CAMBRIDGE TRACTS IN MATHEMATICS

84 *Consequences of Martin's axiom*

D.H. FREMLIN
Lecturer in Mathematics, University of Essex

Consequences of Martin's axiom

The right of the University of Cambridge to print and sell all manner of books was granted by Henry VIII in 1534. The University has printed and published continuously since 1584.

CAMBRIDGE UNIVERSITY PRESS
Cambridge
London New York New Rochelle
Melbourne Sydney

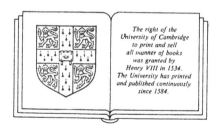

CAMBRIDGE UNIVERSITY PRESS
Cambridge, New York, Melbourne, Madrid, Cape Town, Singapore, São Paulo, Delhi

Cambridge University Press
The Edinburgh Building, Cambridge CB2 8RU, UK

Published in the United States of America by Cambridge University Press, New York

www.cambridge.org
Information on this title: www.cambridge.org/9780521250917

First published 1984
This digitally printed version 2008

A catalogue record for this publication is available from the British Library

Library of Congress Catalogue Card Number: 83-15426

ISBN 978-0-521-25091-7 hardback
ISBN 978-0-521-08954-8 paperback

But there are some special difficulties here, in that the unifying principle of this book is both unusual and relatively modern. Consequently, for instance, certain cardinal numbers, which play an enormous role in the subject, do not have standard names, and their very definitions involve ideas which will be new to most of the people I am trying to reach. So if, looking ahead, you find the symbols $\mathfrak{m}$, $\mathfrak{m}_K$ and $\mathfrak{p}$ both ubiquitous and mysterious, try not to be dismayed, and use the note on page viii as a provisional interpretation of their meaning. Other obscurities may be clarified by a glance at §A1.

A second result of choosing my material on an unconventional criterion is that I have found, even more often than is usual, that customary usages in the different topics I cover sit awkwardly together. The most violent clash is between the terminology for concepts associated with partially ordered sets favoured by specialists in forcing, and the older words more commonly used by other mathematicians; I refer you to 11A for the definitions I have chosen, and to 11G for my reasons. In the sections on measure theory (§§32–3) I have found it convenient to express results in terms of concepts that may be unfamiliar to the general reader; and again in §§42–3 I have based arguments on new definitions. I hope that in all these cases it will be found that the power of the new concepts justifies the effort required to master them. Elsewhere, I have tried to limit the difficulties by avoiding certain phrases, and accepting the extra verbosity of circumlocutions.

An allied point is the range of preliminary material that is necessary to a full understanding of this book. The audience I am trying to address consists of all those who are doing research in any of the topics dealt with here; which covers a substantial proportion of pure mathematicians. Correspondingly, only an exceptionally wide-ranging knowledge of abstract pure mathematics is likely to encompass all the miscellaneous information which turns out to be relevant. I have therefore written out a long list (Appendix A) of definitions and theorems which I use, with references or proofs. I have enjoyed immensely learning enough of each topic to be able to understand the applications of Martin's axiom in it, and I hope that you will have some of the same pleasure.

The majority of the theorems in this book depend for their significance, if not for their truth, on special axioms, and it is therefore necessary to establish certain codes. In many cases it is possible to regard the results as properties of the special cardinals $\mathfrak{m}$, $\mathfrak{m}_K$ and $\mathfrak{p}$; in which case questions about applicability of the results resolve themselves into questions about the possible values of these cardinals. In other cases, we need to make specific assumptions about these values. When this happens I shall write '**Theorem** $[\mathfrak{m}_K > \omega_1]$...' to mean that the theorem in question is true if

$\mathfrak{m}_K$ is greater than ω_1, but may (so far as I know) be false otherwise. Frequently the special assumption is needed for only part of a theorem, in which case I write '**Proposition** (*a*)...(*b*) [$\mathfrak{p} = \mathfrak{c}$]...' to mean that part (*a*) is a theorem of ZFC (Zermelo–Fraenkel set theory, together with the axiom of choice), but that part (*b*) is to be proved under the extra axiom that $\mathfrak{p} = \mathfrak{c}$. Sometimes I need a compound axiom, e.g. '[$\mathfrak{p} = \mathfrak{c} = \omega_2$]'. All the combinations which I express in this way are known to be relatively consistent with ZFC; when I discuss one of the stronger, and potentially more dangerous, axioms, I shall use more explicit formulations.

An obvious question to ask of every result proved in this way is whether the special axiom is really necessary, or whether the result is actually a theorem of ZFC. For the great majority of the theorems in this book, it is known that something more than ZFC is needed, though of course they rarely need the whole strength of the axiom which I appeal to. In those cases where in my view there is serious doubt that any special axiom is needed, I have indicated this as a 'problem' at the end of the appropriate section. For the rest, I have not attempted to give proper references to the alternative models in which the results are known to be false, but these are often to be found with the original versions of the theorems, listed as 'sources'. Equally, one can ask of any pair of undecidable results whether one implies the other. Taken to its logical extreme, this would be an impossibly complicated subject. Many individual enquiries, of course, are of great interest; but here also I have not felt able to give more than a few isolated comments on the known answers and outstanding problems.

In each section of Chapters 1–4 I have included a paragraph headed 'Sources', a brief list of the articles containing the earliest results known to me which are recognizably versions of the theorems I give. As well as being an attempt to give credit to the appropriate authors, these references may be of some use in filling in background, because the peculiar principle on which I have selected the material of this book means that many results are deprived of their natural context; and while I hope that they are all visibly beautiful, it may not always be clear why the questions dealt with are important.

The basic structure of this book is a division into three chapters (Chapters 2–4) according to the cardinal involved; the fundamental relations being that $\mathfrak{m} \leq \mathfrak{m}_K \leq \mathfrak{p}$, so that the results of each chapter are relevant to those after it. Within each chapter, the division is, somewhat arbitrarily, by subject. Apart from this, Chapter 1 comprises definitions of $\mathfrak{m}$, $\mathfrak{m}_K$ and $\mathfrak{p}$, with their most important alternative forms; Appendix A consists of background information; and Appendix B lists questions which Martin's axiom is known *not* to resolve.

1

The three cardinals

11 $\mathfrak{m}$, $\mathfrak{m}_K$ and $\mathfrak{p}$

I define the cardinals $\mathfrak{m}$, $\mathfrak{m}_K$ and $\mathfrak{p}$ in terms of which the rest of this book is expressed [11B, 11D], with their most important relationships [11C, 11E].

11A Partially ordered sets

Recall that a **partially ordered set** is a set P with a relation $\leq$ such that for $p, q, r \in P$

$$p \leq q \;\&\; q \leq r \Rightarrow p \leq r;$$

$$p \leq q \;\&\; q \leq p \Leftrightarrow p = q.$$

If P is a partially ordered set, an **up-antichain** in P is a set $R \subseteq P$ such that no two distinct elements of R have a common upper bound in P. P is **upwards-ccc** if every up-antichain in P is countable. A set $R \subseteq P$ is **upwards-linked** if every pair of elements of R has an upper bound in P. P satisfies **Knaster's condition upwards** if, whenever $R \subseteq P$ is uncountable, there is an uncountable upwards-linked $R' \subseteq R$. A set $R \subseteq P$ is **upwards-directed** if any two elements of R have an upper bound in R (so that every non-empty finite subset of R has an upper bound in R). A set $Q \subseteq P$ is **cofinal** with P if for every $p \in P$ there is a $q \in Q$ with $p \leq q$.

Since the inverse relation $\geq$ is also a partial ordering on P, all these concepts have inverted equivalents. Thus a set $Q \subseteq P$ is **coinitial** if for every $p \in P$ there is a $q \in Q$ such that $q \leq p$, and I shall speak of partially ordered sets which are **downwards-ccc** or satisfy **Knaster's condition downwards**, and subsets which are **down-antichains** or **downwards-linked** or **downwards-directed**, as the occasion arises.

A **chain** in a partially ordered set is a totally ordered subset. Finally, a **weak antichain** in P is a set $A \subseteq P$ such that $x \not< y$ for any $x, y \in A$.

11B Three formulae

(*a*) For cardinals κ, let $\mathrm{MA}(\kappa)$ be the statement

whenever P is a non-empty upwards-ccc partially ordered set and $\mathcal{Q}$ is a family of cofinal subsets of P with $\#(\mathcal{Q}) \leq \kappa$, then there is an upwards-directed subset of P meeting every member of $\mathcal{Q}$.

(*b*) Let MAK(κ) be the statement

whenever P is a non-empty partially ordered set satisfying Knaster's condition upwards, and $\mathcal{Q}$ is a family of cofinal subsets of P with $\#(\mathcal{Q}) \leq \kappa$, then there is an upwards-directed subset of P meeting every member of $\mathcal{Q}$.

(*c*) Let P(κ) be the statement

whenever $\mathcal{A}$ is a family of subsets of $\mathbf{N}$ such that $\#(\mathcal{A}) < \kappa$ and $A_0 \cap \ldots \cap A_n$ is infinite whenever $A_0, \ldots, A_n \in \mathcal{A}$, then there is an infinite set $I \subseteq \mathbf{N}$ such that $I \backslash A$ is finite for every $A \in \mathcal{A}$.

11C **Theorem**

(*a*) MA(ω) is true.
(*b*) MA(κ) $\Rightarrow$ MAK(κ) for every cardinal κ.
(*c*) MAK(κ) $\Rightarrow$ P(κ^+) for every cardinal κ.
(*d*) P($\mathfrak{c}^+$) is false.

Proof (*a*) Let P be a non-empty partially ordered set and $\mathcal{Q}$ a countably infinite family of cofinal subsets of P. Enumerate $\mathcal{Q}$ as $\langle Q_n \rangle_{n \in \mathbf{N}}$ and choose $\langle p_n \rangle_{n \in \mathbf{N}}$ inductively so that

$p_0 \in P$

given p_n, then $p_{n+1} \in Q_n$ and $p_{n+1} \geq p_n$.

Then $\{p_n : n \in \mathbf{N}\}$ is upwards-directed and meets every member of $\mathcal{Q}$.
If $\mathcal{Q}$ is finite the same idea works more quickly.

(*b*) We need only observe that a partially ordered set which satisfies Knaster's condition upwards is upwards-ccc.

(*c*) Assume MAK(κ). Since P(κ^+) is plainly true for every finite κ, we may suppose that κ is infinite. Let $\mathcal{A} \subseteq \mathcal{P}\mathbf{N}$ be such that $\#(\mathcal{A}) < \kappa^+$ (i.e. $\#(\mathcal{A}) \leq \kappa$) and $\bigcap \mathcal{A}_0$ is infinite for every finite $\mathcal{A}_0 \subseteq \mathcal{A}$. If $\mathcal{A} = \emptyset$, there is nothing to prove. Otherwise, take

$$\mathcal{A}^* = \{\bigcap \mathcal{A}_0 : \emptyset \neq \mathcal{A}_0 \subseteq \mathcal{A}, \mathcal{A}_0 \text{ finite}\},$$

so that $\mathcal{A}^*$ is a downwards-directed family of infinite subsets of $\mathbf{N}$. Set

$$P = \{(I, A) : I \subseteq \mathbf{N} \text{ is finite, } A \in \mathcal{A}^*\}.$$

Order P by saying

$$(I, A) \leq (J, B) \text{ if } I \subseteq J \subseteq I \cup A \text{ and } B \subseteq A.$$

This is a partial order on P. For each finite $I \subseteq \mathbf{N}$, $P_I = \{(I, A) : A \in \mathcal{A}^*\}$

is an upwards-directed subset of P, because $(I, A) \leq (I, A \cap B)$ whenever $A, B \in \mathscr{A}^*$; so P must satisfy Knaster's condition upwards, because any uncountable $R \subseteq P$ must meet some P_I is an uncountable set, and now $R \cap P_I$ is upwards-linked.

Now , for each $n \in \mathbf{N}$, set

$$Q_n = \{(I, A) : (I, A) \in P, \exists m \in I, m \geq n\}.$$

Then Q_n is cofinal with P. **P** If $(I, A) \in P$, then A is infinite, so there is an $m \in A$ with $m \geq n$; now $(I \cup \{m\}, A\} \in Q_n$ and $(I, A) \leq (I \cup \{m\}, A)$ in P. **Q**

Next, for $C \in \mathscr{A}$, set

$$Q_C = \{(I, A) : (I, A) \in P, A \subseteq C\}.$$

Then Q_C is also cofinal with P. **P** If $(I, A) \in P$, then $(I, A \cap C) \in Q_C$ and $(I, A) \leq (I, A \cap C)$ in P. **Q**

Accordingly MAK(κ) assures us that there is an upwards-directed $R \subseteq P$ meeting every Q_n and every Q_C, since $\#(\{Q_n : n \in \mathbf{N}\} \cup \{Q_C : C \in \mathscr{A}\}) = \max(\omega, \#(\mathscr{A})) \leq \kappa$. Set

$$K = \bigcup\{I : (I, A) \in R\}.$$

Then K is infinite. **P** Let $n \in \mathbf{N}$. Then R meets Q_n; say $(I, A) \in R \cap Q_n$. There is an $m \in I$ such that $m \geq n$; now $m \in K$. As n is arbitrary, K is infinite. **Q**

Finally, if $C \in \mathscr{A}$, $K \setminus C$ is finite. **P** $R \cap Q_C \neq \varnothing$; say $(I_0, A_0) \in R \cap Q_C$. Let (I, A) be any member of R. As R is upwards-directed, there is a $(J, B) \in R$ which is a common upper bound of (I, A) and (I_0, A_0). Now $I \subseteq J \subseteq I_0 \cup A_0$, so $I \setminus C \subseteq I \setminus A_0 \subseteq I_0$. As (I, A) is arbitrary, $K \setminus C \subseteq I_0$ is finite. **Q**

As $\mathscr{A}$ is arbitrary, P(κ^+) is true.

(*d*) This is obvious; try $\mathscr{A}$ any non-principal ultrafilter on $\mathbf{N}$.

11D $\mathfrak{m}, \mathfrak{m}_K, \mathfrak{p}$ and Martin's axiom

(*a*) From 11C*b*–*d* we see that MA($\mathfrak{c}$) is false. So we may define a cardinal $\mathfrak{m}$ by saying that $\mathfrak{m}$ is the least cardinal such that MA($\mathfrak{m}$) is false. Of course we now have MA(κ) false for every $\kappa \geq \mathfrak{m}$; so, for a cardinal κ, MA(κ)$\Leftrightarrow \kappa < \mathfrak{m}$.

Similarly, $\mathfrak{m}_K$ is to be the least cardinal such that MAK($\mathfrak{m}_K$) is false, so that MAK(κ)$\Leftrightarrow \kappa < \mathfrak{m}_K$. Finally, $\mathfrak{p}$ is to be the least cardinal such that $P(\mathfrak{p}^+)$is false, so that $P(\kappa) \Leftrightarrow \kappa \leq \mathfrak{p}$.

(*b*) Now we can restate Theorem 11C in the form

$\omega_1 \leq \mathfrak{m}$ [11C*a*];

$\mathfrak{m} \leq \mathfrak{m}_K$ [11C*b*];

$\mathfrak{m}_K \leq \mathfrak{p}$ [11C*c*];

$\mathfrak{p} \leq \mathfrak{c}$ [11C*d*].

(*c*) Martin's axiom, 'MA', is the statement '$\mathfrak{m} = \mathfrak{c}$' i.e. 'MA($\kappa$) $\forall \kappa < \mathfrak{c}$'. Since $\omega_1 \leq \mathfrak{m} \leq \mathfrak{c}$, the continuum hypothesis implies MA.

11E **Theorem**

MA $\not\Rightarrow$ CH; in fact $\mathfrak{m} = \mathfrak{c}$ is consistent with any consistent assignment of cardinal powers such that $2^\kappa \leq \mathfrak{c}$ for every $\kappa < \mathfrak{c}$.

Proof This is quite outside the scope of this book. Proofs are given in SOLOVAY & TENNENBAUM 71; JECH 71; BURGESS 77; JECH 78; COHEN 78; KUNEN 80; BAUMGARTNER *a*.

Remark The condition '$2^\kappa \leq \mathfrak{c}\ \forall \kappa < \mathfrak{c}$' is forced by 21C.

11F **Exercise**

Let P be the partially ordered set $\bigcup_{n \in \mathbf{N}} \omega_1^n$, ordered by saying that $\sigma \leq \tau$ if τ extends σ. For each $\xi \in \omega_1$ set $Q_\xi = \{\sigma : \exists i, \sigma(i) = \xi\}$. Show that each Q_ξ is cofinal with P but that there is no upwards-directed $R \subseteq P$ meeting every Q_ξ. Hence show that if we omit the phrase 'upwards-ccc' in the statement of MA(κ), we obtain a statement equivalent to '$\kappa \leq \omega$'.

11G **Notes on terminology**

It will already be clear to anyone who has studied this subject before that my terminology differs from that of other authors more than is strictly necessary, and a word of explanation is perhaps in order.

(*a*) Because the earliest work in this subject was done by specialists in mathematical logic, a number of words crept in ('filter', 'generic', 'dense') which for the majority of mathematicians are merely confusing; so I do not use them, preferring the words I learnt as an undergraduate ('directed', 'cofinal').

(*b*) It is not customary to distinguish, for example, between 'upwards-ccc' and 'downwards-ccc' partially ordered sets; most authors fix on one of these (usually, I think, 'downwards-') and call it simply 'ccc'. The difficulty with this is that many partially ordered sets come to us with a natural orientation. (Consider, for instance, 11F above; or the cases in which our partially ordered set P is a collection of sets ordered by $\subseteq$, as in 32B and 33E below.) There is no logical difficulty in declaring, if necessary, that '$A \leq B$ iff $B \subseteq A$'; but I find that following the subsequent arguments is like drinking a glass of water while hanging upside-down. I think it easier to take the trouble to use a language which can itself perform the necessary inversions. Every axiom, definition or theorem which I give involving partially ordered sets will have a corresponding inverted form; and when appropriate I shall use axioms or theorems which have been stated in one form, in the other form, without comment, as they should be instantly recognizable.

(*c*) The letters 'ccc' are an acronym for 'countable chain condition'; which is unfortunate, as the condition really refers to antichains. (Matters are not improved by the fact that we shall have occasion to deal with ordered sets in which all *chains* are countable.) For this reason some authors (notably MARTIN & SOLOVAY 70) use 'cac', 'countable antichain condition'. Regrettably this remains very much a minority choice, and I feel that I am taking up quite enough minority positions already.

(*d*) The word 'antichain' itself is not perfectly happy, because it is perhaps more natural to use it for what I am calling a 'weak antichain', and this is quite often done. But I think that the majority of authors (in the present context) use it to mean either an 'up-' or a 'down-' antichain.

(*e*) The cardinals $\mathfrak{m}$ and $\mathfrak{m}_K$ have not been handled in this way before; the practice has been to work with the formulae MA(κ) and MA. I have chosen to use $\mathfrak{m}$, $\mathfrak{m}_K$ and $\mathfrak{p}$ throughout this book for three reasons. (i) In general, they make for elegance, conciseness and flexibility. Consider, for instance, the restatement of Theorem 11C as '$\omega_1 \leq \mathfrak{m} \leq \mathfrak{m}_K \leq \mathfrak{p} \leq \mathfrak{c}$'. (ii) The accident that the accepted meanings of MA(κ) and P(κ) follow different conventions makes these awkward to use together. (iii) In recent years a good deal of interest has been taken in the relationships between the various consequences of assuming that $\mathfrak{p} = \mathfrak{c}$. It seems that an effective way of discussing many of these is to define further special cardinals lying between $\mathfrak{p}$ and $\mathfrak{c}$ and then to investigate the properties of these cardinals. (See DOUWEN 83 and the notes to §24. I use the same device in §B1.) The notation I have chosen is therefore well adapted to certain types of more detailed enquiry, though it is too early to be sure that this structure will continue to seem appropriate.

(*f*) 'Knaster's condition' has usually (following KNASTER 45) been called 'property (K)'. I think it has turned out sufficiently important to be given a proper name.

11H Sources

KNASTER 45 for (a special manifestation of) Knaster's condition. ROTHBERGER 48 considers the axiom $\mathfrak{p} > \omega_1$. MARTIN & SOLOVAY 70 consider axioms A_κ and S_κ equivalent respectively to $\kappa < \mathfrak{m}$ and $\kappa < \mathfrak{p}$, and show that $A_\kappa \Rightarrow S_\kappa$. BOOTH 70 and SILVER 70 also give $\mathrm{MA} \Rightarrow \mathfrak{p} = \mathfrak{c}$. (Martin's axiom actually arose as the common centre of the proof by SOLOVAY & TENNENBAUM 71 that Souslin's hypothesis is consistent, and a number of other consistency results of Solovay.) JECH 71 for 11F. KUNEN & TALL 79 for the axiom MAK (meaning '$\mathrm{MAK}(\kappa)\forall\kappa < \mathfrak{c}$', or $\mathfrak{m}_K = \mathfrak{c}$).

Notes and comments

I have spelt out the proof that $\mathfrak{m}_K \leq \mathfrak{p}$ in full detail, because it is not only one of the most beautiful arguments of the theory, but also an archetype. The idea of the partially ordered set P is that each (I, A) in P is in some sense an attempt at the set K we are seeking; I is the set of numbers 'already' put into K, while A is the set of numbers which will be allowed in future. (Observe that $I \subseteq K \subseteq I \cup A$ for every $(I, A) \in R$.) Many of the partially ordered sets to which we shall apply Martin's axiom can be thought of as sets of pairs (t, T) where t is a kind of approximation to the object we are seeking, and T is a condition, or set of conditions, which are to be satisfied henceforth as the approximation improves. The upwards-directed set R is now a consistent family of approximations sufficiently large so that the object it defines has all the properties we need.

There is one point, of course, where the proof that $\mathfrak{m}_K \leq \mathfrak{p}$ is far from typical: the proof that the partially ordered set P is upwards-ccc is very easy. Usually, I suppose, the 'ccc' requirement is the most irksome condition to check; for this reason I have presented 11F to show that it really is essential.

The theorem I have not proved, that $\mathfrak{m} > \omega_1$ is (relatively) consistent with ZFC [11E], is of course the whole point of Martin's axiom. If you glance through the pages ahead, I think you will agree with me that the most interesting theorems are those declared as consequences of $\mathfrak{m} > \omega_1$ (or $\mathfrak{m}_K > \omega_1$ or $\mathfrak{p} > \omega_1$, of course); that the many theorems invoking no special axiom, but just a requirement '$\#(X) \leq \mathfrak{p}$' or '$\#(X) < \mathfrak{m}$' are less remarkable; and that theorems depending on MA or $\mathfrak{p} = \mathfrak{c}$ are, with some notable exceptions, of a more conventional type, being consequences of the continuum hypothesis.

At the same time, it is the case that in most of the usual models of $\mathfrak{c} > \omega_1$ (e.g. P.J. Cohen's original one) we have $\mathfrak{m} = \omega_1$. Once we have established that $\mathfrak{m}$ is not fixed to either ω_1 or $\mathfrak{c}$, it is natural to ask what are the possible values of $\mathfrak{m}$, $\mathfrak{m}_K$ and $\mathfrak{p}$. The constraints known to me are (i) $\omega_1 \leq \mathfrak{m} \leq \mathfrak{m}_K \leq \mathfrak{p} \leq \mathfrak{c}$ [11C]; (ii) if $\mathfrak{m} > \omega_1$ then $\mathfrak{m}_K = \mathfrak{m}$ [41Cc]; (iii) $\mathrm{cf}(\mathfrak{m}_K) > \omega$ [31K]; (iv) $\mathfrak{p}$ is regular [21K]; (v) $2^\kappa = \mathfrak{c}$ whenever $\omega \leq \kappa < \mathfrak{p}$

[21C]. As far as I am aware there are no others which are not consequences of these. In particular, $\omega_1 = \mathfrak{m} < \mathfrak{m}_K$ and $\mathfrak{m}_K < \mathfrak{p}$ are both possible. For a fuller discussion of these points see §B1.

A rough census suggests that the axioms which are most often used are, in order, $\mathfrak{m} > \omega_1$ or MA(ω_1); $\mathfrak{m}_K > \omega_1$; $\mathfrak{p} > \omega_1$ or P(ω_2); and $\mathfrak{p} = \mathfrak{c}$ or P($\mathfrak{c}$) or P, occasionally called "Booth's lemma'. All of these are consequences of $\mathfrak{m} = \mathfrak{c} > \omega_1$ or MA +not-CH. I have already suggested that to begin with it is reasonable, for definiteness, to fix on the axiom $\mathfrak{m} = \mathfrak{c} = \omega_2$.

In §14 I show that $\mathfrak{p}$, like $\mathfrak{m}$ and $\mathfrak{m}_K$, can be defined in terms of directed sets meeting cofinal sets. This makes the inequality $\mathfrak{m}_K \leq \mathfrak{p}$ obvious, and suggests a variety of other cardinals ($\mathfrak{m}_{\sigma\text{-linked}}$, $\mathfrak{m}_{\text{pc}(\omega_1)}$, $\mathfrak{m}_{\text{countable}}$) as described in B1A–B1D. But there is no natural limit to this kind of subdivision, and I have not found a way to break Chapter 2 into balanced fragments, so I have settled for just the three cardinals $\mathfrak{m}$, $\mathfrak{m}_K$ and $\mathfrak{p}$.

Several authors have sought axioms similar to MA which would operate at levels above $\mathfrak{c}$. I shall not try to deal with these questions in this book. Results on these lines may be found in ANTONOVSKII & CHUDNOVSKY 76, HERINK 77, SHELAH 78, TALL 79, BAUMGARTNER *a*, TALL *b*, WEISS 81*b* and WEISS 83.

12 Boolean algebras and topological spaces

A number of results concerning these structures and their relations with partially ordered sets are so fundamental to the work of this book that I have brought them forward rather than relegating them to Appendix A. The first part of this section should be either easy or familiar or both; the hard work begins with Theorems 12E–F, which are needed in §13. I then give the Δ-system lemma [12H], which will be used repeatedly below, and one of its first consequences [12I], with some further theorems on product spaces which will be useful to us.

12A Boolean algebras

[HALMOS 63, SIKORSKI 64] (*a*) A **Boolean algebra** is a ring $\mathfrak{A}$ with multiplicative identity 1 such that $a^2 = a$ for every $a \in \mathscr{A}$. (Following Sikorski but not Halmos, I allow $1 = 0$.) In this case $\mathfrak{A}$ is commutative and $a + a = 0$ for every $a \in \mathfrak{A}$. If $\mathfrak{A}$ and $\mathfrak{B}$ are Boolean algebras, a function $\varphi : \mathfrak{A} \to \mathfrak{B}$ is a **Boolean homomorphism** if it is a ring homomorphism and $\varphi(1) = 1$.

(*b*) If $\mathfrak{A}$ is any Boolean algebra, it can be expressed (essentially uniquely) as the algebra of open-and-closed sets in a zero-dimensional compact Hausdorff space Z, the **Stone space** of $\mathfrak{A}$, writing $E + F = E \triangle F$

and $E.F = E \cap F$ for $E, F \subseteq Z$. Z can be regarded as the set of Boolean homomorphisms from $\mathfrak{A}$ to $\mathbf{Z}_2$. [SIKORSKI 64, §8; HALMOS 63, §18]

If $\mathfrak{A}$ and $\mathfrak{B}$ are Boolean algebras with Stone spaces Z and W respectively, there is a canonical correspondence between Boolean homomorphisms $\varphi:\mathfrak{A}\to\mathfrak{B}$ and continuous functions $f:W\to Z$, given by identifying φE with $f^{-1}[E]$ (if we think of the algebras as consisting of open-and-closed sets) or by identifying $f(t)$ with $t\circ\varphi$ (if we think of the Stone spaces as consisting of Boolean homomorphisms). φ is surjective iff f is injective, and φ is injective iff f is surjective. So W can be embedded in Z iff $\mathfrak{B}$ is a quotient of $\mathfrak{A}$, and Z is a continuous image of W iff $\mathfrak{A}$ can be embedded in $\mathfrak{B}$. [SIKORSKI 64, §11; HALMOS 63, §20]

(*c*) Corresponding to this representation of $\mathfrak{A}$ we have a partial order $\subseteq$ on $\mathfrak{A}$, which can be defined directly from the ring structure by writing $a\subseteq b$ iff $ab=a$. I shall normally (unless, as in 12B, this is incompatible with another structure) write $a\cap b$ for $\inf\{a,b\}=ab$, $a\cup b$ for $\sup\{a,b\}=a+b+ab$, and $a\backslash b$ for $a+ab$.

(*d*) A Boolean algebra $\mathfrak{A}$ is **Dedekind (σ)-complete** if every (countable) subset of $\mathfrak{A}$ has a supremum and an infimum. $\mathfrak{A}$ is Dedekind complete iff its Stone space is extremally disconnected. [SIKORSKI 64, 22.4; HALMOS 63, §21, Theorem 10]

(*e*) If $\mathfrak{A}$ and $\mathfrak{B}$ are Boolean algebras, a Boolean homomorphism $\varphi:\mathfrak{A}\to\mathfrak{B}$ is **(sequentially) order-continuous** (or a '(σ-) complete homomorphism') if $\inf\varphi[C]=0$ in $\mathfrak{B}$ whenever C is a (countable) downwards-directed set in $\mathfrak{A}$ with $\inf C=0$; equivalently, if φ preserves all (countable) suprema and infima which exist in $\mathfrak{A}$.

(*f*) If $\mathfrak{A}$ is a Boolean algebra, a **subalgebra** of $\mathfrak{A}$ is a subring $\mathfrak{B}$ which contains 1. $\mathfrak{B}$ is a **regular** subalgebra if the identity map $\mathfrak{B}\to\mathfrak{A}$ is order-continuous i.e. whenever $C\subseteq\mathfrak{B}$ and $\inf C=0$ in $\mathfrak{B}$ then $\inf C=0$ in $\mathfrak{A}$. $\mathfrak{B}$ is **order-dense** in $\mathfrak{A}$ if $a=\sup\{b:b\in\mathfrak{B}, b\subseteq a\}$ for every $a\in\mathfrak{A}$; equivalently, if $\mathfrak{B}\backslash\{0\}$ is coinitial with $\mathfrak{A}\backslash\{0\}$. An order-dense subalgebra must be regular.

(*g*) If $\mathfrak{A}$ is any Boolean algebra, there is an (essentially unique) Dedekind complete Boolean algebra $\hat{\mathfrak{A}}$ in which $\mathfrak{A}$ is embedded as an order-dense subalgebra; I shall speak of $\hat{\mathfrak{A}}$ as 'the' Dedekind completion of $\mathfrak{A}$. [SIKORSKI 64, §35; HALMOS 63, §21; see also 12B]

(*h*) If $\mathfrak{A}$ is a Boolean algebra, a subalgebra $\mathfrak{B}$ of $\mathfrak{A}$ is **order-closed** if $\sup C\in\mathfrak{B}$ whenever $C\subseteq\mathfrak{B}$ and $\sup C$ is defined in $\mathfrak{A}$. (Sikorski uses the phrase 'complete subalgebra'.) I will say that a set $A\subseteq\mathfrak{A}$ **weakly generates**

$\mathfrak{A}$ if $\mathfrak{A}$ is the smallest order-closed subalgebra of itself including A, and that $\mathfrak{A}$ is **weakly countably generated** if it is generated in this sense by a countable set. (Sikorski uses the phrase 'generate completely' for my 'weakly generate'.)

(*i*) If $\mathfrak{A}$ is a Boolean algebra and $\mathscr{I}$ is an ideal of $\mathfrak{A}$ (in the ring-theoretic sense), then $\mathfrak{A}/\mathscr{I}$ is a Boolean algebra. $\mathscr{I}$ is called **κ-additive** if whenever $C \subseteq \mathscr{I}$ and $\#(C) < \kappa$, then C is bounded above in $\mathscr{I}$. $\mathscr{I}$ is a **σ-ideal** if whenever $C \subseteq \mathscr{I}$ is countable and $\sup C$ exists in $\mathfrak{A}$, then $\sup C \in \mathscr{I}$. (Thus an ω_1-additive ideal is always a σ-ideal, and a σ-ideal in a Dedekind σ-complete algebra is ω_1-additive.) $\mathscr{I}$ is a σ-ideal iff the canonical map from $\mathfrak{A}$ to $\mathfrak{A}/\mathscr{I}$ is sequentially order-continuous [HALMOS 63, §13]. If $\mathfrak{A}$ is Dedekind σ-complete and $\mathscr{I}$ is a σ-ideal in $\mathfrak{A}$, then $\mathfrak{A}/\mathscr{I}$ is Dedekind σ-complete.

(*j*) An element a of a Boolean algebra $\mathfrak{A}$ is an **atom** if $a \neq 0$ and $b \subseteq a \Rightarrow b = 0$ or $b = a$. $\mathfrak{A}$ is **atomless** if it has no atoms.

(*k*) If $\mathfrak{A}$ is a Boolean algebra, a set $A \subseteq \mathfrak{A}$ is **free** if $\inf B \not\subseteq \sup C$ whenever B, C are disjoint finite subsets of A.

12B Regular open sets

Let X be any topological space. (We are going to use the ideas here on spaces which are not even T_1.) An open set $G \subseteq X$ is **regular** if it is the interior of some closed set; equivalently, if $G = \operatorname{int} \bar{G}$. The set $\mathscr{G}$ of regular open subsets of X can be given a Boolean algebra structure by writing

$$G.H = G \cap H;$$

$$G + H = \operatorname{int}(\overline{G \cup H}) \setminus (\overline{G \cap H}).$$

I call $\mathscr{G}$ the **regular open algebra** of X. The ordering induced by this Boolean algebra structure is precisely $\subseteq$. The lattice operations are not exactly what we should hope for; $\inf\{G, H\}$ is $G \cap H$ but $\sup\{G, H\}$ is $\operatorname{int}(\overline{G \cup H})$. Complementation in the Boolean algebra $\mathscr{G}$ is given by $G \mapsto X \setminus \bar{G}$. $\mathscr{G}$ is Dedekind complete, with

$$\inf \mathscr{H} = \operatorname{int}(\overline{\textstyle\bigcap \mathscr{H}}), \quad \sup \mathscr{H} = \operatorname{int}(\overline{\textstyle\bigcup \mathscr{H}})$$

for any non-empty $\mathscr{H} \subseteq \mathscr{G}$.

Any open-and-closed set is a regular open set. If X is zero-dimensional, then $\mathscr{G}$ can be identified with the Dedekind completion of the algebra $\mathscr{E}$ of open-and-closed subsets of X (because $\mathscr{E}$ is order-dense in $\mathscr{G}$). X is extremally disconnected iff $\mathscr{G} = \mathscr{E}$.

$\mathscr{G}$ can be identified with the quotient Boolean algebra $\mathfrak{A}/\mathscr{J}$ where $\mathfrak{A}$ is the algebra of subsets of X with nowhere-dense boundaries and $\mathscr{J}$ is the ideal of nowhere dense sets [SIKORSKI 64, §10, Example D]. If X is a Baire space then $\mathscr{G}$ can be identified with the quotient $\mathscr{B}/\mathscr{M}$ where $\mathscr{B}$ is the algebra of Borel sets and $\mathscr{M}$ is the ideal of meagre Borel sets. [SIKORSKI 64, §22, Example G]

12C **Up- and down-topologies**

Let P be a partially ordered set. The family $\{\{q:q\geq p\}:p\in P\}$ is a topology base on P; I shall call the topology it generates the **up-topology**, and the open sets for this topology **up-open**. The up-topology is always T_0 but only in trivial cases is it T_1. Observe that $Q\subseteq P$ is cofinal iff it is dense for the up-topology.

Similarly P has a **down-topology** with basic **down-open** sets $\{q:q\leq p\}$.

12D **The ccc**

(*a*) A topological space X is **ccc** if every disjoint family of open sets in X is countable. (See A4C*a*.)

(*b*) A Boolean algebra $\mathfrak{A}$ is **ccc** if every disjoint set in $\mathfrak{A}$ is countable. (A set $A\subseteq\mathfrak{A}$ is **disjoint** if $a\cap b=0$ for all distinct $a,b\in A$.) If $\mathfrak{A}$ is ccc then for every $A\subseteq\mathfrak{A}$ there is a countable $B\subseteq A$ with the same upper bounds as A. [HALMOS 63, §14, Lemma 1]

(*c*) Observe that

(i) a partially ordered set is upwards-ccc iff its up-topology is ccc;
(ii) a topological space is ccc iff its regular open algebra is ccc;
(iii) a Boolean algebra is ccc iff its Stone space is ccc;
(iv) a Boolean algebra $\mathfrak{A}$ is ccc iff $(\mathfrak{A}\setminus\{0\},\subseteq\}$ is downwards-ccc iff $(\mathfrak{A}\setminus\{1\},\subseteq)$ is upwards-ccc;
(v) a topological space $(X,\mathfrak{T})$ is ccc iff $\mathfrak{T}\setminus\{\emptyset\}$ is downwards-ccc;
(vi) a continuous image of a ccc topological space is ccc [A4Dc].

(*d*) If $\mathfrak{A}$ is a Boolean algebra, $\mathscr{J}$ an ideal of $\mathfrak{A}$, and κ is an infinite cardinal, then $\mathscr{J}$ is **κ-saturated** if whenever $A\subseteq\mathfrak{A}\setminus\mathscr{J}$ has cardinal κ, there are distinct $a,b\in A$ with $a\cap b\notin\mathscr{J}$. Observe that $\mathscr{J}$ is ω_1-saturated iff the quotient algebra $\mathfrak{A}/\mathscr{J}$ is ccc iff $\mathfrak{A}\setminus\mathscr{J}$ is downwards-ccc. If $\mathscr{J}$ is a σ-ideal in a Dedekind σ-complete Boolean algebra, it will be ω_1-saturated iff there is no uncountable disjoint set $A\subseteq\mathfrak{A}\setminus\mathscr{J}$. In this case $\mathfrak{A}/\mathscr{J}$ will be Dedekind complete (being ccc and Dedekind σ-complete).

12E **Theorem**

Every ccc Boolean algebra $\mathfrak{A}$ of cardinal less than or equal to $\mathfrak{c}$

can be expressed as a regular subalgebra of a Dedekind complete weakly countably generated ccc Boolean algebra.

Proof (*a*) Express $\mathfrak{A}$ as the algebra of open-and-closed sets in a zero-dimensional compact Hausdorff space Z. Let $\mathscr{A}$ be an almost-disjoint family of infinite subsets of $\mathbf{N}$ with $\#(\mathscr{A}) = \mathfrak{c}$ [A3A*d*]. As $\#(\mathfrak{A}) \leq \mathfrak{c}$, there is an injection $E \mapsto A_E : \mathfrak{A} \to \mathscr{A}$. Set $W = Z \times [\mathbf{N}]^{<\omega}$.

Let $\mathscr{R}$ be the ideal of $\mathscr{P}\mathbf{N}$ generated by the finite sets and $\mathscr{A}$. If $R \in \mathscr{R}$, then

$$\mathscr{E}_R = \{E : E \in \mathfrak{A}, A_E \cap R \text{ is infinite}\} \cup \{Z\}$$

must be finite, because $\mathscr{A}$ is almost-disjoint, so we can set

$$F(R) = \bigcap \mathscr{E}_R \in \mathfrak{A}.$$

Now if $R \in \mathscr{R}$ and $C \in [R]^{<\omega}$, write

$$U(C, R) = \{(z, I) : z \in F(R), I \in [\mathbf{N}]^{<\omega}, I \cap R = C\} \subseteq W.$$

(*b*) The family

$$\mathscr{U} = \{U(C, R) : R \in \mathscr{R}, C \in [R]^{<\omega}\}$$

is a topology base on W. **P** (i) $\bigcup \mathscr{U} \supseteq U(\emptyset, \emptyset) = W$. (ii) If $(z, I) \in U(C, R) \cap U(D, S)$, then $\mathscr{E}_{R \cup S} = \mathscr{E}_R \cup \mathscr{E}_S$, $F(R \cup S) = F(R) \cap F(S)$ so $(z, I) \in U(C \cup D, R \cup S) \subseteq U(C, R) \cap U(S, D)$. (The point is that $(C \cup D) \cap R = I \cap R = C$.) **Q**

(*c*) Accordingly we can give W the topology generated by $\mathscr{U}$. Under this topology, W is ccc. **P** Let $\mathscr{G}$ be an uncountable family of non-empty open sets in W. For each $G \in \mathscr{G}$ we can find $R_G \in \mathscr{R}$, $C_G \in [R_G]^{<\omega}$ such that $\emptyset \neq U(C_G, R_G) \subseteq G$. Now there is a $C \in [\mathbf{N}]^{<\omega}$ such that $\mathscr{G}_1 = \{G : G \in \mathscr{G}, C_G = C\}$ is uncountable. For each $G \in \mathscr{G}_1$, $F(R_G)$ is a non-zero member of $\mathfrak{A}$; so (as $\mathfrak{A}$ is ccc) there are distinct G, $H \in \mathscr{G}_1$ such that $F(R_G) \cap F(R_H) \neq \emptyset$. Now if $z \in F(R_G) \cap F(R_H)$, $(z, C) \in U(C, R_G) \cap U(C, R_H) \subseteq G \cap H$. **Q**

(*d*) So if $\mathscr{G}$ is the regular open algebra of W, $\mathscr{G}$ is a ccc Dedekind complete Boolean algebra. Now $\pi_1 : W \to Z$ is continuous, because

$$\pi_1^{-1}[E] = \bigcup \{U(C, A_E) : C \in [A_E]^{<\omega}\}$$

is open for each $E \in \mathfrak{A}$, and $\mathfrak{A}$ is a base for the topology of Z. Consequently $\pi_1^{-1}[E]$ is open-and-closed in W for each $E \in \mathfrak{A}$, and belongs to $\mathscr{G}$. Write $\varphi E = \pi_1^{-1}[E]$; then $\varphi : \mathfrak{A} \to \mathscr{G}$ is an embedding of the algebra $\mathfrak{A}$ in $\mathscr{G}$, because the lattice operations of $\mathscr{G}$ agree with $\cup$ and $\cap$ on the open-and-closed sets of W. Finally

$$\pi_1[U(C,R)] = F(R)$$

is open for each $U(C,R) \in \mathscr{U}$, so π_1 is an open mapping and $\varphi[\mathfrak{A}]$ is a regular subalgebra of $\mathscr{G}$. **P** If $\emptyset \neq \mathscr{D} \subseteq \mathfrak{A}$ and $\inf \mathscr{D} = \emptyset$ in $\mathfrak{A}$, then $\text{int}(\bigcap \mathscr{D}) = \emptyset$. Now if $G \subseteq W$ is a non-empty open set, $\pi_1[G]$ is a non-empty open set, and $\pi_1[G] \not\subseteq \bigcap \mathscr{D}$; thus $G \not\subseteq \bigcap \{\pi_1^{-1}[E] : E \in \mathscr{D}\}$ and $\inf \varphi[\mathscr{D}] = \emptyset$ in $\mathscr{G}$. **Q**

(*e*) For $n \in \mathbf{N}$, write

$$\begin{aligned} W_n &= \{(z,I) : (z,I) \in W, n \notin I\} \\ &= U(\emptyset, \{n\}) = W \setminus U(\{n\}, \{n\}) \in \mathscr{G}. \end{aligned}$$

For non-empty $R \in \mathscr{R}$, consider $U(\emptyset, R)$. This is $\text{int}(\bigcap_{n \in R} W_n)$. **P** $U(\emptyset, R) \subseteq \bigcap_{n \in R} W_n$ and $U(\emptyset, R)$ is open, so $U(\emptyset, R) \subseteq \text{int}(\bigcap_{n \in R} W_n)$. On the other hand, if $(z,I) \in \text{int}(\bigcap_{n \in R} W_n)$, there are $S \in \mathscr{R}$, $D \in [S]^{<\omega}$ such that $(z,I) \in U(D,S) \subseteq \bigcap_{n \in R} W_n$. Now we see that $z \in F(S)$ and that $J \cap R = \emptyset$ whenever $J \cap S = D$. So $R \subseteq S \setminus D$ and $\mathscr{E}_R \subseteq \mathscr{E}_S$ and $F(R) \supseteq F(S)$, so $z \in F(R)$. Also $I \cap R = \emptyset$, so $(z,I) \in U(\emptyset, R)$. As (z,I) is arbitrary, $\text{int}(\bigcap_{n \in R} \mathrm{W}_n) \subseteq U(\emptyset, R)$. **Q**

Accordingly $U(\emptyset, R) \in \mathscr{G}$ and $U(\emptyset, R)$ belongs to the order-closed subalgebra $\mathfrak{B}$ of $\mathscr{G}$ generated by $\{W_n : n \in \mathbf{N}\}$.

(*f*) Observe finally that if $R \in \mathscr{R}$, $C \in [R]^{<\omega}$ then $F(R \setminus C) = F(R)$, so

$$U(C,R) = U(\emptyset, R \setminus C) \setminus \bigcup_{n \in C} W_n \in \mathscr{G}.$$

This shows that $\mathscr{U} \subseteq \mathfrak{B}$. It follows at once that every element of $\mathscr{G}$ is the supremum of a family in $\mathfrak{B}$ and so $\mathfrak{B} = \mathscr{G}$ and $\mathscr{G}$ is weakly countably generated.

Thus $\varphi : \mathfrak{A} \to \mathscr{G}$ is the required embedding.

12F **Theorem**

Let $\mathfrak{A}$ be a weakly countably generated Dedekind complete ccc Boolean algebra. Then $\mathfrak{A}$ is isomorphic to some quotient $\mathscr{B}(\{0,1\}^{\mathbf{N}})/\mathscr{I}$ where $\mathscr{B}(\{0,1\}^{\mathbf{N}})$ is the σ-algebra of Borel sets of $\{0,1\}^{\mathbf{N}}$ and $\mathscr{I}$ is an ω_1-saturated σ-ideal of $\mathscr{B}$.

Proof Start by expressing $\mathfrak{A}$ as the algebra of open-and-closed sets in an extremally disconnected compact Hausdorff space Z, i.e. the regular open sets of Z. Let $\langle W_n \rangle_{n \in \mathbf{N}}$ be a sequence in $\mathfrak{A}$ which weakly generates $\mathfrak{A}$. Define $f : Z \to \{0,1\}^{\mathbf{N}}$ by writing $(ft)(n) = (\chi W_n)(t)$. Then f is continuous and induces a sequentially order-continuous Boolean homomorphism

$\varphi:\mathscr{B}(\{0,1\}^{\mathbf{N}})\rightarrow\mathscr{B}(Z)$, given by $\varphi E=f^{-1}[E]$. Because Z is a Baire space, we can identify $\mathfrak{A}$ with $\mathscr{B}(Z)/\mathscr{M}$, where $\mathscr{M}$ is the σ-ideal of meagre Borel sets in Z; let $\psi:\mathscr{B}(Z)\rightarrow\mathfrak{A}$ be the corresponding homomorphism. Now $\psi\varphi:\mathscr{B}(\{0,1\}^{\mathbf{N}})\rightarrow\mathfrak{A}$ is a sequentially order-continuous Boolean homomorphism; its kernel $\mathscr{I}=\varphi^{-1}[\mathscr{M}]$ is a σ-ideal in $\mathfrak{B}(\{0,1\}^{\mathbf{N}})$. Next, $\psi\varphi$ is a surjection, because $\mathfrak{B}=(\psi\varphi)[\mathscr{B}(\{0,1\}^{\mathbf{N}})]$ is a subalgebra of $\mathfrak{A}$ closed under countable suprema and infima and containing $W_n=(\psi\varphi)(\{t: t(n)=1\})$ for every $n\in\mathbf{N}$. Because $\mathfrak{A}$ is ccc, $\mathfrak{B}$ is order-closed in $\mathfrak{A}$ and must be $\mathfrak{A}$ itself. Accordingly $\psi\varphi$ induces an isomorphism between $\mathscr{B}/\mathscr{I}$ and $\mathfrak{A}$. Finally, $\mathscr{I}$ is ω_1-saturated because $\mathscr{B}/\mathscr{I}=\mathfrak{A}$ is ccc.

12G **Definition**

An indexed family $\langle I_\alpha\rangle_{\alpha\in A}$ is a **Δ-system** with **root** I if $I_\alpha\cap I_\beta=I$ for all distinct $\alpha,\beta\in A$. I will call it a **constant-size** Δ-system if $\#(I_\alpha\backslash I)=\#(I_\beta\backslash I)$ for all $\alpha,\beta\in A$.

I will also say that a collection $\mathscr{J}$ of sets is a Δ-system with root I if $J\cap K=I$ for all distinct $J,K\in\mathscr{J}$, and that it is constant-size if $\#(J\backslash I)$ is constant.

12H **Δ-system lemma**

(*a*) Let $\langle I_\alpha\rangle_{\alpha\in A}$ be an uncountable indexed family of finite sets. Then there is an uncountable $B\subseteq A$ such that $\langle I_\alpha\rangle_{\alpha\in B}$ is a constant-size Δ-system.

(*b*) Let $\mathscr{A}$ be an uncountable set of finite sets. Then there is an uncountable constant-size Δ-system $\mathscr{J}\subseteq\mathscr{A}$.

Proof (*a*) As A is uncountable and every I_α is finite, there is an $n\in\mathbf{N}$ such that $C=\{\alpha:\#(I_\alpha)=n\}$ is uncountable. Set

$$\mathscr{J}=\{J:\{\alpha:\alpha\in C, J\subseteq I_\alpha\}\text{ is uncountable}\}.$$

As $\#(J)\leq n$ for every $J\in\mathscr{J}$, $\mathscr{J}$ has a maximal element I say. Set $D=\{\alpha:\alpha\in C, I\subseteq I_\alpha\}$. Then D is uncountable; but if $x\notin I$, then $I\cup\{x\}\notin\mathscr{J}$, and $\{\alpha:\alpha\in D, x\in I_\alpha\}$ must be countable. Accordingly, if Y is any countable set disjoint from I,

$$\{\alpha:\alpha\in D, Y\cap I_\alpha\neq\emptyset\}$$

is countable.

Choose $\langle\alpha(\xi)\rangle_{\xi<\omega_1}$ inductively in D so that

$$I_{\alpha(\xi)}\cap(\textstyle\bigcup_{\eta<\xi}I_{\alpha(\eta)}\backslash I)=\emptyset, \alpha(\xi)\neq\alpha(\eta)\,\forall\eta<\xi$$

for each $\xi<\omega_1$; this will be possible because at any stage all but countably many members of D will be available to us as possible $\alpha(\xi)$. Set $B=\{\alpha(\xi):\xi<\omega_1\}$; then $\langle I_\beta\rangle_{\beta\in B}$ is a constant-size Δ-system with root I.

(*b*) The alternative form follows at once from (*a*).

12I Theorem

Let $\langle X_\iota\rangle_{\iota\in I}$ be a family of topological spaces such that $\prod_{\iota\in J}X_\iota$ is ccc for every finite $J\subseteq I$. Then $X=\prod_{\iota\in I}X_\iota$ is ccc.

Proof Let $\langle G_\xi\rangle_{\xi<\omega_1}$ be any family of non-empty open sets in X. For each $\xi<\omega_1$ we can find a finite $J(\xi)\subseteq I$ and open sets $G_{\xi\iota}\subseteq X_\iota$, for $\iota\in J(\xi)$, such that

$$\varnothing\neq G'_\xi=\{x:x\in X, x(\iota)\in G_{\xi\iota}\ \forall\iota\in J(\xi)\}\subseteq G_\xi.$$

By the Δ-system lemma, there is a an uncountable $C\subseteq\omega_1$ such that $\langle J(\xi)\rangle_{\xi\in C}$ is a Δ-system; let J be its root. Now $\prod_{\iota\in J}X_\iota$ is ccc, so there are distinct $\xi,\eta\in C$ such that

$$\prod_{\iota\in J}G_{\xi\iota}\cap\prod_{\iota\in J}G_{\eta\iota}\neq\varnothing.$$

In this case however we can find an $x\in X$ such that

$$\begin{aligned} &x(\iota)\in G_{\xi\iota}\cap G_{\eta\iota} && \text{for } \iota\in J\\ &x(\iota)\in G_{\xi\iota} && \text{for } \iota\in J(\xi)\setminus J\\ &x(\iota)\in G_{\eta\iota} && \text{for } \iota\in J(\eta)\setminus J \end{aligned}$$

and now $x\in G_\xi\cap G_\eta$. As $\langle G_\xi\rangle_{\xi<\omega_1}$ is arbitrary, X is ccc.

12J Corollary

Let Z be a ccc topological space, and $\langle X_\iota\rangle_{\iota\in I}$ a family of separable topological spaces. Then $Z\times\prod_{\iota\in I}X_\iota$ is ccc.

Proof By Theorem 12I, we need only consider finite sets I; in this case $X=\prod_{\iota\in I}X_\iota$ is also separable. Let $Y\subseteq X$ be a countable dense set. If $\langle G_\xi\rangle_{\xi<\omega_1}$ is any family of non-empty open sets in $Z\times X$, there are non-empty open sets $E_\xi\subseteq Z$, $H_\xi\subseteq X$ such that $E_\xi\times H_\xi\subseteq G_\xi$ for each ξ. As Y is dense in X, every H_ξ must meet Y; as Y is countable, there is a $y\in Y$ such that $C=\{\xi:y\in H_\xi\}$ is uncountable. Now, because Z is ccc, there must be distinct ξ, $\eta\in C$ such that $E_\xi\cap E_\eta\neq\varnothing$; in which case $G_\xi\cap G_\eta\neq\varnothing$.

12K Theorem

Let $\langle X_\iota\rangle_{\iota\in I}$ be a family of separable topological spaces with $\#(I)\leq\mathfrak{c}$. Then $X=\prod_{\iota\in I}X_\iota$ is separable.

Proof If one of the X_ι is empty, so is X, and we're done. Otherwise, let $f:I\to[0,1]$ be an injection, and for $\iota\in I$ choose a function $g_\iota:\mathbf{N}\to X$ such that $g_\iota[\mathbf{N}]$ is dense in X_ι. Define $\varphi:\mathbf{N}^{[0,1]}\to X$ by

$$(\varphi h)(\iota)=g_\iota(h(f(\iota)))\ \forall\,\iota\in I,\ h\in\mathbf{N}^{[0,1]}.$$

Then φ is continuous (if $\mathbf{N}$ is given the discrete topology and $\mathbf{N}^{[0,1]}$ the product topology), and $\varphi[\mathbf{N}^{[0,1]}]=\prod_{\iota\in I}g_\iota[\mathbf{N}]$ is dense in X. Let $D\subseteq\mathbf{N}^{[0,1]}$ be the set of functions $h:[0,1]\to\mathbf{N}$ such that h is of bounded variation and continuous at all irrational points. Members of D can jump only at rational points and can make only finitely many jumps, so D is countable. Clearly D is dense in $\mathbf{N}^{[0,1]}$ so $\varphi[D]$ is a countable dense set in X.

12L **Sources**

MARCZEWSKI 47 for 12K. MARTIN & SOLOVAY 70 for 12E–F. JUHÁSZ 71 for 12I.

12M **Exercises**

(*a*) We say that a topological space X satisfies **Knaster's condition** if whenever $\mathscr{G}$ is an uncountable family of open sets in X there is an uncountable $\mathscr{H}\subseteq\mathscr{G}$ such that $G\cap H\neq\varnothing$ for all $G,H\in\mathscr{H}$. Formulate a definition of 'Knaster's condition' for Boolean algebras so that results corresponding to those of 12D*c* and 12E will be true. Show that the product of any family of topological spaces satisfying Knaster's condition satisfies Knaster's condition, and that the product of a ccc topological space with one satisfying Knaster's condition is ccc. [MARCZEWSKI 47; COMFORT & NEGREPONTIS 82, 7.4; 41E below.]

(*b*) Let X be a compact Hausdorff space, $\mathscr{G}$ its regular open algebra, and Z the Stone space of $\mathscr{G}$. Then there is a continuous surjection (the *Gleason map*) $\varphi:Z\to X$ defined by saying that if $G\in\mathscr{G}$ and $z\in\hat{G}$, the open-and-closed subset of Z corresponding to G, then $\varphi(z)\in\bar{G}$ in X. (Compare A7H*c*.)

(*c*) Let X be a zero-dimensional compact Hausdorff space and $\mathscr{E}$ its algebra of open-and-closed sets. Then there is a natural one-to-one correspondence between open sets in X and ideals in $\mathscr{E}$. If $\mathscr{J}\lhd\mathscr{E}$ and $G=\bigcup\mathscr{J}\subseteq X$, then $X\setminus G$ is (homeomorphic to) the Stone space of $\mathscr{E}/\mathscr{J}$. (Compare 26A.)

Notes and comments

The relationships between partially ordered sets, Boolean algebras and compact Hausdorff spaces set out in 12B–D form part of the machinery of forcing; but we shall not in this book be concerned with such applications. The proof of 12E is a direct translation of that given by MARTIN & SOLOVAY 70, and readers who are readier to make friends with strange partially ordered sets than with strange topologies may prefer the original. The basic open sets $U(C,R)$ may be thought of as forcing conditions; a downwards-directed family of $U(C,R)$ corresponds to upwards-directed R, C and 'forces' a set I which has finite intersection with each R.

Recall that the Borel structure of $\{0,1\}^{\mathbf{N}}$ is shared by any uncountable Borel set in any Polish space [A5E*c*], so that there is nothing special about $\{0,1\}^{\mathbf{N}}$ in 12F; $\mathbf{R}$ or $[0,1]$, for instance would do just as well. Note that if $\mathcal{J}$ is an ω_1-saturated σ-ideal of $\mathcal{B}=\mathcal{B}(\{0,1\}^{\mathbf{N}})$, then $\mathcal{B}/\mathcal{J}$ will be weakly countably generated and Dedekind complete. Thus 12F is a complete representation theorem for weakly countably generated Dedekind complete ccc Boolean algebras; but its value is limited by the absence of a useful classification of the ω_1-saturated σ-ideals of $\mathcal{B}$.

One of the original applications of Martin's axiom was to prove that the product of ccc topological spaces is ccc [41E]. Without special axioms, this need not be true; but we can prove some interesting and useful partial results [12I–K, 12M*a*]. I have taken the trouble to spell both forms of the Δ-system lemma out at length because it will be useful later to have both immediately to hand.

13 Other definitions of $\mathfrak{m}$

In a portmanteau theorem [13A] I give the most important alternative formulations of MA(κ): the compact-Hausdorff-space form [13A(iii)], the Boolean-algebra form [13A(v) and 13C*c*(ii)] and the ideals-of-Borel-sets form [13A(vi)], with a couple of other versions. It is worth noting that there are unobvious variations on the partially-ordered-set form given in 11B*a* [13A(ii)].

13A Theorem

For a cardinal κ, the following are equivalent.

(i) $\kappa<\mathfrak{m}$.

(ii) If P is a non-empty upwards-ccc partially ordered set with $\#(P)\leq\kappa$, and $\mathcal{Q}$ is a family of up-open cofinal subsets of P with $\#(\mathcal{Q})\leq\kappa$, then there is an upwards-linked subset of P meeting every member of $\mathcal{Q}$.

(iii) If X is a non-empty ccc compact Hausdorff space, it is not the union of κ or fewer nowhere-dense sets.

(iv) If X is a non-empty ccc compact Hausdorff space and $w(X)\leq\kappa$, then X is not the union of κ or fewer nowhere-dense sets.

(v) Let $\mathfrak{A}$ be a ccc Boolean algebra, not $\{0\}$, and $\mathscr{A}$ a family of subsets of $\mathfrak{A}$ with $\#(\mathscr{A}) \leq \kappa$. Then there is a Boolean homomorphism $\varphi: \mathfrak{A} \to \mathbf{Z}_2$ such that, for every $A \in \mathscr{A}$, either there is an $a \in A$ with $\varphi(a) = 1$, or there is an upper bound b of A with $\varphi(b) = 0$.

(vi) Let $\mathscr{J}$ be a proper ω_1-saturated σ-ideal of the σ-algebra of Borel subsets of $\{0, 1\}^{\mathbf{N}}$. Then $\{0, 1\}^{\mathbf{N}}$ is not the union of κ or fewer members of $\mathscr{J}$.

(vii) Let X be a non-empty compact Hausdorff space of weight less than or equal to κ, Y a ccc compact Hausdorff space, and $f: X \to Y$ an irreducible continuous surjection. Then there is a $u \in Y$ such that $f^{-1}[\{u\}]$ is a singleton set.

Proof I show that (i) $\Rightarrow$ (iii) $\Rightarrow$ (v) $\Rightarrow$ (ii) $\Rightarrow$ (i); (v) $\Leftrightarrow$ (vi); and (iii) $\Rightarrow$ (vii) $\Rightarrow$ (iv) $\Rightarrow$ (iii). Since it is easy to see that (i)–(vii) are all true for all finite κ, I assume throughout the proof that $\kappa \geq \omega$.

(*a*) **(i) $\Rightarrow$ (iii)** Assume that $\kappa < \mathfrak{m}$. Let X be a non-empty ccc compact Hausdorff space and $\mathscr{E}$ a family of nowhere-dense subsets of X with $\#(\mathscr{E}) \leq \kappa$. Let P be $\{G: G \subseteq X \text{ open}, G \neq \varnothing\}$, ordered by $\subseteq$. Then P is non-empty and downwards-ccc [12D*c*(v)]. For $E \in \mathscr{E}$ set

$$Q_E = \{G: G \in P, \bar{G} \cap E = \varnothing\}.$$

Because X is regular and E is nowhere dense, Q_E is coinitial with P. Now $\mathscr{Q} = \{Q_E: E \in \mathscr{E}\}$ is a family of fewer than $\mathfrak{m}$ coinitial subsets of P, so by the definition of $\mathfrak{m}$ (inverted) there is a downwards-directed set $R \subseteq P$ meeting every Q_E. R has the finite intersection property, so

$$F = \bigcap\{\bar{G}: G \in R\} \neq \varnothing.$$

But $F \cap \bigcup \mathscr{E} = \varnothing$, because R meets every Q_E. So $X \neq \bigcup \mathscr{E}$. As X and $\mathscr{E}$ are arbitrary, (iii) is true.

(*b*) **(iii) $\Rightarrow$ (v)** Assume (iii). Let $\mathfrak{A}$ be a non-trivial ccc Boolean algebra and $\mathscr{A}$ a family of subsets of $\mathfrak{A}$ with $\#(\mathscr{A}) \leq \kappa$. Express $\mathfrak{A}$ as the algebra of open-and-closed subsets of its Stone space Z. For $A \subset \mathscr{A}$ set $F_A = \overline{\bigcup A} \setminus \bigcup A \subseteq Z$. Then F_A is a closed nowhere-dense set. As Z is ccc [12D*c*(iii)], $Z \neq \bigcup_{A \in \mathscr{A}} F_A$ (using (iii)). Let $t \in Z \setminus \bigcup_{A \in \mathscr{A}} F_A$. Then t corresponds to a Boolean homomorphism $\hat{t}: \mathfrak{A} \to \mathbf{Z}_2$ defined by writing $\hat{t}(E) = (\chi E)(t)$ for $E \in \mathfrak{A}$. If $A \in \mathscr{A}$, then either $t \in \bigcup A$ and $\hat{t}(E) = 1$ for some $E \in A$, or $t \notin \overline{\bigcup A}$ and $\hat{t}(E) = 1$ for some $E \in \mathfrak{A}$ with $E \cap \bigcup A = \varnothing$; in the latter case, $\hat{t}(Z \setminus E) = 0$ and $Z \setminus E$ is an upper bound for A. Thus $\hat{t}$ is the homomorphism required by (v).

(*c*) **(v) $\Rightarrow$ (ii)** Assume (v). Let P be a non-empty upwards-ccc partially ordered set, and $\mathscr{Q}$ a family of cofinal subsets of P with $\#(\mathscr{Q}) \leq \kappa$.

Give P its up-topology [12C], and let $\mathscr{G}$ be the regular open algebra of P. Then $\mathscr{G}$ is ccc [12D*c*].

For $p \in P$ let $G_p = \operatorname{int}\overline{\{q : q \geq p\}} \in \mathscr{G}$. For $Q \in \mathscr{Q}$ write $A_Q = \{G_q : q \in Q\}$. By (v), there is a Boolean homomorphism $\varphi : \mathscr{G} \to \mathbf{Z}_2$ such that

$$\forall Q \in \mathscr{Q}, \text{ either } \exists q \in Q, \varphi(G_q) = 1$$
$$\text{or } \exists G \in \mathscr{G}, G_q \subseteq G \; \forall q \in Q, \varphi(G) = 0.$$

But for any $Q \in \mathscr{Q}$, $\bigcup A_Q \supseteq Q$ is dense in P, so the only member of $\mathscr{G}$ which includes every member of A_Q is P itself; thus the second alternative can never occur, and for every $Q \in \mathscr{Q}$ there is a $q \in Q$ with $\varphi(G_q) = 1$.

Set $R = \{p : p \in P, \varphi(G_p) = 1\}$. Then R meets every member of $\mathscr{Q}$. If p and q belong to R, then $\varphi(G_p \cap G_q) = 1$, so $G_p \cap G_q \neq \emptyset$; consequently $G_p \cap \overline{\{r : r \geq q\}} \neq \emptyset$, $G_p \cap \{r : r \geq q\} \neq \emptyset$, and similarly $\{r : r \geq p\} \cap \{r : r \geq q\} \neq \emptyset$. As p and q are arbitrary, this means that R is upwards-linked, as required for (ii).

(*d*) **(ii)$\Rightarrow$(i)** Assume (ii), and let P be any non-empty upwards-ccc partially ordered set, with a family $\mathscr{Q}$ of not more than κ cofinal subsets.

Our first task is to cut P down to not more than κ elements, which we can do as follows. Let S_0 be a singleton subset of P. Given $S_n \subseteq P$ with $\#(S_n) \leq \kappa$, let S_{n+1} be such that

(α) $S_n \subseteq S_{n+1} \subseteq P$;
(β) $\forall p \in S_n$, $Q \in \mathscr{Q} \, \exists q \in Q \cap S_{n+1}$ such that $q \geq p$;
(γ) whenever $I \subseteq S_n$ is a finite set with an upper bound in P, it has an upper bound in S_{n+1};
(δ) $\#(S_{n+1}) \leq \max(\omega, \#(S_n), \#(\mathscr{Q})) \leq \kappa$.

Set $\tilde{P} = \bigcup_{n \in \mathbf{N}} S_n$. $\tilde{P}$ is non-empty because S_0 is non-empty; $\#(\tilde{P}) \leq \kappa$, by ($\delta$); $\tilde{P} \cap Q$ is cofinal with $\tilde{P}$ for every $Q \in \mathscr{Q}$, by (β); and a subset of $\tilde{P}$ is an up-antichain in $\tilde{P}$ iff it is an up-antichain in P, by (γ), so $\tilde{P}$ is upwards-ccc.

For $Q \in \mathscr{Q}$, set $\tilde{Q} = \{p : p \in \tilde{P}, \exists q \in Q, p \geq q\}$, so that $\tilde{Q}$ is up-open and cofinal with $\tilde{P}$. Next, for each pair p, q of elements of $\tilde{P}$, set

$$R_{pq} = \{r : r \in \tilde{P} \text{ and either } r \geq p, r \geq q$$
$$\text{or } \{r, p\} \text{ has no upper bound in } \tilde{P}$$
$$\text{or } \{r, q\} \text{ has no upper bound in } \tilde{P}\}.$$

Clearly R_{pq} is up-open in $\tilde{P}$. Also it is cofinal with $\tilde{P}$. **P** If $r \in \tilde{P}$, then either $r \in R_{pq}$ or $\{r, p\}$ has an upper bound r_1 in $\tilde{P}$. In the latter case, either $r_1 \in R_{pq}$ or $\{r_1, q\}$ has an upper bound $r_2 \in \tilde{P}$. But in this last case r_2 is an upper bound of $\{p, q\}$ so belongs to R_{pq}. Thus we have either $r \in R_{pq}$ or $r \leq r_1 \in R_{pq}$ or $r \leq r_2 \in R_{pq}$. **Q**

So by (ii) there is an upwards-linked $\tilde{R} \subseteq \tilde{P}$ meeting every $\tilde{Q}$ and every R_{pq}. Now $\tilde{R}$ is upwards-directed. **P** Let $p, q \in \tilde{R}$. Then $\tilde{R}$ meets R_{pq}; say $r \in \tilde{R} \cap R_{pq}$. Since $\{r, p\}$ and $\{r, q\}$ must both have upper bounds in $\tilde{P}$, we see that r is the required upper bound of $\{p, q\}$ in $\tilde{R}$. **Q**

Accordingly, if we set

$$R = \{p : p \in P, \exists r \in \tilde{R}, p \leq r\}$$

we see that R is upwards-directed and that R meets every $Q \in \mathcal{Q}$ because $\tilde{R}$ meets every $\tilde{Q}$. As P and $\mathcal{Q}$ are arbitrary, $\kappa < \mathfrak{m}$.

(*e*) **(v)** $\Rightarrow$ **(vi)** Assume (v). Let $\mathcal{J}$ be a proper ω_1-saturated σ-ideal of $\mathcal{B}$, the algebra of Borel sets in $\{0,1\}^{\mathbf{N}}$. For any $\Phi \subseteq \mathcal{B}^{\mathbf{N}}$, let $\mathcal{S}(\Phi)$ be the smallest collection of subsets of $\{0,1\}^{\mathbf{N}}$ such that

every open-and-closed set belongs to $\mathcal{S}(\Phi)$;
the complement of any set in $\mathcal{S}(\Phi)$ belongs to $\mathcal{S}(\Phi)$;
if $\langle S_n \rangle_{n \in \mathbf{N}} \in \Phi \cap \mathcal{S}(\Phi)^{\mathbf{N}}$ then $\bigcup_{n \in \mathbf{N}} S_n \in \mathcal{S}(\Phi)$.

Then we see that $\bigcup \{\mathcal{S}(\Phi) : \Phi \subseteq \mathcal{B}^{\mathbf{N}} \text{ is countable}\}$ is a σ-algebra of sets so must be $\mathcal{B}$ itself, and we can choose for each $B \in \mathcal{B}$ a countable $\Phi_B \subseteq \mathcal{B}^{\mathbf{N}}$ such that $B \in \mathcal{S}(\Phi_B)$.

Let $\mathcal{A} \subseteq \mathcal{J}$, $\#(\mathcal{A}) \leq \kappa$. Let $\mathfrak{A}$ be the quotient algebra $\mathcal{B}/\mathcal{J}$; then $\mathfrak{A}$ is non-zero, ccc and Dedekind complete. Applying (v) to the family

$$\{\{E_n^{\cdot} : n \in \mathbf{N}\} : \langle E_n \rangle_{n \in \mathbf{N}} \in \bigcup_{A \in \mathcal{A}} \Phi_A\}$$

of subsets of $\mathfrak{A}$, there is a Boolean homomorphism $\varphi : \mathfrak{A} \to \mathbf{Z}_2$ such that

$$\varphi(\sup_{n \in \mathbf{N}} E_n^{\cdot}) = \sup_{n \in \mathbf{N}} \varphi E_n^{\cdot} \; \forall \langle E_n \rangle_{n \in \mathbf{N}} \in \bigcup_{A \in \mathcal{A}} \Phi_A.$$

Set $H_n = \{s : s(n) = 1\} \in \mathcal{B}$ for each $n \in \mathbf{N}$, and define $t \in \{0,1\}^{\mathbf{N}}$ by writing

$$t(n) = \varphi(H_n^{\cdot}) \; \forall n \in \mathbf{N}.$$

Consider

$$\mathcal{S} = \{E : E \in \mathcal{B}, \varphi(E^{\cdot}) = (\chi E)(t)\}.$$

Then $\mathcal{S}$ is a subalgebra of $\mathcal{B}$ containing every H_n, so contains all the open-and-closed sets. Moreover, if $A \in \mathcal{A}$ and $\langle E_n \rangle_{n \in \mathbf{N}} \in \mathcal{S}^{\mathbf{N}} \cap \Phi_A$, then

$$\varphi((\bigcup_{n \in \mathbf{N}} E_n)^{\cdot}) = \varphi(\sup_{n \in \mathbf{N}} E_n^{\cdot}) = \sup_{n \in \mathbf{N}} \varphi(E_n^{\cdot})$$

(because $\langle E_n \rangle_{n \in \mathbf{N}} \in \Phi_A$)

$$= \sup_{n \in \mathbf{N}} (\chi E_n)(t) = (\chi(\bigcup_{n \in \mathbf{N}} E_n))(t)$$

so that $\bigcup_{n \in \mathbf{N}} E_n \in \mathcal{S}$. So $\mathcal{S} \supseteq \mathcal{S}(\Phi_A)$ and $A \in \mathcal{S}$.

But in this case $(\chi A)(t) = \varphi(A^{\cdot}) = \varphi(0) = 0$, for every $A \in \mathcal{A}$. Thus $t \notin \bigcup \mathcal{A}$ and $\bigcup \mathcal{A} \neq \{0,1\}^{\mathbf{N}}$, as required by (vi).

(*f*) **(vi)⇒(v)** Assume (vi). Let $\mathfrak{A}$ be a non-zero ccc Boolean algebra and $\mathscr{A}$ a family of subsets of $\mathfrak{A}$ with $\#(\mathscr{A})\le\kappa$. For each $A\in\mathscr{A}$, set $A'=A\cup\{b:a\cap b=0\ \forall a\in A\}$. Then $\sup A'=1$ in $\mathfrak{A}$, so there is a sequence $\langle b_n^A\rangle_{n\in\mathbf{N}}$ in A' such that $\sup_{n\in\mathbf{N}}b_n^A=1$ [12D*b*]. Of course (vi) implies that $\kappa<\mathfrak{c}$ (since there are non-trivial proper ω_1-saturated σ-ideals of Borel sets; e.g. the ideal of meagre Borel sets). So the subalgebra $\mathfrak{B}$ of $\mathfrak{A}$ generated by $\{b_n^A:A\in\mathscr{A},\ n\in\mathbf{N}\}$ has cardinal less than $\mathfrak{c}$.

Consequently $\mathfrak{B}$ can be embedded as a regular subalgebra of $\mathscr{B}/\mathscr{I}$ for some ω_1-saturated σ-ideal $\mathscr{I}$ of $\mathscr{B}$, the algebra of Borel sets in $\{0,1\}^{\mathbf{N}}$ [12E–F]. For each $n\in\mathbf{N}$, $A\in\mathscr{A}$, choose $E_n^A\in\mathscr{B}$ such that $(E_n^A)^{\cdot}=b_n^A$. Let $\mathscr{B}_1$ be the smallest subalgebra of $\mathscr{B}$ including

$$\{E_n^A:n\in\mathbf{N},A\in\mathscr{A}\}\cup\{\textstyle\bigcup_{n\in\mathbf{N}}E_n^A:A\in\mathscr{A}\}.$$

Then $\#(\mathscr{B}_1)\le\kappa$. By (vi), $\bigcup(\mathscr{B}_1\cap\mathscr{I})$ cannot be $\{0,1\}^{\mathbf{N}}$; take $t\in\{0,1\}^{\mathbf{N}}\setminus\bigcup(\mathscr{B}_1\cap\mathscr{I})$. Let $\mathscr{J}$ be the ideal of $\mathscr{B}/\mathscr{I}$ generated by $\{E^{\cdot}:E\in\mathscr{B}_1,t\notin E\}$. Then $\mathscr{J}$ is a proper ideal because if $E\in\mathscr{B}_1$ and $t\notin E$ then $\{0,1\}^{\mathbf{N}}\setminus E\notin\mathscr{I}$, i.e. $E^{\cdot}\ne1$ in $\mathscr{B}/\mathscr{I}$. So $\mathscr{J}\cap\mathfrak{B}$ is a proper ideal of $\mathfrak{B}$. Once again regarding $\mathfrak{B}$ as a subalgebra of $\mathfrak{A}$, let $\mathscr{J}^*$ be a maximal ideal of $\mathfrak{A}$ including $\mathscr{J}\cap\mathfrak{B}$, and let $\varphi:\mathfrak{A}\to\mathbf{Z}_2$ be the Boolean homomorphism with kernel $\mathscr{J}^*$.

If $A\in\mathscr{A}$, then $\sup_{n\in\mathbf{N}}b_n^A=1$ in $\mathfrak{A}$. As $\mathfrak{B}$ is a subalgebra of $\mathfrak{A}$, $\sup_{n\in\mathbf{N}}b_n^A=1$ in $\mathfrak{B}$. As $\mathfrak{B}$ is a regular subalgebra of $\mathscr{B}/\mathscr{I}$, $\sup_{n\in\mathbf{N}}b_n^A=1$ in $\mathscr{B}/\mathscr{I}$. Accordingly $\{0,1\}^{\mathbf{N}}\setminus\bigcup_{n\in\mathbf{N}}E_n^A$ belongs to $\mathscr{I}$; as it also belongs to $\mathscr{B}_1$, $t\notin\{0,1\}^{\mathbf{N}}\setminus\bigcup_{n\in\mathbf{N}}E_n^A$ and there is some $n\in\mathbf{N}$ such that $t\in E_n^A\in\mathscr{B}_1$. Now $(\{0,1\}^{\mathbf{N}}\setminus E_n^A)^{\cdot}=1\setminus b_n^A\in\mathscr{J}$ and $\varphi(1\setminus b_n^A)=0$, i.e. $\varphi b_n^A=1$.

This means that either there is an $a\in A$ with $\varphi a=1$, or there is a $b\in A'\setminus A$ with $\varphi b=1$; in which case $1\setminus b$ is an upper bound of A and $\varphi(1\setminus b)=0$. Thus φ is a homomorphism of the kind required by (v).

(*g*) **(iii)⇒(vii)** Assume (iii), and let X, Y and f be as in the statement of (vii). If $G,H\subseteq X$ are disjoint open sets, $E_{GH}=\overline{f[G]}\cap\overline{f[H]}$ is nowhere dense in Y. **P** Set $U=f^{-1}[\operatorname{int}E_{GH}]$. Then $f[G\cap U]\subseteq E_{GH}\subseteq\overline{f[H]}$, so $\overline{f[X\setminus(G\cap U)]}=\overline{f[X\setminus(G\cap U)]}=f[X]=Y$. Because f is irreducible, $G\cap U=\varnothing$, and $f[G]$ does not meet $\operatorname{int}E_{GH}$. But $E_{GH}\subseteq\overline{f[G]}$, so $\operatorname{int}E_{GH}$ must be $\varnothing$. **Q**

Let $\mathscr{U}$ be a base for the topology of X, with $\#(\mathscr{U})\le\kappa$. Set $\mathscr{A}=\{(G,H):G,H\in\mathscr{U},G\cap H=\varnothing\}$; then $\#(\mathscr{A})\le\kappa$, so by (iii) there is a $u\in Y\setminus\bigcup_{(G,H)\in\mathscr{A}}E_{GH}$. As X is Hausdorff and $u\notin\overline{f[G]}\cap\overline{f[H]}$ for any pair $(G,H)\in\mathscr{A}$, $f^{-1}[\{u\}]$ cannot have more than one element, and must be a singleton, as required by (vii).

(*h*) **(vii)⇒(iv)** Assume (vii). Let Z be a non-empty ccc compact Hausdorff space of weight less than or equal to κ, and $\mathscr{E}$ a family of nowhere-dense sets in Z with $\#(\mathscr{E}) \leq \kappa$. Set $X = Z \times \{0,1\}^{\mathscr{E}}$. Define an equivalence relation $\sim$ on X by writing

$$(z,u) \sim (z',u') \text{ if } z = z' \text{ and } \forall\, E \in \mathscr{E}, \text{ either } u(E) = u'(E) \text{ or } z \in \bar{E}.$$

Let Y be $X/\sim$ and give Y the quotient topology. It is easy to check that Y is Hausdorff (because $\sim$ is closed in $X \times X$, or otherwise). Because Z has weight less than or equal to κ and $\#(\mathscr{E}) \leq \kappa$, $w(X) \leq \kappa$; because Z is ccc, so is X [12J] and therefore so is Y. Let $f: X \to Y$ be the quotient map.

f is irreducible. **P** If F is a proper closed subset of X, let $(z,u) \in X \setminus F$. Then there is an open set $G \subseteq Z$ and a finite set $\mathscr{C} \in \mathscr{E}$ such that

$$(z,u) \in G \times \{v : v \restriction \mathscr{C} = u \restriction \mathscr{C}\} \subseteq X \setminus F.$$

Now $G \nsubseteq \overline{\bigcup \mathscr{C}}$, because each member of $\mathscr{C}$ is nowhere dense; let $w \in G \setminus \overline{\bigcup \mathscr{C}}$. ? If $f(w,u) \in f[F]$, there is a $(w',u') \in F$ such that $(w',u') \sim (w,u)$. We must have $w' = w \in G \setminus \overline{\bigcup \mathscr{C}}$. So $u'(E) = u(E)$ for every $E \in \mathscr{C}$. But this means that $(w',u') \notin F$, by the choice of G and $\mathscr{C}$. **X** Thus $f[F] \neq Y$. As F is arbitrary, f is irreducible. **Q**

By (vii), there is a $y \in Y$ such that $f^{-1}[\{y\}]$ is a singleton $\{(z,u)\}$ say. If $E \in \mathscr{E}$, define $u_E \in \{0,1\}^{\mathscr{E}}$ by writing

$$u_E(E) = 1 - u(E), \quad u_E(E') = u(E') \forall E' \in \mathscr{E} \setminus \{E\}.$$

Then $f(z,u_E) \neq y$, so $(z,u_E) \nsim (z,u)$ and $z \notin \bar{E}$. As E is arbitrary, $z \notin \bigcup \mathscr{E}$. As Z and $\mathscr{E}$ are arbitrary, (iv) is true.

(*i*) **(iv)$\Rightarrow$(iii)** Assume (iv). Let X be any non-empty ccc compact Hausdorff space, and $\mathscr{E}$ a family of nowhere-dense sets in X with $\#(\mathscr{E}) \leq \kappa$. Because X is ccc and completely regular, we can choose for each $E \in \mathscr{E}$ a continuous function $f_E : X \to [0,1]$ such that

$$E \subseteq F_E = f_E^{-1}[\{0\}], \quad \operatorname{int} F_E = \emptyset$$

[A5B*e*]. Define $f: X \to [0,1]^{\mathscr{E}}$ by writing $(ft)(E) = f_E(t)$ for $E \in \mathscr{E}$ and $t \in X$, and set $Z = f[X]$. Then Z is ccc [A4D*c*], compact, Hausdorff and $w(Z) \leq \kappa$ [A4D*a*]. Also $f[F_E]$ has empty interior in Z for each E because $F_E = f^{-1}[f[F_E]]$ has empty interior in X. So, by (iv), $Z \neq \bigcup_{E \in \mathscr{E}} f[F_E]$ and $X \neq \bigcup \mathscr{E}$. This proves (iii).

13B Sources

MARTIN & SOLOVAY 70 for the equivalence of 13A(i), 13A(v) (in the form of 13C*c*(ii)) and 13A (vi). JUHÁSZ 71 for 13A(iii)$\Leftrightarrow$(v). BENNETT & MCLAUGHLIN 76 for 13A(ii)$\Rightarrow$(i). MALYHIN 77 for 13A(vii).

13C **Exercises**

(*a*) Formulate statements which will be equivalent to $\kappa < \mathfrak{m}_K$ and will correspond to 13A(ii)–(vii) above. [Use 12M*a*. See KUNEN & TALL 79].

(*b*) Find (i) a non-empty compact Hausdorff space which is the union of ω_1 nowhere-dense subsets (ii) a proper σ-ideal $\mathcal{J}$ of the algebra of Borel sets of $\{0,1\}^{\mathbf{N}}$ such that $\{0,1\}^{\mathbf{N}}$ is the union of ω_1 members of $\mathcal{J}$. [Hints for (ii): A5G*c* or 21L.]

(*c*) Show that the following statements are equivalent to '$\kappa < \mathfrak{m}$'.

(i) Let X be a non-empty ccc topological space and $\mathcal{B}$ a π-base for the topology of X. Let $\mathcal{H}$ be a family of dense open sets in X with $\#(\mathcal{H}) \le \kappa$. Then there is a downwards-directed $\mathcal{R} \subseteq \mathcal{B}$ such that every member of $\mathcal{H}$ includes a member of $\mathcal{R}$.

(ii) Let $\mathfrak{A}$ be a non-zero Dedekind complete ccc Boolean algebra and $\mathcal{A} \subseteq \mathcal{P}\mathfrak{A}$, $\#(\mathcal{A}) \le \kappa$. Then there is a Boolean homomorphism $\varphi = \mathfrak{A} \to \mathbf{Z}_2$ such that $\varphi(\sup A) = \sup \varphi[A]$ for every $A \in \mathcal{A}$.

(*d*) Let $\mathrm{MA}^{\mathrm{Lus}}$ be the statement

$\mathfrak{c}$ is regular and whenever $\mathcal{J}$ is an ω_1-saturated σ-ideal of the Borel algebra $\mathcal{B}$ of $\{0,1\}^{\mathbf{N}}$ containing all the singleton sets, there is a $(\mathcal{B}, \mathcal{J}, \mathfrak{c})$-Lusin set.

[For the definition of 'Lusin set', see A3E.] Show that $\mathrm{MA} \Leftrightarrow \mathrm{MA}^{\mathrm{Lus}}$. [For $\Rightarrow$, use A3F*a* and 13A(vi), with the help of A5E*c*. See MILLER 79.]

(e) Show that $\kappa < \mathfrak{m}$ implies (and is therefore equivalent to) each of the following.

(i) Let X be an analytic space, $\mathcal{J}$ a proper ω_1-saturated σ-ideal of the algebra of Borel subsets of X. Then X is not the union of κ or fewer members of $\mathcal{J}$.

[Hint: first consider Polish X, using 13A(vi) and A5E*c*; then other metrizable X, using A5G*c*; and finally the general case, using A5F*e*.]

(ii) Let X be a compact Hausdorff space, $\mathcal{J}$ a proper ω_1-saturated σ-ideal of the algebra of Borel subsets of X. Suppose that for every open $G \subseteq X$ there is an F_σ set $H \subseteq G$ such that $G \setminus H \in \mathcal{J}$. Then X is not the union of κ or fewer members of $\mathcal{J}$.

[Hint: adapt the argument of 13A prf *e*]

13D **Further results**

(*a*) STARK 80 gives a result, equivalent to MA, in the theory of models for infinitary logics; it is an extension of the Barwise completeness theorem (Barwise 75, §III.5.6). The problem is to find necessary and sufficient conditions for the existence of a model for a theory T in a language L in which infinite conjunctions and disjunctions of propositions are allowed. Barwise gave such conditions, in terms of the existence of models for enough fragments of T, for countable T. Stark extends this to theories T of cardinal less than $\mathfrak{c}$ such that an adequate family F of consistent finite extensions of T is upwards-ccc. The idea is to use an upwards-directed subset of F to determine which of the other sentences of L are to be true in the model. Stark shows also that every partially ordered set can be used to define an appropriate theory and accordingly that his theorem is actually equivalent to MA.

(*b*) The following statement is equivalent to $\mathfrak{m} > \omega_1$:

Let P be an uncountable upwards-ccc partially ordered set. Then P has an uncountable upwards-directed subset.

[Fremlin *b*. Compare H of 41L below.]

(*c*) If $\mathfrak{p} > \omega_1$ then $\mathfrak{m}$ is equal to the cardinal $\mathfrak{l}$ of 41L below. [Fremlin *b*. See 41P*a*]

Notes and comments

Part (*a*) of the proof of 13A is our second example of a proof using MA(κ); and it is of course very easy, since the partially ordered set we need is directly in front of us and is ccc by hypothesis. For anyone who is using this book to learn about Martin's axiom for the first time, I should like to recommend, as further practice, writing out proofs that $\kappa < \mathfrak{m}$ implies the statement of 13C*c*(i), and (just a little less obvious) a direct proof of 13A(i) $\Rightarrow$(v).

Note that, for $\kappa = \omega$, 13A(iii) is nothing but Baire's theorem for compact Hausdorff spaces, 13A(v) is the Rasiowa–Sikorski theorem, and 13A(vi) is trivial; in none of these, of course, is the ccc involved, just as it is not needed in MA(ω). Let me again recommend that you take the trouble to work out examples which are not ccc [13C*b*].

13A(iii) is of course one of the most important properties of $\mathfrak{m}$, although in this book it is overshadowed by its extension in 43F*a*. The fact that it characterizes $\mathfrak{m}$ is practically never used in the formal sense, since the usual consistency proofs work directly with MA(κ). But one of the things I am trying to emphasize in this book is that the cardinal $\mathfrak{m}$ is one you might encounter even if you think you are interested only in general topology or Boolean algebras.

Observe that 13A(vi) is *not* the same thing as saying that if $\mathcal{J}$ is an ω_1-saturated σ-ideal of the Borel algebra of $\{0,1\}^{\mathbb{N}}$, then $\mathcal{J}$ is κ^+-additive; in fact there is always an ω_1-saturated σ-ideal which is not ω_2-additive [Martin & Solovay 70, p. 160]. However we shall see below [22B, 32F] that ideals of negligible and meagre sets in second-countable spaces are $\mathfrak{m}$-additive. 13A(vi) has natural extensions to analytic spaces [13C*e*(i)] and to compact Hausdorff spaces [13C*e*(ii)]. The regularity condition in the latter seems to be the essential common ingredient of the applications to negligible sets [32B] and to meagre sets [13A(iii)].

I think that V.I. Malyhin's formulation 13A(vii) is charming; I know of no other reason to take it seriously. Of course each element of 13A prf *g*–*h* is an important argument.

13C*c* gives some elementary variants of 13A. $\mathrm{MA}^{\mathrm{Lus}}$ [13C*d*] lies a little deeper. Of course 43E and 43F*a* also define $\mathfrak{m}$. In 41L I discuss a proposition *K* which implies that $\mathfrak{m}=\mathfrak{m}_K$ and a cardinal $\mathfrak{l}$ such that $\mathfrak{m}=\min(\mathfrak{p},\mathfrak{l})$.

14 Other definitions of $\mathfrak{p}$

The main result of this section is M.G. Bell's theorem that $\mathfrak{p}=\mathfrak{c}$ implies $\mathrm{MA}_{\sigma\text{-cent}}$, 'Martin's axion for σ-centered partially ordered sets'. The machinery of §§12–13 now provides further definitions of $\mathfrak{p}$, such as a separable-compact-Hausdorff-space form [14C(iv)]. The proof of Bell's theorem depends on one of the most important properties of $\mathfrak{p}$ [14B]. I give also F. Rothberger's equivalent form of the axiom $\mathfrak{p}>\omega_1$ [14D].

14A **σ-centered sets**

(*a*) Let *P* be a partially ordered set. A subset *R* of *P* is **upwards-centered** if every non-empty finite subset of *R* is bounded above in *P*. (Thus an upwards-directed set is upwards-centered, and an upwards-centered set is upwards-linked.) *P* is **σ-centered upwards** if it is expressible as a countable union of upwards-centered sets.

Similarly, *P* is **σ-centered downwards** if it is expressible as a countable union of downwards-centered sets.

(*b*) A topological space $(X,\mathfrak{T})$ is σ-centered if $\mathfrak{T}\setminus\{\emptyset\}$ is σ-centered downwards [A4Q–R]. Now a partially ordered set P is σ-centered upwards iff it is σ-centered for its up-topology [12C]. **P** Write $U_p=\{q:q\geq p\}$, $\mathcal{U}=\{U_p:p\in P\}$, so that $\mathcal{U}$ is a base for the up-topology $\mathfrak{T}$. A set $R\subseteq P$ is upwards-centered iff $\{U_p:p\in R\}$ has the finite intersection property, so that *P* is σ-centered upwards iff $\mathcal{U}$ is σ-centered downwards. But by A4Q*b* $\mathcal{U}$ is σ-centered downwards iff P is σ-centered under $\mathfrak{T}$. **Q**

(*c*) Let us write $\mathrm{MA}_{\sigma\text{-cent}}(\kappa)$ for the statement whenever P is a non-empty partially ordered set which is σ-centered upwards, and $\mathcal{Q}$ is a family of cofinal subsets of P with $\#(\mathcal{Q}) \leq \kappa$, there is an upwards-directed subset of P meeting every member of $\mathcal{Q}$.

[Compare 11B*a*–*b*.]

14B **Lemma**

Let $A \subseteq \mathbf{N}^{\mathbf{N}}$ be a set with $\#(A) < \mathfrak{p}$. Then there is a $g \in \mathbf{N}^{\mathbf{N}}$ such that $\{n : g(n) < f(n)\}$ is finite for every $f \in A$.

Proof Enumerate A as $\langle f_\xi \rangle_{\xi < \kappa}$ where $\kappa < \mathfrak{p}$. For infinite $C \subseteq \mathbf{N}$ let $g_C \in \mathbf{N}^{\mathbf{N}}$ be the enumeration of C in ascending order. Choose inductively sets $\langle C(\xi) \rangle_{\xi \leq \kappa}$ in $\mathbf{N}$ such that

$C(0) = \mathbf{N}$;

if $C(\xi)$ is infinite, where $\xi < \kappa$, then $C(\xi + 1)$ is to be an infinite subset of $C(\xi)$ such that $g_{C(\xi+1)}(n) \geq f_\xi(n) + n$ for every $n \in \mathbf{N}$;

if $\langle C(\eta) \rangle_{\eta < \xi}$ is such that there is an infinite set C with $C \backslash C(\eta)$ finite for each $\eta < \xi$, where $\xi \leq \kappa$ is a non-zero limit ordinal, then $C(\xi)$ is to be such a set.

Now we see by induction on ξ that every $C(\xi)$ is infinite and that $C(\xi) \backslash C(\eta)$ is finite whenever $\eta \leq \xi \leq \kappa$; at limit ordinals ξ we must apply $\mathrm{P}(\mathfrak{p})$ to the family $\{C(\eta) : \eta < \xi\}$ to see that $C(\xi)$ can be chosen according to the recipe offered.

In particular $C(\kappa) \backslash C(\xi + 1)$ is finite for every $\xi < \kappa$, so that

$$\{n : g_{C(\kappa)}(n) < f_\xi(n)\} \subseteq \{n : g_{C(\kappa)}(n) + n < g_{C(\xi+1)}(n)\}$$

is finite for every $\xi < \kappa$. So $g_{C(\kappa)}$ is the function we seek.

14C **Bell's theorem**

For any cardinal κ, the following are equivalent.

(i) $\kappa < \mathfrak{p}$.

(ii) $\mathrm{MA}_{\sigma\text{-cent}}(\kappa)$ [14A*c*].

(iii) Let P be a non-empty partially ordered set which is σ-centered upwards and has cardinal less than or equal to κ. If $\mathcal{Q}$ is a family of up-open cofinal subsets of P with $\#(\mathcal{Q}) \leq \kappa$, there is an upwards-linked subset of P meeting every member of $\mathcal{Q}$.

(iv) Let X be a non-empty separable compact Hausdorff space. Then X is not the union of κ or fewer nowhere-dense sets.

Proof (*a*) (i)⇒(iii) Assume that $\kappa < \mathfrak{p}$, and let P be a non-empty partially ordered set which is σ-centered upwards and has cardinality not more than κ. Let $\mathcal{Q}$ be a family of up-open cofinal subsets of P with $\#(\mathcal{Q}) \leq \kappa$.

Let $\langle P_n \rangle_{n\in\mathbf{N}}$ be a sequence of upwards-centered subsets of P covering P; we may suppose that no P_n is empty. If some P_n meets every member of $\mathcal{Q}$, then it is itself an upwards-linked set of the kind we seek. So let us suppose that there are $Q_n^* \in \mathcal{Q}$ such that $P_n \cap Q_n^* = \emptyset$ for each $n \in \mathbf{N}$.

For each $p \in P$, $Q \in \mathcal{Q}$ set

$$A(p, Q) = \{n : \exists q \in P_n \cap Q, q \geq p\}.$$

Then we find that if $p_0, \ldots, p_k \in P_r$ and $Q_0, \ldots, Q_k \in \mathcal{Q}$, $\bigcap_{i\leq k} A(p_i, Q_i)$ is infinite. **P** Let $m \in \mathbf{N}$. As P_r is upwards-centered, there is a $p \in P$ such that $p_i \leq p$ for every $i \leq k$. Now because each Q_i, Q_j^* is cofinal and up-open,

$$Q = \bigcap_{i\leq k} Q_i \cap \bigcap_{j\leq m} Q_j^*$$

is cofinal with P; let $q \in Q$, $q \geq p$. Then $q \in P_n$ for some $n \in \mathbf{N}$. As $q \in Q_j^*$ for each $j \leq m$, $q \notin P_j$ for $j \leq m$, and $n > m$. Now $n \in A(p_i, Q_i)$ for each $i \leq k$. As m is arbitrary, $\bigcap_{i\leq k} A(p_i, Q_i)$ is infinite. **Q**

Accordingly (since $\#(P \times \mathcal{Q}) \leq \max(\kappa, \omega) < \mathfrak{p}$) there is an infinite $I_r \subseteq \mathbf{N}$ such that $I_r \setminus A(p, Q)$ is finite whenever $p \in P_r$ and $Q \in \mathcal{Q}$. Write $\mathbf{N}^{(\mathbf{N})} = \bigcup_{n\in\mathbf{N}} \mathbf{N}^n$ and for $\sigma \in \mathbf{N}^n$, $i \in \mathbf{N}$ let $\sigma^\frown i$ be the member of $\mathbf{N}^{n+1}$ such that $\sigma^\frown i \restriction n = \sigma$ and $(\sigma^\frown i)(n) = i$. Define $h : \mathbf{N}^{(\mathbf{N})} \to \mathbf{N}$ by saying that

$$h(\emptyset) = 0;$$

$$\langle h(\sigma^\frown i) \rangle_{i\in\mathbf{N}} \text{ enumerates } I_{h(\sigma)} \text{ in ascending order, } \forall \sigma \in \mathbf{N}^{(\mathbf{N})}.$$

Now choose $p(Q, \sigma)$, for $Q \in \mathcal{Q}$ and $\sigma \in \mathbf{N}^{(\mathbf{N})}$, so that

$$p(Q, \sigma) \in P_{h(\sigma)}\ \forall Q \in \mathcal{Q}, \sigma \in \mathbf{N}^{(\mathbf{N})};$$

if $h(\sigma^\frown i) \in A(p(Q, \sigma), Q)$ then $p(Q, \sigma^\frown i) \in Q$ and $p(Q, \sigma^\frown i) \geq p(Q, \sigma)$.

Since $I_{h(\sigma)} \setminus A(p(Q, \sigma), Q)$ is finite, we can find $f_Q(\sigma) \in \mathbf{N}$ such that $h(\sigma^\frown i) \in A(p(Q, \sigma), Q)$ for every $i \geq f_Q(\sigma)$, $Q \in \mathcal{Q}$, $\sigma \in \mathbf{N}^{(\mathbf{N})}$.

Because $\mathbf{N}^{(\mathbf{N})}$ is countable, and $\#(\mathcal{Q}) < \mathfrak{p}$, we can apply 14B to see that there is an $f : \mathbf{N}^{(\mathbf{N})} \to \mathbf{N}$ such that

$$\{\sigma : f(\sigma) < f_Q(\sigma)\} \text{ is finite } \forall Q \in \mathcal{Q}.$$

Define $g \in \mathbf{N}^{\mathbf{N}}$ by

$$g(n) = f(g \restriction n)\ \forall n \in \mathbf{N}.$$

For each $Q \in \mathcal{Q}$, $\{n : g(n) < f_Q(g \restriction n)\}$ is finite; let $m(Q)$ be such that $g(n) \geq f_Q(g \restriction n)$ for $n \geq m(Q)$; we see that $h(g \restriction n+1) \in A(p(Q, g \restriction n), Q)$ if $n \geq m(Q)$, so that $p(Q, g \restriction n+1) \in Q$ and $p(Q, g \restriction n+1) \geq p(Q, g \restriction n)$ if $n \geq m(Q)$.

Set

$$R = \{p(Q, g \upharpoonright n) : Q \in \mathcal{Q}, n \geq m(Q)\}.$$

To see that this R serves, we have to check the following. (i) If $Q \in \mathcal{Q}$, $p(Q, g \upharpoonright m(Q) + 1) \in Q \cap R$, so R meets Q. (ii) If q, $q' \in R$, express them as $p(Q, g \upharpoonright n)$, $p(Q', g \upharpoonright n')$ where $n \geq m(Q)$, $n' \geq m(Q')$. Suppose that $n \geq n'$. Then $p(Q', g \upharpoonright n) \geq p(Q', g \upharpoonright n')$. But $p(Q', g \upharpoonright n)$ and $p(Q, g \upharpoonright n)$ both belong to $P_{h(g \upharpoonright n)}$ so have a common upper bound in P, which is now a common upper bound of q, q'. Thus R is upwards-linked and meets every member of $\mathcal{Q}$. As P and $\mathcal{Q}$ are arbitrary, (iii) is true.

(*b*) **(iii)$\Rightarrow$(ii)** The argument is identical to that of 13A prf *d*. The only point that needs alteration is that it is no longer sufficient to know that $\tilde{P}$ is ccc; we need to know that it is σ-centered. But by condition (γ) in 13A prf *d*, a subset of $\tilde{P}$ will be upwards-centered in $\tilde{P}$ iff it is upwards-centered in P. So if $\langle P_n \rangle_{n \in \mathbb{N}}$ is a sequence of upwards-centered sets covering P, then $\langle P_n \cap \tilde{P} \rangle_{n \in \mathbb{N}}$ is a cover of $\tilde{P}$ by sets which are upwards-centered in $\tilde{P}$, and $\tilde{P}$ is σ-centered upwards.

(*c*) **(ii)$\Rightarrow$(i)** Use the argument of 11C*c* above, observing that the partially ordered set

$$P = \{(I, A) : I \in [\mathbb{N}]^{<\omega}, A \in \mathcal{A}^*\}$$

constructed there is σ-centered upwards, being the union of countably many upwards-directed sets P_I.

(*d*) **(ii)$\Rightarrow$(iv)** Use the argument of 13A prf *a*, observing that since X is now separable, $\mathfrak{T} \setminus \{\emptyset\}$ is σ-centered downwards.

(*e*) **(iv)$\Rightarrow$(iii)** Assume (iv), and let P be a partially ordered set which is σ-centered upwards. Let $\mathcal{Q}$ be a family of cofinal subsets of P with $\#(\mathcal{Q}) \leq \kappa$. Give P its up-topology; let $\mathcal{G}$ be the algebra of regular open sets in P and Z the Stone space of $\mathcal{G}$. Then Z is separable [14A*b* and A4R].

For $p \in P$, let $V_p \subseteq Z$ be the open-and-closed set corresponding to $\operatorname{int} \overline{\{q : q \geq p\}} \in \mathcal{G}$. For $Q \in \mathcal{Q}$, set $H_Q = \bigcup_{p \in Q} V_p$, and observe that H_Q is dense in Z because Q is cofinal with P. By (iv) there is a $z \in \bigcap_{Q \in \mathcal{Q}} H_Q$. Set $R = \{p : z \in V_p\}$; then R meets every $Q \in \mathcal{Q}$; and if $p, q \in R$ then $V_p \cap V_q \neq \emptyset$ so p and q have a common upper bound in P (just as in the last paragraph of 13A prf *c*). Thus R is an upwards-linked set as required by (iii).

14D Theorem

The following statement is true iff $\mathfrak{p} > \omega_1$:

(*) whenever $\langle A_\xi \rangle_{\xi < \omega_1}$ is a family of infinite subsets of $\mathbf{N}$ such that $A_\xi \setminus A_\eta$ is finite whenever $\eta \le \xi$, there is an infinite set $A \subseteq \mathbf{N}$ such that $A \setminus A_\xi$ is finite whenever $\xi < \omega_1$.

Proof Evidently $\mathfrak{p} > \omega_1$ implies (*). For the rest of this proof, therefore, I assume (*) and seek to prove that $\mathfrak{p} > \omega_1$. Let us write $A \subseteq_{\text{ess}} B$ to mean that $A \setminus B$ is finite.

(*a*) If $F \subseteq \mathbf{N}^{\mathbf{N}}$ has $\#(F) \le \omega_1$, there is a $g \in \mathbf{N}^{\mathbf{N}}$ such that $\{n : f(n) > g(n)\}$ is finite for every $f \in F$. **P** Use the argument of 14B above. **Q**

(*b*) Suppose that $\mathscr{C} \subseteq \mathscr{P}\mathbf{N}$, $\mathscr{B} \subseteq \mathscr{P}\mathbf{N}$, $\#(\mathscr{C}) \le \omega_1$, $\mathscr{B}$ is countable, and that $C \subseteq_{\text{ess}} B$ for every $C \in \mathscr{C}$, $B \in \mathscr{B}$. Then there is a $D \subseteq \mathbf{N}$ such that $C \subseteq_{\text{ess}} D \subseteq_{\text{ess}} B$ for every $C \in \mathscr{C}$, $B \in \mathscr{B}$. **P** If $\mathscr{B} = \emptyset$, take $D = \mathbf{N}$. Otherwise, let $\langle B_n \rangle_{n \in \mathbf{N}}$ be a sequence running over $\mathscr{B}$; set $B'_n = \bigcap_{i \le n} B_i$; then $C \subseteq_{\text{ess}} B'_n$ for every $C \in \mathscr{C}$, $n \in \mathbf{N}$. If there is some $n \in \mathbf{N}$ such that $B'_n \subseteq_{\text{ess}} B'_m$ for every $m \in \mathbf{N}$, we can take $D = B'_n$. Otherwise, there is a strictly increasing sequence $\langle n(k) \rangle_{k \in \mathbf{N}}$ such that $D_k = B'_{n(k)} \setminus B'_{n(k+1)}$ is infinite for each $k \in \mathbf{N}$. For $k \in \mathbf{N}$, $C \in \mathscr{C}$, $D_k \cap C \subseteq C \setminus B'_{n(k+1)}$ is finite; let $f_C(k) = \sup(D_k \cap C)$. By (*a*) above there is a $g \in \mathbf{N}^{\mathbf{N}}$ such that $\{k : f_C(k) > g(k)\}$ is finite for each $C \in \mathscr{C}$. Set

$$D = B'_{n(0)} \setminus \bigcup_{k \in \mathbf{N}} \{i : i \in D_k, i > g(k)\}.$$

It is easy to check that this D will serve. **Q**

(*c*) Now we are ready to prove that $\omega_1 < \mathfrak{p}$. Let $\mathscr{A} \subseteq \mathscr{P}\mathbf{N}$ be a set such that $\#(\mathscr{A}) \le \omega_1$ and $\bigcap \mathscr{A}_0$ is infinite for every finite $\mathscr{A}_0 \subseteq \mathscr{A}$. If $\mathscr{A}$ is countable, we can use the fact that $\omega < \mathfrak{p}$. Otherwise, enumerate $\mathscr{A}$ as $\langle A_\xi \rangle_{\xi < \omega_1}$, and set

$$\mathscr{A}^* = \{\bigcap \mathscr{A}_0 : \mathscr{A}_0 \subseteq \mathscr{A} \text{ is finite, not } \emptyset\}.$$

Choose $\langle B_\xi \rangle_{\xi < \omega_1}$ inductively so that

$B_\xi \subseteq_{\text{ess}} B_\eta$ if $\eta \le \xi$;

$B_{\xi+1} \subseteq A_\xi$;

$B_\xi \cap A$ is infinite $\forall \xi < \omega_1$, $A \in \mathscr{A}^*$;

as follows. $B_0 = \mathbf{N}$. Given B_ξ satisfying the inductive hypothesis, set $B_{\xi+1} = B_\xi \cap A_\xi$. Given $\langle B_\eta \rangle_{\eta < \xi}$ where ξ is a non-zero countable limit ordinal, then for each $A \in \mathscr{A}^*$ we have an essentially decreasing family $\langle B_\eta \cap A \rangle_{\eta < \xi}$ of infinite sets, so there is an infinite $C^\xi_A \subseteq A$ such that $C^\xi_A \subseteq_{\text{ess}} B_\eta$ for every $\eta < \xi$. By (*b*) above there is a $B_\xi \subseteq \mathbf{N}$ such that $C^\xi_A \subseteq_{\text{ess}} B_\xi \subseteq_{\text{ess}} B_\eta$ for every $A \in \mathscr{A}^*$, $\eta < \xi$, so that $B_\xi \cap A$ is infinite for every $A \in \mathscr{A}^*$, and the induction continues.

Now by (*) there is an infinite $I \subseteq \mathrm{N}$ such that $I \subseteq_{\text{ess}} B_{\xi+1} \subseteq A_\xi$ for every $\xi < \omega_1$. As $\mathscr{A}$ is arbitrary, $\mathrm{P}(\omega_2)$ is true and $\mathfrak{p} > \omega_1$.

14E Sources

ROTHBERGER 48 for 14B (with $\#(A) = \omega_1$) and 14D. KUNEN 68 for 14B in general. DOUWEN 77*a* for 14C(ii)$\Leftrightarrow$(iv). BELL 81*b* for 14C(i)$\Rightarrow$(ii).

14F Exercise

Formulate a Boolean-algebra definition of $\mathfrak{p}$. Are there definitions of $\mathfrak{p}$ corresponding to 13A(vi)–(vii) above?

14G Problem

Is there a version of (*) in 14D that can be used to define $\mathfrak{p}$ completely? [See DOUWEN 77*a*, DOUWEN 83.]

Notes and comments

The proof I give of 14B is taken from ROTHBERGER 48; there are more elegant proofs, but this is the argument needed for 14D. It will be clear that everything in (*b*)–(*e*) of the proof of 14C is taken from Theorem 13A; the different method in part (*e*) is due to my omission of a Boolean-algebra form of $\kappa < \mathfrak{p}$, on the grounds that the appropriate class of Boolean algebras is not of much independent interest [but see 41N*g*]. The new idea in 14C is in part (*a*) of the proof, and I shall therefore refer to 14C(i)$\Rightarrow$(ii) as 'Bell's theorem'.

As is to be expected from the fact that $\mathrm{P}(\kappa^+) \Rightarrow \mathrm{MA}_{\sigma\text{-cent}}(\kappa)$ is harder to prove than its converse, the great majority of the properties of $\mathfrak{p}$ in the next chapter are going to be proved from the fact that it is the least cardinal for which $\mathrm{MA}_{\sigma\text{-cent}}(\mathfrak{p})$ is false. There would have been a greater symmetry and elegance, perhaps, in using $\mathrm{MA}_{\sigma\text{-cent}}(\kappa)$ for the definition of $\mathfrak{p}$, in line with the definitions of $\mathfrak{m}$ and $\mathfrak{m}_K$. I chose the historic definition from $\mathrm{P}(\kappa)$ partly in order to maintain links with the past and partly in order to have a real theorem in §11.

It will become clear that $\mathfrak{p} > \omega_1$ is already a very powerful axiom, so I think it worth while to give an alternative form [14D], even though this does not seem to provide a characterization of $\mathfrak{p}$ [14G]. (This is also a way of paying tribute to F. Rothberger, who had the courage to take $\mathfrak{p} > \omega_1$ seriously twenty years before techniques for proving it consistent appeared.) The attempts I have made to find similar forms for $\mathfrak{p} > \kappa$ have all come to grief on Hausdorff's gap [21L], which shows that there is no simple-minded way to lift part (*b*) of the proof of 14D to higher cardinals.

Among the results in Chapter 2 which are given as properties of $\mathfrak{p}$, but clearly also define it, it is perhaps worth mentioning 21A, 24J and 26C*a*. See also 24N*n*.

2

When $\mathfrak{p} > \omega_1$

This chapter is devoted to results which involve the cardinal $\mathfrak{p}$. The definition of $\mathfrak{p}$ in terms of families of subsets of $\mathbf{N}$ makes it plain that we must expect to be limited to contexts in which countable sets play a dominant role. Thus in Theorem 21A, the underlying set X must be countable; in §§22–23, we deal mainly with second-countable spaces; in §24, we have separable spaces; and in §25 we work with spaces which have associated second-countable topologies. It is, however, worth noting that several of the arguments can reach surprisingly far. Thus 21G can apply to arbitrary subsets of $\mathfrak{c}^2$ and 24K does not mention any restriction on cardinality, though of course such a restriction is present.

21 Combinatorics

The axiom $\mathfrak{p} = \mathfrak{c}$ is, of course, a tool for finding (or, rather, an excuse for declaring the existence of) subsets of $\mathbf{N}$; and it is natural to give priority to its set-theoretic consequences, even though the distinction between these and its topological consequences is not always sharp. I begin with a portmanteau theorem [21A] which embodies most of the straightforward ways of using the principle $P(\kappa)$ itself. I use this to prove first that $2^\kappa \leq \mathfrak{c}$ for $\kappa < \mathfrak{p}$ [21C] and then to give conditions sufficient to make every subset of $X \times Y$ attainable in two steps from sequences of 'rectangles' [21G]. I conclude with remarks on four further topics: a lemma in the partition calculus [21I], R.B. Jensen's principle $\diamondsuit$ [21J], the possible values of $\mathfrak{p}$ [21K] and Hausdorff's gap [21L].

21A **Theorem**

Let X be a countable set, and $\mathscr{A}$, $\mathscr{B}$, $\mathscr{C}$ and $\mathscr{D}$ four families of subsets of X such that

(i) $A \cap C$ is finite $\forall A \in \mathscr{A}$, $C \in \mathscr{C}$;

(ii) $B \setminus \bigcup \mathscr{A}_0$ is infinite $\forall B \in \mathscr{B}$, finite $\mathscr{A}_0 \subseteq \mathscr{A} \setminus \{B\}$;

(iii) $D \setminus \bigcup \mathscr{C}_0$ is infinite $\forall D \in \mathscr{D}$, finite $\mathscr{C}_0 \subseteq \mathscr{C} \setminus \{D\}$;

(iv) $\mathscr{A}$ is countable; $\mathscr{B}$, $\mathscr{C}$, and $\mathscr{D}$ have cardinals less than $\mathfrak{p}$.

Then there is an $I \subseteq X$ such that

$A \setminus I$ is finite if $A \in \mathscr{A}$;

$B\backslash I$ is non-empty if $B \in \mathscr{B}$, infinite if $B \in \mathscr{B}\backslash\mathscr{A}$;
$C \cap I$ is finite if $C \in \mathscr{C}$;
$D \cap I$ is non-empty if $D \in \mathscr{D}$, infinite if $D \in \mathscr{D}\backslash\mathscr{C}$.

Proof Let $\mathscr{E}$ be the set

$$\{K \triangle \bigcup \mathscr{A}_0 : K \in [X]^{<\omega}, \mathscr{A}_0 \in [\mathscr{A}]^{<\omega}\}.$$

Note that $\mathscr{E}$ is countable. For $E \in \mathscr{E}$, set

$$\mathscr{B}_E = \{B : B \in \mathscr{B},\ B\backslash E \text{ is finite}\}.$$

By condition (ii), if $E \triangle \bigcup \mathscr{A}_0$ is finite where $\mathscr{A}_0 \in [\mathscr{A}]^{<\omega}$, $\mathscr{B}_E = \mathscr{B} \cap \mathscr{A}_0$. Consequently $\mathscr{B}_{E \cup F} = \mathscr{B}_E \cup \mathscr{B}_F$ for all $E, F \in \mathscr{E}$, and $\mathscr{B}_E = \emptyset$ if E is finite.

Let $\mathscr{G}$ be the set

$$\{K \triangle \bigcup \mathscr{C}_0 : K \in [X]^{<\omega}, \mathscr{C}_0 \in [\mathscr{C}]^{<\omega}\}.$$

For $G \in \mathscr{G}$, set

$$\mathscr{D}_G = \{D : D \in \mathscr{D},\ D\backslash G \text{ is finite}\}.$$

Using condition (iii) this time, we see that $\mathscr{D}_{G \cup H} = \mathscr{D}_G \cup \mathscr{D}_H$ for all $G, H \in \mathscr{G}$, and $\mathscr{D}_G = \emptyset$ if G is finite.

(*b*) Let P be the set

$$\{(E, G) : E \in \mathscr{E}, G \in \mathscr{G}, E \cap G = \emptyset, E \cap D \neq \emptyset\ \forall D \in \mathscr{D}_G,$$
$$G \cap B \neq \emptyset\ \forall B \in \mathscr{B}_E\}.$$

Order P by saying that

$$(E, G) \leq (F, H) \text{ if } E \subseteq F,\ G \subseteq H.$$

Observe that if (E, G) and (E, H) belong to P, then $(E, G \cup H) \in P$; as $\mathscr{E}$ is countable, P is the union of countably many upwards-directed sets, so is certainly σ-centered upwards. $(\emptyset, \emptyset) \in P$ so P is not empty.

(*c*) For $A \in \mathscr{A}\backslash\mathscr{B}$, set

$$Q_A = \{(E, G) : (E, G) \in P, A\backslash E \text{ is finite}\}.$$

Then Q_A is cofinal with P. **P** Let $(E, G) \in P$. By condition (i), $A \cap G$ is finite. Set $F = E \cup A\backslash G$. Then $\mathscr{B}_F = \mathscr{B}_E \cup \mathscr{B}_{A\backslash G} = \mathscr{B}_E$, so $(F, G) \in P$ and $(E, G) \leq (F, G) \in Q_A$. **Q**

(*d*) For $A \in \mathscr{A} \cap \mathscr{B}$, set

$$R_A = \{(E, G) : (E, G) \in P, A\backslash E \text{ is finite}, A \cap G \neq \emptyset\}.$$

Then R_A is cofinal with P. **P** Let $(E, G) \in P$. If $A\backslash E = \emptyset$, then $A \in \mathscr{B}_E$ so

(E, G) itself belongs to R_A. Otherwise, take $x \in A \setminus E$. As before, $A \cap G$ is finite. Set

$$F = E \cup A \setminus (G \cup \{x\}), \quad H = G \cup \{x\}.$$

Then $\mathscr{B}_F = \mathscr{B}_E \cup \{A\}$ and $\mathscr{D}_H = \mathscr{D}_G$, so $(F, H) \in P$ and $(E, G) \leq (F, H) \in R_A$. **Q**

(*e*) For $B \in \mathscr{B} \setminus \mathscr{A}$, $n \in \mathbf{N}$ set

$$S_{Bn} = \{(E, G):(E, G) \in P,\ \#(G \cap B) \geq n\}.$$

Then S_{Bn} is cofinal with P. **P** Let $(E, G) \in P$. Then $B \setminus E$ is infinite, by (ii). Let $K \in [B \setminus E]^n$, and set $H = G \cup K$. Then $\mathscr{B}_H = \mathscr{B}_G$ so $(E, H) \in P$ and $(E, G) \leq (E, H) \in S_{Bn}$. **Q**

(*f*) For $C \in \mathscr{C} \setminus \mathscr{D}$, set

$$T_C = \{(E, G):(E, G) \in P,\ C \setminus G \text{ is finite}\}.$$

Then T_C is cofinal with P by the argument of (*c*) above with the coordinates reversed. For $C \in \mathscr{C} \cap \mathscr{D}$, set

$$U_C = \{(E, G):(E, G) \in P, C \setminus G \text{ is finite},\ C \cap E \neq \emptyset\};$$

for $D \in \mathscr{D} \setminus \mathscr{C}$, $n \in \mathbf{N}$ set

$$V_{Dn} = \{(E, G):(E, G) \in P,\ \#(D \cap E) \geq n\}.$$

Then the sets U_C, V_{Dn} are all cofinal with P by the arguments in (*d*), (*e*) above.

(*g*) Set

$$\begin{aligned}\mathfrak{Q} = \{Q_A : A \in \mathscr{A} \setminus \mathscr{B}\} &\cup \{R_A : A \in \mathscr{A} \cap \mathscr{B}\}\\ &\cup \{S_{Bn} : B \in \mathscr{B} \setminus \mathscr{A}, n \in \mathbf{N}\} \cup \{T_C : C \in \mathscr{C} \setminus \mathscr{D}\}\\ &\cup \{U_C : C \in \mathscr{C} \cap \mathscr{D}\} \cup \{V_{Dn} : D \in \mathscr{D} \setminus \mathscr{C}, n \in \mathbf{N}\}.\end{aligned}$$

Then $\#(\mathfrak{Q}) \leq \max(\omega, \#(\mathscr{B}), \#(\mathscr{C}), \#(\mathscr{D})) < \mathfrak{p}$, so by Bell's theorem [14C] there is an upwards-directed set $W \subseteq P$ which meets every member of $\mathfrak{Q}$. Set $I = \bigcup\{E:(E, G) \in W\}$. Then $I \cap G = \emptyset$ whenever $(E, G) \in W$, because W is upwards-directed (if (E_1, G_1) and (E_2, G_2) both belong to W, they have a common upper bound (E, G), so that $E_1 \cap G_2 \subseteq E \cap G = \emptyset$). Now each of the six requirements imposed on I corresponds to one of the six families of sets met by W, and I is the set we are looking for.

21B Corollaries

(*a*) No non-principal ultrafilter on $\mathbf{N}$ has a base of cardinal less than $\mathfrak{p}$.

(*b*) Any infinite maximal almost-disjoint family in $[\mathbf{N}]^\omega$ has cardinal greater than or equal to $\mathfrak{p}$.

(*c*) If $\mathscr{E} \subseteq [\mathbf{N}]^{\omega}$ is almost disjoint, $\#(\mathscr{E}) < \mathfrak{p}$, and $\mathscr{D} \subseteq \mathscr{E}$, there is an $I \subseteq \mathbf{N}$ such that $\mathscr{D} = \{E : E \in \mathscr{E},\ E \cap I \text{ is infinite}\}$.

Proof (*a*) This is in fact immediate from the definition of $\mathfrak{p}$; if $\mathscr{A}$ is a filter base on $\mathbf{N}$ of cardinal less than $\mathfrak{p}$, not containing a finite set, there is an infinite $I \subseteq \mathbf{N}$ with $I \setminus A$ finite for every $A \in \mathscr{A}$. If we take $J \subseteq I$ such that J and $I \setminus J$ are both infinite, then neither J nor $\mathbf{N} \setminus J$ can be in the filter generated by $\mathscr{A}$, so that $\mathscr{A}$ does not generate an ultrafilter.

(*b*) If $\mathscr{C}$ is an almost-disjoint family in $[\mathbf{N}]^{\omega}$ such that $\omega \leq \#(\mathscr{C}) < \mathfrak{p}$, then no finite subset of $\mathscr{C}$ can have cofinite union in $\mathbf{N}$, so by 21A with $\mathscr{A} = \mathscr{B} = \emptyset$, $\mathscr{D} = \{\mathbf{N}\}$ there is an infinite $I \subseteq \mathbf{N}$ almost disjoint from every member of $\mathscr{C}$.

(*c*) Apply 21A with $\mathscr{A} = \mathscr{B} = \emptyset$, $\mathscr{C} = \mathscr{E} \setminus \mathscr{D}$.

21C **Theorem**

If $\kappa < \mathfrak{p}$ then $2^{\kappa} \leq \mathfrak{c}$.

Proof There is an almost-disjoint family $\mathscr{A}$ of infinite subsets of $\mathbf{N}$ with $\#(\mathscr{A}) = \mathfrak{c}$ [A3A*d*]. As $\kappa < \mathfrak{c}$ [11D*b*], there is an injection $f : \kappa \to \mathscr{A}$. For $I \subseteq \mathbf{N}$, set $D_I = \{\xi : \xi < \kappa, I \cap f(\xi) \text{ is infinite}\}$. By 21B*c*, with $\mathscr{E} = f[\kappa]$, the map $I \mapsto D_I : \mathscr{P}\mathbf{N} \to \mathscr{P}\kappa$ is surjective. So

$$2^{\kappa} = \#(\mathscr{P}\kappa) \leq \#(\mathscr{P}\mathbf{N}) = \mathfrak{c}.$$

21D **Corollary**

(*a*) $\#([\mathfrak{c}]^{<\mathfrak{p}}) = \mathfrak{c}$. (*b*) $\mathrm{cf}(\mathfrak{c}) \geq \mathfrak{p}$.

Proof (*a*) For any $\kappa < \mathfrak{p}$,

$$\#([\mathfrak{c}]^{\kappa}) \leq \#(\mathfrak{c}^{\kappa}) = \#(\mathscr{P}(\mathbf{N} \times \kappa)) \leq 2^{\max(\omega,\kappa)} = \mathfrak{c}$$

by 21C. So $\#([\mathfrak{c}]^{<\mathfrak{p}}) \leq \sum_{\kappa < \mathfrak{p}} \#([\mathfrak{c}]^{\kappa}) = \mathfrak{c}$.

(*b*) König's theorem [A2C].

21E **Corollary**

$[\mathfrak{p} = \mathfrak{c}]$ (*a*) $\#([\mathfrak{c}]^{<\mathfrak{c}}) = \mathfrak{c}$. (*b*) $\mathfrak{c}$ is regular.

21F **Lemma**

(*a*) Let $S \subseteq [\mathfrak{p}]^2$. Then there is a family $\langle A_{\xi} \rangle_{\xi < \mathfrak{p}}$ of subsets of $\mathbf{N}$ such that

$$S = \{\{\eta, \xi\} : \eta, \xi < \mathfrak{p}, \eta \neq \xi, A_{\eta} \cap A_{\xi} \text{ is infinite}\}.$$

(*b*) Let $S \subseteq \mathfrak{p}^2$. Then there are families $\langle A_{\xi} \rangle_{\xi < \mathfrak{p}}$, $\langle B_{\xi} \rangle_{\xi < \mathfrak{p}}$ of subsets of **N** such that

$$S = \{(\eta, \xi) : \eta, \xi < \mathfrak{p},\ A_\eta \cap B_\xi \text{ is infinite}\}.$$

Proof (*a*) I construct $\langle A_\xi \rangle_{\xi < \mathfrak{p}}$ inductively. In other to have room for manoeuvre at the inductive step, I need an inductive hypothesis in three parts:

(i) if $\eta < \zeta < \xi$ then $A_\eta \cap A_\zeta$ is infinite iff $\{\eta, \zeta\} = S$;
(ii) if $I \in [\xi]^{<\omega}$ then $\mathbf{N} \setminus \bigcup_{\eta \in I} A_\eta$ is infinite;
(iii) if $I \in [\xi]^{<\omega}$ and $\zeta \in \xi \setminus I$ then $A_\zeta \setminus \bigcup_{\eta \in I} A_\eta$ is infinite.

(Of course (ii) is redundant except when $\xi < \omega$.) Now suppose that I have $\langle A_\eta \rangle_{\eta < \xi}$ satisfying (i)–(iii), where $\xi < \mathfrak{p}$. In order to continue the induction, I need to arrange the following properties for A_ξ:

(i′) if $\eta < \xi$ and $\{\eta, \xi\} \in S$ then $A_\xi \cap A_\eta$ is infinite;
(i″) if $\eta < \xi$ and $\{\eta, \xi\} \notin S$ then $A_\xi \cap A_\eta$ is finite;
(ii′) if $I \in [\xi]^{<\omega}$ then $(\mathbf{N} \setminus \bigcup_{\eta \in I} A_\eta) \setminus A_\xi$ is infinite;
(iii′) if $I \in [\xi]^{<\omega}$ then $A_\xi \setminus \bigcup_{\eta \in I} A_\eta$ is infinite;
(iii″) if $I \in [\xi]^{<\omega}$ and $\zeta \in \xi \setminus I$ then $(A_\zeta \setminus \bigcup_{\eta \in I} A_\eta) \setminus A_\xi$ is finite.

So I try to apply 21A with $\mathscr{A} = \varnothing$,

$$\mathscr{B} = \{\mathbf{N} \setminus \textstyle\bigcup_{\eta \in I} A_\eta : I \in [\xi]^{<\omega}\} \cup \{A_\zeta \setminus \bigcup_{\eta \in I} A_\eta : I \in [\xi]^{<\omega}, \zeta \in \xi \setminus I\},$$
$$\mathscr{C} = \{A_\eta : \eta < \xi, \{\eta, \xi\} \notin S\},$$
$$\mathscr{D} = \{A_\eta : \eta < \xi, \{\eta, \xi\} \in S\} \cup \{\mathbf{N} \setminus \textstyle\bigcup_{\eta \in I} A_\eta : I \in [\xi]^{<\omega}\}.$$

For this to work, I need conditions on the cardinalities of $\mathscr{B}$, $\mathscr{C}$ and $\mathscr{D}$ and on the overlapping of members of $\mathscr{C}$ and $\mathscr{D}$. First, $\mathscr{B}$, $\mathscr{C}$ and $\mathscr{D}$ all have cardinals less than or equal to $\max(\omega, \#(\xi)) < \mathfrak{p}$. Next, I need to know that

(α) every member of $\mathscr{B}$ is infinite;
(β) if $\mathscr{C}_0 \subseteq \mathscr{C}$ is finite and $D \in \mathscr{D}$ then $D \setminus \bigcup \mathscr{C}_0$ is infinite;

and both of these follow at once from (ii) and (iii) of the inductive hypothesis. So 21A does apply and the set it provides will serve for A_ξ. Thus the induction continues.

(*b*) Take any disjoint sets C, $D \subseteq \mathfrak{p}$ of cardinal $\mathfrak{p}$ and enumerate them as $\langle \gamma(\xi) \rangle_{\xi < \mathfrak{p}}$, $\langle \delta(\xi) \rangle_{\xi < \mathfrak{p}}$ respectively. Set

$$T = \{\{\gamma(\eta), \delta(\xi)\} : (\eta, \xi) \in S\} \subseteq [\mathfrak{p}]^2,$$

and use (*a*) to find a family $\langle E_\alpha \rangle_{\alpha<\mathfrak{p}}$ of subsets of $\mathbf{N}$ such that $T = \{\{\alpha,\beta\}:\alpha,\beta < \mathfrak{p}, \alpha \neq \beta, E_\alpha \cap E_\beta$ is infinite$\}$. Now set

$$A_\xi = E_{\gamma(\xi)}, B_\xi = E_{\delta(\xi)}.$$

21G **Theorem**

Let X and Y be sets. If

either (i) $\#(X) < \mathfrak{p}$

or (ii) $\#(Y) < \mathfrak{p}$

or (iii) $\#(X) = \#(Y) = \mathfrak{p}$,

then for every $S \subseteq X \times Y$ there are sequences $\langle C_n \rangle_{n\in\mathbf{N}}, \langle D_n \rangle_{n\in\mathbf{N}}$ of subsets of X, Y respectively such that $S = \bigcap_{n\in\mathbf{N}} \bigcup_{m\geq n} C_m \times D_m$.

Proof (*a*) Let us take first the case (i). As in 21C, take an almost-disjoint family $\langle A_x \rangle_{x\in X}$ of infinite subsets of $\mathbf{N}$. For each $y \in Y$ use 21B*c* to choose set $I_y \subseteq \mathbf{N}$ such that $A_x \cap I_y$ is infinite if $(x, y) \in S$, finite if $(x, y) \notin S$. Set

$$C_n = \{x : n \in A_x\}, D_n = \{y : n \in I_y\},$$

so that

$$S = \{(x, y) : \forall n \in \mathbf{N}\ \exists m \geq n, m \in A_x \cap I_y\} = \bigcap_{n\in\mathbf{N}} \bigcup_{m\geq n} C_m \times D_m.$$

(*b*) Of course (ii) proceeds similarly. For (iii), we can without losing generality suppose that $X = Y = \mathfrak{p}$. Given $S \subseteq \mathfrak{p}^2$, let $\langle A_\xi \rangle_{\xi<\mathfrak{p}}$ and $\langle B_\xi \rangle_{\xi<\mathfrak{p}}$ be families of subsets of $\mathbf{N}$ chosen by 21F*b*; set

$$C_n = \{\xi : n \in A_\xi\}, D_n = \{\xi : n \in B_\xi\}$$

and continue as in (*a*).

21H **Corollary**

Let X be a set and $\mathscr{A}$ a family of subsets of X. If

either (i) $\#(X) < \mathfrak{p}$

or (ii) $\#(\mathscr{A}) < \mathfrak{p}$

or (iii) $\#(X) = \#(\mathscr{A}) = \mathfrak{p}$,

then there is a sequence $\langle C_n \rangle_{n\in\mathbf{N}}$ of subsets of X such that every member of $\mathscr{A}$ is expressible as $\bigcap_{n\in\mathbf{N}} \bigcup_{m\in I\setminus n} C_m$ for some $I \subseteq \mathbf{N}$.

Proof Consider $S = \{(x, A) : x \in A \in \mathscr{A}\} \subseteq X \times \mathscr{A}$. By 21G, there are sequences $\langle C_n \rangle_{n\in\mathbf{N}}, \langle \mathscr{D}_n \rangle_{n\in\mathbf{N}}$ in $\mathscr{P}X$, $\mathscr{P}\mathscr{A}$ respectively such that $S =$

$\bigcap_{n\in\mathbf{N}} \bigcup_{m\geq n} C_m \times \mathscr{D}_m$. Now if $A \in \mathscr{A}$, let I be $\{n : A \in \mathscr{D}_n\}$; we shall have

$$A = \{x : (x, A) \in S\} = \bigcap_{n\in\mathbf{N}} \bigcup_{m\geq n} \{x : (x, A) \in C_m \times \mathscr{D}_m\}$$
$$= \bigcap_{n\in\mathbf{N}} \bigcup_{m\in I\setminus n} C_m.$$

21I **Lemma**

Suppose that $\omega < \mathrm{cf}(\kappa) \leq \kappa < \mathfrak{p}$. Let $\mathscr{S}$ be a finite cover of $\mathbf{N} \times \kappa$. Then there are an infinite $I \subseteq \mathbf{N}$, a $C \subseteq \kappa$ with $\#(C) = \kappa$, and an $S \in \mathscr{S}$ such that $I \times C \subseteq S$.

Proof Enumerate $\mathscr{S}$ as $\langle S_i \rangle_{i\leq m}$. For each $\xi < \kappa$ set $S_i^\xi = \{n : (n, \xi) \in S_i\}$. Let $\mathscr{F}$ be any non-principal ultrafilter on $\mathbf{N}$. For each $\xi < \kappa$, $\{S_i^\xi : i \leq m\}$ is a finite cover of $\mathbf{N}$, so there is an $i(\xi)$ such that $S_{i(\xi)}^\xi \in \mathscr{F}$. Let $j \leq m$ be such that $C_0 = \{\xi : i(\xi) = j\}$ has cardinal κ. Applying the definition of $\mathfrak{p}$ to $\{S_j^\xi : \xi \in C_0\}$, we see that there is an infinite $I_0 \subseteq \mathbf{N}$ such that $I_0 \setminus S_j^\xi$ is finite for each $\xi \in C_0$. Because $\mathrm{cf}(\kappa) > \omega$, there is a $k \in \mathbf{N}$ such that $C = \{\xi : \xi \in C_0, I_0 \setminus S_j^\xi \subseteq k\}$ has cardinal κ; set $I = I_0 \setminus k$ and see that $I \times C \subseteq S_j \in \mathscr{S}$.

21J **Diamond, club and stick**

These are combinatorial principles which are consequences of the axiom of constructibility; my only concern here is to show that they are incompatible with Martin's axiom except in the trivial circumstances, i.e. when the continuum hypothesis is true. Their statements are:

$\diamondsuit$ [R.B. Jensen]: There is an indexed family $\langle A_\xi \rangle_{\xi<\omega_1}$ of sets such that $\{\xi : \xi < \omega_1, A \cap \xi = A_\xi\}$ is stationary in ω_1 for every $A \subseteq \omega_1$.

$\clubsuit$ [Ostaszewski 76]: There is a ladder system $\langle \theta(\zeta, n) \rangle_{n\in\mathbf{N}, \zeta\in\Omega}$ on ω_1 such that whenever $A \subseteq \omega_1$ is uncountable there is a $\zeta \in \Omega$ such that $\theta(\zeta, n) \in A$ for every $n \in \mathbf{N}$.

† [Broverman, Ginsburg, Kunen & Tall 78]: There is a family $\mathscr{A}$ of infinite subsets of ω_1 such that $\#(\mathscr{A}) = \omega_1$ and every uncountable subset of ω_1 includes some member of $\mathscr{A}$.

Proposition

(*a*) $\diamondsuit \Leftrightarrow \clubsuit + \mathrm{CH}$.

(*b*) $\clubsuit \Rightarrow$ †; $\mathrm{CH} \Rightarrow$ †.

(*c*) † $\Rightarrow \mathfrak{p} = \omega_1$.

Proof For (*a*), see Ostaszewski 76. (*b*) is obvious (since $\#([\omega_1]^\omega) = \mathfrak{c}$). For (*c*), we regard $\mathscr{P}\omega_1$ as a compact Hausdorff space homeomorphic to $\{0, 1\}^{\omega_1}$; by 12K, it is separable. Now if $\mathscr{A} \subseteq \mathscr{P}\omega_1$ is a family of infinite sets as in †, we can set

$$E_A = \{C : A \subseteq C \subseteq \omega_1\},\ F_\xi = \mathscr{P}\xi$$

for $A \in \mathscr{A}$ and $\xi < \omega_1$. All these are closed nowhere-dense sets in $\mathscr{P}\omega_1$, and they cover $\mathscr{P}\omega_1$. So $\omega_1 \geq \mathfrak{p}$ by 14C(i)$\Rightarrow$(iv).

Remark I omit the proof of (*a*) because, while it may not be obvious that $\diamondsuit \Rightarrow \clubsuit$, it is surely obvious that $\diamondsuit \Rightarrow$ † so that $\diamondsuit \Rightarrow \mathfrak{p} = \omega_1$.

21K **Proposition**
$\mathfrak{p}$ is regular.

Proof Suppose that $\omega \leq \kappa \leq \mathfrak{p}$ and that $\mathrm{cf}(\kappa) = \lambda < \mathfrak{p}$. Let $\mathscr{A} \subseteq \mathscr{P}\mathbf{N}$ be a non-empty family of sets such that $\#(\mathscr{A}) \leq \kappa$ and $\bigcap \mathscr{A}'$ is infinite for every finite $\mathscr{A}' \subseteq \mathscr{A}$. Set

$$\mathscr{A}^* = \{\bigcap \mathscr{A}' : \varnothing \neq \mathscr{A}' \in [\mathscr{A}]^{<\omega}\}.$$

Then $\#(\mathscr{A}^*) \leq \kappa$ so $\mathscr{A}^*$ can be expressed as $\bigcup_{\xi<\lambda} \mathscr{A}_\xi$ where $\#(\mathscr{A}_\xi) < \kappa$ for every $\xi < \lambda$.

We can choose inductively an indexed family $\langle E_\xi \rangle_{\xi<\lambda}$ of subsets of $\mathbf{N}$ such that

$E_\xi \cap A$ is infinite for every $A \in \mathscr{A}^*$, $\xi < \lambda$;
$E_\xi \setminus E_\eta$ is finite whenever $\eta \leq \xi$;
$E_\xi \setminus A$ is finite whenever $A \in \mathscr{A}_\xi$, $\zeta < \lambda$.

P Given $\langle E_\eta \rangle_{\eta<\xi}$, where $\xi < \lambda$, then for each $\zeta < \lambda$ set

$$\mathscr{B}_\zeta = \{E_\eta : \eta < \xi\} \cup \mathscr{A}_\xi \cup \mathscr{A}_\zeta.$$

$\#(\mathscr{B}_\zeta) < \kappa \leq \mathfrak{p}$ and $\bigcap \mathscr{B}'$ is infinite for every finite $\mathscr{B}' \subseteq \mathscr{B}_\zeta$. So we can choose an infinite set $D_\zeta \subseteq \mathbf{N}$ such that

$D_\zeta \setminus E_\eta$ is finite $\forall \eta < \xi$, $D_\zeta \setminus A$ is finite $\forall A \in \mathscr{A}_\xi \cup \mathscr{A}_\zeta$.

Now apply 21A with

$$\mathscr{C} = \{\mathbf{N} \setminus E_\eta : \eta < \xi\} \cup \{\mathbf{N} \setminus A : A \in \mathscr{A}_\xi\},$$
$$\mathscr{D} = \{D_\zeta : \zeta < \lambda\}$$

to see that there is an $E_\xi \subseteq \mathbf{N}$ such that

$E_\xi \setminus E_\eta$ is finite $\forall \eta < \xi$, $E_\xi \setminus A$ is finite $\forall A \in \mathscr{A}_\xi$, $E_\xi \cap D_\zeta$ is infinite $\forall \zeta < \lambda$.

If $A \in \mathscr{A}^*$ there is a $\zeta < \lambda$ such that $A \in \mathscr{A}_\zeta$; now $D_\zeta \setminus A$ is finite and $E_\xi \cap D_\zeta$ is infinite, so $E_\xi \cap A$ is infinite. Thus the induction continues. **Q**

Because $\lambda < \mathfrak{p}$, there is an infinite $E \subseteq \mathbf{N}$ such that $E \setminus E_\xi$ is finite for every $\xi < \lambda$. In this case $E \backslash A$ is finite for every $A \in \bigcup_{\xi < \lambda} \mathscr{A}_\xi \supseteq \mathscr{A}$. As $\mathscr{A}$ is arbitrary, $\kappa < \mathfrak{p}$. As κ is arbitrary, $\mathrm{cf}(\mathfrak{p}) = \mathfrak{p}$ and $\mathfrak{p}$ is regular.

21L **Hausdorff's gap: proposition**

(*a*) There are indexed families $\langle A_\xi \rangle_{\xi < \omega_1}$, $\langle B_\xi \rangle_{\xi < \omega_1}$ of subsets of $\mathbf{N}$ such that

if $\eta < \xi < \omega_1$, then $A_\eta \backslash A_\xi$ and $B_\eta \backslash B_\xi$ and $A_\xi \cap B_\xi$ are finite, $A_\xi \backslash A_\eta$ and $B_\xi \backslash B_\eta$ are infinite;

if $\xi < \omega_1$ and $r \in \mathbf{N}$, then $\{\eta : \eta < \xi, B_\eta \cap A_\xi \subseteq r\}$ is finite.

(*b*) There is now no $H \subseteq \mathbf{N}$ such that $A_\xi \backslash H$ and $B_\xi \cap H$ are both finite for every $\xi < \omega_1$.

Proof (*a*) Construct A_ξ, B_ξ inductively; we must include, as part of the inductive hypothesis, the requirement

$$\mathbf{N} \backslash (A_\xi \cup B_\xi) \text{ is infinite } \forall \xi < \omega_1.$$

(i) Start with $A_0 = B_0 = \varnothing$.

(ii) Given A_ξ and B_ξ, take three infinite disjoint subsets C, D and E of $\mathbf{N} \backslash (A_\xi \cup B_\xi)$; set $A_{\xi+1} = A_\xi \cup C$ and $B_{\xi+1} = B_\xi \cup D$. Now

$$A_\eta \backslash A_{\xi+1} \subseteq A_\eta \backslash A_\xi, \quad B_\eta \backslash B_{\xi+1} \subseteq B_\eta \backslash B_\xi, \quad A_{\xi+1} \cap B_{\xi+1} = A_\xi \cap B_\xi$$

are finite for $\eta \leq \xi$;

$$A_{\xi+1} \backslash A_\eta \supseteq C \backslash (A_\eta \backslash A_\xi), \quad B_{\xi+1} \backslash B_\eta \supseteq D \backslash (B_\eta \backslash B_\xi),$$
$$\mathbf{N} \backslash (A_{\xi+1} \cup B_{\xi+1}) \supseteq E$$

are infinite for $\eta \leq \xi$; and finally

$$\{\eta : \eta \leq \xi, B_\eta \cap A_{\xi+1} \subseteq r\} \subseteq \{\xi\} \cup \{\eta : \eta < \xi, B_\eta \cap A_\xi \subseteq r\}$$

is finite for each $r \in \mathbf{N}$.

(iii) Given $\langle A_\eta \rangle_{\eta < \xi}$ and $\langle B_\eta \rangle_{\eta < \xi}$ where ξ is a non-zero countable limit ordinal, choose a strictly increasing sequence $\langle \zeta(n) \rangle_{n \in \mathbf{N}}$ of ordinals with supremum ξ, starting with $\zeta(0) = 0$. Take $k_n \in \mathbf{N}$ such that $k_n \supseteq A_{\zeta(n)} \cap \bigcup_{i \leq n} B_{\zeta(i)}$ and set $A'_\xi = \bigcup_{n \in \mathbf{N}} (A_{\zeta(n)} \backslash k_n)$. Then we have $A_{\zeta(n)} \backslash A'_\xi \subseteq k_n$ for every $n \in \mathbf{N}$, so $A_\eta \backslash A'_\xi$ is finite for each $\eta < \xi$; it follows that $A'_\xi \backslash A_\eta$ is infinite for each $\eta < \xi$; also

$$A'_\xi \cap B_{\zeta(m)} = \bigcup_{n \in \mathbf{N}} A_{\zeta(n)} \cap B_{\zeta(m)} \backslash k_n = \bigcup_{n < m} A_{\zeta(n)} \cap B_{\zeta(m)} \backslash k_n$$

is finite for each $m \in \mathbf{N}$.

Next, for each $n \in \mathbf{N}$,

$$J_n = \{\eta : \zeta(n) \leq \eta < \zeta(n+1), B_\eta \cap A'_\xi \subseteq n\}$$

$$\subseteq \{\eta : \eta < \zeta(n+1), B_\eta \cap (A_{\zeta(n+1)} \setminus k_{n+1}) \subseteq n\}$$
$$\subseteq \{\eta : \eta < \zeta(n+1), B_\eta \cap A_{\zeta(n+1)} \subseteq \max(n, k_{n+1})\}$$

is finite. It follows that if $J = \bigcup_{n\in\mathbf{N}} J_n$ then $J \cap \eta$ is finite for every $\eta < \xi$.

For $\eta \in J \setminus \{0\}$, set

$$C_\eta = B_\eta \setminus \bigcup \{B_{\zeta(i)} : i \in \mathbf{N}, \zeta(i) < \eta\},$$

so that C_η is infinite, and set

$$j(\eta) = \min(C_\eta \setminus n) \text{ if } \eta \in J_n \setminus \{0\}$$

(recalling that $\langle J_n \setminus \{0\} \rangle_{n\in\mathbf{N}}$ is disjoint and covers $J \setminus \{0\}$). Try

$$A_\xi = A'_\xi \cup \{j(\eta) : \eta \in J \setminus \{0\}\}.$$

Because $A_\xi \supseteq A'_\xi$, $A_\eta \setminus A_\xi$ is finite and $A_\xi \setminus A_\eta$ is infinite for every $\eta < \xi$. Also

$$A_\xi \cap B_{\zeta(n)} \subseteq (A'_\xi \cap B_{\zeta(n)}) \cup \{j(\eta) : \eta \in J \setminus \{0\}, \eta \leq \zeta(n)\}$$

because $j(\eta) \in C_\eta \subseteq \mathbf{N} \setminus B_{\zeta(n)}$ if $\eta > \zeta(n)$. So $A_\xi \cap B_{\zeta(n)}$ is finite for every $n \in \mathbf{N}$. We now at last arrive at the point of the construction. Given $r \in \mathbf{N}$,

$$\{\eta : \eta < \xi, B_\eta \cap A_\xi \subseteq r\} = \bigcup_{n\in\mathbf{N}} \{\eta : \zeta(n) \leq \eta < \zeta(n+1), B_\eta \cap A_\xi \subseteq r\}$$
$$= \bigcup_{n\in\mathbf{N}} K^r_n \text{ say.}$$

Now for each $n \in \mathbf{N}$,

$$K^r_n \subseteq \{\eta : \zeta(n) \leq \eta < \zeta(n+1), B_\eta \cap A'_\xi \subseteq r\}$$
$$\subseteq \{\eta : \zeta(n) \leq \eta < \zeta(n+1), B_\eta \cap A_{\zeta(n+1)} \subseteq \max(r, k_{n+1})\}$$

is finite. While for $n \geq r$, $K^r_n \subseteq J_n$. But if $\eta \in J_n \setminus \{0\}$, then $j(\eta) \in A_\xi \cap B_\eta \setminus n$, so $K^r_n \subseteq \{0\}$ if $n \geq r$. Thus $\bigcup_{n\in\mathbf{N}} K^r_n$ is finite i.e. $\{\eta : \eta < \xi, B_\eta \cap A_\xi \subseteq r\}$ is finite, for every $r \in \mathbf{N}$.

We still have to choose B_ξ. Since $A_\xi \cap B_{\zeta(n+1)}$ is finite for every $n \in \mathbf{N}$, $(\mathbf{N} \setminus A_\xi) \setminus \bigcup_{i \leq n} B_{\zeta(i)}$ is always infinite, and there is an infinite $E \subseteq \mathbf{N} \setminus A_\xi$ such that $E \cap B_{\zeta(n)}$ is finite for every $n \in \mathbf{N}$ (this is a kind of application of $\mathrm{P}(\omega_1)$). Set $B_\xi = \mathbf{N} \setminus (A_\xi \cup E)$. Then $B_{\zeta(n)} \setminus B_\xi \subseteq (B_{\zeta(n)} \cap A_\xi) \cup (B_{\zeta(n)} \cap E)$ is finite for each $n \in \mathbf{N}$, so $B_\eta \setminus B_\xi$ is finite and $B_\xi \setminus B_\eta$ is infinite for each $\eta < \xi$; while $A_\xi \cap B_\xi = \emptyset$ is finite and $\mathbf{N} \setminus (A_\xi \cup B_\xi) = E$ is infinite. Thus the induction continues.

(*b*) **?** Suppose, if possible, that there is such an H. Then there is an $n \in \mathbf{N}$ such that

$$C = \{\xi : B_\xi \cap H \subseteq n\}$$

is uncountable. Take $\xi \in C$ such that $C \cap \xi$ is infinite. Then $A_\xi \setminus H$ is finite;

say $A_\xi \setminus H \subseteq m$. We see now that

$$\{\eta : \eta < \xi, B_\eta \cap A_\xi \subseteq \max(m, n)\} \supseteq C \cap \xi$$

is infinite, which we are supposing to be impossible. **X**

21M **Sources**

HAUSDORFF 36 for 21L. ROTHBERGER 48 for 21C (with $\kappa = \omega_1$). KUNEN 68 for 21F*b*, 21G(iii). MARTIN & SOLOVAY 70 for 21A (in part), 21B*c*, 21C and 21E*a*. HECHLER 71 for 21B*b*. ERDÖS & SHELAH 72 for more of 21A. ELLENTUCK & RUCKER 72 for 21B*a*. BAUMGARTNER & HAJNAL 73 for 21I. BROVERMAN, GINSBURG, KUNEN & TALL 78 for 21J*c*. SZYMAŃSKI *a* for 21K.

21N **Exercises**

(*a*) Let $\langle D_n \rangle_{n \in \mathbf{N}}$ be a sequence in $\mathscr{P}\mathbf{N}$, and $\mathscr{C} \subseteq \mathscr{P}\mathbf{N}$; suppose that $\#(\mathscr{C}) < \mathfrak{p}$, and that $\sup_{n \in \mathbf{N}} \#(D_n \setminus \bigcup \mathscr{C}_0) = \omega$ for every finite $\mathscr{C}_0 \subseteq \mathscr{C}$. Then there is an $I \subseteq \mathbf{N}$ such that $I \cap C$ is finite for every $C \in \mathscr{C}$ and $\sup_{n \in \mathbf{N}} \# (I \cap D_n) = \omega$. [Hint: take the partially ordered set P of 21A with $\mathscr{A} = \emptyset$ and different V_{Dn}.] Formulate an elaboration of 21A which will include this result as a special case. [BOOTH 70, Theorem 4.10.]

(*b*) If $\mathscr{A} \subseteq [\mathbf{N}]^\omega$, $\#(\mathscr{A}) < \mathfrak{p}$, and $f \in \mathbf{N}^\mathbf{N}$, there is a strictly increasing $g \in \mathbf{N}^\mathbf{N}$ such that $g(n) \geq f(n)$ for every $n \in \mathbf{N}$ and $g^{-1}[A]$ is infinite for every $A \in \mathscr{A}$. [Hint: take P to be $\{g : \exists n \in \mathbf{N}, g \in \mathbf{N}^n, g(i) \geq f(i) \forall i < n, g(i) < g(j) \forall i < j < n\}$.]

(*c*) Let $\mathscr{C} \subseteq \mathscr{P}\mathbf{N}$ be such that $\#(\mathscr{C}) < \mathfrak{p}$ and $C \setminus \bigcup \mathscr{C}_0$ is infinite whenever $\mathscr{C}_0 \in [\mathscr{C}]^{<\omega}$ and $C \in \mathscr{C} \setminus \mathscr{C}_0$. Let $f : \mathscr{C} \to \mathbf{N}$ be any function. Then there is an $I \subseteq \mathbf{N}$ such that $f(C) \leq \#(I \cap C) < \omega$ for every $C \in \mathscr{C}$. [Cf. HECHLER 71, Lemma 9.2.]

(*d*) Let $\mathscr{C}$ be an almost-disjoint family of subsets of $\mathbf{Q}$, with $\omega \leq \#(\mathscr{C}) < \mathfrak{p}$. Then there is a dense set $I \subseteq \mathbf{Q}$ such that $I \cap C$ is finite for every $C \in \mathscr{C}$. [$\mathfrak{p} = \mathfrak{c}$] There is a maximal almost-disjoint family of infinite subsets of $\mathbf{Q}$ such that every member of the family is dense in $\mathbf{Q}$. [HECHLER 71.]

(*e*) Let $S \subseteq [\mathfrak{p}]^{<\omega}$ be such that $\emptyset \in S$ and $I \subseteq J \in S \Rightarrow I \in S$. Then there is a family $\langle A_\xi \rangle_{\xi < \mathfrak{p}}$ in $\mathscr{P}\mathbf{N}$ such that

$$S = \{I : I \in [\mathfrak{p}]^{<\omega}, \bigcap_{\eta \in I} A_\eta \text{ is infinite}\}.$$

[Hint: choose the A_ξ inductively, requiring $\bigcap_{\eta \in I} A_\eta \setminus \bigcup_{\eta \in J} A_\eta$ to be infinite whenever $I \in S$ and $J \in [\mathfrak{p} \setminus I]^{<\omega}$.] So S is expressible as $\bigcap_{n\in\mathbf{N}} \bigcup_{m\geq n} [C_m]^{<\omega}$ for some sequence $\langle C_m \rangle_{m\in\mathbf{N}}$ of subsets of $\mathfrak{p}$.

(*f*) $[\mathfrak{p} > \omega_1]$ (i) Let $A \subseteq \{0,1\}^{\omega_1}$ be infinite. Then there are an infinite $B \subseteq A$ and an uncountable $C \subseteq \omega_1$ such that $x \restriction C = y \restriction C$ for all $x, y \in B$. [Hint: 21I. See RUDIN 75, p. 30.] (ii) Let $\mathscr{A} \subseteq \mathscr{P}\omega_1$ be infinite. Then there is an uncountable $C \subseteq \omega_1$ such that one of $\{A : A \in \mathscr{A},\ C \subseteq A\}$ or $\{A : A \in \mathscr{A},\ C \cap A = \emptyset\}$ is infinite.

(*g*) Let $\mathscr{A}$ be a family of infinite sets with $\#(\mathscr{A}) < \mathfrak{p}$. Then there is a set B such that $A \cap B \neq \emptyset$ and $A \setminus B \neq \emptyset$ for every $A \in \mathscr{A}$. [Take X such that $\#(X) \leq \mathfrak{c}$ and $A \cap X$ is infinite for every $A \in \mathscr{A}$, and use the method of 21J*c*. See ERDÖS, HAJNAL & MÁTÉ 73, Lemma 4.4.]

(*h*) $[\mathfrak{p} = \mathfrak{c}]$ There is a set $S \subseteq [\mathfrak{c}]^2$ such that if $A \subseteq \mathfrak{c}$ is infinite then $\{\xi : \{\eta, \xi\} \in S\ \forall \eta \in A\}$ and $\{\xi : \{\eta, \xi\} \notin S\ \forall \eta \in A\}$ both have cardinals less than $\mathfrak{c}$. [Enumerate $[\mathfrak{c}]^\omega$ as $\langle A_\xi \rangle_{\xi<\mathfrak{c}}$. Use 21N*g* to choose $B \subseteq \xi$ such that neither B nor $\xi \setminus B$ includes A_η for any $\eta < \xi$. See LAVER 75 and 41O*c*.]

(*i*) A **Boolean group** is an abelian group G such that $x + x = 0$ for every $x \in G$. Let G be a Boolean group and $\mathscr{A}$ a family of infinite subsets of G with $\#(\mathscr{A}) < \mathfrak{p}$. Then there is a homomorphism $f : G \to \mathbf{Z}_2$ such that f is not constant on any member of $\mathscr{A}$. [Express $(G, +)$ as $([X]^{<\omega}, \triangle)$ and use the idea of 21J*c*/21N*g*. See DOUWEN 80.]

(*j*) Suppose that $\#(X) < \mathfrak{p}$ and that $\mathscr{A}$ is a family of infinite subsets of X with $\#(\mathscr{A}) < \mathfrak{p}$. Then X is expressible as $\bigcup_{n\in\mathbf{N}} X_n$ where no X_n includes any member of $\mathscr{A}$. [Hint: seek a suitable subset of $X \times \mathbf{N}$, using the method of 21J*c*/21N*g*. See WEISS 83, 8.12.]

(*k*) There is an almost-disjoint family $\mathscr{A}$ of infinite subsets of $\mathbf{N}$ such that (i) $\#(\mathscr{A}) = \omega_1$ (ii) if $\mathscr{B}, \mathscr{C} \subseteq \mathscr{A}$ are uncountable and disjoint then there is no $H \subseteq \mathbf{N}$ such that $B \cap H$ and $C \setminus H$ are finite for all $B \in \mathscr{B}$, $C \in \mathscr{C}$. [Take $\mathscr{A} = \{A_\xi : \xi < \omega_1\}$ where $\{\eta : \eta < \xi, A_\eta \cap A_\xi \subseteq k\}$ is finite for all $\xi < \omega_1$, $k \in \mathbf{N}$. See DOUWEN 83, 4.1.]

(*l*) If κ is a regular cardinal, there are families $\langle A_\xi \rangle_{\xi<\kappa^+}$, $\langle B_\xi \rangle_{\xi<\kappa^+}$ of subsets of κ such that (i) all the A_ξ and B_ξ are non-stationary, (ii) if $\eta < \xi < \kappa^+$ then $\#(A_\eta \setminus A_\xi) < \kappa$, $\#(B_\eta \setminus B_\xi) < \kappa$, $\#(A_\xi \cap B_\xi) < \kappa$ and $\#(A_\xi \setminus A_\eta) = \kappa$, (iii) if $\xi < \kappa^+$ and $I \in [\xi]^{<\kappa}$ then $\#(B_\xi \setminus \bigcup_{\eta\in I} B_\eta) = \kappa$, (iv) if $\xi < \kappa^+$ and $\alpha < \kappa$ then $\#(\{\eta : \eta < \xi,\ B_\eta \cap A_\xi \subseteq \alpha\}) < \kappa$. Now there is no $H \subseteq \kappa$ such that $\#(A_\xi \setminus H) < \kappa$ and $\#(B_\xi \cap H) < \kappa$ for every $\xi < \kappa^+$. [See BALCAR, FRANKIEWICZ & MILLS 80, Lemma 3.5; HERINK 77, Theorem 2.7.]

(*m*) Let P be a particularly ordered set which is σ-centered upwards and has cardinal less than $\mathfrak{p}$. Then P is expressible as a countable union of upwards-directed sets. [Use the technique of 31A.]

21O Further results

(*a*) An ordinal α is **indecomposable** if $\mathrm{otp}(\alpha\setminus\xi) = \alpha$ for every $\xi < \alpha$; equivalently, if α is not the ordinal sum of two smaller ordinals. For each non-zero indecomposable ordinal $\alpha < \omega_1$, there is an ultrafilter $\mathscr{F}$ on α such that if $\mathscr{A} \subseteq \mathscr{F}$ and $\#(\mathscr{A}) < \mathfrak{p}$ there is an $I \subseteq \alpha$ such that $\mathrm{otp}(I) = \alpha$ and $\sup(I\setminus A) < \alpha$ for every $A \in \mathscr{A}$. [LAVER 75.]

(*b*) Suppose that $\omega < \mathrm{cf}(\kappa) \leq \kappa < \mathfrak{p}$ and that $\alpha < \omega_1$ is indecomposable. Let $\mathscr{S}$ be a finite cover of $\alpha \times \kappa$. Then there are an $I \subseteq \alpha$, a $C \subseteq \kappa$ and an $S \in \mathscr{S}$ such that $\mathrm{otp}(I) = \alpha$, $\#(C) = \kappa$ and $I \times C \subseteq S$. [BAUMGARTNER & HAJNAL 73; LAVER 75.]

(*c*) [$\mathfrak{p} > \omega_1$] Let α be the ordinal power $\omega_1^{\omega+2}$ and β the ordinal power ω^2 [definition: A2A*b*]. If $f:\alpha \to \beta$ is any function, there is an $A \subseteq \alpha$ such that $\mathrm{otp}(A) = \alpha$ but $\mathrm{otp}(f[A]) < \beta$. [LARSON 79.]

(*d*) **Automorphisms of Boolean algebras** For a Boolean algebra $\mathfrak{A}$, write $\mathrm{Aut}(\mathfrak{A})$ for its group of automorphisms. (i) If $\mathfrak{A}$ has infinitely many atoms and $\#(\mathfrak{A}) < \mathfrak{p}$, then $\mathrm{Aut}(\mathscr{P}\mathbf{N})$ can be embedded into $\mathrm{Aut}(\mathfrak{A})$ [MCKENZIE & MONK 75], there is a surjective Boolean homomorphism from $\mathfrak{A}$ to $\mathfrak{A}$ which is not injective [LOATS 79], and there is an injective Boolean homomorphism from $\mathfrak{A}$ to $\mathfrak{A}$ which is not surjective. (ii) [$\mathfrak{p} = \mathfrak{c}$] If $\#(\mathrm{Aut}(\mathfrak{A})) = \omega$ then $\#(\mathfrak{A}) \geq \mathfrak{c}$ [MCKENZIE & MONK 75]. (iii) [$\mathfrak{p} = \mathfrak{c}$] If $\omega \leq \kappa \leq \mathfrak{c}$ there is a subalgebra $\mathfrak{A}$ of $\mathscr{P}\kappa$ such that $[\kappa]^{<\omega} \subseteq \mathfrak{A}$, $\#(\mathfrak{A}) = \mathfrak{c}$, a Boolean homomorphism from $\mathfrak{A}$ to itself is injective iff it is surjective, and every member of $\mathrm{Aut}(\mathfrak{A})$ moves only finitely many atoms of $\mathfrak{A}$. [LOATS & ROITMAN 81.]

(*e*) [$\mathfrak{p} = \mathfrak{c} > \omega_1$ + $\mathfrak{c}$ a successor cardinal] There is a Boolean algebra $\mathfrak{A}$ with $\#(\mathfrak{A}) = \mathfrak{c}$, such that (i) if B and C are countable subsets of $\mathfrak{A}$ with $b \subseteq c$ for every $b \in B$ and $c \in C$ there is an $a \in \mathfrak{A}$ with $b \subseteq a \subseteq c$ for every $b \in B$ and $c \in C$ (ii) there is no surjective Boolean homomorphism from a Dedekind σ-complete Boolean algebra onto $\mathfrak{A}$. [DOUWEN & MILL 80.]

(*f*) [$\mathfrak{p} = \mathfrak{c}$] There is a countable extremally disconnected abelian Hausdorff topological group which is not discrete. [LOUVEAU 72, MALYHIN 75*b*.]

(*g*) [$\mathfrak{p} = \mathfrak{c}$] There is a surjection $f:\mathscr{P}\mathbf{N} \to \mathscr{P}\mathbf{N}$, not the identity, such that $f(A) \supseteq A$ and $f(A \cup B) = f(A) \cup f(B)$ for all $A, B \subseteq \mathbf{N}$. [PRICE 79.]

(*h*) (i) If $\kappa < \mathfrak{p}$ there is a $C \subseteq \kappa$ such that every subset of κ is constructible from C and a subset of $\mathbf{N}$. [MARTIN & SOLOVAY 70.] (ii) $[\mathfrak{p} > \omega_1]$ There is a set $A \subseteq \mathbf{N}$ such that $\omega_1 = \omega_1^{L(A)}$ iff every subset of ω_1 is constructible from a subset of $\mathbf{N}$. [ibid.] (iii) $[\mathfrak{p} > \omega_1]$ If on the other hand $\omega_1^{L(A)} < \omega_1$ for every $A \subseteq \mathbf{N}$, then ω_1 is a Mahlo cardinal in $L(A)$ for every $A \subseteq \mathbf{N}$. [KANOVEI 79; see also HARRINGTON & SHELAH *a*.]

(*i*) Results on similar topics may also be found in BURGESS 78, GÖBEL & WALD 80, HECHLER 72*b* and MALYHIN 75*a*.

21P **Problem**

Are special axioms really needed in 21O*e–g*?

Notes and comments

Theorem 21A covers most of the extensions of $\mathrm{P}(\kappa)$ which I have seen; not all, because further variations using the idea of 21N*a* are possible. I have written 21A out as a consequence of Bell's theorem; it is of course not hard to work directly from the definition of $\mathfrak{p}$, but I think that the proof here makes it clearer what we are really doing. Each (E, G) in the partially ordered set P is a kind of provisional commitment, declaring that we mean to have $E \subseteq I \subseteq X \setminus G$ when we've finished; as usual, we use a form of Martin's axiom to pick out a consistent family W of such commitments. Further refinements which operate by using the same set P but expanding $\mathcal{Q}$ should not now be hard to find.

There is an asymmetry in 21A; the family $\mathcal{A}$ must be countable, while $\mathcal{B}$, $\mathcal{C}$ and $\mathcal{D}$ are allowed to have any cardinals less than $\mathfrak{p}$. To show that this is necessary, I have included a description of Hausdorff's gap [21L], where we see that 21A becomes false if we allow $\#(\mathcal{A}) = \#(\mathcal{C}) = \omega_1$ even if $\mathcal{B}$ and $\mathcal{D}$ are empty. (This can be regarded as a further warning that only ccc partially ordered sets will work.) Using $\mathfrak{m}$ rather than $\mathfrak{p}$, the other countability restriction here (on the set X) can be relaxed, at the cost of some further complications; see 42I below.

A simpler result which uses $\mathrm{MA}_{\sigma\text{-cent}}(\kappa)$ in much the same way as 21A is in 21N*b*. (But note that the partially ordered set in 21N*b* is actually countable; see B1B.) The corollaries of 21A in 21B are typical of properties of cardinals less than $\mathfrak{p}$ which can be false for other cardinals less than $\mathfrak{c}$. For more about almost-disjoint families, see 21N*c–d*. I use 21B*c* for 21C–E and parts (i) and (ii) of 21G–H; it is also the basis of some striking model-theoretic consequences of $\mathfrak{p} > \omega_1$ (see 21O*h*, and also 23N*g*).

Theorem 21C and its corollaries 21D–E form one of the characteristic consequences of axiom P; both parts of 21E are frequently used. The results in 21F–H constitute a pleasing and important pattern. I have presented 21G(i)–(ii) and 21H(i)–(ii) as if they were perfectly symmetrical; but if you look at the proof of 21G you will see that the C_n of part (*a*) are chosen before looking at the set *S*, while the D_n in part (*a*), and both the C_n and the D_n in part (*b*), are defined in a way that depends on *S*. So there are occasions on which the C_n of 21G(i) or 21H(i) can be chosen with extra properties. This is what happens, in effect, in 23B below. Applications of 21G–H are in 23F, 23M*e* and 23N*i*. 21N*e* is an elaboration of 21F*a*.

21I is a kind of partition theorem which will be extremely useful in Chapter 4 (usually, of course, with $\kappa = \omega_1$); elementary applications are in 21N*f*. There is an interesting extension in 21O*b*, which uses 21O*a* in place of the definition of $\mathfrak{p}$. I include 21J only because $\diamondsuit$ is so important that it's worth recording that most of the things you use it to prove imply that $\mathfrak{p} = \omega_1$; but the method of proof which I use in 21J*c* is important to us now, because it gives a method of obtaining special subsets of uncountable sets, as in 21N*f*–*j*. Perhaps I ought to remark here that 21A cannot be approached directly by this method; by B1G below, it is consistent to assume that the union of $\mathfrak{p}$ meagre sets in $\mathscr{P}\mathbf{N}$ is always meagre.

The $\diamondsuit$ of 21J is '$\diamondsuit_{\omega_1}$'. It is worth noting that '$\diamondsuit_{\mathfrak{c}}$' is *not* inconsistent with $\mathfrak{m} = \mathfrak{c} > \omega_1$ [B2H].

For the results concerning $\mathfrak{m}$ and $\mathfrak{m}_K$ which correspond to 21K, see 31K, 41C*d* and B2A. There is another consequence of 21L in 25K; variations of 21L are in 21N*k*–*l* and 26K*c*.

As this book progresses it will become apparent that there are interesting and subtle differences between results involving directed sets and those involving centered sets; 21N*m* may help to illuminate some of these, though I shall not use it directly.

22 Category and measure

The work of this section is based on two facts: if $\kappa < \mathfrak{p}$, then the union of κ meagre sets in $\mathbf{R}$ is meagre, and the union of κ closed negligible sets is negligible. Actually something much stronger is true: if we have κ nowhere-dense sets $E_\xi \subseteq \mathbf{R}$, then there is a sequence of nowhere-dense sets F_n such that every E_ξ is included in some F_n. This is what I call the '$(<\mathfrak{p}, \omega)$-covering property' [22A]. Naturally, there are extensive generalizations of the arguments used, and since this is general topology I follow current practice in this subject, by giving results in the greatest convenient generality. However, the reader who has no use for 'countable π-bases' and the like will lose nothing of importance by taking every topological space to be separable and metrizable, and every measure space to be a subspace of **R** with Lebesgue measure.

I start with the $(<\mathfrak{p},\omega)$-covering property for nowhere-dense sets [22B–D]. In 22E I give a lemma for constructing nowhere-dense sets, with a simple corollary. 22G–H deal with the $(<\mathfrak{p},\omega)$-covering property for closed negligible sets; 22I–J with the same for totally bounded sets. 22K shows how the axiom $\mathfrak{p}=\mathfrak{c}$ makes a variety of directed sets have a simple final structure. In 22L I discuss the relationship between $\mathfrak{p}$ and the measurable cardinal problem.

22A Definition

Let κ be a cardinal. I shall say that a family $\mathscr{A}$ of sets has the **(κ,ω)-covering property** if whenever $\mathscr{A}_1\subseteq\mathscr{A}$ and $\#(\mathscr{A}_1)\leq\kappa$ there is a countable $\mathscr{A}_0\subseteq\mathscr{A}$ such that every member of $\mathscr{A}_1$ is included in a member of $\mathscr{A}_0$. If $\mathscr{A}$ has the (λ,ω)-covering property for every $\lambda<\kappa$, I shall say that $\mathscr{A}$ has the **$(<\kappa,\omega)$-covering property**.

Observe that if $\mathscr{A}$ has the $(<\kappa,\omega)$-covering property then the σ-ideal of sets generated by $\mathscr{A}$ is κ-additive.

22B Theorem

If X is a topological space with a countable π-base, then the family of nowhere-dense sets in X has the $(<\mathfrak{p},\omega)$-covering property.

Proof If $X=\emptyset$ this is trivial. Otherwise, let $\mathscr{U}$ be a countable π-base for the topology of X, and let $\langle U_n\rangle_{n\in\mathbf{N}}$ be a sequence in $\mathscr{U}$ such that each member of $\mathscr{U}$ recurs infinitely often in the sequence. Let $\mathscr{E}$ be a family of nowhere-dense sets in X with $\#(\mathscr{E})<\mathfrak{p}$. For $E\in\mathscr{E}$ set

$$C_E=\{n:n\in\mathbf{N},E\cap U_n\neq\emptyset\},$$

and for $U\in\mathscr{U}$ set

$$D_U=\{n:n\in\mathbf{N},U_n\subseteq U\}.$$

If $\mathscr{E}_0\subseteq\mathscr{E}$ is finite and $U\in\mathscr{U}$, then $\bigcup\mathscr{E}_0$ is nowhere dense, so $U\setminus\overline{\bigcup\mathscr{E}_0}$ is non-empty and there is a $V\in\mathscr{U}$ such that $V\subseteq U\setminus\overline{\bigcup\mathscr{E}_0}$; now

$$D_U\setminus\bigcup\{C_E:E\in\mathscr{E}_0\}\supseteq\{n:U_n=V\},$$

which is infinite. Accordingly we can apply 21A with $\mathscr{A}=\mathscr{B}=\emptyset$, $\mathscr{C}=\{C_E:E\in\mathscr{E}\}$, $\mathscr{D}=\{D_U:U\in\mathscr{U}\}$ and obtain a set $I\subseteq\mathbf{N}$ such that $I\cap C_E$ is finite for each $E\in\mathscr{E}$ and $I\cap D_U$ is infinite for each $U\in\mathscr{U}$. Set $G_m=$

$\bigcup_{n\in I\setminus m} U_n$, $F_m = X\setminus G_m$ for each $m\in\mathbf{N}$. If $U\in\mathscr{U}$ and $m\in\mathbf{N}$ there is an $n\in D_U\cap I\setminus m$; now $U_n\subseteq U\cap G_m$ so $U\cap G_m$ is not empty; as U is arbitrary, G_m is dense and F_m is nowhere dense. If $E\in\mathscr{E}$ there is an m such that $C_E\cap I\subseteq m$; now $E\cap G_m=\varnothing$ and $E\subseteq F_m$. Thus $\langle F_m\rangle_{m\in\mathbf{N}}$ is a sequence of nowhere-dense sets such that each member of $\mathscr{E}$ is included in at least one of them, as required.

22C Corollary

Let X be a topological space with a countable π-base. Then

(*a*) the union of fewer than $\mathfrak{p}$ meagre sets is meagre;

(*b*) the union of fewer than $\mathfrak{p}$ sets with the Baire property again has the Baire property.

Proof (*a*) is immediate from 22B. For (*b*), let $\mathscr{E}$ be a family of subsets of X with the Baire property, with $\#(\mathscr{E})<\mathfrak{p}$. For each $E\in\mathscr{E}$, choose an open set G_E such that $G_E\triangle E$ is meagre. Then

$$(\bigcup\mathscr{E})\triangle(\bigcup_{E\in\mathscr{E}}G_E)\subseteq\bigcup_{E\in\mathscr{E}}(E\triangle G_E)$$

is meagre, so $\bigcup\mathscr{E}$ has the Baire property.

22D Corollary

$[\mathfrak{p}=\mathfrak{c}]$ Let X be a topological space with a countable π-base in which singleton sets are nowhere dense. Then X has a $\mathfrak{c}$-Lusin subset.

Proof By 22B, no non-meagre set can be covered by fewer than $\mathfrak{c}$ meagre sets. Also X is ccc because $c(X)\leq\pi(X)\leq\omega$. So A3F*b* applies and X has a $\mathfrak{c}$-Lusin subset.

22E Proposition

Let X be a separable metric space without isolated points, and $x\in X$. Let $\mathscr{C}$ be a family of subsets of X with $\#(\mathscr{C})<\mathfrak{p}$. Then there is a closed nowhere-dense set $F\subseteq X$ such that $x\in F$ and, for every $C\in\mathscr{C}$, either $F\cap C\neq\varnothing$ or $F\cap\bar{C}=\varnothing$.

Proof Let $\mathscr{U}$ be a countable base for the topology of X, and let $\mathscr{U}^*$ be the set of finite unions of members of $\mathscr{U}$. Set

$$P=\{(I,G):G\in\mathscr{U}^*, x\in I\in[G]^{<\omega}\}.$$

Say that $(I,G)\leq(J,H)$ if $I\subseteq J$ and $H\subseteq G$. As $\mathscr{U}^*$ is countable, P is σ-centered upwards.

For each $U \in \mathscr{U} \setminus \{\emptyset\}$ set

$$Q_U = \{(I, G) : (I, G) \in P, U \setminus \bar{G} \neq \emptyset\}.$$

Then Q_U is cofinal with P. **P** Let $(I, G) \in P$. As X has no isolated points, $U \nsubseteq I$. Let $y \in U \setminus I$ and let V be an open neighbourhood of y such that $I \cap \bar{V} = \emptyset$. Now take an $H \in \mathscr{U}^*$ such that $I \subseteq H \subseteq G \setminus \bar{V}$ and observe that $(I, G) \leq (I, H) \in Q_U$. **Q**

For each $C \in \mathscr{C}$ set

$$Q'_C = \{(I, G) : (I, G) \in P, I \cap C \neq \emptyset \text{ or } \bar{G} \cap \bar{C} = \emptyset\}.$$

Then Q'_C is cofinal with P. **P** Let $(I, G) \in P$. If $C \cap G \neq \emptyset$, take $y \in C \cap G$ and see that $(I, G) \leq (I \cup \{y\}, G) \in Q'_C$. If $C \cap G = \emptyset$, take $H \in \mathscr{U}^*$ such that $I \subseteq H \subseteq \bar{H} \subseteq G$ and observe that $(I, G) \leq (I, H) \in Q'_C$. **Q**

By Bell's theorem [14C], there is an upwards-directed set $R \subseteq P$ meeting every Q_U and every Q'_C. Set $F = \bigcap\{\bar{G} : (I, G) \in R\}$. Because R is upwards-directed, $F \supseteq I$ whenever $(I, G) \in R$; in particular, $x \in F$. If $U \in \mathscr{U} \setminus \{\emptyset\}$, then R meets Q_U, so $U \setminus F \neq \emptyset$; thus F is nowhere dense. Finally, if $C \in \mathscr{C}$, then R meets Q'_C in (I, G) say; if $I \cap C \neq \emptyset$, then $F \cap C \neq \emptyset$; if $\bar{G} \cap \bar{C} = \emptyset$, then $F \cap \bar{C} = \emptyset$.

22F Corollary

[$\mathfrak{p} = \mathfrak{c}$] There is a set $S \subseteq \mathbf{R}^2$ with nowhere-dense vertical sections such that if $A \subseteq \mathbf{R}$ and $\#(A) = \mathfrak{c}$ then there are distinct $s, t \in A$ with $(s, t) \in S$.

Proof Enumerate $\mathbf{R}$ as $\langle s_\xi \rangle_{\xi < \mathfrak{c}}$ and $[\mathbf{R}]^\omega$ as $\langle C_\xi \rangle_{\xi < \mathfrak{c}}$. For each $\xi < \mathfrak{c}$ use 22E to find a nowhere-dense set $S_\xi \subseteq \mathbf{R}$ such that $s_\xi \in S_\xi$ and if $\eta \leq \xi$ and $S_\xi \cap \bar{C}_\eta \neq \emptyset$ then $S_\xi \cap C_\eta \neq \emptyset$. Set

$$S = \{(s_\xi, t) : \xi < \mathfrak{c}, t \in S_\xi\}.$$

Then the vertical sections of S are the S_ξ and are nowhere dense. If $A \subseteq \mathbf{R}$ and $\#(A) = \mathfrak{c}$, let C be a countable dense subset of A; let $\eta < \mathfrak{c}$ be such that $C = C_\eta$; let $\xi \geq \eta$ be such that $s_\xi \in A \setminus C$. Then $S_\xi \cap C \neq \emptyset$, by construction; if $t \in S_\xi \cap C$ then s_ξ, t are distinct members of A and $(s_\xi, t) \in S$.

22G Theorem

Let X be a second-countable topological space and μ a σ-finite Borel measure on X. Then the family of closed negligible sets in X has the $(<\mathfrak{p}, \omega)$-covering property.

Proof Let $\langle X_n \rangle_{n \in \mathbf{N}}$ be an increasing sequence of Borel sets of finite measure covering X. Let $\mathscr{U}$ be a countable base for the topology of X, and $\mathscr{U}^* = \{\bigcup \mathscr{U}_0 : \mathscr{U}_0 \in [\mathscr{U}]^{<\omega}\}$. Let $\langle G_n \rangle_{n\in\mathbf{N}}$ be a sequence in $\mathscr{U}^*$ such that each member of $\mathscr{U}^*$ recurs infinitely often in the sequence.

Let $\mathscr{E}$ be a family of closed negligible sets in X with $\#(\mathscr{E}) < \mathfrak{p}$. For $m\in\mathbf{N}$, set

$$D_m = \{n : n\in\mathbf{N}, \mu(X_m \setminus G_n) \le 2^{-m}\}.$$

For $E\in\mathscr{E}$, set

$$C_E = \{n : n\in\mathbf{N}, G_n \cap E \neq \varnothing\}.$$

If $m\in\mathbf{N}$ and $\mathscr{E}_0 \subseteq \mathscr{E}$ is finite, then $F = \bigcup \mathscr{E}_0$ is a closed negligible set, so

$$\mathscr{G} = \{G : G\in\mathscr{U}^*, G\cap F = \varnothing\}$$

is upwards-directed and has union $X\setminus F$. As $\mathscr{G}$ is countable,

$$\mu X_m = \mu(X_m \setminus F) = \sup\{\mu(X_m \cap G) : G\in\mathscr{G}\}$$

and there is a $G\in\mathscr{G}$ such that $\mu(X_m\setminus G) \le 2^{-m}$. Now

$$D_m \setminus \bigcup\{C_E : E\in\mathscr{E}_0\} \supseteq \{n : G_n = G\}$$

is infinite.

So we can apply 21A to find an $I\subseteq\mathbf{N}$ such that $C_E\cap I$ is finite for every $E\in\mathscr{E}$ and $D_m \cap I$ is infinite for every $m\in\mathbf{N}$. Set

$$F_n = X\setminus \bigcup_{i\in I, i\ge n} G_i.$$

For each $E\in\mathscr{E}$, there is an $n\in\mathbf{N}$ such that $I\cap C_E \subseteq n$; now $E\cap G_i = \varnothing$ whenever $i\in I$ and $i \ge n$, so $E\subseteq F_n$. For each $m\in\mathbf{N}$, $n\in\mathbf{N}$ there is an $i\in D_m\cap I$ such that $i\ge n$; now $F_n \cap G_i = \varnothing$ so $\mu(X_m\cap F_n) \le \mu(X_m\setminus G_i) \le 2^{-m}$; as m is arbitrary, $\mu F_n = 0$. Thus $\{F_n : n\in\mathbf{N}\}$ is the countable family of closed negligible sets which we need.

22H **Corollary**

(*a*) Let X be a second-countable space, μ a quasi-Radon measure on X. If $\mathscr{E}$ is a family of closed sets in X with $\#(\mathscr{E}) < \mathfrak{p}$, then $\bigcup\mathscr{E}$ is measurable and

$$\mu(\bigcup\mathscr{E}) = \sup\{\mu(\bigcup\mathscr{E}_0) : \mathscr{E}_0 \in [\mathscr{E}]^{<\omega}\} \le \sum_{E\in\mathscr{E}} \mu E.$$

If $\mathscr{E}$ is upwards-directed, then $\mu(\bigcup\mathscr{E}) = \sup\{\mu E : E\in\mathscr{E}\}$; if $\mathscr{E}$ is disjoint, then $\mu(\bigcup\mathscr{E}) = \sum_{E\in\mathscr{E}} \mu E$.

(*b*) If $A\subseteq\mathbf{R}$ and $\#(A) < \mathfrak{p}$ then A is Lebesgue negligible.

(*c*) If (X, Σ, μ) is any atomless σ-finite measure space and $\#(X) < \mathfrak{p}$ then $\mu X = 0$.

(*d*) If X is any Hausdorff space and $A \subseteq X$ is a set of cardinal less than $\mathfrak{p}$, then A is universally negligible.

Proof (*a*) Recall that a quasi-Radon measure on a second-countable space must be σ-finite [A7B*j*]. So there is a countable $\mathscr{E}_1 \subseteq \mathscr{E}$ such that $\mu(E \setminus \bigcup \mathscr{E}_1) = 0$ for every $E \in \mathscr{E}$ [A6G*a*]. Define a Borel measure ν on X by writing

$$\nu E = \mu(E \setminus \bigcup \mathscr{E}_1)$$

for every Borel set $E \subseteq X$. Then $\nu E \leq \mu E$ for every Borel set E, so ν is also σ-finite [using A7B*c*]. Also $\nu E = 0$ for every $E \in \mathscr{E}$, so by 22G there is a sequence $\langle F_n \rangle_{n \in \mathbf{N}}$ of closed ν-negligible sets covering $\bigcup \mathscr{E}$. Now $\bigcup \mathscr{E}_1 \subseteq \bigcup \mathscr{E} \subseteq \bigcup_{n \in \mathbf{N}} F_n$ and

$$\mu(\bigcup_{n \in \mathbf{N}} F_n \setminus \bigcup \mathscr{E}_1) = \nu(\bigcup_{n \in \mathbf{N}} F_n) = 0;$$

as (X, μ) is complete, $\bigcup \mathscr{E}$ is μ-measurable and

$$\begin{aligned}\mu(\bigcup \mathscr{E}) = \mu(\bigcup \mathscr{E}_1) &= \sup\{\mu(\bigcup \mathscr{E}_0) : \mathscr{E}_0 \in [\mathscr{E}_1]^{<\omega}\} \\ &\leq \sup\{\mu(\bigcup \mathscr{E}_0) : \mathscr{E}_0 \in [\mathscr{E}]^{<\omega}\} \leq \mu(\bigcup \mathscr{E}).\end{aligned}$$

It follows at once that $\mu(\bigcup \mathscr{E}) \leq \sum_{E \in \mathscr{E}} \mu E$. If $\mathscr{E}$ is upwards-directed, then $\sup\{\mu(\bigcup \mathscr{E}_0) : \mathscr{E}_0 \in [\mathscr{E}]^{<\omega}\} = \sup_{E \in \mathscr{E}} \mu E$; if $\mathscr{E}$ is disjoint, then

$$\sup\{\mu(\bigcup \mathscr{E}_0) : \mathscr{E}_0 \in [\mathscr{E}]^{<\omega}\} = \sum_{E \in \mathscr{E}} \mu E.$$

(*b*) This is immediate either from (*a*) or from 22G itself.

(*c*) Let (X, Σ, μ) be an atomless σ-finite measure space, with $\#(X) < \mathfrak{p}$. Then there is a Borel set $I \subseteq \mathbf{R}$ and a function $f : X \to I$ which is inverse-measure-preserving for μ and the restriction ν of Lebesgue measure to the Borel sets of I [A6I*b*]. By (*b*), $\nu^*(f[X]) = 0$ so that μX must be 0.

(*d*) This follows at once from (*c*), or by repeating the argument for (*c*).

22I **Proposition**

Let (X, ρ) be a separable metric space. Then the family of totally bounded subsets of X has the $(<\mathfrak{p}, \omega)$-covering property.

Proof If X is itself totally bounded, there is nothing to prove. Otherwise, let $\langle x_i \rangle_{i \in \mathbf{N}}$ enumerate a countable dense subset of X. A set $A \subseteq X$ is totally bounded iff

$$\forall n \in \mathbf{N}\ \exists m \in \mathbf{N} \text{ such that } A \subseteq \bigcup_{i \leq m} V(x_i, 2^{-n})$$

where $V(x,\delta) = \{y:\rho(y,x) \le \delta\}$. If $\mathscr{E}$ is a collection of totally bounded subsets of X and $\#(\mathscr{E}) < \mathfrak{p}$, then for each $E \in \mathscr{E}$ there is a function $f_E:\mathbf{N} \to \mathbf{N}$ such that

$$E \subseteq \bigcup_{i \le f_E(n)} V(x_i, 2^{-n}) \, \forall n \in \mathbf{N}.$$

By 14B there is an $f:\mathbf{N} \to \mathbf{N}$ such that $\{n:f(n) < f_E(n)\}$ is finite for each $E \in \mathscr{E}$. Set

$$F_k = \bigcap_{n \ge k} \bigcup_{i \le f(n)} V(x_i, 2^{-n}) \, \forall k \in \mathbf{N};$$

then each F_k is totally bounded and each member of $\mathscr{E}$ is included in some F_k.

22J Corollary

Let X be a Polish space, and let $\mathscr{I}$ be the σ-ideal of $\mathscr{P}X$ generated by the compact sets. Then $\mathscr{I}$ is $\mathfrak{p}$-additive.

22K Scales

Let P be a partially ordered set. An **up-scale** in P is a well-ordered cofinal subset of P of minimal cardinality; call its cardinal the **rank** of the scale. Observe that if P has any up-scale, then all its up-scales have the same rank.

Proposition

[$\mathfrak{p} = \mathfrak{c}$] The following partially ordered sets have up-scales of rank either 1 or $\mathfrak{c}$:

(*a*) the meagre sets in any space with a countable π-base;
(*b*) the negligible F_σ sets in a second-countable space with a quasi-Radon measure;
(*c*) the K_σ sets in a Polish space;
(*d*) $(\mathbf{N}^{\mathbf{N}}, \preccurlyeq)$ where $f \preccurlyeq g$ if either $f = g$ or $\{n:g(n) \le f(n)\}$ is finite.

Proof (*a*)–(*c*) are all consequences of the following fact: if $\mathscr{A} \subseteq \mathscr{P}X$ has the $(<\mathfrak{c},\omega)$-covering property and is closed under countable unions, then every $\mathscr{A}_1 \in [\mathscr{A}]^{<\mathfrak{c}}$ has an upper bound in $\mathscr{A}$. Consequently, if $\#(\mathscr{A}) \le \mathfrak{c}$, then $\mathscr{A}$ has an up-scale which is of rank 1 if $\bigcup \mathscr{A} \in \mathscr{A}$ and of rank $\mathfrak{c}$ otherwise. Now 22B, 22G, 22I/22J provide the relevant facts. Of course (*d*) follows from 14B by a similar argument. [See also 33I*b*.]

22L Atomlessly measurable cardinals: theorem

(*a*) [$\mathfrak{p} > \omega_1$] Let $(X, \mathscr{P}X, \mu)$ be a probability space in which every subset of X is measurable. Then $L^1(X)$ is separable.

(*b*) [$\mathfrak{p} > \omega_1$] Let $\bar{\mu}$ be a measure, extending Lebesgue measure, which has domain $\mathscr{P}([0,1])$. Then there is an $A \subseteq [0,1]$ such that $\bar{\mu}A = 1$ and $\#(A) < \mathfrak{c}$.

(*c*) [$\mathfrak{p} = \mathfrak{c}$] There is no atomlessly measurable cardinal.

Proof (*a*) ? Suppose, if possible, otherwise. Then there is an $s > 0$ and a family $\langle E_\xi \rangle_{\xi < \omega_1}$ in $\mathscr{P}X$ such that $\mu(E_\xi \triangle E_\eta) \geq 3s$ whenever $\eta \neq \xi$. Now by 21H(ii) there is a sequence $\langle C_n \rangle_{n \in \mathbf{N}}$ in $\mathscr{P}X$ such that every E_ξ is expressible as $\bigcap_{n \in \mathbf{N}} \bigcup_{i \in I(\xi) \setminus n} C_i$ for some $I(\xi) \subseteq \mathbf{N}$. For each $\xi < \omega_1$ there are $k(\xi)$, $m(\xi) \in \mathbf{N}$ such that $\mu(E_\xi \triangle F_\xi) \leq s$, where

$$F_\xi = \bigcap_{n \leq k(\xi)} \bigcup_{i \in m(\xi) \cap I(\xi) \setminus n} C_i.$$

But since there are only countably many pairs $(k, m \cap I)$, there must be distinct $\eta, \xi < \omega_1$ such that $F_\eta = F_\xi$, in which case $\mu(E_\eta \triangle E_\xi) \leq 2s$. **X**

(*b*) By (*a*), $L^1 = L^1([0,1], \bar{\mu})$ is separable, and $\#(L^1) = \mathfrak{c}$. Let $\langle f_\xi \rangle_{\xi < \mathfrak{c}}$ be a family of $\bar{\mu}$-integrable functions such that $L^1 = \{f^{\cdot}_\xi : \xi < \mathfrak{c}\}$, and enumerate $[0,1]$ as $\langle s_\xi \rangle_{\xi < \mathfrak{c}}$. Choose $f: [0,1] \to [0,1]$ such that

$$f(s_\xi) \neq f_\eta(s_\xi) \text{ if } \eta \leq \xi.$$

Then (because $\bar{\mu}$ measures every set) f is integrable and $f^{\cdot} \in L^1$. Let $\eta < \mathfrak{c}$ be such that $f^{\cdot} = f^{\cdot}_\eta$; then $\bar{\mu}\{s : f(s) = f_\eta(s)\} = 1$. But also $f(s_\xi) \neq f_\eta(s_\xi)$ for $\xi \geq \eta$, so $\bar{\mu}\{s_\xi : \xi < \eta\} = 1$.

(*c*) If $\mathfrak{c} = \omega_1$ this is a result of S. Ulam; see A6P*e*(ii). If $\mathfrak{p} = \mathfrak{c} > \omega_1$ it follows from (*b*) and 22G/H.

22M Sources

ROTHBERGER 48 for 22H*b* (with $\#(A) = \omega_1$). MARTIN & SOLOVAY 70 for 22C*a* (with $X = \mathbf{R}$) and 22L*c*. HECHLER 72*a* for 22F. BOOTH 74 for another version of 22H*b*. SHOENFIELD 75 for 22C*b*.

22N Exercises

(*a*) Let X be a non-empty topological space with a countable π-base, and $\mathscr{G}$ a collection of open sets in X with the finite intersection property. Suppose that $\#(\mathscr{G}) < \mathfrak{p}$. Then there is a decreasing sequence $\langle H_n \rangle_{n \in \mathbf{N}}$ of non-empty open sets in X such that every member of $\mathscr{G}$ includes some H_n. [Compare 32I.]

(*b*) (i) The Stone space Z of the regular open algebra of $\mathbf{R}$ has a countable π-base but is not first-countable. (ii) $\chi(z, Z) \geq \mathfrak{p}$ for every $z \in Z$. [Use 22N*a*.]

(*c*) Suppose $\kappa < \mathfrak{p}$. Let X be a non-empty absolute G_κ space with a countable π-base and no isolated points. Then X has a non-empty compact perfect subset. [Show that X has a dense Cech-complete subspace. This result can be strengthened; see 43Q*c*.]

(*d*) Let X be a Hausdorff space with $w(X) < \mathfrak{p}$ and $\pi(X) \leq \omega$. Then (i) X has a metrizable comeagre subset Y (ii) $\{x:\psi(x, X) \leq \omega\}$ is comeagre in X. [Hint for (i): take a countable π-base $\mathscr{U}$ and for open $V \subseteq X$ set

$$G_V = \bigcup\{U : U \in \mathscr{U}, U \subseteq V\} \cup (X \setminus \bar{V}).$$

Set $Y = \bigcap_{V \in \mathscr{V}} G_V$ for a base $\mathscr{V} \supseteq \mathscr{U}$.]

(*e*) Let $\kappa < \mathfrak{p}$, and let $\langle A_\xi \rangle_{\xi < \kappa}$ be an almost-disjoint family in $\mathscr{P}\mathbf{N}$. Define $f:\mathscr{P}\mathbf{N} \to \mathscr{P}\kappa$ by setting $f(A) = \{\xi : A \cap A_\xi \text{ is infinite}\}$ for each $A \subseteq \mathbf{N}$. Then f is a surjection and $f^{-1}[E]$ is Borel in $\mathscr{P}\mathbf{N}$ for every open-and-closed set $E \subseteq \mathscr{P}\kappa$. Consequently $f^{-1}[E]$ has the strong Baire property for each Borel set $E \subseteq \mathscr{P}\kappa$. [See MAULDIN 78, Theorem 4.3.]

(*f*) [$\mathfrak{p} = \mathfrak{c}$] There are $\mathfrak{c}$-Lusin sets $A, B \subseteq \mathbf{R}$ such that A, B are both linear spaces over $\mathbf{Q}$, $A + B = \mathbf{R}$, and $A \cap B = \{0\}$. [Take A, B to be the linear spaces over $\mathbf{Q}$ generated by $\{s_\xi : \xi < \mathfrak{c}\}$ and $\{t_\xi : \xi < \mathfrak{c}\}$ respectively, where s_ξ, t_ξ are chosen by an inductive process similar to that of A3F*a*. See ERDÖS, KUNEN & MAULDIN 81, Theorem 7.]

(*g*) Let X be a Hausdorff space with a countable π-base and no isolated points. Let $\mathscr{E} \subseteq \mathscr{P}X$ be such that $\#(\mathscr{E}) < \mathfrak{p}$ and $\bigcap_{E \in \mathscr{E}} \bar{E} \neq \emptyset$. Then there is a nowhere-dense set $Y \subseteq X$ which meets every member of $\mathscr{E}$. [Let $\mathscr{U}^*$ be a countable π-base for the topology of X which is closed under finite unions, and pick $x \in \bigcap_{E \in \mathscr{E}} \bar{E}$. Let P be $\{(I, G) : G \in \mathscr{U}^*, x \in I \in [X \setminus \bar{G}]^{<\omega}\}$; say $(I, G) \leq (J, H)$ if $I \subseteq J$ and $G \subseteq H$. Compare 22E.]

(*h*) [$\mathfrak{p} = \mathfrak{c}$] (i) There is a set $A \subseteq [0, 1]^2$ such that the horizontal sections of A are negligible and meagre and the vertical sections are conegligible and comeagre in $[0, 1]$. (ii) There is a function $f:[0, 1] \to [0, 1]$ such that $\{s : f(s) = g(s)\}$ is negligible and meagre for every Borel measurable function $g:[0, 1] \to [0, 1]$.

(*i*) [$\mathfrak{p} = \mathfrak{c}$] (i) There is an injective function $f:[0, 1] \to [0, 1]$ such that $\#(f^{-1}[F]) < \mathfrak{c}$ for every nowhere-dense set $F \subseteq [0, 1]$. Now there is no extension of Lebesgue measure on $[0, 1]$ for which f is measurable. (ii) There is a bijection $f:[0, 1] \to [0, 1]$ such that the graph of f is universally negligible in $[0, 1]^2$.

(*j*) Let X be an atomless compact Radon measure space. (i) If $A \subseteq X$ and $\#(A) < \mathfrak{p}$ there is a negligible F_σ set including A. [Hint: find a countable family $\mathscr{U}$ of open sets such that $\mu X = \sup\{\mu U : U \in \mathscr{U}, x \notin \bar{U}\}$

for every $x\in X$, and use the method of 22B/22G.] (ii) [$\mathfrak{p}>\omega_1$] If the compact negligible sets of X are metrizable then X is metrizable. [Hint: show that X^2 is hereditarily Lindelöf.]

(*k*) The family of compact subsets of $\mathbf{Q}$ does not have the (ω_1,ω)-covering property. [Recall that any countable compact set can be derived down to $\varnothing$ in countably many steps: see 44H prf *f*.]

(*l*) (i) Let $M\subseteq\mathbf{N}^{\mathbf{N}}$ be the set of unbounded monotonic sequences. If $A\subseteq M$ and $\#(A)<\mathfrak{p}$ there is a $g\in M$ such that $\lim_{n\to\infty}[g(n)/f(n)]=0$ for every $f\in A$. [Hint: 14B.] (ii) If $A\subseteq c_0(\mathbf{N})$ and $\#(A)<\mathfrak{p}$ there is a $g\in c_0(\mathbf{N})$ such that $\lim_{n\to\infty}[f(n)/g(n)]=0$ for every $f\in A$.

(*m*) (i) If X is a metric space, the family of uniformly equicontinuous sets in $C(X)$ has the $(<\mathfrak{p},\omega)$-covering property. (ii) If X is any set and $\langle f_n\rangle_{n\in\mathbf{N}}$ any sequence in $\mathbf{R}^X$ then $\{A:A\subseteq X,\langle f_n\restriction A\rangle_{n\in\mathbf{N}}$ is uniformly convergent$\}$ has the $(<\mathfrak{p},\omega)$-covering property. (iii) If X is any σ-finite measure space, the family of weakly compact sets in $L^1(X)$ has the $(<\mathfrak{p},\omega)$-covering property. [Use 22N*l*(ii). For (ii), compare Shinoda 73, §1.]

(*n*) The family of compact subsets of $\mathbf{R}^2$ with Lebesgue negligible vertical sections has the $(<\mathfrak{p},\omega)$-covering property. [Hint: show that a compact set with negligible vertical sections belongs to the ideal $\mathscr{E}$ of 32R*d*, and then use the method of 22G.]

(*o*) [$\mathfrak{p}>\omega_1$] Let (X,μ) be a probability space such that for every $f:X\to\mathbf{R}$ there is a measure $\bar{\mu}$ extending μ for which f is measurable. Then $L^1(X)$ is separable. [Use the argument of 22L*a*. See Prikry *a*.]

(*p*) Let (X,ρ) be a separable metric space; write $V(x,s)=\{y:\rho(y,x)\le s\}$. (i) If $A\subseteq X$ and $\#(A)<\mathfrak{p}$ then A has strong measure 0. [Let D be a countable dense subset of X. Given a sequence $\langle s_n\rangle_{n\in\mathbf{N}}$ in $]0,1]$, let P be $\bigcup_{n\in\mathbf{N}}D^n$ and for $x\in A$ set $Q_x=\{\langle x_i\rangle_{i<n}: x\in\bigcup_{i<n}V(x_i,s_i)\}$. Compare 33B.] (ii) If A is a $\mathfrak{p}$-Lusin set in X then $f[A]$ has strong measure 0 in Y whenever f is a function with the Baire property from X to another separable metric space Y. (iii) [$\mathfrak{p}=\mathfrak{c}$] There is a set of strong measure 0 in $[0,1]$ which has cardinal $\mathfrak{c}$. [See Kunen 68, Theorem 14.6; also 22N*r*(iii).]

(*q*) Let X be a separable metric space. A set $A\subseteq X$ has **Rothberger's property** [Laver 77] if for every family $\langle s_n^x\rangle_{n\in\mathbf{N},x\in A}$ in $]0,1]$ there is a sequence $\langle x(n)\rangle_{n\in\mathbf{N}}$ in A such that $A\subseteq\bigcup_{n\in\mathbf{N}}V(x(n),s_n^{x(n)})$. (i) A has Rothberger's property iff whenever Y is a metric space and $f:A\to Y$ is continuous then $f[A]$ has strong measure 0 in Y. [Recall that A is paracompact.] (ii) Any set of cardinal less than $\mathfrak{p}$ has Rothberger's property.

(iii) A $\mathfrak{p}$-Lusin set always has Rothberger's property. [See LAVER 77, Theorem 1.4; compare SHINODA 73, p. 123, Proposition. See also 33I*a*.]

(*r*) Let X be a separable metric space, A and Y subsets of X. Say A is **$<\kappa$-concentrated** around Y if $\#(A\setminus G) < \kappa$ for every open $G \supseteq Y$. Say A is **concentrated** if it is $<\omega_1$-concentrated around some countable set. (i) If Y is of strong measure 0 and A is $<\mathfrak{p}$-concentrated around Y then A is of strong measure 0. [Hint: use 22N*p*(i).] (ii) [$\mathfrak{p} > \omega_1$] Every concentrated set is countable. [Hint: 23B.] (iii) [$\mathfrak{p} = \mathfrak{c}$] There is an $A \subseteq \mathbf{R}$ which is of strong measure 0 but is not $<\mathfrak{c}$-concentrated around any set of cardinal less than $\mathfrak{c}$. [Hint: enumerate $[\mathbf{R}]^{<\mathfrak{c}}$ as $\langle B_\xi\rangle_{\xi<\mathfrak{c}}$, and $]0,1]^{\mathbf{N}}$ as $\langle\langle s_n^\xi\rangle_{n\in\mathbf{N}}\rangle_{\xi<\mathfrak{c}}$. Choose inductively $A_\xi \in [\mathbf{R}]^{<\mathfrak{c}}$, non-empty perfect sets $F_\xi \subseteq \mathbf{R}$, dense open sets $G_\xi \subseteq \mathbf{R}$ such that (α) $F_\xi \subseteq \bigcap_{\eta<\xi} G_\eta \setminus B_\xi$ [using 22Nc], (β) $A_\xi \subseteq \bigcap_{\eta<\xi} G_\eta \setminus \bigcup_{\eta<\xi} A_\eta$, $A_\xi \cap F_\xi$ is dense in F_ξ, and $A_\xi \cap F_\eta \neq \emptyset\ \forall \eta < \xi$ (possible because $F_\eta \cap G_\zeta$ is dense in F_η for all $\eta, \zeta < \xi$), (γ) $G_\xi \supseteq \bigcup_{\eta\leq\xi} A_\eta$ and $G_\xi \subseteq \bigcup_{n\in\mathbf{N}} V(x_n, s_n^\xi)$ for some $\langle x_n\rangle_{n\in\mathbf{N}}$ (using 22N*p*(i)). Try $A = \bigcup_{\eta<\mathfrak{c}} A_\eta$. See LAVER 77.] (iv) [$\mathfrak{p} = \mathfrak{c}$] There is an $A \subseteq \mathbf{R}$ which has Rothberger's property but is not $<\mathfrak{c}$-concentrated around any set of cardinal less than $\mathfrak{c}$. (v) [$\mathfrak{p} = \mathfrak{c}$] There is an $A \subseteq \mathbf{R}\setminus\mathbf{Q}$ which is $<\mathfrak{c}$-concentrated around $\mathbf{Q}$ but does not have Rothberger's property. [Take a homeomorphism $f:\mathbf{N}^{\mathbf{N}} \to \mathbf{R}\setminus\mathbf{Q}$ and seek a set $B \subseteq \mathbf{N}^{\mathbf{N}}$ such that $f[B]$ is $<\mathfrak{c}$-concentrated around $\mathbf{Q}$ but B is not of strong measure 0. 14B may help.]

22O Further results

(*a*) [$\mathfrak{p} = \mathfrak{c}$] Let μ be Lebesgue measure on $[0,1]$ and $\mathscr{L}^0 \subseteq \mathbf{R}^{[0,1]}$ the set of μ-measurable functions. Let $K \subseteq \mathscr{L}^0$ be a countable set. Then the following are equivalent: (i) $\bar{K} \subseteq \mathscr{L}^0$, where $\bar{K}$ is the closure of K in $\mathbf{R}^{[0,1]}$; (ii) whenever $\mu E > 0$ and $s < t$ in $\mathbf{R}$, there is a $k \geq 1$ such that

$$\mu^{2k}\{\langle u_i\rangle_{i<2k} : u_i \in E\ \forall i < 2k, \exists f \in K, f(u_i) \leq s \text{ if } i < k,$$
$$f(u_i) \geq t \text{ if } k \leq i < 2k\} < (\mu E)^{2k},$$

where μ^{2k} is Lebesgue measure on $[0,1]^{2k}$. [See FREMLIN & TALAGRAND 79, TALAGRAND *a*.]

(*b*) [$\mathfrak{p} = \mathfrak{c}$] Let X be a perfect probability space, $\langle f_n\rangle_{n\in\mathbf{N}}$ a pointwise bounded sequence of measurable real-valued functions on X for which every cluster point in $\mathbf{R}^X$ is measurable. If f is any cluster point of $\langle f_n\rangle_{n\in\mathbf{N}}$ in $\mathbf{R}^X$, it is the limit almost everywhere of a subsequence of $\langle f_n\rangle_{n\in\mathbf{N}}$. [FREMLIN & TALAGRAND 79.]

(*c*) [$\mathfrak{p} = \mathfrak{c}$] Let X be a perfect probability space, $\varphi: X \to \ell^\infty(\mathbf{N})$ a bounded scalarly measurable function. Then φ has a Pettis integral. [Fremlin & Talagrand 79, Talagrand *a*.]

(*d*) [$\mathfrak{p} > \omega_1$] Let (X, μ) be a probability space and $K \subseteq \mathbf{R}^X$ a set which is compact for the usual topology of $\mathbf{R}^X$ and consists entirely of measurable functions. Suppose moreover that $\mu\{x : f(x) \neq g(x)\} > 0$ for all distinct $f, g \in K$. Then K is metrizable. [Fremlin *a*; see also A8E and Talagrand 80*a*, Talagrand *a*.]

(*e*) [$\mathfrak{p} = \mathfrak{c}$] There is a Banach space E with the Radon–Nikodým property and a bounded scalarly measurable function $f : [0, 1] \to E$ which has no Pettis integral. [Edgar 80.]

(*f*) [$\mathfrak{p} = \mathfrak{c}$] Every separable infinite-dimensional Banach space has a universally measurable hyperplane which is not closed. [Talagrand 80*c*. Compare 220*l*.]

(*g*) Let $\mathfrak{F}$ be a non-empty family of filters on $\mathbf{N}$ such that each member of $\mathfrak{F}$ is a non-meagre set in $\mathscr{P}\mathbf{N}$, and $\#(\mathfrak{F}) < \mathfrak{p}$. Then $\bigcap \mathfrak{F}$ is non-meagre. [Talagrand 80*a*.]

(*h*) Say that a set $E \subseteq [0, 2\pi]$ is a **U-set** if whenever $\langle a_n \rangle_{n \in \mathbf{N}}$, $\langle b_n \rangle_{n \geq 1}$ are real sequences such that $\frac{1}{2}a_0 + \sum_{n \geq 1} a_n \cos n\theta + b_n \sin n\theta = 0$ for every $\theta \in [0, 2\pi] \backslash E$, then all the a_n and b_n are 0 [Zygmund 59]. Then the union of fewer than $\mathfrak{p}$ closed U-sets is a U-set. [Holščevnikova 81.]

(*i*) Let X be a topological space. Let Γ be the following infinite game for two players. White chooses a point $x_0 \in X$; Black chooses a neighbourhood G_0 of x_0; White chooses $x_1 \in X$; Black chooses a neighbourhood G_1 of x_1; and so on. White wins if $\bigcup_{n \in \mathbf{N}} G_n = X$; otherwise Black wins. If X is separable and metrizable and $\omega < \#(X) < \mathfrak{p}$ then Γ is undetermined. [Galvin 78.]

(*j*) If $\kappa < \mathfrak{p}$ and $\mathrm{cf}(\kappa) = \omega$, and X is a Hausdorff space with $hc(X) = \kappa$, then X has a discrete subset of cardinal κ. [Kunen & Roitman 77; Juhász 80*b*, 4.4.]

(*k*) [$\mathfrak{p} = \mathfrak{c}$] Let X be a completely regular Hausdorff space such that not every continuous real-valued function on X is bounded. Let βX be the Stone–Cech compactification of X. Then there is an $x \in \beta X \backslash X$ such that $x \notin \bar{A}$ for every countable nowhere-dense $A \subseteq \beta X \backslash \{x\}$. [Mill 82.]

(*l*) [$\mathfrak{p} = \mathfrak{c}$] There is a **medial limit** i.e. a functional $F \in (\ell^\infty(\mathbf{N}))'$ such that $\|F\| = 1$, $F(x) = \lim_{n \to \infty} x(n)$ whenever this exists, and F is universally measurable for $\mathfrak{T}_s(\ell^\infty, \ell^1)$. Now if X is any probability space and $\langle f_n \rangle_{n \in \mathbf{N}}$ is any bounded sequence of measurable functions on X, the limit function $x \mapsto F(\langle f_n(x) \rangle_{n \in \mathbf{N}})$ is measurable. [See NORMANN 76, HOFFMANN-JØRGENSEN 78.]

(*m*) Let X be a probability space and $E \subseteq L^1(X)$ a closed linear subspace. (i) If $d(E) < \mathfrak{p}$ then E is weakly compactly generated. (ii) [$\mathfrak{p} > \omega_1$] If E is not separable then it has a non-separable weakly compact subset. [ROSENTHAL 74.]

(*n*) Let F be a Banach lattice with an order-continuous norm and $F \subseteq E$ a closed linear subspace such that the unit ball of F' is not sequentially compact for $\mathfrak{T}_s(F', F)$. Then there is a complemented linear subspace of F isomorphic (as linear topological space) to $\ell^1(\mathfrak{p})$. [FIGIEL, GHOUSSOUB & JOHNSON 81.]

22P Problems

(*a*) I do not know whether special axioms are really necessary for the results given in 22N*h*(ii), 22O*a*–*f* and 22O*k*–*l*.

(*b*) Is $\mathfrak{p} > \omega_1$ enough to ensure that there is no atomlessly measurable cardinal?

Notes and comments

I have had to invent the phrase '$(<\mathfrak{p}, \omega)$-covering property' because, while the phenomenon is well known, I have not seen it brought out in this way before . Once we start looking for it, of course, it is liable to appear anywhere where we have a $\mathfrak{p}$-additive ideal. Examples which I have collected are 22B, 22G, 22I, 22N*m* and 22N*n*; also 23A*a* can be regarded as a $(<\mathfrak{p}, \omega)$-covering property for the set of closed sets disjoint from $\bigcup \mathscr{K}$.

I regret that the statement of Theorem 22B contains two phrases ('countable π-base' and '$(<\mathfrak{p}, \omega)$-covering property') which are liable to be off-putting to many; but the proof is virtually identical to the natural proof of its most important corollary, that the union of fewer than $\mathfrak{p}$ meagre sets in **R** is meagre [22C*a*]. The same argument yields 22N*a*. I include 22N*b* to show that 'countable π-base' really is weaker than 'countable base' and that stating 22B–D in full generality enables us to apply them to some extremely non-metrizable spaces. 22N*d* is another corollary of 22B which is of interest primarily if we look at non-metrizable compact spaces; 22N*c*, however, has some sort of content even in the metrizable case. 22N*e* is a further glance at the idea which I used for 21C above.

22D is the second appearance (the first was in 13C*d*) of one of the *leitmotivs* of this book; every time we find ourselves with an ω_1-saturated

σ-ideal we can try to prove that it is $\mathfrak{c}$-additive and/or look for a corresponding $\mathfrak{c}$-Lusin set. There are many ways of refining the arguments of A3F in special cases; 22N*f* is one that can be done here.

22E, and its corollary 22F, are lightweight; I spell them out because here also we have results which awake echoes elsewhere (compare 22E with 22N*g* and 23M*h*, and 22F with 42E). Technically the proofs of 22E and 22N*g* are related to those of 21A and 25A in that we use a carefully chosen set P with a natural order on it and prove that it is σ-centered by looking at a countable set built into the definition. 22E can also be regarded as a kind of strong Baire theorem for a Pixley–Roy topology; see 43P*n*.

In 22G–H we have more results of the first importance. I can see four points here which the non-specialist may feel need explanation. First, I talk about 'second-countable spaces with σ-finite measures' when you are very likely interested only in Lebesgue measure on **R**. In this case, you have only to make the appropriate translations as you read; I am wasting none of your time, because the proof is not altered in any way by the demands of extra generality. Second, I once again talk about the $(<\mathfrak{p},\omega)$-covering property when what I am really trying to say is that the union of fewer than $\mathfrak{p}$ closed negligible sets is negligible. I do this to bring out the characteristic nature of $\mathfrak{p}$; if you look ahead to Theorem 32E, you will see that the arguments there work quite differently. Third, having glanced at 32E–F, you will realize that with the other cardinal we can get a better result, and you have some grounds for irritation at being asked to consider the partial form. This is one of the very few occasions (22N*p*(i)/33B is another) where my decision to classify the results of this book by the cardinals which they involve causes some repetition of material. My justification is that 22H*b* is important enough (consider, for instance, its consequence 22L*c*) to merit taking a bit of trouble to observe that it is a property of $\mathfrak{p}$ rather than of $\mathfrak{m}$ (in fact, of course, we can do better still [B1B]). Fourth, I expect you know little and care less about 'quasi-Radon measures'. In the present context this terminology has nothing useful to contribute; on a second-countable space a quasi-Radon measure is virtually the same thing as the completion of a Borel measure [A7B*e*(iii)]. But there are, I believe, good reasons for using the concept in §32, as I try to explain in the notes to that section; and, finding that I need complete measures in 22H*a*, I use the same phrase here.

22H is, of course, elementary. However, the shift in apparent generality on moving to 22H*c*—*d* is so great that it is as well to have them laid out explicitly even though they are really only repetitions of 22H*b*. Among the exercises, 22N*h*–*i* are examples of the kind of elementary, but sometimes useful, constructions which 22B/22H make possible. 22N*j*(i) is a kind of strengthening of 22H*d*. The point of 22N*j*(ii) is that it is false if CH is true [KUNEN 81]. Note also that, despite 22G and 22N*j*(i), the family of compact negligible subsets of a Radon measure space cannot in general be expected to have the $(<\mathfrak{p}, \omega)$-covering property [32P*b*].

22I–K are more easy ideas. I note that 22I relies only on 14B, and is therefore essentially weaker than 22B or 22G. In a non-Polish analytic space, the family of compact sets need not have the $(<\mathfrak{p}, \omega)$-covering property [22N*k*]; but I do not know whether it always has the weaker property that the union of fewer than $\mathfrak{p}$ compact sets is always included in a K_σ set [23O*b*; see also 23H*a*]. I do not think that the results of 22K are important; they show just that some questions which are obscure in the absence of special axioms are settled fairly tidily by $\mathfrak{p} = \mathfrak{c}$.

22L is much more interesting. There are many important questions outstanding about atomlessly measurable cardinals, starting with the question of whether they can exist at all [see B2D]. Certainly there are several reasons why they are incompatible with Martin's axiom; 22L*c* and 22N*i*(i) are two of them. Moreover, even assuming that atomlessly measurable cardinals are admissible in ZFC, it is not clear whether they can have small Maharam type; for instance, it could well be a theorem of ZFC that any extension of Lebesgue measure to the whole of $\mathscr{P}([0, 1])$ must have a non-separable L^1 space. If this is so, then 22L*b* becomes vacuous and 22L*c* a consequence of 22L*a*, answering 22P*b*. See also 22N*o* and 23K.

22N*p*–*r* deal with three more notions of 'small' set: strong measure 0, Rothberger's property, and 'concentration'. A countable set is concentrated; a concentrated set has Rothberger's property; a set with Rothberger's property has strong measure 0; and a set of strong measure 0 is universally negligible. LAVER 77 describes a model in which every set of strong measure 0 is countable. A fourth notion of 'small' set is one in which the player Black of 22O*i* has no winning strategy; this must have Rothberger's property.

In 22O I present several results which can be proved using $\mathfrak{p} = \mathfrak{c}$ but for which there is no good reason, so far as I know, to believe that special axioms are actually necessary. Of course this is one of the dangers of enthusiasm for something like Martin's axiom; having proved a result using it, we may lose the impulse to find out what happens without it. When the result is indeed undecidable, this is a regrettable, but sometimes forgivable, kind of specialization; when the result is in fact provable in ZFC, we have merely made fools of ourselves. (A number of respected mathematicians have pointed out, for instance, that using CH or MA we can find a universally negligible set which is not of strong measure 0; but as it happens such a set always exists [GRZEGOREK 81].) In 22P*a* I list those items which seem to me most likely to be true in 'real' mathematics. My guess is that at least one of these does not need MA. For further details concerning some of them, see FREMLIN & TALAGRAND 79.

22O*a*–*f* can all be proved from 22H*a*; I do not know that they should be regarded as consequences of it. (22O*b*–*c* are derivable from 22O*a*.) With the exception of 22O*h* (which depends on 22C and 22H) and 22O*m* (which is associated with 22N*m*(iii)), the other results of 22O are not connected with the work of this section except, very vaguely, by subject; their proofs are direct from the definition of $\mathfrak{p}$, $MA_{\sigma\text{-cent}}$ or 14B.

23 Descriptive set theory

In this section I continue the work of §22 with results which use more precise descriptions of the sets involved than 'Baire property' or 'measurable'. I begin with the Rothberger–Silver theorem on Q-sets [23B] and some miscellaneous corollaries of results in §§21–22 [23F–H]; then I give the remarkable results of D.A. Martin and R.M. Solovay on the structure of coanalytic and PCA sets [23J], with another of their theorems on ω_1-saturated ideals [23K].

23A **Theorem**

(*a*) Let X be a second-countable space, $\mathscr{K}$ a family of compact subsets of X, and $\mathscr{E}$ a family of closed subsets of X. Suppose that $\#(\mathscr{K}) < \mathfrak{p}$, $\#(\mathscr{E}) < \mathfrak{p}$ and $\bigcup\mathscr{K} \cap \bigcup\mathscr{E} = \emptyset$. Then there is an F_σ set $H \subseteq X$ such that $\bigcup\mathscr{E} \subseteq H$ and $\bigcup\mathscr{K} \cap H = \emptyset$.

(*b*) If X is regular, then there is also an F_σ set $H' \subseteq X$ such that $\bigcup\mathscr{K} \subseteq H'$ and $H' \cap H = \emptyset$.

Proof (*a*) Let $\mathscr{U}$ be a countable base for the topology of X, and $\mathscr{U}^* = \{\bigcup\mathscr{U}_0 : \mathscr{U}_0 \in [\mathscr{U}]^{<\omega}\}$. Let $\langle G_n \rangle_{n\in\mathbf{N}}$ be a sequence in $\mathscr{U}^*$ such that each member of $\mathscr{U}^*$ recurs infinitely often in the sequence.

For $E \in \mathscr{E}, K \in \mathscr{K}$ set

$$C_E = \{n : E \cap G_n \neq \emptyset\},\ D_K = \{n : K \subseteq G_n\}.$$

If $\mathscr{E}_0 \subseteq \mathscr{E}$ is finite and $K \in \mathscr{K}$, then $\bigcup\mathscr{E}_0$ is closed and disjoint from K; so (because K is compact) there is a $G \in \mathscr{U}^*$ such that $K \subseteq G$ and $G \cap \bigcup\mathscr{E}_0 = \emptyset$, and now

$$D_K \setminus \bigcup\{C_E : E \in \mathscr{E}_0\} \supseteq \{n : G_n = G\}$$

is infinite. So by 21A there is an $I \subseteq \mathbf{N}$ such that

$$I \cap D_K \text{ is infinite } \forall K \in \mathscr{K},$$

$$I \cap C_E \text{ is finite } \forall E \in \mathscr{E}.$$

Set $F_n = X \setminus \bigcup_{m \in I, m \geq n} G_m$, $H = \bigcup_{n \in \mathbb{N}} F_n$.

(*b*) The idea is to replace $\mathscr{K}$ and $\mathscr{E}$ in the argument above by $\{F_n : n \in \mathbb{N}\}$ and $\mathscr{K}$. To do this, we need to be able to replace $\mathscr{U}^*$ by a countable family $\mathscr{G}$ of open sets such that, if $n \in \mathbb{N}$ and $\mathscr{K}_0 \subseteq \mathscr{K}$ is finite, there is a $G \in \mathscr{G}$ such that $F_n \subseteq G$ and $G \cap \bigcup \mathscr{K}_0 = \emptyset$. Try

$$\mathscr{G} = \{X \setminus \bar{G} : G \in \mathscr{U}^*\}.$$

Because X is regular, this works.

23B Corollary

If X is a second-countable T_1 space and $\#(X) < \mathfrak{p}$, then X is a **Q-space** i.e. every subset of X is F_σ and G_δ.

Proof For $E \subseteq X$, apply 23A*a* with $\mathscr{E} = \{\{x\} : x \in E\}$, $\mathscr{K} = \{\{x\} : x \in X \setminus E\}$ to see that E is F_σ. Similarly, $X \setminus E$ is F_σ so E is also G_δ.

23C Corollary

$[\mathfrak{p} = \mathfrak{c}]$ Let X be a second-countable T_1 space without isolated points. Then X has a $\mathfrak{c}$-Lusin subset. If Y is a $\mathfrak{c}$-Lusin subset of X, then every subset of Y of cardinal less than $\mathfrak{c}$ is relatively F_σ in Y.

Proof By 22D, X has a $\mathfrak{c}$-Lusin subset. If Y is such a set and $A \in [Y]^{<\mathfrak{c}}$, then A is meagre in X, by 22C*a*; let H be a meagre F_σ set in X including A. As Y is $\mathfrak{c}$-Lusin in X, $\#(H \cap Y) < \mathfrak{c}$. Now A is relatively F_σ in $H \cap Y$, by 23B; as $H \cap Y$ is relatively F_σ in Y, so is A.

23D Corollary

Let X be a separable metric space. Let $\mathscr{E}$ be a disjoint family of subsets of X such that $\#(\mathscr{E}) < \mathfrak{p}$ and each member of $\mathscr{E}$ is expressible as the union of fewer than $\mathfrak{p}$ compact sets. Then there is a disjoint family $\langle H_E \rangle_{E \in \mathscr{E}}$ of F_σ sets in X such that $E \subseteq H_E$ for every $E \in \mathscr{E}$.

Proof Set $Y = \bigcup \mathscr{E}$. If $E \in \mathscr{E}$, then both E and $Y \setminus E$ are expressible as unions of fewer than $\mathfrak{p}$ compact sets, because $\mathfrak{p}$ is regular [21K]; so by 23A*a* there is an F_σ set C_E including E and disjoint from $Y \setminus E$, i.e. such that $Y \cap C_E = E$. Now set

$$Z = \bigcup \{C_E \cap C_F : E, F \in \mathscr{E}, E \neq F\}.$$

Then Z is a union of fewer than $\mathfrak{p}$ closed sets and $Y \cap Z = \emptyset$. By 23A*b* there is an F_σ set $H' \subseteq X$ such that $Y \subseteq H'$ and $Z \cap H' = \emptyset$. Set $H_E = H' \cap C_E$ for each $E \in \mathscr{E}$.

23E Sub-second-countable spaces

Corollary 23B can be used in some surprising contexts through the following elementary idea. Let us say that a topological space $(X, \mathfrak{T})$ is **sub-second-countable** if there is a second-countable T_1 topology $\mathfrak{S} \subseteq \mathfrak{T}$. Clearly, if $(X, \mathfrak{T})$ is sub-second-countable and $\#(X) < \mathfrak{p}$, then X is a Q-space for $\mathfrak{S}$ and therefore for $\mathfrak{T}$.

(i) All cometrizable spaces are sub-second-countable [25B]. (ii) If X is a Moore space and $d(X) \leq \mathfrak{c}$, it is sub-second-countable [A4P*a*]. (iii) In particular, if X is metrizable and $d(X) \leq \mathfrak{c}$, then it is sub-second-countable. So any metric space of cardinal less than $\mathfrak{p}$ (whether separable or not) is a Q-space.

For other examples see 23M*c* and 23N*j*.

23F Corollary to 21H

$[\mathfrak{p} = \mathfrak{c}]$ There is a sequence $\langle C_n \rangle_{n \in \mathbf{N}}$ of subsets of $\mathbf{R}$ such that every projective set in $\mathbf{R}$ is expressible as $\bigcap_{n \in \mathbf{N}} \bigcup_{m \in I \setminus n} C_m$ for some $I \subseteq \mathbf{N}$.

Proof For there are only $\mathfrak{c}$ projective subsets of $\mathbf{R}$ [A5G*a*].

23G Corollary to 22C

$[\mathfrak{p} > \omega_1]$ Let X be a Polish space. Then every PCA or CPCA set in X has the strong Baire property.

Proof Recall that every PCA set A is expressible as the union of ω_1 Borel sets [A5G*c*] and therefore has the Baire property, by 22C*b*. Repeating the argument on $A \cap F$ for an arbitrary closed set $F \subseteq X$, we see that A has the strong Baire property. It follows at once that $X \setminus A$ also has the strong Baire property.

23H Corollary to 22J

Let X be a K-analytic space.

(*a*) If $A \subseteq X$ and $\#(A) < \mathfrak{p}$ there is a K_σ set in X including A.

(*b*) $[\mathfrak{p} > \omega_1]$ If $A \subseteq X$ is uncountable there is a compact $K \subseteq X$ such that $K \cap A$ is uncountable.

Proof (*a*) As X is K-analytic there is an usco-compact relation $R \subseteq \mathbf{N}^{\mathbf{N}} \times X$ with $\pi_2[R] = X$. If $A \subseteq X$ and $\#(A) < \mathfrak{p}$ there is a $B \subseteq \mathbf{N}^{\mathbf{N}}$ such that $\#(B) < \mathfrak{p}$ and $R[B] \supseteq A$. By 22J, there is a sequence $\langle K_n \rangle_{n \in \mathbf{N}}$ of compact subsets of $\mathbf{N}^{\mathbf{N}}$ covering B. Now $\langle R[K_n] \rangle_{n \in \mathbf{N}}$ is a sequence of compact subsets of X [A5C*b*] which covers A.

(*b*) Follows at once from (*a*), since A has a subset of cardinal ω_1.

23I **Proposition**

Let X and Y be uncountable Polish spaces, $A \subseteq X$ and $B \subseteq Y$ subsets with $\#(A) = \#(B) < \mathfrak{p}$. Then any bijection $f: A \to B$ can be extended to a Borel isomorphism $g: X \to Y$.

Proof It is enough to consider the case $X = Y = \{0,1\}^{\mathbf{N}}$, since all uncountable Polish spaces have the same Borel structure [A5E*c*]. By 23B, every subset of A is relatively Borel in A. So we can find Borel sets $E_n \subseteq X$ such that

$$A \cap E_n = \{x : x \in A, (fx)(n) = 1\}.$$

Define $h_1: X \to Y$ by

$$(h_1 x)(n) = (\chi E_n)(x) \;\forall n \in \mathbf{N}, x \in X.$$

Then h_1 is Borel measurable and extends f. Similarly, there is a Borel measurable $h_2: Y \to X$ extending f^{-1}. Let E, F be Borel sets in X, Y such that $A \subseteq E$, $B \subseteq F$ and $X \backslash E$ and $Y \backslash F$ are both uncountable. (To see that such exist, if A is uncountable, observe that A itself is not Borel, so that $X \backslash A$ is uncountable, and we can apply 23B to $A \subseteq A \cup A'$ where $A' \subseteq X \backslash A$ has cardinal ω_1.) Set

$$D = \{(x, y) : x \in E, y \in F, h_1(x) = y, h_2(y) = x\}.$$

Then D is a Borel set, the graph of a bijection $g_1 : \pi_1[D] \to \pi_2[D]$, and g_1 extends f. Since $\pi_1 : D \to X$ is injective, $\pi_1[D]$ is a Borel set [A5E*b*(i)]; similarly $\pi_2[D]$ is a Borel set, and g_1 is a Borel isomorphism between $\pi_1[D]$ and $\pi_2[D]$ [A5E*b*(ii)–(iii)]. Now $X \backslash \pi_1[D]$ and $Y \backslash \pi_2[D]$ are both uncountable Borel sets in Polish spaces, so there is a Borel isomorphism $g_2 : X \backslash \pi_1[D] \to Y \backslash \pi_2[D]$ [A5E*c* again]. Finally, let $g: X \to Y$ be the common extension of g_1 and g_2; this is the function we seek.

23J **Theorem**

[$\mathfrak{p} > \omega_1$] If one of the following is true, so are the others.

(i) There is a Polish space X and a PCA set ($= \Sigma_2^1$ set) $A \subseteq X$ with $\#(A) = \omega_1$.

(ii) Every set of cardinal ω_1 in every Polish space is coanalytic.

(iii) If X is a Polish space, a set $A \subseteq X$ is PCA iff it is expressible as the union of ω_1 Borel sets.

(iv) There is a Polish space X and an uncountable coanalytic set $A \subseteq X$ which has no uncountable compact subset.

(v) There is a Polish space X and an uncountable PCA set $A \subseteq X$ which has no uncountable compact subset.

Proof (*a*) **(i)$\Rightarrow$(ii)** Suppose that X is Polish and that $A \subseteq X$ is a PCA set of cardinal ω_1. Then there is a Polish space Y, a coanalytic set $B \subseteq Y$, and a continuous surjection $f: B \to A$. Let F be the graph of f; then F is coanalytic in $Y \times X$ [A5G*b*(iii)]. By Kondô's theorem [A5H], there is a coanalytic set $C \subseteq F$ such that $\pi_2[C] = A$ and $\pi_2: C \to A$ is injective, so that $\#(C) = \omega_1$.

Now if Z is any Polish space, $D \subseteq Z$ and $\#(D) = \omega_1$, there is a Borel isomorphism $h: Z \to X \times Y$ such that $D = h^{-1}[C]$, by 23I. So D is coanalytic [A5G*b*(iv)].

(*b*) **(ii)$\Rightarrow$(iii)** Let X be Polish. If $A \subseteq X$ is PCA it is certainly the union of ω_1 Borel sets [A5G*c*]. On the other hand, assuming (ii), suppose that $A \subseteq X$ is expressible as the union of ω_1 Borel sets. There is an analytic set $R \subseteq \mathbf{N}^{\mathbf{N}} \times X$ such that every Borel set in X is a vertical section of R [A5F*g*]. So there is a $B \subseteq \mathbf{N}^{\mathbf{N}}$ such that $\#(B) = \omega_1$ and $R[B] = A$. By (ii), B is coanalytic so $R \cap (B \times X)$ is PCA [A5G*c*] and $A = \pi_2[R \cap (B \times X)]$ is PCA.

(*c*) Finally, (iii)$\Rightarrow$(i) is trivial; (ii)$\Rightarrow$(iv) because $\omega_1 < \mathfrak{p} \leq \mathfrak{c}$, so that no uncountable compact set in a Polish space can have cardinal ω_1; (iv)$\Rightarrow$(v) is trivial; and (v)$\Rightarrow$(i) because a PCA set A which has no uncountable compact subset must be expressible as the union of ω_1 Borel sets, none of which have uncountable compact subsets, so all of them are countable [A5F*f*], and $\#(A) = \omega_1$.

Terminology Following MARTIN & SOLOVAY 70, I shall call the assertion (ii) above **L**. See also 23N*g*.

23K **Proposition**

Suppose that $\kappa < \mathfrak{p}$. Then the following are equivalent: (i) there is an ω_1-saturated σ-ideal $\mathscr{I}$ of the algebra $\mathscr{B}$ of Borel sets in $\{0,1\}^{\mathbf{N}}$, containing all singletons, such that not every subset of $\{0,1\}^{\mathbf{N}}$ of cardinal κ is included in a member of $\mathscr{I}$; (ii) there is a proper ω_1-saturated σ-ideal of $\mathscr{P}\kappa$ containing all singleton sets.

Proof **(i)$\Rightarrow$(ii)** Let $A \subseteq \{0,1\}^{\mathbf{N}}$ be a set of cardinal κ not included in any member of $\mathcal{I}$. Set $\mathcal{I}^* = \{A \cap I : I \in \mathcal{I}\}$. By 23B, $\mathcal{P}A = \{A \cap B : B \in \mathcal{B}\}$ so $\mathcal{I}^*$ is a non-trivial σ-ideal of $\mathcal{P}A$. Also, if $\langle A_\xi \rangle_{\xi < \omega_1}$ is a disjoint family in $\mathcal{P}A$, there are $B_\xi \in \mathcal{B}$ such that $A_\xi = B_\xi \cap A$ for each ξ; now set

$$B'_\xi = B_\xi \setminus \bigcup_{\eta < \xi} B_\eta \in \mathcal{B},$$

and see that some B'_ξ must belong to $\mathcal{I}$, so that $A_\xi \in \mathcal{I}^*$. Thus $\mathcal{I}^*$ is ω_1-saturated.

(ii)⇒(i) Take a set $A \subseteq \{0,1\}^{\mathbf{N}}$ of cardinal κ, and let $\mathcal{I}^*$ be a non-trivial ω_1-saturated σ-ideal of $\mathcal{P}A$. Let $\mathcal{I}$ be $\{I : I \in \mathcal{B}, I \cap A \in \mathcal{I}^*\}$; then $\mathcal{I}$ is a non-trivial ω_1-saturated σ-ideal of $\mathcal{B}$, and A is not included in any member of $\mathcal{I}$.

23L Sources

ROTHBERGER 48 for 23B (with $\#(X) = \omega_1$). MARTIN & SOLOVAY 70 for 23B in general (attributed to J. Silver), 23G, 23J and 23K. MATHIAS, OSTASZEWSKI & TALAGRAND 78 for 23H*b*. PRZYMUSIŃSKI 81 for 23A.

23M Exercises

(*a*) [$\mathfrak{p} = \mathfrak{c}$] Let X be a second-countable T_1 space. If $Y \subseteq X$ is a $\mathfrak{c}$-Lusin set and μ is an atomless quasi-Radon measure on X, then $\mu Y = 0$. [Hint: show that for any $\varepsilon > 0$ there is a dense open set G with $\mu G \le \varepsilon$; now use A7B*k* and 22H*c*. Hence there is a universally negligible $Y \subseteq [0,1]$ with $\#(Y) = \mathfrak{c}$; compare 22N*p*(iii), 22N*r*(iii).]

Every relatively Borel subset of Y is of the form $Y \cap (G \triangle M)$, where G is open and M is F_σ in X, so is relatively $G_{\delta\sigma}$. [Use 23C.] If $A \subseteq Y$ has the Baire property in Y then it is relatively Borel in Y. If X is Polish and uncountable, then no countable dense set in Y can be relatively G_δ in Y, so that $\Sigma^0_2(Y) \neq \Sigma^0_3(Y) = \mathcal{B}(Y)$. [See MILLER 79, Theorem 17; also 22N*q*(iii).]

(*b*) [$\mathfrak{p} = \mathfrak{c}$] Let X be a second-countable T_1 space, $\mathcal{B}(X)$ its Borel σ-algebra, $\mathcal{I}$ an ω_1-saturated σ-ideal of $\mathcal{B}(X)$. Suppose that (through 13C*d* or otherwise) there is a $(\mathcal{B}, \mathcal{I}, \mathfrak{c})$-Lusin subset Y of X. (i) If $A \subseteq X$ is obtainable by Souslin's operation from Borel sets, then there are $E \in \mathcal{B}(X)$ and $I \in \mathcal{I}$ such that $A \triangle E \subseteq I$. [Use A3G*b*.] (ii) $\mathcal{B}(Y)$ is closed under Souslin's operation. [Use the arguments of 23C. See MILLER 79, Theorem 46.] (iii) If every member of $[Y]^{<\mathfrak{c}}$ is included in some member of $\mathcal{I}$, then $[Y]^{<\mathfrak{c}}$ is an ω_1-saturated σ-ideal of $\mathcal{B}(Y)$, and $\bigcup \mathcal{E} \in \mathcal{B}(Y)$ whenever $\mathcal{E} \in [\mathcal{B}(Y)]^{<\mathfrak{c}}$. [Use the idea of 22C*b*/22H*a*.] (iv) If not every member of $[Y]^{<\mathfrak{c}}$ is included in some member of $\mathcal{I}$, then there is a $\kappa < \mathfrak{c}$ and a proper ω_1-saturated σ-ideal of $\mathcal{P}\kappa$ containing singletons. [See 23K, B2D.]

(*c*) (i) If X is a Hausdorff continuous image of a second-countable space (e.g. if X is an analytic space), then X is sub-second-countable. (ii) Any subspace of a sub-second-countable space is sub-second-countable. (iii) A countable product of sub-second-countable spaces is sub-second-countable.

(*d*) (i) If X is a first-countable space without isolated points, there are disjoint dense subsets $A, B \subseteq X$. (ii) A first-countable Q-space without isolated points in meagre in itself. (iii) A metrizable Baire space without isolated points has cardinal greater than or equal to $\mathfrak{p}$. [FLEISSNER & KUNEN 78, Corollary 3.2. See B3I*b*–B3J.]

(*e*) $[\mathfrak{p} = \mathfrak{c}]$ There is a separable metric space Y such that $\#(Y) = \mathfrak{c}$ and every $A \subseteq Y$ with $\#(A) < \mathfrak{c}$ is both F_σ and G_δ. [Use 21Eb and 21H(iii). See MAULDIN 78 and SHINODA 73, §5.]

(*f*) Let X and Y be separable metric spaces with $\#(X) < \mathfrak{p}$, and $f: X \to Y$ any function. Then f is **JR-piecewise continuous**, i.e. there is a sequence $\langle F_n \rangle_{n \in \mathbf{N}}$ of closed subsets of X, covering X, such that $f \restriction F_n$ is continuous for each $n \in \mathbf{N}$. [Express f as $g \restriction X$ for some Borel function $g: \hat{X} \to \hat{Y}$, where $\hat{X}$ is the completion of X, and apply 23H*a* to the graph of f. Compare SHINODA 73, p. 120, Proposition 2.]

(*g*) Let X be an analytic space and $\mathscr{E}$ a cover of X by closed sets with $\#(\mathscr{E}) < \mathfrak{p}$. Then $\mathscr{E}$ has a countable subcover. [First consider Polish X, using the method of 44F. See STERN 78, Theorem 3.]

(*h*) (α) $[\mathfrak{p} > \omega_1]$ If $\mathscr{E}$ is a cover of $\mathbf{R}$ by nowhere-dense sets, there is an $\mathscr{E}' \subseteq \mathscr{E}$ such that $\bigcup \mathscr{E}'$ is not analytic. [Use 23M*g*.] (β) $[\mathfrak{p} = \mathfrak{c}]$ There is a family $\mathscr{E}$ of compact Lebesgue negligible subsets of $\mathbf{R}$ such that $\bigcup \mathscr{E} = \mathbf{R}$ and $\bigcup \mathscr{E}'$ is Lebesgue measurable and has the strong Baire property for every $\mathscr{E}' \subseteq \mathscr{E}$. [Enumerate $\mathbf{R}$ as $\langle t_\xi \rangle_{\xi < \mathfrak{c}}$ and $[\mathbf{R}]^\omega$ as $\langle C_\xi \rangle_{\xi < \mathfrak{c}}$. Adapt 22E or 25I*f* to construct compact negligible sets E_ξ such that $t_\xi \in E_\xi$ and if $\eta < \xi$ and $\bar{C}_\eta \cap E_\xi \neq \emptyset$ then $C_\eta \cap E_\xi \neq \emptyset$. Now show that if $A \subseteq \mathfrak{c}$ and $F = \mathbf{R} \setminus \bigcup_{\xi \in A} E_\xi$ then $\#(\{\xi : \xi \in A,\ F \cap E_\xi \neq \emptyset\} < \mathfrak{c}$.] ($\gamma$) $[\mathfrak{p} = \mathfrak{c} = \omega_2 + \mathbf{L}]$. There is a cover $\mathscr{D}$ of $\mathbf{R}$ by nowhere-dense sets of cardinal less than or equal to ω_1 such that $\bigcup \mathscr{D}'$ is coanalytic for every $\mathscr{D}' \subseteq \mathscr{D}$. [Use ($\beta$) and 23J(ii).]

23N **Further results**

(*a*) **Ramsey sets in $[\mathbf{N}]^\omega$. Let** $\mathfrak{S}$ be the usual Polish topology on $X = [\mathbf{N}]^\omega$, regarded as a G_δ set in $\mathscr{P}\mathbf{N} \cong \{0, 1\}^{\mathbf{N}}$. For $A \in X$ and $n \in \mathbf{N}$ write $U(n, A) = \{B : B \in [A]^\omega, A \cap n \subseteq B\}$, and let $\mathfrak{T}$ be the topology on X generated by $\{U(n, A) : A \in X, n \in \mathbf{N}\}$.

(i) Say that $Y \subseteq X$ is **completely Ramsey** If for every $A \in X$ and $n \in \mathbf{N}$ there is a $B \in U(n, A)$ such that $U(n, B)$ is either included in Y or disjoint from Y. Then Y is completely Ramsey iff it has the Baire property for $\mathfrak{T}$. [ELLENTUCK 74.]

(ii) The union of fewer than $\mathfrak{p}$ $\mathfrak{T}$-nowhere-dense sets is $\mathfrak{T}$-nowhere dense. The union of fewer than $\mathfrak{p}$ completely Ramsey sets is completely Ramsey. [$\mathfrak{p} > \omega_1$] If $Y \subseteq X$ is PCA or CPCA for $\mathfrak{S}$, then Y is completely Ramsey. [SILVER 70.]

(iii) Every $\mathfrak{S}$-analytic subset of X is completely Ramsey. [SILVER 70, ELLENTUCK 74.]

(iv) Let $\mathscr{F}$ be a non-principal ultrafilter on $\mathbf{N}$. Say that a set $Y \subseteq X$ is **$\mathscr{F}$-Ramsey** if there is a $B \in \mathscr{F}$ such that either $[B]^\omega \subseteq Y$ or $[B]^\omega \cap Y = \emptyset$. If $\mathscr{F}$ is a Ramsey ultrafilter, every $\mathfrak{S}$-analytic set is $\mathscr{F}$-Ramsey [LOUVEAU 74, MATHIAS 77]. [$\mathfrak{p} = \mathfrak{c} > \omega_1$] There is a Ramsey ultrafilter $\mathscr{F}$ such that every $\mathfrak{S}$-PCA set is $\mathscr{F}$-Ramsey [MATHIAS 77]. [$\mathfrak{p} = \mathfrak{c} > \omega_1$ +L] There is a Ramsey ultrafilter $\mathscr{F}$ such that not every $\mathfrak{S}$-PCA set is $\mathscr{F}$-Ramsey [MATHIAS 77].

(*b*) [$\mathfrak{p} > \omega_2$] If there is a two-valued-measurable cardinal, then every PCPCA set ($= \Sigma_3^1$ set) in a Polish space has the strong Baire property. [MARTIN & SOLOVAY 70. See also 32Q*h*.]

(*c*) [$\mathfrak{p} > \omega_1$] Let X be a Polish space and $A \subseteq X$ an analytic non-Borel set. Then there is a compact $K \subseteq X$ such that $K \cap A$ is not Borel. [MATHIAS, OSTASZEWSKI & TALAGRAND 78.]

(*d*) [$\mathfrak{p} > \omega_1$] If X is a K-analytic space and every compact subset of X is metrizable, then X is analytic. [FREMLIN 77*a*.]

(*e*) If X is an analytic space and $\mathscr{E}$ is a partition of X into $G_{\delta\sigma}$ sets with $\#(\mathscr{E}) < \mathfrak{p}$, then $\mathscr{E}$ is countable. [FREMLIN & SHELAH 79. Compare 44G.]

(*f*) [$\mathfrak{p} = \mathfrak{c}$] For any ordinal α with $1 \leq \alpha \leq \omega_1$, there is a separable metric space X such that $\alpha = \min\{\xi : \mathscr{B}(X) = \Sigma_\xi^0(X)\}$ and $\mathscr{B}(X)$ is closed under Souslin's operation, where $\mathscr{B}(X)$ is the algebra of Borel subsets of X. [MILLER 79; see 23M*a*, 32P*f*.]

(*g*) [$\mathfrak{p} > \omega_1$] The statement **L** of 23J is equivalent to (vi) there is an $A \subseteq \mathbf{N}$ such that $\omega_1 = \omega_1^{L(A)}$ (see 21O*h*(ii)). [MARTIN & SOLOVAY 70.]

(*h*) [$\mathfrak{p} > \omega_1$] Let X, Y and Z be metric spaces. (i) If $\mathscr{E} \subseteq \mathscr{P}X$ is a point-finite family of sets such that $\bigcup \mathscr{E}'$ is Borel for every $\mathscr{E}' \subseteq \mathscr{E}$, then there is a $\zeta < \omega_1$ such that $\bigcup \mathscr{E}' \in \Sigma_\zeta^0(X)$ for every $\mathscr{E}' \subseteq \mathscr{E}$. (ii) If $f: X \to Y$ is a Borel measurable function, there is a $\zeta < \omega_1$ such that $f^{-1}[G] \in \Sigma_\zeta^0(X)$

for every open $G \subseteq Y$. (iii) If $R \subseteq X \times Y$ has compact vertical sections and $R^{-1}[G]$ is Borel for every open $G \subseteq Y$, then R has a Borel measurable selector. (iv) If $f: X \to Y$ and $g: X \to Z$ are Borel measurable, then $x \mapsto (fx, gx): X \to Y \times Z$ is Borel measurable. [FREMLIN, HANSELL & JUNNILA *a*.]

(*i*) Let E be a Banach space such that $d(E) \leq \mathfrak{p}$. Then there is a separable metric space X such that E can be isometrically embedded in the subspace of $\ell^\infty(X)$ consisting of pointwise limits of sequences of continuous functions. [MAULDIN 77.]

(*j*) (i) If X is a T_1 space, $\#(X) \leq \mathfrak{c}$, and the topology of X has a σ-point-finite base, then X is sub-second-countable [PRZYMUSIŃSKI 81]. (ii) If X is a regular Hausdorff space, then X is a metacompact Moore space iff open sets in X are F_σ and the topology of X has a σ-point-finite base. (iii) If X is a regular Hausdorff space with a σ-point-finite base and $\#(X) < \mathfrak{p}$ then X is a metacompact Moore space. [FLEISSNER & REED 78.]

(*k*) Results on similar topics may also be found in LOUVEAU 80 and JAYNE & ROGERS 80*a*.

23O Problems

(*a*) Is any special axiom necessary for 23N*h*? [See FREMLIN, HANSELL & JUNNILA *a*.]

(*b*) Let X be a Polish space, $E \subseteq X$ a Borel set, $\mathscr{K}$ a family of compact subsets of E with $\#(\mathscr{K}) < \mathfrak{p}$. Is there always a K_σ set H with $\bigcup \mathscr{K} \subseteq H \subseteq E$? What if E is $F_{\sigma\delta}$?

(*c*) [$\mathfrak{p} > \omega_1$] Does **L** [23J] imply that the union of ω_1 compact sets in **R** is always coanalytic?

(*d*) [$\mathfrak{m} = \mathfrak{c} > \omega_1$] Let $A \subseteq \mathbf{N}^{\mathbf{N}}$ be such that $A \cap K$ is Borel for every compact $K \subseteq \mathbf{N}^{\mathbf{N}}$. Does it follow that A is Borel?

Notes and comments

The proof of 23A is nearly identical to the natural proof of 23B (see MARTIN & SOLOVAY 70); the only extra step called for is the switch from $\mathscr{U}$ to $\mathscr{U}^*$. Applications of 23A are in 23D and 25D–E. Note the asymmetry of 23A, which is forced by the proof; I do not have examples to show that this is really the right shape for the result, and perhaps in favourable circumstances something better can be proved; this may depend on the answer to 23O*b*.

23C and 23M*a*–*b* are an investigation into the properties of generalized

Lusin sets when $\mathfrak{p} = \mathfrak{c}$. The same ideas are used in 23N*f*. 23E, 23M*c*–*d* and 23N*j* show that 23B reaches further than one might think. 23F–H spell out easy consequences of earlier results; 23H*b* is the basis of 23N*c*–*d* and will also be useful in §44. 23I makes rather unscrupulous use of classical descriptive set theory; it is possible to give sharper descriptions of the Borel isomorphism *g*.

Further classical descriptive set theory (this time reaching to Kondô's theorem) yields the remarkable equivalences of Theorem 23J. Perhaps I should explain here what the consistency status of **L** is. The ordinary models of $\mathfrak{m} = \mathfrak{c} > \omega_1$ constructed by the Solovay–Tennenbaum method from other familiar models all satisfy (vi) of 23N*g*, so have **L** true. To obtain a model in which $\mathfrak{p} > \omega_1$ but **L** is false, we need to start from one with a weakly compact cardinal [B3G]. The present consensus is that these large cardinal axioms are fairly safe, so that **L** is probably truly undecided by MA +not-CH. Further consequences of $\mathfrak{p} = \mathfrak{c} > \omega_1$ +**L** are in 23M*h*(γ) and 23N*a*(iv). Perhaps there are yet other forms for **L** [23O*c*].

23M*g* is useful and 23N*e* is interesting; both can be proved in stronger forms if we use $\mathfrak{m}$ rather than $\mathfrak{p}$, which is what I do in 44F–G. It is a theorem of ZFC that any uncountable Polish space can be expressed as the union of a strictly increasing family $\langle G_\xi \rangle_{\xi < \omega_1}$ of G_δ sets, and can therefore be partitioned into just ω_1 $F_{\sigma\delta}$ sets; this is quite easy to prove from the existence of Hausdorff's gap [21L; see Hausdorff 36]. But 23N*e* shows that if $\mathfrak{p} > \omega_1$ then [0, 1] cannot be partitioned into just ω_1 G_δ sets. The point of 23M*h*(α) is that if the continuum hypothesis is true there is a cover $\mathscr{E}$ of **R** by countable compact sets such that $\bigcup \mathscr{E}'$ is F_σ for every $\mathscr{E}' \subseteq \mathscr{E}$ [Juhász, Kunen & Rudin 76]; the argument I suggest for 23M*h*(β) is an adaptation of theirs.

23N*a*(iii), the 'Mathias–Silver theorem', is famous for having been first proved using $\mathfrak{m} > \omega_1$ (via 23N*a*(ii)) and then observing that by Shoenfield's absoluteness theorem [Drake 74, 5.7.15; Moschovakis 80, 8F.10] this can't have made any difference. My own view is that any result proved in this way is going to have a better proof of a conventional kind. 23N*b* depends on a theorem of Martin *a* that if there is a two-valued-measurable cardinal then every PCPCA set is expressible as the union of ω_2 Borel sets. 23N*h* uses (perhaps unnecessarily) 21H(ii).

24 Sequential convergence

In this section I have tried to collect those results which depend in one way or another on the power of P(κ) to provide us with convergent sequences; 24A is typical. I give some straightforward results on the products of not-too-many spaces all of which have plenty of convergent sequences [24B–C], a less straightforward one of the same kind on sequentially continuous functions [24E], and two theorems on sequential compactness [24F–G]. In 24G–L I deal with countably compact spaces. 24K–L are especially interesting because (alone in this section) they depend on the continuum hypothesis being false.

24A **Proposition**

Let X be a topological space and $A \subseteq X$ a countable set. Suppose that $x \in \bar{A}$ and that either (i) $\chi(x, X) < \mathfrak{p}$ or (ii) $\psi(\overline{\{x\}}, X) < \mathfrak{p}$ and X is regular and A is relatively countably compact in X or (iii) $hL(X) < \mathfrak{p}$ and X is Hausdorff and A is relatively countably compact in X. Then there is a sequence in A converging to x.

Proof If there is a constant sequence in A converging to x, stop here. Otherwise, $U \cap A$ must be infinite for every neighbourhood U of x. (i) If $\chi(x, X) < \mathfrak{p}$, let $\mathscr{U}$ be a base of neighbourhoods of x with $\#(\mathscr{U}) < \mathfrak{p}$. Then $\{U \cap A : U \in \mathscr{U}\}$ is a downwards-directed family of infinite sets, so the definition of $\mathfrak{p}$ shows that there is an infinite $I \subseteq A$ such that $I \setminus U$ is finite for every $U \in \mathscr{U}$. Now any sequence enumerating I will converge to x. (ii) In the next case, because X is regular, $\overline{\{x\}} = \bigcap \mathfrak{N}$ where $\mathfrak{N}$ is the set of open neighbourhoods of x. As $\psi(\overline{\{x\}}, X) < \mathfrak{p}$, there is a family $\mathscr{U} \subseteq \mathfrak{N}$ such that $\#(\mathscr{U}) < \mathfrak{p}$ and $\bigcap \mathscr{U} = \overline{\{x\}}$. For each $U \in \mathscr{U}$ choose $V_U \in \mathfrak{N}$ such that $\bar{V}_U \subseteq U$. Once again, there is a sequence $\langle x_n \rangle_{n \in \mathbf{N}}$ in A such that $\{n : x_n \notin V_U\}$ is finite for every $U \in \mathscr{U}$. But now, if $W \in \mathfrak{N}$, $W \supseteq \bigcap_{U \in \mathscr{U}} \bar{V}_U$, which contains every cluster point of $\langle x_n \rangle_{n \in \mathbf{N}}$; as A is relatively countably compact, $\{n : x_n \notin W\}$ must be finite, and $\langle x_n \rangle_{n \in \mathbf{N}} \to x$. (iii) For the last case, because X is Hausdorff, $\{x\} = \bigcap_{U \in \mathfrak{N}} \bar{U}$; because $L(X \setminus \{x\}) \leq hL(X) < \mathfrak{p}$, there is a $\mathscr{U} \in [\mathfrak{N}]^{<\mathfrak{p}}$ such that $\{x\} = \bigcap_{U \in \mathscr{U}} \bar{U}$; now we can argue as in (ii) with $V_U = U$.

24B **Proposition**

(*a*) The product of fewer than $\mathfrak{p}$ sequentially compact spaces is sequentially compact.

(*b*) The product of $\mathfrak{p}$ or fewer sequentially compact spaces is countably compact.

Proof (*a*) Let $\langle X_\xi \rangle_{\xi < \kappa}$ be a family of sequentially compact spaces, where $\kappa < \mathfrak{p}$, and let $\langle x_n \rangle_{n \in \mathbf{N}}$ be a sequence in $X = \prod_{\xi < \kappa} X_\xi$. Choose infinite sets $I(\xi) \subseteq \mathbf{N}$ inductively, for $\xi \leq \kappa$, as follows. $I(0) = \mathbf{N}$. Given $I(\xi)$, where $\xi < \kappa$, let $I(\xi + 1)$ be an infinite subset of $I(\xi)$ such that $\langle x_n(\xi) \rangle_{n \in I(\xi+1)}$ has a limit in X_ξ. Given $\langle I(\eta) \rangle_{\eta < \xi}$, where $\xi \leq \kappa$ is a non-zero limit ordinal, such that

$I(\zeta)\setminus I(\eta)$ is finite whenever $\eta \leq \zeta < \xi$, use the definition of $\mathfrak{p}$ to find an infinite set $I(\xi)$ such that $I(\xi)\setminus I(\eta)$ is finite for every $\eta < \xi$.

Now $\langle x_n \rangle_{n\in I(\kappa)}$ is a convergent subsequence of $\langle x_n \rangle_{n\in \mathbf{N}}$.

(*b*) Proceed as above, but with $\kappa \leq \mathfrak{p}$, up to the point where $\langle I(\xi) \rangle_{\xi<\kappa}$ have been chosen. If $\kappa = \mathfrak{p}$, we can no longer be sure of an infinite set $I(\kappa)$. But if for each $\xi < \kappa$ we choose a limit $x(\xi)$ of $\langle x_n(\xi) \rangle_{n\in I(\xi+1)}$, then x will be a cluster point of $\langle x_n \rangle_{n\in\mathbf{N}}$ in X.

24C **Proposition**

The product of fewer than $\mathfrak{p}$ sequentially separable spaces is sequentially separable.

Proof Let $\langle X_\iota \rangle_{\iota\in I}$ be a family of sequentially separable spaces, with $\#(I) < \mathfrak{p}$, and set $X = \prod_{\iota\in I} X_\iota$. If any X_ι is empty, so is X, and we're done. Otherwise, for each $\iota\in I$ there is a non-empty countable set $D_\iota \subseteq X$ such that each point of X_ι is the limit of a sequence in D_ι. Give each D_ι its discrete topology, and consider $\prod_{\iota\in I} D_\iota$. By 12K, this is separable; let D be a countable dense subset; then for any finite $J \subseteq I$ and family $\langle y_\iota \rangle_{\iota\in J} \in \prod_{\iota\in J} D_\iota$, there is a $z\in D$ such that $z(\iota) = y_\iota$ for each $\iota\in J$.

Let $x\in X$. For each $\iota\in I$ there is a sequence $\langle y_n^\iota \rangle_{n\in\mathbf{N}}$ in D_ι which converges to $x(\iota)$ in X. For each $\iota\in I$, $n\in\mathbf{N}$ set

$$E_n^\iota = \{z : z\in D, \exists m \geq n, z(\iota) = y_m^\iota\}.$$

Then $\mathscr{E} = \{E_n^\iota : \iota\in I, n\in\mathbf{N}\}$ has the finite intersection property, and $\#(\mathscr{E}) < \mathfrak{p}$. So there is a sequence $\langle z_j \rangle_{j\in\mathbf{N}}$ in D such that $\{j : j\in\mathbf{N}, z_j \notin E_n^\iota\}$ is finite for each $\iota\in I$, $n\in\mathbf{N}$; and now $\langle z_j \rangle_{j\in\mathbf{N}} \to x$. As x is arbitrary, X is sequentially separable.

24D **Lemma**

(*a*) Let $\langle X_\xi \rangle_{\xi<\kappa}$ be a family of first-countable topological spaces, wih product X, and let Y be a regular topological space. Suppose that $f : X \to Y$ is a sequentially continuous function such that whenever $x\in X$ and V is a neighbourhood of $f(x)$ in Y, there is a countable $I \subseteq \kappa$ such that $f(z)\in V$ whenever $z\in X$ and $z\restriction I = x\restriction I$. Then f is continuous.

(*b*) Let Y be a regular topological space and κ a cardinal such that every sequentially continuous $f : \{0,1\}^\kappa \to Y$ is continuous. Then if $\langle X_\xi \rangle_{\xi<\kappa}$ is any family of separable first-countable spaces, and $X = \prod_{\xi<\kappa} X_\xi$, every sequentially continuous $f : X \to Y$ is continuous.

Proof (*a*) Let $x\in X$. For each $\xi < \kappa$ let $\langle U_n^\xi \rangle_{n\in\mathbf{N}}$ be a decreasing sequence such that $\{U_n^\xi : n \in \mathbf{N}\}$ is a base of neighbourhoods of $x(\xi)$ in X_ξ. Let V be a closed neighbourhood of $f(x)$ in Y.

? Suppose, if possible, that $f^{-1}[V]$ is not a neighbourhood of x. Choose inductively sequences $\langle z_n \rangle_{n \in \mathbf{N}}$, $\langle I_n \rangle_{n \in \mathbf{N}}$, $\langle \xi_{nm} \rangle_{n,m \in \mathbf{N}}$, $\langle J_n \rangle_{n \in \mathbf{N}}$ as follows. $J_n = \{\xi_{ij} : i,j < n\}$. Given that J_n is a finite subset of κ, take $z_n \in X$ such that $z_n(\xi) \in U_n^\xi$ for $\xi \in J_n$ but $f(z_n) \notin V$; take a countable non-empty set $I_n \subseteq \kappa$ such that if $z \in X$ and $z \upharpoonright I_n = z_n \upharpoonright I_n$ then $f(z) \notin V$; and let $\langle \xi_{nm} \rangle_{m \in \mathbf{N}}$ be a sequence running over I_n. Continue.

Write $I = \bigcup_{n \in \mathbf{N}} J_n = \bigcup_{n \in \mathbf{N}} I_n$. For each $n \in \mathbf{N}$, define $z_n' \in X$ by writing

$$z_n'(\xi) = z_n(\xi) \text{ if } \xi \in I,$$
$$z_n'(\xi) = x(\xi) \text{ if } \xi \in \kappa \setminus I.$$

Then, by the choice of I_n, $f(z_n') \notin V$. But observe that

$$z_n'(\xi) \in U_n^\xi \; \forall \xi \in J_n \cup (\kappa \setminus I),$$

so that $\langle z_n' \rangle_{n \in \mathbf{N}} \to x$, although $\langle f(z_n') \rangle_{n \in \mathbf{N}} \not\to f(x)$. **X**

As x and V are arbitrary, and Y is regular, f is continuous.

(*b*) If any X_ξ is empty, or if $\kappa \le \omega$, then X is first-countable, and the result is trivial. Otherwise, for each $\xi < \kappa$, choose a sequence $\langle t_n^\xi \rangle_{n \in \mathbf{N}}$ in X_ξ such that $\{t_n^\xi : n \in \mathbf{N}\}$ is dense. Suppose that $f : X \to Y$ is sequentially continuous.

The condition of (*a*) is satisfied. **P** Let $x \in X$ and let V be a closed neighbourhood of $f(x)$ in Y. For each $n \in \mathbf{N}$, $\xi < \kappa$ define $\varphi_n^\xi : 2^n \to X$ by

$$\varphi_n^\xi(0) = x(\xi), \quad \varphi_n^\xi(i) = t_{i-1}^\xi \text{ for } 0 < i < 2^n,$$

and let $\varphi_n : (2^n)^\kappa \to X$ be given by

$$\varphi_n(w) = \langle \varphi_n^\xi(w(\xi)) \rangle_{\xi < \kappa} \; \forall w \in (2^n)^\kappa.$$

Then φ_n is continuous (giving each copy of 2^n the discrete topology) so $f\varphi_n : (2^n)^\kappa \to Y$ is sequentially continuous. As $(2^n)^\kappa$ is homeomorphic to $\{0\}$ or $\{0, 1\}^\kappa$, $f\varphi_n$ is continuous, by our hypothesis. Writing **0** for the member of $(2^n)^\kappa$ with constant value 0, $f\varphi_n(\mathbf{0}) = f(x) \in \operatorname{int} V$, so there is a finite $I_n \subseteq \kappa$ such that $f\varphi_n(w) \in V$ whenever $w \in (2^n)^\kappa$ and $w(\xi) = 0$ for every $\xi \in I_n$. Consider $I = \bigcup_{n \in \mathbf{N}} I_n$.

If $z \in X$ and $z \upharpoonright I = x \upharpoonright I$, then for each $\xi \in \kappa \setminus I$ choose a sequence $\langle m(\xi, j) \rangle_{j \in \mathbf{N}}$ in $\mathbf{N}$ such that $\langle t_{m(\xi,j)}^\xi \rangle_{j \in \mathbf{N}} \to z(\xi)$. Define $w_{nj} \in (2^n)^\kappa$ by

$$w_{nj}(\xi) = m(\xi, j) + 1 \text{ if } \xi \in \kappa \setminus I \text{ and } m(\xi, j) + 1 < 2^n,$$
$$w_{nj}(\xi) = 0 \qquad \text{otherwise.}$$

Then (by the choice of I_n) $f\varphi_n(w_{nj}) \in V$ for all $n, j \in \mathbf{N}$. Now, for each $j \in \mathbf{N}$,

$\langle \varphi_n(w_{nj}) \rangle_{n\in\mathbf{N}} \to z_j$ where

$$z_j(\xi) = t^{\xi}_{m(\xi,j)} \text{ if } \xi\in\kappa\backslash I,$$
$$z_j(\xi) = x(\xi) \quad \text{if } \xi\in I.$$

So $f(z_j)\in V$ for each $j\in\mathbf{N}$. But $\langle z_j \rangle_{j\in\mathbf{N}} \to z$, so $f(z)\in V$. As x and V are arbitrary, and Y is regular, the condition of (*a*) is satisfied. **Q**

So f is continuous.

24E **Theorem**

Suppose that $\langle X_\xi \rangle_{\xi<\kappa}$ is a family of separable first-countable topological spaces, Y is a regular topological space, and $f:\prod_{\xi<\kappa} X_\xi \to Y$ is sequentially continuous. If *either* (*a*) $\kappa < \mathfrak{p}$ *or* (*b*) $\kappa = \mathfrak{p} = \mathfrak{c}$, then f is continuous.

Proof (*a*) Let us begin with the case $\kappa < \mathfrak{p}$. By 12K, $X = \prod_{\xi<\kappa} X_\xi$ is separable, and $\chi(X) \le \max(\omega, \kappa) < \mathfrak{p}$. Let $A \subseteq X$ be a countable dense set. If $x\in X$ and V is an open neighbourhood of $f(x)$ in Y, let W be an open neighbourhood of $f(x)$ such that $\bar{W} \subseteq V$. Consider $B = A\backslash f^{-1}[W]$. By 24A(i), every point of $\bar{B}$ is the limit of a sequence in B. So

$$f[\bar{B}] \subseteq \overline{f[B]} \subseteq Y\backslash W,$$

and $x\notin\bar{B}$. As A is dense in X, $X\backslash\bar{B} \subseteq \overline{A\backslash B}$. Again, every point of $\overline{A\backslash B}$ is the limit of a sequence in $A\backslash B$, so

$$f[X\backslash\bar{B}] \subseteq f[\overline{A\backslash B}] \subseteq \overline{f[A\backslash B]} \subseteq \bar{W} \subseteq V.$$

Thus $x\in X\backslash\bar{B} \subseteq f^{-1}[V]$ and $f^{-1}[V]$ is a neighbourhood of x. As x and V are arbitrary, f is continuous.

(*b*) Now suppose that $\kappa = \mathfrak{p} = \mathfrak{c}$. By 24D*b*, we need consider only the case in which every X_ξ is $\{0,1\}$. It is convenient to change the index set from $\mathfrak{c}$ itself to a $\mathfrak{c}$-Lusin subset C of $\mathbf{R}$, so that we have a sequentially continuous function $f:\{0,1\}^C \to Y$; such a set C exists by 22D.

? Suppose, if possible, that the condition of 24D*a* is not satisfied; that there is an $x\in X = \{0,1\}^C$ and a neighbourhood V of $f(x)$ in Y such that

$$\forall \text{ countable } I \subseteq C\ \exists z\in X \text{ such that } z\upharpoonright I = x\upharpoonright I, f(z)\notin V.$$

Let W be an open neighbourhood of $f(x)$ such that $\bar{W} \subseteq V$. Enumerate $\mathbf{Q}$ as $\langle q_n \rangle_{n\in\mathbf{N}}$ and set

$$B^n_m = C\cap\,]q_n, q_n + 2^{-m}[\ \forall m, n\in\mathbf{N}.$$

Now we can choose inductively a sequence $\langle m(n) \rangle_{n\in\mathbf{N}}$ such that

(*) $\forall$ countable $I \subseteq C \exists z \in X$ such that

$$z \upharpoonright I \cup \bigcup_{i<n} B^i_{m(i)} = x \upharpoonright I \cup \bigcup_{i<n} B^i_{m(i)},\ f(z) \notin \bar{W}.$$

P For $n=0$ this is just the original supposition concerning x and V. **?** Suppose, if possible, that the inductive step to $n+1$ fails. Then we have $\langle m(i) \rangle_{i<n}$ satisfying (*), but no possible choice for $m(n)$; that is to say, writing D_n for $\bigcup_{i<n} B^i_{m(i)}$,

$$\forall r \in \mathbf{N}\ \exists \text{ countable } I_r \subseteq C \text{ such that, for } z \in X,$$

$$z \upharpoonright I_r \cup D_n \cup B^n_r = x \upharpoonright I_r \cup D_n \cup B^n_r \Rightarrow f(z) \in \bar{W}.$$

Set $I = \bigcup_{r \in \mathbf{N}} I_r$. By the inductive hypothesis there is a $z \in X$ such that

$$z \upharpoonright I \cup D_n = x \upharpoonright I \cup D_n,\ f(z) \notin \bar{W}.$$

For $r \in \mathbf{N}$, define z_r by

$$z_r(s) = x(s) \text{ if } s \in B^n_r,\ z(s) \text{ otherwise}.$$

Then

$$z_r \upharpoonright I_r \cup D_n \cup B^n_r = x \upharpoonright I_r \cup D_n \cup B^n_r,$$

so that $f(z_r) \in \bar{W}$ by the choice of I_r. But because $\langle B^n_r \rangle_{r \in \mathbf{N}} \downarrow \emptyset$, $z = \lim_{r\to\infty} z_r$, so $f(z) = \lim_{r\to\infty} f(z_r) \in \bar{W}$, contrary to the choice of z. **X** Thus the induction continues. **Q**

Set $D = \bigcup_{n \in \mathbf{N}} B^n_{m(n)}$. Then D is the intersection of C with a dense open set in $\mathbf{R}$ so $\#(C \setminus D) < \mathfrak{c}$, because C is $\mathfrak{c}$-Lusin. Consider $Z = \{z : z \in X, z \upharpoonright D = x \upharpoonright D\}$. Then Z is homeomorphic to $\{0,1\}^{C \setminus D}$ and $f \upharpoonright Z$ is sequentially continuous. By part (*a*) above, $f \upharpoonright Z$ is continuous. So there is a finite $I \subseteq C \setminus D$ such that

$$z \in Z,\ z \upharpoonright I = x \upharpoonright I \Rightarrow f(z) \in W.$$

By the choice of $\langle m(n) \rangle_{n \in \mathbf{N}}$, there is for each $n \in N$ a $z_n \in X$ such that

$$z_n \upharpoonright I \cup \bigcup_{i<n} B^i_{m(i)} = x \upharpoonright I \cup \bigcup_{i<n} B^i_{m(i)},\quad f(z_n) \notin \bar{W}.$$

Consider $\langle z_n \upharpoonright C \setminus D \rangle_{n \in \mathbf{N}}$. As $\{0,1\}^{C \setminus D}$ is sequentially compact [24B*a*], there is a strictly increasing sequence $\langle n(k) \rangle_{k \in \mathbf{N}}$ such that $\langle z_{n(k)} \upharpoonright C \setminus D \rangle_{k \in \mathbf{N}}$ is convergent. Since $\langle z_{n(k)} \upharpoonright D \rangle_{k \in \mathbf{N}} \to x \upharpoonright D$, $z = \lim_{k\to\infty} z_{n(k)}$ exists in X; and $z \in Z$ and $z \upharpoonright I = x \upharpoonright I$. In this case, $f(z) = \lim_{k\to\infty} f(z_{n(k)}) \notin W$; contrary to the choice of I. **X**

Thus we see that f satisfies the condition of 24D*a*, and is continuous.

24F **Proposition**

Let X be a compact Hausdorff space such that $\#(X) < 2^{\mathfrak{p}}$. Then X is sequentially compact.

Proof Given a sequence $\langle x_n \rangle_{n\in\mathbf{N}}$ in X, apply A4J*b* to its set K of cluster points to see that some $x \in K$ has $\chi(x, K) < \mathfrak{p}$. Now if $Z = K \cup \{x_n : n\in\mathbf{N}\} = \overline{\{x_n : n\in\mathbf{N}\}}$, $\chi(x, Z) < \mathfrak{p}$ so that we can apply 24A to find a subsequence of $\langle x_n \rangle_{n\in\mathbf{N}}$ converging to x. (Formally speaking, it will be necessary to distinguish cases according to whether $(n : x_n = x\}$ is finite or infinite.)

24G **Proposition**

Let X be a countably compact topological space. If $\mathrm{hL}(X) < \mathfrak{p}$ (in particular, if $\#(X) < \mathfrak{p}$), then X is sequentially compact.

Proof Let $\langle x_n \rangle_{n\in\mathbf{N}}$ be a sequence in X. For $I \subseteq \mathbf{N}$ set

$$F(I) = \bigcap_{n\in\mathbf{N}} \overline{\{x_i : i \in I \setminus n\}}.$$

Then there is an $I \in [\mathbf{N}]^\omega$ such that $F(J) = F(I)$ whenever $J \in [I]^\omega$. **P?** Otherwise, choose $\langle I_\xi \rangle_{\xi<\mathfrak{p}}$ in $[\mathbf{N}]^\omega$ inductively, as follows. $I_0 = \mathbf{N}$. Given $I_\xi \in [\mathbf{N}]^\omega$, where $\xi < \mathfrak{p}$, choose $I_{\xi+1} \in [I_\xi]^\omega$ such that $F(I_{\xi+1}) \neq F(I_\xi)$. Given $\langle I_\eta \rangle_{\eta<\xi}$, where $\xi < \mathfrak{p}$ is a non-zero limit ordinal, such that $I_\eta \setminus I_\zeta$ is finite whenever $\zeta \leq \eta < \xi$, choose $I_\xi \in [\mathbf{N}]^\omega$ such that $I_\xi \setminus I_\eta$ is finite for every $\eta < \xi$; this is possible by the definition of $\mathfrak{p}$. This construction ensures that $F(I_\xi) \subseteq F(I_\eta)$ whenever $\eta < \xi < \mathfrak{p}$. Set $Z = \bigcap_{\xi<\mathfrak{p}} F(I_\xi)$. As each $F(I_\xi)$ is closed, and $L(X \setminus Z) \leq hL(X) < \mathfrak{p}$, there must be an $A \in [\mathfrak{p}]^{<\mathfrak{p}}$ such that $Z = \bigcap_{\xi\in A} F(I_\xi)$. Because $\mathfrak{p}$ is regular [21K], $\zeta = \sup A < \mathfrak{p}$, and $F(I_\zeta) = Z \subseteq F(I_{\zeta+1})$. **XQ**

Now, as X is countably compact, there is an $x \in F(I)$, and x is a cluster point of every subsequence of $\langle x_n \rangle_{n\in I}$; i.e. $\langle x_n \rangle_{n\in I} \to x$. As $\langle x_n \rangle_{n\in\mathbf{N}}$ is arbitrary, X is sequentially compact.

24H **Theorem**

Let X be a separable countably compact topological space and $\mathscr{G}$ an open cover of X with $\#(\mathscr{G}) < \mathfrak{p}$. Then there is a finite $\mathscr{G}_0 \subseteq \mathscr{G}$ such that $\bigcup \mathscr{G}_0$ is dense in X.

Proof If X is finite, this is trivial. Otherwise, let $\langle x_n \rangle_{n\in\mathbf{N}}$ enumerate a countable dense subset of X. For $G \in \mathscr{G}$ set $A_G = \{n : x_n \notin \bar{G}\}$. **?** If $X \neq \overline{\bigcup \mathscr{G}_0}$ for any finite $\mathscr{G}_0 \subseteq \mathscr{G}$, then $\bigcap \{A_G : G \in \mathscr{G}_0\}$ is infinite for every finite $\mathscr{G}_0 \subseteq \mathscr{G}$, so there is an infinite $I \subseteq \mathbf{N}$ such that $I \setminus A_G$ is finite for every $G \in \mathscr{G}$. Now $\langle x_n \rangle_{n\in I}$ has no cluster point in $\bigcup \mathscr{G} = X$; but X is supposed to be countably compact. **X**

24I **Corollary**

Let X be a regular separable countably compact space. If $\hat{L}(X) \leq \mathfrak{p}$ (in particular, if either $\#(X) < \mathfrak{p}$ or $w(X) < \mathfrak{p}$), then X is compact.

Proof If $\mathscr{G}$ is an open cover of X, then $\mathscr{H} = \{H : H \text{ open}, \exists G \in \mathscr{G}, \bar{H} \subseteq G\}$ is an open cover of X, because X is regular. Since $\hat{L}(X) \le \mathfrak{p}$, there is an $\mathscr{H}_1 \subseteq \mathscr{H}$, covering X, with $\#(\mathscr{H}_1) < \mathfrak{p}$. By Theorem 24H, there is a finite $\mathscr{H}_0 \subseteq \mathscr{H}_1$ such that $X = \overline{\bigcup \mathscr{H}_0} = \bigcup\{\bar{H} : H \in \mathscr{H}_0\}$. Now there is a finite $\mathscr{G}_0 \subseteq \mathscr{G}$ such that $\bigcup \mathscr{G}_0 \supseteq X$. As $\mathscr{G}$ is arbitrary, X is compact.

24J **Theorem**

If X is a non-empty regular separable countably compact space, it is not the union of fewer than $\mathfrak{p}$ meagre sets.

Proof It is enough to show that $\varnothing$ is not the intersection of fewer than $\mathfrak{p}$ dense open sets in X. Let P be the set of non-empty open subsets of X. Then P is σ-centered downwards because X is separable. Let $\mathscr{G}$ be a non-empty family of dense open sets in X with $\#(\mathscr{G}) < \mathfrak{p}$. For $G \in \mathscr{G}$ set $Q_G = \{H : H \in P, \bar{H} \subseteq G\}$. Then Q_G is coinitial with P because X is regular and G is dense. By Bell's theorem [14C], there is a downwards-directed set $R \subseteq P$ meeting every Q_G; we may suppose that $\#(R) < \mathfrak{p}$. Now $\bigcap\{\bar{H} : H \in R\} \subseteq \bigcap \mathscr{G}$. **?** If $\bigcap\{\bar{H} : H \in R\} = \varnothing$, then $\{X \setminus \bar{H} : H \in R\}$ is an open cover of X of cardinal less than $\mathfrak{p}$. By 24H there is a finite $R_0 \subseteq R$ such that $\bigcup\{X \setminus \bar{H} : H \in R_0\}$ is dense. But now $\bigcap R_0 = \varnothing$, which is impossible, as R is downwards-directed in P. **X** So $\bigcap \mathscr{G} \ne \varnothing$, as required.

24K **Weiss' theorem**

$[\mathfrak{p} > \omega_1]$ Let X be a regular countably compact space in which every open set is F_σ. Then X is compact.

Proof (*a*) I show first that X is hereditarily ccc. **P?** If not, there is an uncountable $A \subseteq X$ such that $x \notin \overline{A \setminus \{x\}}$ for every $x \in A$ [A4E*b*]. Now $A = \bar{A} \cap G$ for some open $G \subseteq X$. As G is F_σ, there is a closed $F \subseteq G$ such that $A \cap F$ is infinite. But in this case $A \cap F = \bar{A} \cap F$ has no cluster point in X, contradicting the hypothesis that X is countably compact. **XQ**

(*b*) If $Y \subseteq X$ is a hereditarily separable subspace, it is Lindelöf. **P?** Otherwise, there is a cover $\mathscr{G}$ of Y by sets which are open in X which has no countable subcover. Choose $\langle x_\xi \rangle_{\xi < \omega_1}$ in Y, $\langle G_\xi \rangle_{\xi < \omega_1}$ in $\mathscr{G}$ inductively so that

$$x_\xi \in Y \setminus \bigcup_{\eta < \xi} G_\eta, \; x_\xi \in G_\xi \in \mathscr{G}$$

for each $\xi < \omega_1$. Choose open sets H_ξ such that $x_\xi \in H_\xi \subseteq \bar{H}_\xi \subseteq G_\xi$ for each ξ. As $H = \bigcup_{\xi < \omega_1} H_\xi$ is open, therefore F_σ, and contains every x_ξ, there is a closed $F \subseteq H$ such that $C = \{\xi : x_\xi \in F\}$ is uncountable. Now $F \cap Y$ is separable, so $\overline{F \cap Y}$ is separable and countably compact, and

$\overline{F \cap Y} \subseteq F \subseteq H = \bigcup_{\xi < \omega_1} H_\xi$. By 24H there is a finite $D \subseteq \omega_1$ such that $\overline{F \cap Y} \subseteq \overline{\bigcup_{\xi \in D} H_\xi} \subseteq \bigcup_{\xi \in D} G_\xi$. But now there is a $\zeta \in C$ such that $\zeta > \sup D$, and $x_\zeta \in F \cap Y \setminus \bigcup_{\xi \in D} G_\xi$. **XQ**

(*c*) It follows that X is Lindelöf. **P?** Otherwise, there is a cover $\mathscr{G}$ of X by open sets which has no countable subcover. As in (*b*), we can find families $\langle x_\xi \rangle_{\xi < \omega_1}$ in X, $\langle G_\xi \rangle_{\xi < \omega_1}$ in $\mathscr{G}$ such that

$$x_\xi \in G_\xi \setminus \bigcup_{\eta < \xi} G_\eta \; \forall \xi < \omega_1.$$

Now $Y = \{x_\xi : \xi < \omega_1\}$ is not Lindelöf, so cannot be hereditarily separable; let $C \subseteq \omega_1$ be such that $Z = \{x_\xi : \xi \in C\}$ is not separable. We can choose $\langle \alpha(\xi) \rangle_{\xi < \omega_1}$ in C such that

$$x_{\alpha(\xi)} \notin \overline{\{x_\zeta : \zeta \in C, \exists \eta < \xi, \zeta \leq \alpha(\eta)\}} \; \forall \xi < \omega_1$$

In this case $\langle \alpha(\xi) \rangle_{\xi < \omega_1}$ will be strictly increasing and

$$H_\xi = G_{\alpha(\xi)} \setminus \overline{\{x_{\alpha(\eta)} : \eta < \xi\}}$$

is a neighbourhood of $x_{\alpha(\xi)}$ for each ξ. But now $x_{\alpha(\xi)} \notin H_\eta$ for $\eta \neq \xi$, so $\{x_{\alpha(\xi)} : \xi < \omega_1\}$ is not ccc, contradicting (*a*) above. **XQ**

(*d*) Thus X is Lindelöf and countably compact, so must be compact.

24L Corollary

[$\mathfrak{p} > \omega_1$] Let X be a locally compact Hausdorff space in which every open set is F_σ. Let Z be the one-point compactification of X. Then sequentially closed sets in Z are closed; in particular, Z is countably tight.

Proof As X is T_1 and open sets are F_σ, singleton sets in X are G_δ; as X is locally compact and regular, it is first-countable [A4B*d*]. Let $A \subseteq Z$ be a sequentially closed set. Then $A \cap X$ must be closed in X. If ∞ (the member of $Z \setminus X$) belongs to A, then A is certainly closed in Z. If not, then no sequence in A can converge to ∞; so every sequence in A has a cluster point in Z other than ∞, i.e. a cluster point in X; as A is closed in X, it has a cluster point in A. Thus A is countably compact. But also relatively open sets in A are relatively F_σ. So, by 24K, A is compact, therefore closed in Z.

24M Sources

Rothberger 48 for special cases of 24C. Malyhin & Sapirovskii 73 for 24A(i)–(ii), 24F, 24H, and 24I (in part). Booth 74 and Tall 74*b* for 24C. HECHLER 75*a* for 24B and 24I. SHOENFIELD 75 for 24B*a*. ANTONOVSKII & CHUDNOVSKY 76 for 24E. WEISS 78 for 24K. GRUENHAGE 80 and ISMAIL & NYIKOS 80 for 24L.

24N **Exercises**

(*a*) Let $\kappa < \mathfrak{p}$. If $A \subseteq \mathbf{R}^\kappa$ is separable and closed and $\#(A) < \mathfrak{p}$ then A has an isolated point. [Hint: $\bar{A}$ in $(\beta\mathbf{R})^\kappa$ is not the union of fewer than $\mathfrak{p}$ nowhere-dense sets, by 14C(iv).] Hence $\mathbf{Q}$ cannot be embedded as a closed subset of $\mathbf{R}^\kappa$. [HECHLER 75*b*.]

(*b*) Let X be a countably tight regular space and $\mathscr{E}$ a family of closed subsets of X such that $\#(\mathscr{E}) < \mathfrak{p}$ and $\bigcup \mathscr{E}$ is countably compact. Then $\bigcup \mathscr{E}$ is closed. [Hint: use the idea of 24A. See ISMAIL & NYIKOS 80, 1.12.]

(*c*) [$\mathfrak{p} = \mathfrak{c}$] Let X be a compact Hausdorff space in which countably compact sets are closed. Then sequentially closed sets are closed. [Hint: if $A \subseteq X$ is countable then there is a countably compact $F \supseteq A$ with $\#(F) \leq \mathfrak{c}$. So by 24F X is sequentially compact, and sequentially closed sets in X are countably compact. See ISMAIL & NYIKOS 80, 1.24.]

(*d*) [$\mathfrak{p} = \mathfrak{c}$] Let X be a compact Hausdorff space expressible as $\bigcup_{n\in\mathbf{N}} X_n$, where each X_n is closed in X and any sequentially closed subset of any X_n is closed. Then all sequentially closed subsets of X are closed. [Hint: if $F \subseteq X$ is countably compact it is Lindelöf, therefore compact; use 24N*c*. See RANČIN 77, Theorem 2.]

(*e*) Let X be a separable countably compact space and $\mathscr{G}$ a family of open sets in X which has the finite intersection property. If $\#(\mathscr{G}) < \mathfrak{p}$ then $\bigcap\{\bar{G} : G \in \mathscr{G}\} \neq \emptyset$.

(*f*) If X is a separable countably compact regular Hausdorff space, $x \in X$, and $\psi(x, X) < \mathfrak{p}$, then $\chi(x, X) = \psi(x, X)$.

(*g*) Let X be a separable countably compact normal space, $F \subseteq X$ a closed set with $\psi(F, X) < \mathfrak{p}$. Then $\chi(F, X) = \psi(F, X)$. [MALYHIN & SAPIROVSKII 73, Theorem 1.5.]

(*h*) Let X be a separable countably compact normal Hausdorff space. If $F \subseteq X$ is closed and $\hat{L}(F) \leq \mathfrak{p}$ then F is compact.

(*i*) Let X be a topological space such that $\hat{L}(X) \leq \mathfrak{p}$, and $Y \subseteq X$ a countable relatively countably compact set. Then any ultrafilter on X containing Y has a limit in X. [Use the method of 24I. See HECHLER 75*a*.]

(*j*) Let $\langle X_\iota \rangle_{\iota \in I}$ be a family of countably compact spaces such that

$\hat{L}(X) \leq \mathfrak{p}$ for every $\iota \in I$. Then $\prod_{\iota \in I} X_\iota$ is countably compact [Use 24N*i*. See HECHLER 75*a*, 3.3.]

(*k*) (i) Let X be a first-countable T_1 space with $\#(X) < \mathfrak{p}$. Then any countable subset Y of X is G_δ. [Use first-countability to associate with each $f \in \mathbf{N}^{\mathbf{N}}$ an open set $G_f \supseteq Y$. Now use 14B. See BELL & GINSBURG 80.] (ii) Let X be a first-countable regular space expressible as the union of fewer than $\mathfrak{p}$ compact sets. Then X is pseudonormal. [If E and F are disjoint closed sets in X and E is countable, there is a sequence $\langle G_n \rangle_{n \in \mathbf{N}}$ of open sets including E such that $F \cap \bigcap_{n \in \mathbf{N}} \bar{G}_n = \emptyset$. See JUHÁSZ & WEISS 78, 1.4; DOUWEN & WAGE 79, 1.3; DOUWEN 83, 12.2.]

(*l*) **Property wD** Let us say that a T_1 topological space X has **property wD** if whenever $A \subseteq X$ is an infinite closed discrete set, there is an infinite discrete collection of open sets all meeting A.

(i) If X has property wD, and $A \subseteq X$ is relatively countably compact, then $\bar{A}$ is countably compact.

(ii) If X is completely regular and has property wD, and $A \subseteq X$ is such that $\sup_{x \in A} f(x) < \infty$ for every $f \in C(X)$, then $\bar{A}$ is countably compact.

(iii) If X is pseudonormal, it has property wD. [If $\langle x_n \rangle_{n \in \mathbf{N}}$ enumerates a closed discrete set, we can find a disjoint sequence $\langle G_n \rangle_{n \in \mathbf{N}}$ of open sets such that $x_n \in G_n$ for each $n \in \mathbf{N}$. Now we can separate $\{x_n : n \in \mathbf{N}\}$ from $\bigcap_{n \in \mathbf{N}} \overline{\bigcup_{i \geq n} G_i}$ by open sets.]

(iv) [$\mathfrak{p} > \omega_1$] If $\langle X_\iota \rangle_{\iota \in I}$ is any family of perfectly normal Hausdorff spaces then $X = \prod_{\iota \in I} X_\iota$ has property wD. [Use 24K to show that if $A \subseteq X$ is an infinite closed discrete set there is an $\iota \in I$ such that $\overline{\pi_\iota[A]}$ is not countably compact. See VAUGHAN 79, Corollary 4.4.]

(*m*) [$\mathfrak{p} > \omega_1$] Let X be a perfectly normal Hausdorff space. (i) If $A \subseteq X$ is such that $\sup_{x \in A} f(x) < \infty$ for every $f \in C(X)$, then A is relatively compact. [Use 23N*l*(ii) and 24K.] (ii) X is angelic. (iii) If X is locally compact then its one-point compactification is angelic.

(*n*) $\mathfrak{p} > \omega_1$ iff whenever X is a regular separable countably compact space and $L(X) \leq \omega_1$ then X is compact. [Work with (*) of 14D, taking X to be a suitable subset of $\beta\mathbf{N}$. See NYIKOS 81, Theorem 11.]

24O Further results

(*a*) If X and Y are angelic spaces and $\chi(x, A) < \mathfrak{p}$ whenever $x \in A \in [X]^\omega$ then $X \times Y$ is angelic. [ARHANGEL'SKII 76.]

(*b*) I shall say that a topological space X is **Fréchet-sequential** if whenever $x \in \bar{A}$ in X there is a sequence in A converging to x. [$\mathfrak{p} = \mathfrak{c}$] There is a countable normal Hausdorff space X such that X^n is Fréchet-sequential for every $n \in \mathbf{N}$ but $X^{\mathbf{N}}$ is not Fréchet-sequential. [GRUENHAGE 79.]

(*c*) A topological space X is **symmetrizable** if there is a function $\rho: X^2 \to \mathbf{R}$ such that (i) $\rho(x, y) = \rho(y, x) \geq 0 \, \forall x, y \in X$, (ii) $\rho(x, y) = 0$ iff $x = y$, (iii) the open sets of X are precisely those sets G such that

$$\forall x \in G \, \exists \delta > 0 \text{ such that } G \supseteq \{y : \rho(y, x) \leq \delta\}.$$

[$\mathfrak{p} = \mathfrak{c}$] Let X be a symmetrizable regular Hausdorff space with a dense relatively countably compact subset. Then X is a separable Moore space. [BURKE & DAVIS 81.]

(*d*) [$\mathfrak{p} = \mathfrak{c}$] Let $\langle X_\xi \rangle_{\xi < \mathfrak{c}}$ be a family of separable topological spaces such that $\chi(t, X_\xi) < \mathfrak{c}$ for every $t \in X_\xi$, $\xi < \mathfrak{c}$. Then every sequentially continuous function from $\prod_{\xi < \mathfrak{c}} X_\xi$ to a regular space is continuous.

(*e*) [$\mathfrak{p} = \mathfrak{c}$] Suppose that κ is a cardinal and that there is no two-valued-measurable cardinal $\lambda \leq \kappa$. Let $\langle X_\xi \rangle_{\xi < \kappa}$ be a family of separable first-countable topological spaces, Y a regular space, and $f: \prod_{\xi < \kappa} X_\xi \to Y$ a sequentially continuous function. Then f is continuous. [ANTONOVSKII & CHUDNOVSKY 76.]

(*f*) [$\mathfrak{p} = \mathfrak{c}$] There are two Hausdorff spaces X and Y which are initially κ-compact for every $\kappa < \mathfrak{c}$, such that $X \times Y$ is normal but not countably compact. [DOUWEN *b*.]

(*g*) (i) [$\mathfrak{p} = \mathfrak{c}$] There is an infinite countably compact subgroup G of $\mathbf{Z}_2^{\mathfrak{c}}$ such that G has no non-trivial convergent sequences. G can be taken to be dense in $\mathbf{Z}_2^{\mathfrak{c}}$ and either separable or non-separable. (ii) [$\mathfrak{p} = \mathfrak{c}$] There are two countably compact subgroups G, H of $\mathbf{Z}_2^{\mathfrak{c}}$ such that $G \times H$ is not countably compact. (iii) [$\mathfrak{p} = \mathfrak{c} > \omega_1$] There are two initially ω_1-compact subgroups G, H of $\mathbf{Z}_2^{\mathfrak{c}}$ such that $G \times H$ is not countably compact. [DOUWEN 80.]

(*h*) [$\mathfrak{p} = \mathfrak{c} > \omega_1$] If X is a separable, countably compact, hereditarily normal Hausdorff space and every point of X has a hereditarily Lindelöf neighbourhood then X is compact. [NYIKOS *b*.]

(*i*) [$\mathfrak{p} > \omega_1$] Let X be a perfectly normal Hausdorff space. Then $C(X)$, with the topology of uniform convergence on compact subsets of X, is barreled. [See WHEELER 76.]

(*j*) [$\mathfrak{p} = \mathfrak{c}$] Let X be a set of cardinal $\mathfrak{c}$, $Y \subseteq X$ a set of cardinal less than $\mathfrak{c}$, $\mathfrak{S}$ a locally compact locally countable Hausdorff topology on Y. Then there is a sequentially compact, locally compact, locally countable topology $\mathfrak{T}$ on X such that $\mathfrak{S}$ is the subspace topology on Y induced by $\mathfrak{T}$. If $\mathscr{F}$ is a non-principal ultrafilter on X such that either $Y \notin \mathscr{F}$ or there is no limit of $\mathscr{F}$ in Y for $\mathfrak{S}$, then $\mathfrak{T}$ can be constructed in such a way that $\mathscr{F}$ has no limit in X. [Juhász, Nagy & Weiss 79, Douwen 83.]

(*k*) [$\mathfrak{p} = \mathfrak{c}$] There is a family of sequentially compact completely regular Hausdorff spaces with a product which is not countably compact. [Douwen, 83, 13.1.]

(*l*) [$\mathfrak{p} = \mathfrak{c}$] For $n \in \mathbf{N}$, let $\mathfrak{c}^{(n+)}$ be the nth cardinal greater than $\mathfrak{c}$. Then there is a sequentially compact, locally countable, locally compact Hausdorff space X of cardinal $\mathfrak{c}^{(n+)}$. If $2^{\mathfrak{c}} = \mathfrak{c}^{(n+)}$, then X can be constructed in such a way that for every non-principal ultrafilter $\mathscr{F}$ on $\mathbf{N}$ there is a sequence $\langle x_n \rangle_{n \in \mathbf{N}}$ in X such that $\lim_{n \to \mathscr{F}} x_n$ does not exist in X. [Juhász, Nagy & Weiss 79.]

(*m*) [$\mathfrak{p} = \mathfrak{c}$] There is a non-separable ccc compact Hausdorff space X such that $w(X) = \mathfrak{c}$, $\chi(x, X) < \mathfrak{c}$ for every $x \in X$. [Bell 81*a*.]

(*n*) Let X be an extremally disconnected ccc regular Hausdorff space with $\pi(X) \leq \mathfrak{p}$. Then

$$\pi\chi(x, X) = \min\{\pi(G) : G \text{ open, } x \in G\}$$

for every $x \in X$. [Dow 82.]

(*o*) Results on similar topics may also be found in Galvin 77 and Galvin 80.

24P **Problems**

(*a*) Do 24N*c*–*d*, 24O*b*, 24O*g*(ii) and 24O*k* really need special axioms?

(*b*) Is there a version of 24E*b* if $\mathfrak{p} < \mathfrak{c}$?

(*c*) Can 24E*b* be deduced from the result in 22L*c*? [See 24O*e*.]

(*d*) Does 24H characterize $\mathfrak{p}$?

(*e*) [$\mathfrak{m} = \mathfrak{c} > \omega_1$] Let X be a first-countable separable countably compact normal space. Does X have to be compact? [See Nyikos 81.]

Notes and comments

With the exception of Weiss' theorem [24K], 24L, and some bits of 24N–O, all the results of this section can be regarded as saying: if

$\kappa < \mathfrak{p}$, then κ is like ω. They are therefore important only in so far as it is surprising that uncountable cardinals should be like ω in these ways, since with few exceptions [e.g. 24E*b*] they become elementary if we assume the continuum hypothesis.

The common idea in 24A–24C is that we have a countable set A, and wish to find a sequence in A satisfying fewer than $\mathfrak{p}$ conditions; which is exactly what P($\mathfrak{p}$) is for. Indeed, in all these results we find that a direct appeal to the definition of $\mathfrak{p}$ is sufficient; we do not need elaborations like 21A. In so far as any subtlety is required, this is a consequence of the hypotheses having been weakened to near their breaking point; thus in 24A the shift from '$\chi(x, X) < \mathfrak{p}$' to '$\psi(\{x\}, X) < \mathfrak{p}$' has to be compensated for, while in 24C the premises are attenuated to an extent quite unnecessary for its principal applications. Ideas of the same type yield 24N*b* and 24O*a*.

24B–E, 24N*a*, 24N*j*, 24N*l*(iv), 24O*a*–*b*, 24O*d*–*g* and 24O*k* all deal with products. Most of them are to the effect that products are better behaved than they might be; but there are also some counter-examples. It is not clear that the latter need special axioms [24P*a*].

24F is an argument of a different kind; the idea here is that there is not enough room in X for the cluster points of a sequence to be kept properly apart, so that some of them must collapse together into limits of subsequences. The work is done by the Cech-Pospíšil theorem [A4J*b*]. There is an interesting consequences of 24F in 24N*c*–*d* (but see 24P*a*). 24G uses a refinement of the idea of 24A(iii). Observe that if in 24G X is regular or Hausdorff, or $w(X) < \mathfrak{p}$, 24A gives a more precise result.

24H takes up another of the ideas of 24A; note that here a spark of imagination was needed to get the form of the theorem right, since it is not translated by the usual process from the result when $\#(\mathcal{G}) = \omega$. It leads naturally to 24I and to 24J, which is a kind of strengthening of 14C(iv). On taking complements in 24H we get 24N*e*; we can use this to relate $\chi(x, X)$ to $\psi(x, X)$ and (in normal spaces) $\chi(F, X)$ to $\psi(F, X)$ [24N*f*–*g*]. 24N*i* is a slight extension of 24I, and leads to another result on the countable compactness of products [24N*j*]. 24N*h* and 24O*h* use the same ideas; there may be stronger results in this direction to be found [24P*e*]. 24N*l* discusses 'property wD' (a weakened version of R.L. Moore's 'property D'), which is interesting in itself (see Vaughan 79) and with 24N*k* provides a lemma which enables us to build locally countable locally compact topologies inductively [24O*j*]; this is the basis of 24O*k*–*l*.

For any result like 24A–C or 24F–J there is an associated cardinal, the largest cardinal that can be substituted for $\mathfrak{p}$ in the statement of the theorem without destroying its truth. In some cases, such as 24J, this is $\mathfrak{p}$ itself [14C(iv)]. (I do not know if the same is true of 24H; see 24P*d*,

24N*n* and 14G.) In other cases, it is consistent to assume that the relevant cardinal is strictly greater than $\mathfrak{p}$. For instance, we can define $\mathfrak{b}$ to be smallest cardinal of any set $A \subseteq \mathbf{N}^{\mathbf{N}}$ for which there is no $g \in \mathbf{N}^{\mathbf{N}}$ such that $\{n : g(n) < f(n)\}$ is finite for every $f \in A$; and now Lemma 14B becomes just '$\mathfrak{p} \leq \mathfrak{b}$'. (For a model in which $\mathfrak{p} < \mathfrak{b}$ see KUNEN 80, Ex. 8.A4.) The value of this notation is that it provides a succinct expression of interconnexions between the various theorems. Thus if I had written out 22I–J with $\mathfrak{b}$ in place of $\mathfrak{p}$, it would have been plain that they were on a different level from 22B or 22G; and the reappearance of $\mathfrak{b}$ in 23H would have signalled its origin. Similarly, 24N*k*, 24O*j*–*l* and 25N*f* are more precisely located if they are written with $\mathfrak{b}$ instead of with $\mathfrak{p}$. I have not attempted to carry any such programme through in this book, mostly because none of the individual cardinals which have been studied so far is associated with more than a handful of the results I give. (An exception is the additivity of Lebesgue measure; see 33 *notes*.) But this does seem to be an effective language for the investigation of their relationships. For work along these lines, see DOUWEN 83 and VAUGHAN 83.

Weiss' theorem [24K] gives further conditions under which regular countably compact spaces must be compact. This has the distinction that for once there is no cardinal mentioned in the hypotheses. Furthermore, it is not a consequence of the continuum hypothesis; the space of OSTASZEWSKI 76, depending on ♣ [21J], is a counter-example. The spaces of 24K must of course be hereditarily Lindelöf [A4L*d*]; I shall have a great deal more to say about hereditarily Lindelöf spaces in §44. Among the consequences of 24K are 24L (which will be useful later), 24N*l*(iv), 24N*m* and 24O*i*.

24N*a* is close to the subject-matter of this section but the natural proof seems to be direct from 14C(iv) (see also 43Q*c*). 24O*m* is important as a counter-example to some natural conjectures arising out of the work of §43. Finally, 24O*n* bears no relation to anything else here; it is proved directly from 21A.

25 Normal spaces

Martin's axiom has been used in various ways to prove that certain interesting topological spaces are normal. The most powerful method is that of ALSTER & PRZYMUSIŃSKI 76, which I give in 25A–25D; it relies on the concept of 'cometrizability' [25B]. I then describe a string of important examples: Cantor trees [25F], special Aronszajn trees [25G], Pixley-Roy spaces [25H–I], the density topology [25J], and E.K. van Douwen's gap spaces [25K].

25A Theorem

Let $(X, \mathfrak{S})$ be a second-countable topological space and $\mathscr{A}$ and

$\mathscr{B}$ two families of $\mathfrak{S}$-compact subsets of X, both of cardinal less than $\mathfrak{p}$. Write $Y = \bigcup \mathscr{A}, Z = \bigcup \mathscr{B}$. Suppose that $\mathfrak{T}$ is another topology on X, finer than $\mathfrak{S}$, such that

$$\forall A \in \mathscr{A}\ \exists V \in \mathfrak{T} \text{ such that } A \subseteq V \text{ and } \bar{V}^{\mathfrak{S}} \cap Z = \emptyset;$$
$$\forall B \in \mathscr{B}\ \exists V \in \mathfrak{T} \text{ such that } B \subseteq V \text{ and } \bar{V}^{\mathfrak{S}} \cap Y = \emptyset.$$

Then Y and Z can be separated by $\mathfrak{T}$-open sets.

Proof (*a*) Let $\mathscr{U}$ be a countable base for $\mathfrak{S}$, and let $\mathscr{D}$ be the (countable) algebra of subsets of X generated by $\mathscr{U}$. Let P be

$$\{(G, H): G, H \in \mathfrak{T}, \bar{G}^{\mathfrak{S}} \cap Z = \bar{H}^{\mathfrak{S}} \cap Y = \emptyset,$$
$$\exists D \in \mathscr{D} \text{ such that } G \subseteq D \subseteq X \setminus H\}.$$

Order P by saying that

$$(G, H) \leq (G', H') \text{ if } G \subseteq G', H \subseteq H'.$$

(*b*) For each $D \in \mathscr{D}$,

$$\{(G, H): (G, H) \in P, G \subseteq D, H \subseteq X \setminus D\}$$

is upwards-directed. So P is σ-centered upwards.

(*c*) For $A \in \mathscr{A}$, set $Q_A = \{(G, H): (G, H) \in P, A \subseteq G\}$. Then Q_A is cofinal with P. **P** Let $(G, H) \in P$. Then there is a $D \in \mathscr{D}$ such that $G \subseteq D \subseteq X \setminus H$. Also there is a $V \in \mathfrak{T}$ such that $A \subseteq V \subseteq \bar{V}^{\mathfrak{S}} \subseteq X \setminus Z$. Set $\mathscr{U}_0 = \{U : U \in \mathscr{U}, U \cap H = \emptyset\}$. Then $A \subseteq Y \subseteq X \setminus \bar{H}^{\mathfrak{S}} = \bigcup \mathscr{U}_0$. As A is compact for $\mathfrak{S}$, there is a finite $\mathscr{U}_1 \subseteq \mathscr{U}_0$ such that $A \subseteq \bigcup \mathscr{U}_1$. Set $D_1 = D \cup \bigcup \mathscr{U}_1 \in \mathscr{D}$ and $G_1 = G \cup (V \cap \bigcup \mathscr{U}_1) \in \mathfrak{T}$, and see that

$$G \cup A \subseteq G_1 \subseteq D_1 \subseteq X \setminus H, \bar{G}_1^{\mathfrak{S}} \cap Z = \emptyset,$$

so that $(G_1, H) \in P$ and $(G, H) \leq (G_1, H) \in Q_A$. **Q**

(*d*) Similarly, for $B \in \mathscr{B}$,

$$Q_B = \{(G, H): B \subseteq H\}$$

is cofinal with P. By Bell's theorem [14C] there is an upwards-directed set $R \subseteq P$ meeting every Q_A and every Q_B. Set

$$U = \bigcup \{G: (G, H) \in R\}, \quad V = \bigcup \{H: (G, H) \in R\}$$

and see that U and V belong to $\mathfrak{T}$ and separate Y from Z.

25B Cometrizable spaces

By far the commonest situation in which 25A is used is the following. Say a topological space $(X, \mathfrak{T})$ is **cometrizable** if there is a separable metrizable topology $\mathfrak{S}$ on X such that

(i) $\mathfrak{S} \subseteq \mathfrak{T}$;
(ii) each point of X has a base of neighbourhoods for $\mathfrak{T}$ consisting of $\mathfrak{S}$-closed sets.

Note that in this case $\mathfrak{T}$ must be regular, Hausdorff and sub-second-countable [23E].

If $\mathfrak{S}$ can be taken to be Polish, I call $\mathfrak{T}$ **copolish**.

25C **Lemma**

(*a*) Any subspace of a cometrizable space is cometrizable.

(*b*) The product of countably many cometrizable spaces is cometrizable.

(*c*) If $(X,\mathfrak{T})$ is a locally compact space and there is a separable metrizable topology $\mathfrak{S} \subseteq \mathfrak{T}$, then $(X,\mathfrak{T})$ is cometrizable.

(*d*) Let $(X,\mathfrak{T})$ be a cometrizable space and Y and Z two subsets of X such that $\bar{Y} \cap Z = Y \cap \bar{Z} = \varnothing$. Suppose that both Y and Z can be expressed as the union of fewer than $\mathfrak{p}$ compact sets. Then Y and Z can be separated by open sets.

Proof (*a*) (*b*) Assign a separable metrizable topology to the subspace or product space in the obvious way, and check that it works.

(*c*) As $\mathfrak{S}$ is Hausdorff, so is $\mathfrak{T}$. $\mathfrak{T}$ is therefore regular, and each point of X has a base of $\mathfrak{T}$-neighbourhoods consisting of $\mathfrak{T}$-compact sets, which are $\mathfrak{S}$-closed.

(*d*) Let $\mathfrak{S}$ be the auxiliary separable metrizable topology. Let $\mathscr{A}$, $\mathscr{B}$ be families of $\mathfrak{T}$-compact sets such that $\#(\mathscr{A}) < \mathfrak{p}$, $\#(\mathscr{B}) < \mathfrak{p}$, $Y = \bigcup \mathscr{A}$ and $Z = \bigcup \mathscr{B}$. If $x \in A \in \mathscr{A}$ there is a $V \in \mathfrak{T}$ such that $x \in V \subseteq \bar{V}^{\mathfrak{S}} \subseteq X \backslash \bar{Z}^{\mathfrak{T}}$, because $X \backslash \bar{Z}^{\mathfrak{T}}$ is a $\mathfrak{T}$-neighbourhood of x. Because A is $\mathfrak{T}$-compact there is a $V \in \mathfrak{T}$ such that $A \subseteq V \subseteq \bar{V}^{\mathfrak{S}} \subseteq X \backslash Z$. Similarly, if $B \in \mathscr{B}$, there is a $V \in \mathfrak{T}$ such that $B \subseteq V$ and $\bar{V}^{\mathfrak{S}} \cap Y = \varnothing$. Accordingly we can apply Theorem 25A, and Y and Z can be separated by $\mathfrak{T}$-open sets.

25D **Theorem**

Let X be a cometrizable space which is expressible as the union of fewer than $\mathfrak{p}$ compact sets. Then X and all its countable powers are perfectly normal.

Proof Let $\mathfrak{T}$ be the given topology on X and $\mathfrak{S}$ a separable metrizable topology satisfying the conditions of 25B.

(*a*) If $G \in \mathfrak{T}$ then G is expressible as the union of fewer than $\mathfrak{p}$ $\mathfrak{T}$-compact sets. **P** If $K \subseteq X$ is $\mathfrak{T}$-compact, then $\mathfrak{S}$ and $\mathfrak{T}$ agree on K, so K is metrizable for $\mathfrak{T}$. Consequently $K \cap G$ is $\mathfrak{T}$-K_σ. As X is the union of fewer than $\mathfrak{p}$ $\mathfrak{T}$-compact sets, so is G. **Q**

It follows that G is actually $\mathfrak{S}$-F_σ. **P** Both G and $X\setminus G$ are expressible as unions of fewer than $\mathfrak{p}$ $\mathfrak{T}$-compact sets, which are also $\mathfrak{S}$-compact and $\mathfrak{S}$-closed. So by 23A*a* there is an $\mathfrak{S}$-F_σ set E with $G \subseteq E$ and $E \cap (X\setminus G) = \emptyset$, and $G = E$. **Q**

(*b*) Next, X is normal by 25C*d*, because any closed set in X is the union of fewer than $\mathfrak{p}$ compact sets. So X is perfectly normal.

(*c*) The same arguments apply to all finite powers of X, so they also are perfectly normal. But now countable powers of X are also perfectly normal, by A4L*e*.

25E **Proposition**

Let X be a ccc cometrizable space. If every nowhere-dense closed set in X is the union of fewer than $\mathfrak{p}$ compact sets, then X is perfectly normal.

Proof (*a*) X is normal. **P** Let E and F be disjoint closed sets in X. Then ∂E and ∂F are nowhere-dense closed sets so are both expressible as unions of fewer than $\mathfrak{p}$ compact sets, and by 24C*d* they can be separated by open sets G, H say. Now $G' = (G\setminus F) \cup \operatorname{int} E$ and $H' = (H\setminus E) \cup \operatorname{int} F$ are open sets separating E from F. **Q**

(*b*) To show that open sets are F_σ, I adapt the arguments of 25D. Let $G \subseteq X$ be open. Let $\mathscr{H}$ be a maximal disjoint family of open sets such that $\bar{H} \subseteq G$ for every $H \in \mathscr{H}$. Then $\bigcup\mathscr{H}$ is dense in G and $\mathscr{H}$ is countable, because X is ccc. Observe that $G\setminus\bigcup\mathscr{H}$ is a relatively open subset of the nowhere dense closed set $\partial(\bigcup\mathscr{H})$; so that, just as in (*a*) of the proof of 25D, it is expressible as the union of fewer than $\mathfrak{p}$ compact sets. Consequently, by 23A*a*, working with the associated coarser separable metrizable topology of X, there is an F_σ set E such that $G\setminus\bigcup\mathscr{H} \subseteq E$ and $E \cap (\partial(\bigcup\mathscr{H})\setminus G) = \emptyset$. Now $G = (E \cap \overline{\bigcup\mathscr{H}}) \cup \bigcup\{\bar{H} : H \in \mathscr{H}\}$ is F_σ.

25F **Example**

(*a*) Set

$$Z = \{(2^{-m}i, 2^{-m}) : m \in \mathbf{N}, i \in \mathbf{Z}\} \subseteq \mathbf{R}^2,$$

so that Z is discrete and its accumulation points constitute the horizontal axis. Take any $Y \subseteq \mathbf{R} \times \{0\}$ and set $X = Y \cup Z$. Let $\mathfrak{T}$ be the topology on

X generated by

$$\{\{z\}:z\in Z\}\cup\{U_n(y):n\in\mathbf{N}, y\in Y\},$$

where for $y=(s,0)\in Y$

$$U_n(y)=\{y\}\cup\{(2^{-m}i, 2^{-m}):m\geq n, |s-2^{-m}i|\leq 2^{-m}\}.$$

Then $\mathfrak{T}$ is cometrizable (because all the $U_n(y)$ are closed for the Euclidean topology on X), therefore regular and Hausdorff; it is separable (because $X=\bar{Z}$), locally compact (because the $U_n(y)$ are compact for $\mathfrak{T}$) and locally countable, therefore zero-dimensional. Observing that $\{Y\}\cup\{\{z\}:z\in Z\}$ is a countable partition of X into closed discrete sets, we see that X is a Moore space [A4Nf]. But if Y uncountable, X cannot be metrizable, since X is separable and its subspace Y is not separable for the topology induced by $\mathfrak{T}$.

(*b*) If $\#(Y)<\mathfrak{p}$, then $X^{\mathbf{N}}$ is normal, by 25D.

(*c*) [$\mathfrak{p}>\omega_1$] There is a separable, locally compact, locally countable, normal Moore space which is not metrizable.

25G **Example**

Let X be a special Hausdorff Aronszajn tree with the fine tree topology [A3J].

(i) X is a cometrizable, non-metrizable, locally compact, locally countable, non-separable Moore space.

(ii) [$\mathfrak{p}>\omega_1$] X, and all its countable powers, are normal.

Proof (*a*) X is locally compact because it is a tree; Hausdorff because it is a Hausdorff tree; locally countable because all its elements have countable rank; and not metrizable because it is not metalindelöf. **P** Let $\mathscr{G}$ be the set of countable open subsets of X; then $\mathscr{G}$ is a cover of X. **?** If $\mathscr{H}$ is a point-countable open cover of X refining $\mathscr{G}$, then for each $x\in X$ set

$$h(x)=\sup\{\text{rank}(y):y\in\text{St}(x,\mathscr{H})\}<\omega_1.$$

As $\{x:\text{rank}(x)=\xi\}$ is countable for each $\xi<\omega_1$, there is a non-zero limit ordinal $\zeta<\omega_1$ such that $h(x)<\zeta$ whenever $\text{rank}(x)<\zeta$. But now (because the height of X is ω_1) there is a $z\in X$ such that $\text{rank}(z)=\zeta$; there is an $H\in\mathscr{H}$ such that $z\in H$; and there is an $x\in H$ such that $x<z$, in which case $z\in\text{St}(x,\mathscr{H})$, but $\text{rank}(z)>h(x)$. **XQ**

(*b*) Let $\langle A_n\rangle_{n\in\mathbf{N}}$ be a sequence of weak antichains in X covering X. Then each A_n is closed and discrete. As also X is first-countable, X is a Moore space [A4Nf].

(*c*) Let Z be the set of isolated points in X, namely

$\{x : \text{rank}(x) \text{ is a successor or } 0\}$.

For $x \in X$ set $]x, \infty[= \{y : y > x\}$. For each $n \in \mathbf{N}$, I define an open partition $\mathscr{G}_n$ of X as follows. Write

$$H_n = \bigcup\{]x, \infty[: x \in A_n\}, \quad G_n = X \setminus (H_n \cup (A_n \cap Z)),$$
$$\mathscr{G}_n = (\{]x, \infty[: x \in A_n\} \cup \{\{x\} : x \in A_n \cap Z\} \cup \{G_n\}) \setminus \{\varnothing\}.$$

G_n is open in X because $y \in G_n$ whenever $y \leq x \in G_n$, so $\mathscr{G}_n$ is a family of open sets; $\bigcup \mathscr{G}_n = X$ by the choice of G_n; and $\mathscr{G}_n$ is disjoint because A_n is an up-antichain. As $\#(\mathscr{G}_n) \leq \#(X) = \omega_1 \leq \mathfrak{c}$, there is a function $f_n : X \to \mathbf{R}$ such that $\mathscr{G}_n = \{f_n^{-1}[\{s\}] : s \in f_n[X]\}$, and f_n is continuous.

Let $\mathfrak{T}$ be the tree topology on X, and let $\mathfrak{S}$ be the topology induced by $\{f_n : n \in \mathbf{N}\}$. Then $\mathfrak{S} \subseteq \mathfrak{T}$ and $\mathfrak{S}$ is separable and pseudo-metrizable. But also $\mathfrak{S}$ is Hausdorff. **P** Let x and y be distinct members of X; suppose that $x \nleq y$. If $x \in Z$, let n be such that $x \in A_n$; then $\{x\} \in \mathscr{G}_n$ so $f_n(x) \neq f_n(y)$. Otherwise, there is a $z < x$ such that $z \nless y$, because X is a Hausdorff tree. Let n be such that $z \in A_n$; then $x \in]z, \infty[\in \mathscr{G}_n$ and $y \notin]z, \infty[$, so again $f_n(x) \neq f_n(y)$. Thus $\{f_n : n \in \mathbf{N}\}$ separates the points of X and $\mathfrak{S}$ is metrizable. **Q**

By 25C*c*, $\mathfrak{T}$ is cometrizable.

(*d*) This proves (i). If $\mathfrak{p} > \omega_1$, (ii) follows at once, by 25D.

25H Pixley–Roy spaces

Let S be a topological space, X a family of subsets of S. For $A, G \subseteq S$ write

$$U_X(A, G) = \{K : K \in X, A \subseteq K \subseteq G\}.$$

Then $\{U_X(A, G) : A \subseteq S \text{ finite}, G \subseteq S \text{ open}\}$ is a topology base; the topology it generates is the **Pixley–Roy** topology on X.

25I Example

Let S be a separable metric space, X the set of compact subsets of S, Z the set of finite subsets of S. Give X and Z their Pixley–Roy topologies.

(*a*) X is cometrizable and σ-centered [A4Q–R].

(*b*) Z is a dense subspace of X; it is a σ-centered cometrizable hereditarily metacompact Moore space.

(*c*) If S is uncountable, Z is not separable and neither Z nor X

is metrizable. If S has no isolated points, then $\varnothing$ is the only isolated point of either X or Z.

(*d*) If $\#(S) < \mathfrak{p}$ then Z, and all its countable powers, are normal.

(*e*) If S is Polish then X is copolish.

(*f*) If S is Polish then the intersection of fewer than $\mathfrak{p}$ dense open sets in X is dense.

(*g*) [$\mathfrak{p} = \mathfrak{c}$] If S is Polish and uncountable and has no isolated points then $X \setminus \{\varnothing\}$ has a $\mathfrak{c}$-Lusin subset which is perfectly normal.

Proof Let ρ be the metric of S and fix on a countable base $\mathscr{G}$ for the topology of S which is closed under finite unions. Write $\mathfrak{T}_X$, $\mathfrak{T}_Z$ for the Pixley–Roy topologies on X and Z respectively. Observe that $\mathfrak{T}_Z$ is the subspace topology on Z induced by $\mathfrak{T}_X$. For $A \subseteq S$, $n \in \mathbb{N}$ set

$$G_n(A) = \{s : s \in S, \exists t \in A, \rho(s,t) < 2^{-n}\}.$$

(*a*) (i) Let $\mathfrak{S}$ be the Vietoris topology on X. Then $\mathfrak{S}$ is separable and metrizable [A4T*b*–*c*]. If $K \in U \in \mathfrak{T}_X$ then there is a finite $A \subseteq K$ and an open $G \subseteq S$ such that $K \in U_X(A, G) \subseteq U$. Let H be an open set in S such that $K \subseteq H \subseteq \bar{H} \subseteq G$. Then

$$K \in U_X(A, H) \subseteq U_X(A, \bar{H}) \subseteq U.$$

But $U_X(A, \bar{H})$ is closed for $\mathfrak{S}$ and $U_X(A, H) \in \mathfrak{T}_X$. As U is arbitrary, K has a base of $\mathfrak{T}_X$-neighbourhoods consisting of $\mathfrak{S}$-closed sets; as K is arbitrary, $\mathfrak{T}_X$ is cometrizable.

(ii) Write

$$\mathscr{U} = \{U_X(A, G) : G \in \mathscr{G}, A \in [G]^{<\omega}\}.$$

Then $\mathscr{U}$ is a base for $\mathfrak{T}_X$, not containing $\varnothing$, because if $H \subseteq S$ is open and $K \subseteq H$ is compact there is a $G \in \mathscr{G}$ such that $K \subseteq G \subseteq H$. But $\mathscr{U}$ is σ-centered downwards, because $\mathscr{G}$ is countable. So $\mathfrak{T}_X$ must be σ-centered.

(*b*) Clearly Z is dense in X; so Z is a σ-centered cometrizable space [A4R(ii), 25C*a*]. To see that it is a Moore space, observe that $\{U_Z(A, G_n(A)) : n \in \mathbb{N}\}$ is a base of neighbourhoods of A, for any $A \in Z$, so Z is first-countable. Next,

$$\{A : A \in Z, \rho(s,t) \geq 2^{-n} \forall \text{ distinct } s, t \in A\}$$

is closed and discrete in Z, for each $n \in \mathbb{N}$; so Z is the union of a sequence of closed discrete sets, and is a Moore space by A4N*f*.

If $\mathscr{V}$ is any family of open sets in Z, then for each $K \in \bigcup \mathscr{V}$ choose a

$V_K \in \mathscr{V}$ and an open set $G_K \subseteq S$ such that $U_Z(K, G_K) \subseteq V_K$. Then $\mathscr{V}' = \{U_Z(K, G_K): K \in \bigcup \mathscr{V}\}$ is a refinement of $\mathscr{V}$ with $\bigcup \mathscr{V}' = \bigcup \mathscr{V}$, and $\mathscr{V}'$ is point-finite because every $K \in Z$ is finite, and can belong to $U_Z(K', G_{K'})$ only when $K' \subseteq K$. As $\mathscr{V}$ is arbitrary, Z is hereditarily metacompact.

(*c*) If S is uncountable then for any countable $Z_0 \subseteq Z$ there is an $s \in S \setminus \bigcup Z_0$, and $Z_0 \cap U_Z(\{s\}, S) = \emptyset$. So Z is not separable. As it is ccc, it cannot be metrizable. So X is not metrizable either.

If S has no isolated points, then whenever $G \subseteq S$ is open and not empty and $A \subseteq G$ is finite there is an $s \in G \setminus A$, so that A and $A \cup \{s\}$ are two different members of $U_Z(A, G)$. Thus no open set in Z except $U_Z(\emptyset, \emptyset) = \{\emptyset\}$ is a singleton; and it follows that the same is true for X.

(*d*) If $\#(S) < \mathfrak{p}$ then $\#(Z) < \mathfrak{p}$ so all countable powers of Z are normal, by 25D.

(*e*) If S is Polish then $(X, \mathfrak{S})$ is Polish [A4T*c*–*d*] so $(X, \mathfrak{T}_X)$ is copolish.

(*f*) By 43D*d*, X is Martin-complete under $\mathfrak{T}_X$; by 43F*a* it follows that the intersection of fewer than $\mathfrak{m}$ dense open sets in X is dense. But in the present case, because X is σ-centered, we can replace $\mathfrak{m}$ by $\mathfrak{p}$. **P** By (*a*) above, $P = \mathfrak{T}_X \setminus \{\emptyset\}$ is σ-centered downwards. Now let $\preccurlyeq$ be the auxiliary ordering on P constructed by the recipes of 43D*d*–43C*b* to satisfy the conditions of 43A with $\mathrm{U} = \emptyset$. As remarked in 43B*c*, P will still be σ-centered downwards for $\preccurlyeq$. Turn next to the arguments of 43E. If $\mathscr{V}$ is a family of dense open subsets of X with $\#(\mathscr{V}) < \mathfrak{p}$, form

$$Q_V = \{U: U \in P, U \subseteq V\}$$

for each $V \in \mathscr{V}$, and observe that Q_V is coinitial in P for $\preccurlyeq$. By Bell's theorem [14C] there is a $\preccurlyeq$-downwards-directed $R \subseteq P$ meeting every Q_V. As in (*c*) of the proof of 43E, there is an $x \in \bigcap R$, and now $x \in \bigcap \mathscr{V}$. This shows that $\bigcap \mathscr{V} \neq \emptyset$. Of course we can repeat the argument on $P' = Q_W$, for any $W \in P$, to see that $\bigcap \mathscr{V}$ is actually dense. **Q**

(*g*) $[\mathfrak{p} = \mathfrak{c}]$ As $\{U_X(A, G): G \subseteq S \text{ open}, A \in [G]^{<\omega}\}$ is a base for $\mathfrak{T}_X$, $w(X) \leq \mathfrak{c}$. By (*c*) and (*f*), $X \setminus \{\emptyset\}$ satisfies the conditions of A3F*b*, so has a $\mathfrak{c}$-Lusin subset Y. Now Y is cometrizable (because it is a subset of X) and ccc (because it is dense in an open subset of X); and a subset of Y which is nowhere dense in Y is nowhere dense in $X \setminus \{\emptyset\}$, so has cardinal less than $\mathfrak{c}$. By 25E, Y is perfectly normal.

25J The density topology on R

(*a*) Let μ be Lebesgue measure on **R**, Σ its domain, and $\mathfrak{S}$ the usual topology on **R**. For $E \in \Sigma$, $s \in \mathbf{R}$ and $\delta > 0$ write

$$h_\delta(s, E) = \frac{1}{2\delta}\mu(E \cap [s-\delta, s+\delta]), \quad \varphi(E) = \{s : \lim_{\delta \downarrow 0} h_\delta(s, E) = 1\}.$$

Then $\varphi(E) \in \Sigma$ and $\mu(E \triangle \varphi(E)) = 0$ [MUNROE 53, §42; DUNFORD & SCHWARTZ 57, III.12.7; OXTOBY 71, 3.20.] Set

$$\begin{aligned}\mathfrak{T} &= \{\varphi(E) \setminus F : E, F \in \Sigma, \mu F = 0\} \\ &= \{E : E \in \Sigma, E \subseteq \varphi(E)\}.\end{aligned}$$

Then $\mathfrak{T}$ is a topology [OXTOBY 71, 22.5], the **density topology** on **R**, and $\mathfrak{T} \supseteq \mathfrak{S}$. For $E \subseteq \mathbf{R}$, $\mu E = 0$ iff E is nowhere dense for $\mathfrak{T}$, and in this case E is $\mathfrak{T}$-closed [OXTOBY 71, 22.6]. Consequently E is meagre for $\mathfrak{T}$ iff E is discrete for $\mathfrak{T}$ iff $\mu E = 0$. For $E \subseteq \mathbf{R}$, $E \in \Sigma$ iff E has the Baire property for $\mathfrak{T}$ [OXTOBY 71, 22.7], and in this case is F_σ and G_δ for $\mathfrak{T}$ (because there is an $\mathfrak{S}$-F_σ set $H \subseteq E$ with $\mu(E \setminus H) = 0$). $\mathfrak{T}$ is ccc, because any non-empty set in $\mathfrak{T}$ has positive measure. The proof that $\mathfrak{T}$ is a topology shows that μ is τ-additive for $\mathfrak{T}$, so is quasi-Radon for $\mathfrak{T}$.

(*b*) $\mathfrak{T}$ is copolish. **P** [Cf. OXTOBY 71, 22.9] The auxiliary topology is of course $\mathfrak{S}$. Let $s \in G \in \mathfrak{T}$. Let $\langle \delta_n \rangle_{n \in \mathbf{N}}$ be a decreasing sequence of strictly positive numbers, convergent to 0, such that

$$h_\delta(s, G) \geq 1 - 2^{-n} \; \forall \delta \in]0, \delta_n].$$

Choose $\mathfrak{S}$-closed sets $F_n \subseteq G \cap [s - \delta_n, s + \delta_n]$ such that

$$\mu F_n \geq \mu(G \cap [s - \delta_n, s + \delta_n]) - 2^{-n}\delta_{n+1}.$$

Then for $\delta_{n+1} \leq \delta \leq \delta_n$,

$$h_\delta(s, F_n) \geq h_\delta(s, G) - \frac{1}{2\delta}2^{-n}\delta_{n+1} \geq 1 - 3.2^{-n-1}.$$

Set $F = \{s\} \cup \bigcup_{n \in \mathbf{N}} F_n$. Then F is $\mathfrak{S}$-closed and $\lim_{\delta \downarrow 0} h_\delta(s, F) = 1$, so $s \in \mathrm{int}_{\mathfrak{T}} F \subseteq F \subseteq G$. **Q**

(*c*) Suppose that $Y \subseteq \mathbf{R}$ is an additive subgroup which is a $\mathfrak{c}$-Sierpiński set. ($\mathfrak{p} = \mathfrak{c}$ is not sufficient to ensure the existence of such a set, but $\mathfrak{m}_K = \mathfrak{c}$ is; see 32P*f*.) Give Y the topology induced by $\mathfrak{T}$. Then:

(i) $\#(Y) = \mathfrak{c}$. Y is $\mathfrak{T}$-dense in **R** (because it meets every set of positive measure), so is ccc. Y is cometrizable.

(ii) [$\mathfrak{p} = \mathfrak{c}$] For $A \subseteq Y$,

$$\begin{aligned}\mu A = 0 &\Leftrightarrow \#(A) < \mathfrak{c} \text{ [use 22H}b\text{]} \\ &\Leftrightarrow A \text{ is discrete}\end{aligned}$$

$\Leftrightarrow A$ is closed and nowhere dense

$\Leftrightarrow A$ is meagre.

(iii) If $\mathscr{G}$ is any collection of open sets in Y, there is a countable $\mathscr{G}_0 \subseteq \mathscr{G}$ such that $\#(\bigcup\mathscr{G}\setminus\bigcup\mathscr{G}_0) < \mathfrak{c}$ [use A6G*a*], so $\hat{L}(Y) \leq \mathfrak{c}$.

(iv) $[\mathfrak{p} = \mathfrak{c}]$ The union of fewer than $\mathfrak{c}$ meagre sets in Y is nowhere dense.

(v) $[\mathfrak{p} = \mathfrak{c}]$ Y is perfectly normal [25E].

(vi) For $A \subseteq Y$,

A is Borel $\Leftrightarrow A$ is F_σ and G_δ

$\Leftrightarrow A$ is measurable for the subspace measure on Y [A6J]

$\Leftrightarrow A$ has the Baire property in Y.

(vii) Y is homogeneous (because it is a subgroup of $\mathbf{R}$ and $\mathfrak{T}$ is translation-invariant).

(viii) $[\mathfrak{p} = \mathfrak{c}]$ Y^2 is meagre in itself. **P** Because there is an $\mathfrak{S}$-comeagre negligible set $E \subseteq \mathbf{R}$, and $\#(Y \cap E) < \mathfrak{c}$, Y is $\mathfrak{S}$-meagre [22C*a*]. Let $\langle G_n \rangle_{n\in\mathbf{N}}$ be a sequence of $\mathfrak{S}$-dense $\mathfrak{S}$-open sets such that $Y \cap \bigcap_{n\in\mathbf{N}} G_n = \varnothing$. Consider

$$V_n = \{(s,t): s, t \in Y, s + t \in G_n\}.$$

This is $\mathfrak{T} \times \mathfrak{T}$-open in Y^2. But it is also $\mathfrak{T} \times \mathfrak{T}$-dense; for if $U \subseteq Y^2$ is $\mathfrak{T} \times \mathfrak{T}$-open and not empty, there are non-empty $\mathfrak{T}$-open $H_1, H_2 \subseteq Y$ such that $H_1 \times H_2 \subseteq U$. Now $\varnothing \neq \varphi(H_1) + \varphi(H_2) \subseteq \operatorname{int}(\overline{H_1 + H_2}^{\mathfrak{S}})$, because if $h_\delta(s_1, H_1) \geq \frac{2}{3}$ and $h_\delta(s_2, H_2) \geq \frac{2}{3}$ then $[s_1 + s_2 - \delta/3, s_1 + s_2 + \delta/3] \subseteq \overline{H_1 + H_2}^{\mathfrak{S}}$. Thus $H_1 + H_2$ must be somewhere $\mathfrak{S}$-dense, and meets G_n; in which case U meets V_n. On the other hand, $\bigcap_{n\in\mathbf{N}} V_n = \varnothing$ because $\bigcap_{n\in\mathbf{N}} G_n \cap Y = \varnothing$ and Y is closed under addition. **Q**

25K **van Douwen's gap space**

Let $\langle A_\xi \rangle_{\xi<\omega_1}$, $\langle B_\xi \rangle_{\xi<\omega_1}$ be families of subsets of $\mathbf{N}$ as in 21L ('Hausdorff's gap'). Set

$$X = \mathbf{N} \cup \{a_\xi : \xi < \omega_1\} \cup \{b_\xi : \xi < \omega_1\}$$

where the a_ξ, b_ξ are objects distinct from each other and outside $\mathbf{N}$. Say $G \subseteq X$ is open if

(α) $\{\xi : a_\xi \in G\}$ and $\{\xi : b_\xi \in G\}$ are open for the usual topology of ω_1;

(β) whenever $a_\xi \in G$ and $\xi > 0$, there is an $\eta < \xi$ such that $(A_\xi \setminus A_\eta)\setminus G$ is finite; if $a_0 \in G$, then $A_0 \setminus G$ is finite;
(γ) whenever $b_\xi \in G$ and $\xi > 0$, there is an $\eta < \xi$ such that $(B_\xi \setminus B_\eta)\setminus G$ is finite; if $b_0 \in G$, then$B_0 \setminus G$ is finite.

Then the following are true.

(i) X is Hausdorff. **P** $A_\xi \cup \{a_\eta : \eta \le \xi\}$ and

$$(A_\xi \setminus A_\eta) \cup \{a_\zeta : \eta < \zeta \le \xi\},$$

and the similar sets with the bs are open. **Q**

(ii) X is separable (because **N** is dense in X).

(iii) X is locally compact and locally countable (because $A_\xi \cup \{a_\eta : \eta \le \xi\}$ is compact and countable as well as open; proof is by induction on ξ).

(iv) X is countably paracompact (because it is the union of a collection of isolated points and a countably compact set).

(v) X is pseudonormal (because every countable closed subset of X is included in a countable open-and-closed set.)

(vi) (The point:) X is not normal. **P** $E = \{a_\xi : \xi < \omega_1\}$ and $F = \{b_\xi : \xi < \omega_1\}$ are disjoint closed sets. But if $G \subseteq X$ is any open set, then by 21L*b* either there is a first ξ such that $A_\xi \setminus G$ is infinite, so that $a_\xi \notin G$ and $E \nsubseteq G$, or there is a first ξ such that $G \cap B_\xi$ is infinite, so that $b_\xi \in \bar{G}$ and F meets $\bar{G}$. **Q**

25L Sources

Visible steps forward were made in PRZYMUSIŃSKI 73 (25M*h*), PRZYMUSIŃSKI & TALL 74 (25I), FLEISSNER 75 (25G), WHITE 75 (25J*c*(viii)), ALSTER & PRZYMUSIŃSKI 76 (25D, 25F, 25I), PRZYMUSIŃSKI 77 (25B, 25D), TALL 77*b* (25I), JUHÁSZ & WEISS 78 (25A, 25C*d*, 25G, 25I), TALL 78 (25E, 25J) and PRZYMUSIŃSKI 81 (proofs of 24D–E). The original observation that if $\mathfrak{p} > \omega_1$ there is a non-metrizable separable normal Moore space, constructed as in 25M*k*, was made by J. Silver [TALL 77*a*]. 25K comes from DOUWEN 77*b*.

25M Exercises

(*a*) Let X be a separable metrizable space. If X is expressible as the union of fewer than $\mathfrak{p}$ zero-dimensional compact sets, then any disjoint closed sets $E, F \subseteq X$ can be separated by open-and-closed sets. [Let $\mathscr{K}$ be the cover of X by zero-dimensional compact sets. Let $\mathscr{U}^*$ be a countable base for the topology of X closed under finite unions. Set

$$P = \{(G,H): G, H \in \mathcal{U}^*, \overline{G \cup E} \cap \overline{H \cup F} = \emptyset\},$$

$$Q_K = \{(G,H): (G,H) \in P, K \subseteq G \cup H\} \; \forall K \in \mathcal{K}.$$

See BOOTH 71.]

(*b*) There is a locally compact separable Moore space, of cardinal ω_1, which is not normal. [Use 21N*k*.]

(*c*) Let X be a locally compact normal Moore spaee with $d(X) \leq \mathfrak{c}$. Then X is cometrizable. [Use A4P*c*.]

(*d*) Let X be a normal Moore space with $d(X) < \mathfrak{p}$. Then $w(X) < \mathfrak{c}$. [Use A4P*b*. See ALSTER & PRZYMUSIŃSKI 76, Corollary 1.]

(*e*) $[\mathfrak{p} = \mathfrak{c} > \omega_1]$ Let X be a locally compact normal Moore space. (i) If $A \subseteq X$ and $\#(A) < \mathfrak{c}$ then there is an open-and-closed set $E \supseteq A$ with $w(E) < \mathfrak{c}$. [Use 25M*d*.] (ii) If X is connected then $w(X) < \mathfrak{c}$. [See JUHÁSZ & WEISS 78, 4.11.]

(*f*) $[\mathfrak{p} = \mathfrak{c}]$ Let X be a locally compact normal Moore space with $d(X) < \mathfrak{c}$. Then $X^{\mathbf{N}}$ is normal. [See ALSTER & PRZYMUSIŃSKI 76, Corollary 1.]

(*g*) $[\mathfrak{p} = \mathfrak{c} > \omega_1]$ Let X be a locally compact normal Moore space, $n \in \mathbf{N}$, and E and F disjoint closed sets in X^n such that $d(E) < \mathfrak{c}$. Then E and F can be separated by open sets. [Hint: use 25M*e*(i). See JUHÁSZ & WEISS 78, 4.10.]

(*h*) Give $\mathbf{R}$ the **right-facing Sorgenfrey topology** with basic open sets $[s, t[$. Then $\mathbf{R}$ is copolish. Let $X \subseteq \mathbf{R}$ be a set of cardinal less than $\mathfrak{p}$ such that $X = -X$. Then $X^{\mathbf{N}}$ is perfectly normal, but if X is uncountable then X^2 is not collectionwise Hausdorff nor paracompact. [See ALSTER & PRZYMUSIŃSKI 76.]

(*i*) **Cantor trees** (i) Let $W \subseteq \{0,1\}^{\mathbf{N}}$. Let T be

$$W \cup \bigcup_{n \in \mathbf{N}} \{0,1\}^n,$$

ordered by extension of functions. Then T is a Hausdorff tree. The fine tree topology on T is cometrizable, separable, locally countable, Moore. (ii) If $\#(W) < \mathfrak{p}$ then $T^{\mathbf{N}}$ is normal. (iii) Discuss the differences, if any, between this construction and that of 25F.

(*j*) **Bubble spaces** Let $Y \subseteq \mathbf{R} \times \{0\}$ and set $X = Y \cup (\mathbf{R} \times]0, \infty[)$. Give X the topology generated by the Euclidean topology on X and sets of the form

$$\{(s_0, 0)\} \cup \{(s,t): (s - s_0)^2 + (t - t_0)^2 < t_0^2\}$$

where $(s_0, 0) \in Y$ and $t_0 > 0$. Then X is a cometrizable, connected, locally connected Moore space. If $\#(Y) < \mathfrak{p}$ then $X^{\mathbf{N}}$ is normal.

(*k*) Let Z, Y, X, $\mathfrak{T}$ be as in 25F*a*.

(i) Show directly that if E and F are disjoint subsets of Y which are relatively F_σ in Y for the topology $\mathfrak{S}_0$ on Y induced by the usual topology of $\mathbf{R}^2$, then E and F can be separated by $\mathfrak{T}$-open sets.

(ii) Now suppose that $\mathfrak{S}$ is any topology on Y. Write $\mathfrak{T}_{\mathfrak{S}}$ for $\{G : G \in \mathfrak{T}, G \cap Y \in \mathfrak{S}\}$. If $\mathfrak{S}$ has a base consisting of $\mathfrak{S}_0$-F_σ sets, and $\mathfrak{S}$ is Hausdorff, then $\mathfrak{T}_{\mathfrak{S}}$ is Hausdorff. If $\mathfrak{S}$ has a base consisting of sets which are F_σ and G_δ for $\mathfrak{S}_0$, and $\mathfrak{S}$ is regular, then $\mathfrak{T}_{\mathfrak{S}}$ is regular.

(iii) If W is any topological space with $\#(W) < \mathfrak{p}$, then W can be embedded in a sequentially separable space $W \cup K$ in such a way that $W \cup K$ is Hausdorff if W is, regular if W is, normal if W is. [Hint: use 23B. See REED 76.]

(iv) If W is any topological space with $\#(W) \leq \mathfrak{p}$ and $w(W) \leq \mathfrak{p}$, then W can be embedded in a sequentially separable space $W \cup K$ in such a way that $W \cup K$ is Hausdorff if W is, regular if W is. [Hint: use 21H(iii) to embed W suitably into $\mathbf{R} \times \{0\}$. See PRZYMUSIŃSKI 80.]

(*l*) Let X be a cometrizable space. Suppose there is a $\mathfrak{p}$-Lusin subset Y of X. Then Y is hereditarily normal. [Refine the argument of 25E. See TALL 78, Theorem 12.]

(*m*) Let X be a Hausdorff tree with the fine tree topology and Z the set of its isolated points. Suppose that Z is an F_σ set in X; then every branch of X is countable. If, moreover, $\#(X) \leq \mathfrak{c}$, then X is cometrizable; if $\#(X) < \mathfrak{p}$ then X is a normal Q-space. [TODORČEVIĆ *a*. See 41N*c*.]

(*n*) Suppose that in 25I we take $S = \mathbf{R} \backslash \mathbf{Q}$. (i) If $A \subseteq X \backslash \{0\}$ and $\#(A) < \mathfrak{p}$ then A is nowhere dense. (ii) [$\mathfrak{p} = \mathfrak{c}$] If Y is a $\mathfrak{c}$-Lusin subset of $X \backslash \{\emptyset\}$, a set $A \subseteq Y$ is meagre iff it is nowhere dense.

(*o*) Let S be a separable metric space, X the set of totally bounded subsets of S with its Pixley–Roy topology. (i) The intersection of fewer than $\mathfrak{p}$ dense open sets in X is dense. [Use 43O*i*(iii).] (ii) [$\mathfrak{p} = \mathfrak{c}$] If S has no isolated points then $X \backslash \{\emptyset\}$ has a $\mathfrak{c}$-Lusin subset.

25N **Further results**

(*a*) Special Moore spaces, dependent on MA or a variant, are described in ALSTER & PRZYMUSIŃSKI 76, DAVIS, REED & WAGE 76, WAGE 76, EIDSWICK 76, PRZYMUSIŃSKI 77, FLEISSNER 78, DOUWEN & WAGE

79, REED 80, DOUWEN, LUTZER, PELANT & REED 80, NAVY 81 and PRZYMUSIŃSKI 81.

(*b*) Let $S \subseteq [0,1]$ be a set of cardinal less than $\mathfrak{p}$, and let X be $[0,1]$ split on S [definition: A7J]. Then X^2 is hereditarily normal. [NYIKOS 78.]

(*c*) [$\mathfrak{p} = \mathfrak{c}$] There is a first-countable Dowker space. [BELL 81*b*.]

(*d*) [$\mathfrak{m} = \mathfrak{c}$] If X is a separable Moore space and is the union of fewer than $\mathfrak{c}$ compact sets then X^2 is normal iff X is cometrizable. [PRZYMUSIŃSKI 77.]

(*e*) [$\mathfrak{p} = \mathfrak{c} > \omega_1$] There is a separable, sequentially compact, locally compact, locally countable Hausdorff space which is not normal. [VAUGHAN *a*; see also NYIKOS 81.]

(*f*) [$\mathfrak{p} = \mathfrak{c}$] Let $\langle X_n \rangle_{n \in \mathbb{N}}$ be a sequence of first-countable compact Hausdorff spaces. Then their box product is paracompact. [KUNEN 78; see WILLIAMS *a*, 5.7.]

25O **Problems**

(*a*) Do the results of 25N*c* and 25N*e* require special axioms?

(*b*) Is the product of two locally compact normal Moore spaces normal? [REED 75.]

Notes and comments

25A is taken from JUHÁSZ & WEISS 78. This is one of the cases where the most delicate part of the argument is over by the time we have stated the theorem and defined the partially ordered set to be used. It is only marginally more general than the corresponding result of ALSTER & PRZYMUSIŃSKI 76.

There are various ways of using auxiliary weaker topologies to help with the analysis of given topologies. The two that I use in this book are 'cometrizability' [25B] and 'sub-second-countability' [23E]. Both really amount to using the language of general topology to describe the behaviour of a countable family of sets (compare A7B*f*). 25D is a striking result; in its final form it is due to T.C. Przymusiński. 25M*f*–*h* are corollaries of 25D; 25M*d*–*e* are setting-up lemmas for use before 25M*f*; 25M*g* squeezes a little more out (but see 25O*b*). I give 25E for the sake of applications in 25I–J.

25F–I are some of the most important examples of Moore spaces; I see that more than half this section is taken up with examples. I suppose that they were all originally devised as counter-examples, but that now seems quite the wrong way to look at them; do the group theorists call the sporadic simple groups 'counter-examples'? It is the existence of these objects which makes the subject worth spending time on.

25F is one of the simplest members of a larger and well-known family; if you set $Y = \mathbf{R} \times \{0\}$ you may recognize one of the standard non-normal completely regular spaces. A version of this was the first known example of a non-metrizable normal Moore space. When it was found by J. Silver in 1967, it was already known that there is a non-metrizable separable normal Moore space iff there is an uncountable subset of **R** which is a Q-space [HEATH 64, or BENNETT & MCLAUGHLIN 76, Theorem 5.6; see 25Mk(i) and A4P for parts of the argument], which of course is impossible if $\mathfrak{c} < 2^{\omega_1}$. The normal Moore space problem ('is every normal Moore space metrizable') becomes at once "is it consistent to assume that every normal Moore space is metrizable"; it is very likely that the answer is 'yes' [NYIKOS 80, FLEISSNER 83]. Note that the continuum hypothesis is enough to produce a non-separable non-metrizable normal Moore space [FLEISSNER 82, FLEISSNER 83], even though it renders every separable normal Moore space metrizable. Observe also that the examples of Moore spaces in 25F–I are all constructed in ZFC, and that we use Martin's axiom only to prove that they are normal; it makes them better behaved, not worse.

Apart from being the original route to 25F, 25M*k* gives a method for embedding topological spaces into sequentially separable spaces; compare 26H. There is an essentially identical construction, less easy to draw but perhaps more elegant, in 25M*i*; 25M*j* is another way of expressing the same ideas. 25M*j* differs from the others in being locally connected rather than locally countable and locally compact. (A locally connected, locally compact normal Moore space is always metrizable; see 44O*l*).

25G is a descendant of an example of F.B. Jones. Like 25F it is locally compact; unlike 25F it is non-separable (and all its separable subspaces are metrizable). The Pixley–Roy construction [25H] gives rise to a wide variety of spaces; I investigate only two types, the compact-subset spaces and the finite-subset spaces based on separable metric spaces [25I]. (But compare 25M*o*, 31M*i*–*j*, 43O*i* and 43P*n*.) The finite-subset spaces Z of 25I are ccc (unlike 25G) and metacompact (unlike 25F). (A metacompact separable space has to be Lindelöf, so a metacompact separable regular space has to be paracompact, and a metacompact separable Moore space has to be metrizable.) Many other Moore spaces with particular combinations of properties have been constructed; I list some papers in 25N*a*.

An interesting hereditarily normal compact Hausdorff space is that of

25N*b* (which depends on 25M*h*). A much more complex normal space is that of 25N*c*. (It remains unclear whether in the absence of special axioms there are 'small' Dowker spaces. See also B3H.) With the compact-subset spaces of 25I, and the density topology on **R** [25J], we have a rather different phenomenon. These spaces are not normal, but they are ccc and copolish, so that (subject to Martin's axiom) the intersection of fewer than $\mathfrak{c}$ dense open subsets is always dense [43F*a*] and they have $\mathfrak{c}$-Lusin subsets [43P *f*(i)] which are perfectly normal [43P*f*(ii)]. The arguments of §43 assume $\mathfrak{m} = \mathfrak{c}$; but since the spaces we are dealing with here are not merely ccc but σ-centered [25I] or satisfy Knaster's condition [25J], $\mathfrak{p} = \mathfrak{c}$ is sufficient in the former case, and $\mathfrak{m}_K = \mathfrak{c}$ in the latter. Actually, in 25J, we see that a Lusin set for the density topology is just a Sierpiński set for Lebesgue measure; and for the sake of 25J*c*(vii)–(viii) I suggest taking Y to be an additive subgroup of **R**, which involves a slight refinement of its construction. The arguments of 25I suggest that it might be useful to know when Pixley-Roy spaces are ccc; this question is touched on in 31M*i*.

In 25K I give an example of a non-normal space which seems to form a natural boundary to the methods of this section; for example, in 24N*k*(ii) we cannot improve 'pseudonormal' to 'normal'. 25M*b* is a simpler version of the same idea. 25N*e* can be derived from either of these, using the technique of 24O*j*.

I have not seen a proof of 25N*d*; it gives a further link between cometrizability and normality in product spaces. Finally, I include two results where the proofs are unrelated to the work of this section; 25M*a* goes back to Bell's theorem, while 25N*f* uses 14B only.

26 $\beta\mathbf{N}$ and $\mathscr{P}\mathbf{N}/[\mathbf{N}]^{<\omega}$

Many of the combinatorial results of §21 on subsets of **N** have natural alternative expressions in terms of the corresponding open-and-closed subsets of $\beta\mathbf{N}\setminus\mathbf{N}$; see e.g. 26K*a*. The translation of $P(\mathfrak{p})$ itself [26C*a*] leads naturally to some interesting results on 'residual' sets [26C*b*, 26E]. Turning to $\mathscr{P}\mathbf{N}/[\mathbf{N}]^{<\omega}$, I give an important theorem on its subalgebras [26G] with corollaries involving separable spaces [26H] and continuous images of $\beta\mathbf{N}\setminus\mathbf{N}$ [26I].

26A Stone–Cech compactifications

Except in a few exercises and 'further results', I use this concept only for discrete spaces. Let X be a set. Let βX be the set of all ultrafilters on X. For $A \subseteq X$ write

$$\hat{A} = \{\mathscr{F} : \mathscr{F} \in \beta X, A \in \mathscr{F}\}.$$

Then $\{\hat{A}: A \subseteq X\}$ is a topology base on βX; give βX the topology it generates. βX is compact, Hausdorff, extremally disconnected; we can identify it with the Stone space of $\mathscr{P}X$. If Z is any compact Hausdorff space and $f: X \to Z$ is any function, there is a unique continuous function $g: \beta X \to Z$ such that $g(\mathscr{F}_x) = f(x)$ for every $x \in X$, where $\mathscr{F}_x$ is the principal ultrafilter on X generated by $\{x\}$ [ENGELKING 77, §3.6]. If X is infinite, then $\#(\beta X) = 2^\kappa$ where $\kappa = 2^{\#(X)}$ [COMFORT & NEGREPONTIS 74, 7.4].

There is a canonical bijection between open sets $G \subseteq \beta X$ and ideals $\mathscr{I}$ of $\mathscr{P}X$ given by

$$G \leftrightarrow \mathscr{I}_G = \{A: A \subseteq X, \hat{A} \subseteq G\}.$$

Similarly, there is a canonical bijection between non-empty closed sets of βX and filters on X given by

$$F \leftrightarrow \{A: A \subseteq X, \hat{A} \supseteq F\}.$$

If $G \subseteq \beta X$ is open, then $A \mapsto \hat{A} \setminus G$ is a surjective Boolean homomorphism from $\mathscr{P}X$ onto the algebra of open-and-closed subsets of $\beta X \setminus G$, and has kernel $\mathscr{I}_G$. Thus $\beta X \setminus G$ can be identified with the Stone space of the quotient Boolean algebra $\mathscr{P}X/\mathscr{I}_G$. Taking G to be the set $\tilde{X} = \{\mathscr{F}_x : x \in X\}$ of isolated points of βX, we see that $\beta X \setminus X$ is homeomorphic to the Stone space of $\mathscr{P}X/[X]^{<\omega}$.

Observe that the set of uniform ultrafilters on X, being

$$\beta X \setminus \bigcup \{\hat{A}: A \subseteq X,\ \#(A) < \#(X)\}$$

is always a closed subset of βX.

In this context I normally identify x with $\mathscr{F}_x$, so that the distinction between X and $\tilde{X}$ is erased, and $\beta X \setminus X$ becomes the zero-dimensional compact Hausdorff space of non-principal ultrafilters on X. $\hat{A}$ becomes the closure of A in βX, and $\hat{A} \setminus A$ the set of accumulation points of A in βX.

(For Stone–Cech compactifications of other completely regular Hausdorff spaces, see ENGELKING 77, §3.6.)

26B Definition

Let X be a topological space, κ a cardinal. I will say that a set $A \subseteq X$ is **κ-residual** if $X \setminus A$ can be covered by κ or fewer nowhere-dense sets. (Thus a comeagre set is an ω-residual set.)

26C Theorem

(*a*) Let $\mathscr{G}$ be a non-empty family of open sets in $\beta\mathbf{N} \setminus \mathbf{N}$ with $\#(\mathscr{G}) < \mathfrak{p}$, and suppose that $\bigcap \mathscr{G} \neq \emptyset$. Then $\operatorname{int}(\bigcap \mathscr{G}) \neq \emptyset$.

(*b*) If E is a $\mathfrak{p}$-residual set in $\beta\mathbf{N}\backslash\mathbf{N}$, then E is dense in $\beta\mathbf{N}\backslash\mathbf{N}$ and $\#(E)\geq 2^{\mathfrak{p}}$.

Proof (*a*) For $I\subseteq\mathbf{N}$ write I^* for the set of non-principal ultrafilters on $\mathbf{N}$ containing I, i.e. $\bar{I}\backslash I$, where $\bar{I}$ is the closure of I in $\beta\mathbf{N}$.

Fix on any $x\in\bigcap\mathscr{G}$. For each $G\in\mathscr{G}$ there is an $I_G\subseteq\mathbf{N}$ such that $x\in I_G^*\subseteq G$. If $\mathscr{H}\subseteq\mathscr{G}$ is a non-empty finite set, then $x\in\bigcap_{G\in\mathscr{H}}I_G^*=(\bigcap_{G\in\mathscr{H}}I_G)^*$, so $\bigcap_{G\in\mathscr{H}}I_G$ is infinite. By $P(\mathfrak{p})$, there is an infinite $I\subseteq\mathbf{N}$ such that $I\backslash I_G$ is finite for every $G\in\mathscr{G}$. Now $\varnothing\neq I^*\subseteq\bigcap_{G\in\mathscr{G}}I_G^*\subseteq\bigcap\mathscr{G}$. So $\operatorname{int}(\bigcap\mathscr{G})\neq\varnothing$.

(*b*) Take a family $\mathscr{G}$ of dense open subsets of $\beta\mathbf{N}\backslash\mathbf{N}$ such that $\#(\mathscr{G})\leq\mathfrak{p}$ and $\bigcap\mathscr{G}\subseteq E$. Let $\langle G_\xi\rangle_{\xi<\mathfrak{p}}$ be an indexed family running over $\mathscr{G}$. For each open $G\subseteq\beta\mathbf{N}\backslash\mathbf{N}$ and $\xi<\mathfrak{p}$, choose open sets $\psi_0^\xi(G)$, $\psi_1^\xi(G)\subseteq\beta\mathbf{N}\backslash\mathbf{N}$ such that

$$\overline{\psi_0^\xi(G)}\cap\overline{\psi_1^\xi(G)}=\varnothing,\quad \overline{\psi_0^\xi(G)}\cup\overline{\psi_1^\xi(G)}\subseteq G\cap G_\xi,$$

and if $G\neq\varnothing$ then neither $\psi_0^\xi(G)$ nor $\psi_1^\xi(G)$ is empty (here using the facts that G_ξ is dense and $\beta\mathbf{N}\backslash\mathbf{N}$ is regular, Hausdorff and has no isolated points.)

Start from any non-empty open set $H\subseteq\beta\mathbf{N}\backslash\mathbf{N}$. For each $x\in\{0,1\}^{\mathfrak{p}}$ choose inductively open sets H_ξ^x, for $\xi<\kappa$, as follows. $H_0^x=H$. Given H_ξ^x, set $H_{\xi+1}^x=\psi_{x(\xi)}^\xi(H_\xi^x)$. Given $\langle H_\eta^x\rangle_{\eta<\xi}$, where ξ is a limit ordinal and $0<\xi<\mathfrak{p}$, set $H_\xi^x=\operatorname{int}(\bigcap_{<\xi}H_\eta^x)$.

Then we see by induction on ξ that $H_\xi^x\subseteq H_\eta^x$ whenever $\eta<\xi$. It follows that no H_ξ^x is empty; since if ξ is a limit ordinal and $H_\eta^x\neq\varnothing$ for $\eta<\xi$, then $\bigcap_{\eta<\xi}H_\eta^x=\bigcap_{\eta<\xi}\overline{H_\eta^x}\neq\varnothing$ (because $\beta\mathbf{N}\backslash\mathbf{N}$ is compact), and accordingly $H_\xi^x=\operatorname{int}(\bigcap_{\eta<\xi}H_\eta^x)\neq\varnothing$ by (*a*) above. Consequently $F_x=\bigcap_{\xi<\mathfrak{p}}\overline{H_\xi^x}$ is also non-empty. As $F_x\subseteq\overline{H_1^x}\subseteq H$ and $F_x\subseteq\overline{H_{\xi+1}^x}\subseteq G_\xi$ for every $\xi<\kappa$, $F_x\subseteq E\cap H$.

Observe next that if $x\restriction\xi=y\restriction\xi$ and $x(\xi)\neq y(\xi)$, then $H_\xi^x=H_\xi^y$ and $\overline{H_{\xi+1}^x}\cap\overline{H_{\xi+1}^y}=\varnothing$. Accordingly, if x and y are distinct members of $\{0,1\}^{\mathfrak{p}}$, $F_x\cap F_y=\varnothing$. So $\#(E\cap H)\geq\#(\{0,1\}^{\mathfrak{p}})=2^{\mathfrak{p}}$. As H is arbitrary, E is dense and $\#(E)\geq 2^{\mathfrak{p}}$.

26D **Corollary**

(*a*) $\beta\mathbf{N}\backslash\mathbf{N}$ cannot be covered by $\mathfrak{p}$ nowhere-dense sets.

(*b*) The union of fewer than $\mathfrak{p}$ nowhere-dense sets in $\beta\mathbf{N}\backslash\mathbf{N}$ is nowhere dense.

(*c*) If $x\in\beta\mathbf{N}\backslash\mathbf{N}$ then $\chi(x,\beta\mathbf{N})=\chi(x,\beta\mathbf{N}\backslash\mathbf{N})\geq\mathfrak{p}$.

Proof (*a*) is immediate from 26C*b*. (*b*) By 26C*b*, the intersection of fewer than $\mathfrak{p}$ dense open sets is dense; so by 26C*a* it has dense interior. Now take complements. (*c*) Use 26C*a* or 21B*a*.

26E **Theorem**

(*a*) The set of Ramsey ultrafilters on $\mathbf{N}$ is $\mathfrak{c}$-residual in $\beta\mathbf{N}\backslash\mathbf{N}$.

(*b*) The set of $p(\mathfrak{p})$-point ultrafilters on $\mathbf{N}$ is $\mathfrak{c}$-residual in $\beta\mathbf{N}\backslash\mathbf{N}$.

(*c*) The set of $(\omega, \mathfrak{p})$-saturating ultrafilters on $\mathbf{N}$ is $\mathfrak{c}$-residual in $\beta\mathbf{N}\backslash\mathbf{N}$.

(*d*) $[\mathfrak{p} = \mathfrak{c}]$ There are $2^{\mathfrak{c}}$ $(\omega, \mathfrak{c})$-saturating Ramsey $p(\mathfrak{c})$-point ultrafilters on $\mathbf{N}$.

Proof As in 26C, write I^* for the set of non-principal ultrafilters containing I, for each $I \subseteq \mathbf{N}$.

(*a*) For each $r \in \mathbf{N}$ and finite partition $\mathscr{S}$ of $[\mathbf{N}]^r$ set

$$G_{\mathscr{S}} = \bigcup\{I^* : I \subseteq \mathbf{N}, \exists S \in \mathscr{S}, [I]^r \subseteq S\}.$$

Then $G_{\mathscr{S}}$ is open in $\beta\mathbf{N}\backslash\mathbf{N}$. If $J \in [\mathbf{N}]^\omega$ then $\mathscr{S}$ is a finite cover of $[J]^r$, so by Ramsey's theorem [A2J] there is an infinite $I \subseteq J$ and an $S \in \mathscr{S}$ such that $[I]^r \subseteq S$. Now $\varnothing \neq I^* \subseteq J^* \cap G_{\mathscr{S}}$. As J is arbitrary, $G_{\mathscr{S}}$ is dense.

So we see that

$$R = \bigcap\{G_{\mathscr{S}} : \exists r \in \mathbf{N}, \mathscr{S} \text{ is a finite partition of } [\mathbf{N}]^r\}$$

is $\mathfrak{c}$-residual. But R is precisely the set of Ramsey ultrafilters on $\mathbf{N}$.

(*b*) For each $\mathscr{A} \subseteq \mathscr{P}\mathbf{N}$ set

$$U_{\mathscr{A}} = (\beta\mathbf{N}\backslash\mathbf{N})\backslash\partial(\textstyle\bigcap_{A \in \mathscr{A}} A^*),$$

where $\partial(\bigcap_{A \in \mathscr{A}} A^*)$ is the boundary of $\bigcap_{A \in \mathscr{A}} A^*$ in $\beta\mathbf{N}\backslash\mathbf{N}$. Each $U_{\mathscr{A}}$ is a dense open set in $\beta\mathbf{N}\backslash\mathbf{N}$, and $\bigcap\{U_{\mathscr{A}} : \mathscr{A} \in [\mathscr{P}\mathbf{N}]^{<\mathfrak{p}}\}$ is the set of $p(\mathfrak{p})$-point ultrafilters. As $\#([\mathfrak{c}]^{<\mathfrak{p}}) = \mathfrak{c}$ [21D*a*], this set is $\mathfrak{c}$-residual.

(*c*) For each $\mathscr{B} \subseteq \mathscr{P}(\mathbf{N} \times \mathbf{N})$ with $\#(\mathscr{B}) < \mathfrak{p}$, write

$$\begin{aligned} H_{\mathscr{B}} = \bigcup\{I^* : I \subseteq \mathbf{N}, \; & \textit{either} \; \exists \text{ finite } \mathscr{B}_0 \subseteq \mathscr{B} \text{ such that} \\ & I \cap \pi_1[\textstyle\bigcap \mathscr{B}_0] = \varnothing \; \textit{or} \; \exists f : I \to \mathbf{N} \text{ such that} \\ & \Gamma(f)\backslash B \text{ is finite } \forall B \in \mathscr{B}\} \end{aligned}$$

where $\Gamma(f)$ is the graph of f. Then $H_{\mathscr{B}}$ is open. But also $H_{\mathscr{B}}$ is dense. **P** Let $J \in [\mathbf{N}]^\omega$. If there is a finite $\mathscr{B}_0 \subseteq \mathscr{B}$ such that $I_0 = J\backslash\pi_1[\bigcap\mathscr{B}_0]$ is infinite, then $\varnothing \neq I_0^* \subseteq J^* \cap H_{\mathscr{B}}$. Otherwise, $J \cap \pi_1[\bigcap\mathscr{B}_0]$

is infinite for every finite $\mathscr{B}_0 \subseteq \mathscr{B}$. Apply $P(\mathfrak{p})$ to

$$\mathscr{B} \cup \{(J \backslash n) \times \mathbf{N} : n \in \mathbf{N}\}$$

to find an $F \subseteq \mathbf{N} \times \mathbf{N}$ such that $F \backslash B$ is finite for every $B \in \mathscr{B}$ and $I_1 = J \cap \pi_1[F]$ is infinite. Now there is a function $f: I_1 \rightarrow \mathbf{N}$ such that $\Gamma(f) \subseteq F$. So I_1 satisfies the second alternative in the definition of $H_{\mathscr{B}}$ and $\varnothing \neq I_1^* \subseteq J^* \cap H_{\mathscr{B}}$. Thus in either case, J^* meets $H_{\mathscr{B}}$. As J is arbitrary, $H_{\mathscr{B}}$ is dense. **Q**

Now

$$S = \bigcap \{H_{\mathscr{B}} : \mathscr{B} \in [\mathscr{P}(\mathbf{N} \times \mathbf{N})]^{<\mathfrak{p}}\}$$

is $\mathfrak{c}$-residual (because $\#(\mathscr{P}(\mathbf{N} \times \mathbf{N})]^{<\mathfrak{p}}) = \#([\mathfrak{c}]^{<\mathfrak{p}} = \mathfrak{c}$, by 21D*a*); and every member of S is $(\omega, \mathfrak{p})$-saturating.

(*d*) From (*a*)–(*c*), the set of $(\omega, \mathfrak{c})$-saturating Ramsey $p(\mathfrak{c})$-point ultrafilters is $\mathfrak{c}$-residual, being the intersection of three $\mathfrak{c}$-residual sets; so by 26C*b* it has cardinal greater than or equal to $2^{\mathfrak{c}}$. As $\#(\beta\mathbf{N}) = 2^{\mathfrak{c}}$, we have equality.

26F **Proposition**

Any maximal free set in the Boolean algebra $\mathscr{P}\mathbf{N}/[\mathbf{N}]^{<\omega}$ has cardinal $\geq \mathfrak{p}$.

Proof Let $A \subseteq \mathscr{P}\mathbf{N}/[\mathbf{N}]^{<\omega}$ be a free set of cardinal less than $\mathfrak{p}$. For each $a \in A$ choose a representative $I_a \subseteq \mathbf{N}$. Then for any disjoint finite $B, C \subseteq A$, $\inf B \nsubseteq \sup C$ in $\mathscr{P}\mathbf{N}/[\mathbf{N}]^{<\omega}$, so that

$$J(B, C) = \bigcap_{b \in B} I_b \backslash \bigcup_{c \in C} I_c$$

is infinite. By 21A, with $\mathscr{A} = \mathscr{C} = \varnothing$,

$$\mathscr{B} = \mathscr{D} = \{J(B, C) : B \in [A]^{<\omega} \backslash \{\varnothing\}, C \in [A \backslash B]^{<\omega}\},$$

there is an $I \subseteq \mathbf{N}$ such that $\{I^\cdot\} \cup A$ is a larger free set in $\mathscr{P}\mathbf{N}/[\mathbf{N}]^{<\omega}$, so that A was not maximal.

26G **Theorem**

Let $\mathfrak{A}$ be a Boolean algebra of cardinal less than $\mathfrak{p}$. Then $\mathfrak{A}$ can be embedded in $\mathscr{P}\mathbf{N}/[\mathbf{N}]^{<\omega}$.

Proof (*a*) If $\mathfrak{A}$ is finite, this is easy; let us suppose that $\mathfrak{A}$ is infinite. By 12K, $\{0, 1\}^{\mathfrak{A}}$ is separable; let $K \subseteq \{0, 1\}^{\mathfrak{A}}$ be a countable dense set. Let Z be the Stone space of $\mathfrak{A}$; regard Z as the set of Boolean homomorphisms from $\mathfrak{A}$ to $\mathbf{Z}_2$, so that Z is a closed subset of $\{0, 1\}^{\mathfrak{A}}$. Let $\mathscr{U}$ be the family of open-and-closed sets in $\{0, 1\}^{\mathfrak{A}}$; then $\#(\mathscr{U}) < \mathfrak{p}$.

(*b*) Set

$$\mathscr{C} = \{K \cap U : U \in \mathscr{U}, U \cap Z = \varnothing\},$$
$$\mathscr{D} = \{K \cap U : U \in \mathscr{U}, U \cap Z \neq \varnothing\}.$$

If $D \in \mathscr{D}$, $\mathscr{C}_0 \subseteq \mathscr{C}$ is finite then $D \setminus \bigcup \mathscr{C}_0$ belongs to $\mathscr{D}$ and is dense in some non-empty open set of $\{0,1\}^{\mathfrak{A}}$, so is infinite. By 21A there is an infinite $I \subseteq K$ such that $I \cap C$ is finite for $C \in \mathscr{C}$, $I \cap D$ is infinite for $D \in \mathscr{D}$.

(*c*) Enumerate I as $\langle \varphi_n \rangle_{n \in \mathbb{N}}$ and define $\psi : \mathfrak{A} \to \mathscr{P}\mathbb{N}$, $\theta : \mathfrak{A} \to \mathscr{P}\mathbb{N}/[\mathbb{N}]^{<\omega}$ by

$$\psi a = \{n : \varphi_n(a) = 1\}, \quad \theta a = (\psi a)^{\cdot}.$$

Now we see that if $a, b \in \mathfrak{A}$ then

$$\{\varphi : \varphi \in \{0,1\}^{\mathfrak{A}}, \varphi(a \cup b) \neq \max(\varphi a, \varphi b)\}$$

belongs to $\mathscr{U}$ and does not meet Z, so has finite intersection with I; accordingly

$$\{n : n \in \psi(a \cup b) \triangle (\psi a \cup \psi b)\}$$

is finite, and $\theta(a \cup b) = \theta a \cup \theta b$. Similarly, for any $a \in \mathfrak{A}$,

$$\{\varphi : \varphi(1 \setminus a) \neq 1 - \varphi a\}$$

does not meet Z, and $\theta(1 \setminus a) = 1 \setminus \theta a$. So θ is a Boolean homomorphism. Finally, if $a \neq 0$, then

$$\{\varphi : \varphi a = 1\}$$

does meet Z, so has infinite intersection with I; so ψa is infinite and $\theta a \neq 0$.

(*d*) Thus θ is an injective Boolean homomorphism, as required.

26H **Theorem**

Let X be a non-empty topological space of weight less than $\mathfrak{p}$. Then X can be embedded as a subspace of $X \cup K$, where $\#(K) = \omega$, in such a way that the points of K are isolated in $X \cup K$; every point of X is a limit of a sequence in K; $w(X \cup K) = \max(\omega, w(X))$; and moreover

(i) $X \cup K$ is Hausdorff if X is;

(ii) $X \cup K$ is regular if X is;

(iii) $X \cup K$ is completely regular if X is;

(iv) $X \cup K$ is compact if X is;

(v) $X \cup K$ is first-countable if X is.

Proof (*a*) We can assume that $X\cap\mathbf{N}=\varnothing$, and take $K=\mathbf{N}$.

Let $\mathscr{U}$ be a base for the topology of X, containing $\varnothing$. We can take $\mathscr{U}$ to be such that $X\in\mathscr{U}$ and, when ever $U, V\in\mathscr{U}$ are such that there is a family $\langle E_s\rangle_{s\in\mathbf{Q}\cap[0,1]}$ of open sets in X with $E_1\supseteq U$, $\bar{E}_s\subseteq E_t$ whenever $s>t$, and $E_0\subseteq V$, then there is such a family inside $\mathscr{U}$. We can also suppose that $w(X)\le \#(\mathscr{U})\le\max(\omega, w(X))$.

Let $\mathfrak{A}$ be the algebra of subsets of X generated by $\mathscr{U}$; then $\#(\mathfrak{A})<\mathfrak{p}$. By 26G, there is an injective Boolean homomorphism $\theta:\mathfrak{A}\to\mathscr{P}\mathbf{N}/[\mathbf{N}]^{<\omega}$. Give $X\cup\mathbf{N}$ the topology generated by the base

$$\mathscr{U}^*=\{\{n\}:n\in\mathbf{N}\}\cup\{U\cup I:U\in\mathscr{U}, I\subseteq\mathbf{N}, I^\cdot=\theta U\}.$$

Then X is embedded as a subspace of $X\cup\mathbf{N}$. As $\{n\}\in\mathscr{U}^*$ for each $n\in\mathbf{N}$, the points of $\mathbf{N}$ are isolated in $X\cup\mathbf{N}$. Next, $w(X\cup\mathbf{N})\le\#(\mathscr{U}^*)\le\max(\omega, \#(\mathscr{U}))=\max(\omega, w(X))\le w(X\cup\mathbf{N})$, and $w(X\cup\mathbf{N})<\mathfrak{p}$. Since every non-empty set in $\mathscr{U}^*$ meets $\mathbf{N}$ (because $\theta U\ne 0$ if $U\ne\varnothing$), $\mathbf{N}$ is dense in $X\cup\mathbf{N}$; as $\chi(x, X\cup\mathbf{N})\le w(X\cup\mathbf{N})<\mathfrak{p}$ for every $x\in X$, every point of X is a limit of a sequence in $\mathbf{N}$ [24A].

(*b*) If $U\in\mathscr{U}$, $I^\cdot=\theta U$ then $\bar{I}\subseteq\bar{U}\cup I$ in $X\cup\mathbf{N}$. **P** As each point of $\mathbf{N}$ is isolated, $\bar{I}\subseteq X\cup I$. But also, if $y\in X\setminus\bar{U}$, there is a $V\in\mathscr{U}$ such that $y\in V\subseteq X\setminus U$; now $\theta U\cap\theta V=0$, so there is a $J\subseteq\mathbf{N}\setminus I$ such that $J^\cdot=\theta V$, and $y\in V\cup J$ which is open in $X\cup\mathbf{N}$ and does not meet I. So $\bar{I}\subseteq\bar{U}\cup\mathbf{N}$ also, and $\bar{I}\subseteq\bar{U}\cup I$. **Q** (Note that because the points of $\mathbf{N}$ are isolated in $X\cup\mathbf{N}$, the closure of a subset of X is the same whether taken in X or in $X\cup\mathbf{N}$.)

(*c*) If X is Hausdorff, so is $X\cup\mathbf{N}$. **P** Let x, y be distinct points in $X\cup\mathbf{N}$. If one of them belongs to $\mathbf{N}$, then of course they can be separated by open sets. Otherwise, let $U\in\mathscr{U}$ be such that $x\in U$, $y\in X\setminus\bar{U}$. Take $I\subseteq\mathbf{N}$ such that $I^\cdot=\theta U$; then $x\in U\cup I\in\mathscr{U}^*$ and $\overline{U\cup I}=\bar{U}\cup I$ by (*b*) above, so that $y\notin\overline{U\cup I}$. **Q**

(*d*) If X is regular, so is $X\cup\mathbf{N}$. **P** Take $G\subseteq X\cup\mathbf{N}$ open, and $x\in G$. If $x\in\mathbf{N}$ then $x\subset\{x\}\subseteq G$ and $\{x\}$ is open-and-closed. Otherwise, there are $U\in\mathscr{U}$ and $I\subseteq\mathbf{N}$ such that $I^\cdot=\theta U$ and $x\in U\cup I\subseteq G$. Let $V\in\mathscr{U}$ be such that $x\in V\subseteq\bar{V}\subseteq U$; let $J\subseteq I$ be such that $J^\cdot=\theta V$. Then $x\in V\cup J\subseteq\overline{V\cup J}=\bar{V}\cup J\subseteq U\cup I$. **Q**

(*e*) If X is completely regular, so is $X\cup\mathbf{N}$. **P** Take an open $G\subseteq X\cup\mathbf{N}$ and $x\in G$. If $x\in\mathbf{N}$ then of course x and $(X\cup\mathbf{N})\setminus G$ can be separated by a continuous function. If $x\in X$, let $V\in\mathscr{U}$ and $I\subseteq\mathbf{N}$ be such that $I^\cdot=\theta V$ and $x\in V\cup I\subseteq G$. As X is completely regular, there is a continuous function $g:X\to\mathbf{R}$ such that $g(x)>1$ and $g(y)=0$ for $y\in X\setminus V$. Let $U\in\mathscr{U}$ be such that $x\in U$ and $g(y)>1$ for $y\in U$. Writing $E_s=\{y:g(y)>s\}$, we see that $\langle E_s\rangle_{s\in\mathbf{Q}\cap[0,1]}$ satisfy the conditions of (a) above with respect to U, V. So there is a similar family $\langle U_s\rangle_{s\in\mathbf{Q}\cap[0,1]}$ in $\mathscr{U}$. Now, enumerating $\mathbf{Q}\cap[0,1]$ as $\langle s(n)\rangle_{n\in\mathbf{N}}$, we can choose $\langle I_n\rangle_{n\in\mathbf{N}}$ inductively so that

$$I_n^{\cdot}=\theta(U_{s(n)}),\ I_n\subseteq I_m \text{ whenever } s(n)\geq s(m),\ I_n\subseteq I\ \forall n\in\mathbf{N}.$$

In this case we get

$$\overline{U_{s(n)}\cup I_n}\subseteq U_{s(m)}\cup I_m \text{ if } s(n)>s(m).$$

So we can use the $U_{s(n)}\cup I_n$ to construct a continuous function $f:X\cup\mathbf{N}\to[0,1]$ such that $f(x)=1$ and $f(y)=0$ for $y\notin V\cup I$. **Q**

(f) $X\cup\mathbf{N}$ is compact if X is. **P** Let $\mathscr{F}$ be an ultrafilter on $X\cup\mathbf{N}$. If $X\in\mathscr{F}$, then $\mathscr{F}$ has a limit in X. If $\{n\}\in\mathscr{F}$ for some $n\in\mathbf{N}$, then $\mathscr{F}\to n$. Otherwise, consider

$$\mathscr{E}=\{E:E\in\mathfrak{A},\exists I\subseteq\mathbf{N},I\in\mathscr{F},I^{\cdot}=\theta E\}.$$

Because no finite set belongs to $\mathscr{F}$, $\mathscr{E}$ has the finite intersection property; so there is an $x\in\bigcap_{E\in\mathscr{E}}\bar{E}\subseteq X$. Let $U\in\mathscr{U}$ be such that $x\in U$, and let $I\subseteq\mathbf{N}$ be such that $I^{\cdot}=\theta U$. As $X\setminus U\notin\mathscr{E}$, $\mathbf{N}\setminus I\notin\mathscr{F}$ and $I\in\mathscr{F}$. As U and I are arbitrary, $\mathscr{F}\to x$. Thus every ultrafilter on $X\cup\mathbf{N}$ has a limit, and $X\cup\mathbf{N}$ is compact. **Q**

(g) $X\cup\mathbf{N}$ is first-countable if X is. **P** If $x\in X$ and $\mathscr{U}_0\subseteq\mathscr{U}$ is a countable base of neighbourhoods of x for the topology of X, then $\{U\cup I:U\in\mathscr{U}_0,\ I^{\cdot}=\theta U\}$ is a countable base of neighbourhoods of x in $X\cup\mathbf{N}$. **Q**

26I Corollary

Let X be a non-empty compact Hausdorff space with $w(X)<\mathfrak{p}$. Then X is a continuous image of $\beta\mathbf{N}\setminus\mathbf{N}$.

Proof Embed X in $X\cup K$ as in 26H. As $\#(K)=\omega$ there is a surjection $f:\mathbf{N}\to K$. Because $X\cup K$ is compact and Hausdorff, f extends to a continuous map $\hat{f}:\beta\mathbf{N}\to X\cup K$, and

$$\hat{f}[\beta\mathbf{N}]=\overline{f[\mathbf{N}]}=\bar{K}=X\cup K.$$

As the points of K are isolated in $X\cup K$, $\hat{f}^{-1}[K]=\mathbf{N}$ and $\hat{f}[\beta\mathbf{N}\setminus\mathbf{N}]=X$. So $\hat{f}\restriction\beta\mathbf{N}\setminus\mathbf{N}$ is the required surjection.

26J **Sources**

BOOTH 70 for the existence of Ramsey ultrafilters. ELLENTUCK & RUCKER 72 for the existence of $2^{\mathfrak{c}}$ $(\omega, \mathfrak{p})$-saturating ultrafilters. BLASS 73 for the existence of $2^{\mathfrak{c}}$ Ramsey ultrafilters. KUCIA & SZYMAŃSKI 76 for 26C, 26D*a*–*b* and 26E*a*. KUNEN 80 for 26G. DOUWEN & PRZYMUSIŃSKI 80 for the proof of 26G and 26H–I.

26K **Exercises**

(*a*) (i) Let $\mathscr{A}$, $\mathscr{B}$, $\mathscr{C}$, $\mathscr{D}$ be four families of open-and-closed sets in $\beta\mathbf{N}\setminus\mathbf{N}$ such that

$\bigcup\mathscr{A}\cap\bigcup\mathscr{C}=\emptyset$;

no member of $\mathscr{B}$ is covered by $\mathscr{A}$;

no member of $\mathscr{D}$ is covered by $\mathscr{C}$;

$\mathscr{A}$ is countable; $\mathscr{B}$, $\mathscr{C}$, $\mathscr{D}$ have cardinals less than $\mathfrak{p}$.

Then there is an open-and-closed set $I\subseteq\beta\mathbf{N}\setminus\mathbf{N}$ such that

$\bigcup\mathscr{A}\subseteq I$;

$B\not\subseteq I\ \forall B\in\mathscr{B}$;

$I\cap\bigcup\mathscr{C}=\emptyset$;

$D\cap I\neq\emptyset$ if $D\in\mathscr{D}$.

(ii) In particular, if $G\subseteq\beta\mathbf{N}\setminus\mathbf{N}$ is a cozero set, $H\subseteq\beta\mathbf{N}\setminus\mathbf{N}$ is an open set expressible as the union of fewer than $\mathfrak{p}$ closed sets, and $G\cap H=\emptyset$, then $\bar{G}\cap\bar{H}=\emptyset$.

(*b*) Say an open set $G\subseteq\beta\mathbf{N}\setminus\mathbf{N}$ is **MH** if whenever F is a zero set in $\beta\mathbf{N}\setminus\mathbf{N}$ and $F\not\subseteq G$, then $\operatorname{int}F\not\subseteq G$. (i) If $\kappa<\mathfrak{p}$ and $\langle G_\xi\rangle_{\xi<\kappa}$ is an increasing family of MH open sets, then $\bigcup_{\xi<\kappa}G_\xi$ is MH. [Use 26K*a*(ii).] (ii) Translate this into a statement about MH ideals in $\mathscr{P}$N [MATHIAS 77, 9.28].

(*c*) There is a disjoint family $\mathscr{A}$ of open-and-closed sets in $\beta\mathbf{N}\setminus\mathbf{N}$ such that $\#(\mathscr{A})=\omega_1$ and $\overline{\bigcup\mathscr{B}}\cap\overline{\bigcup\mathscr{C}}\neq\emptyset$ whenever $\mathscr{B}$ and $\mathscr{C}$ are uncountable subsets of $\mathscr{A}$. [See 21N*k*.]

(*d*) $[\mathfrak{p}=\mathfrak{c}]$ If $f:\beta\mathbf{N}\setminus\mathbf{N}\to\mathbf{R}$ is any function, there is a dense $A\subseteq\beta\mathbf{N}\setminus\mathbf{N}$ such that $f\restriction A$ is continuous. [Use 26D*a*. See TALL *a*, 3.7; SZYMAŃSKI 80*b*, Corollary 7.]

(*e*) $[\mathfrak{p}=\mathfrak{c}]$ Let $S\subseteq(\beta\mathbf{N}\setminus\mathbf{N})^2$ be an equivalence relation such that the equivalence classes of S are nowhere dense and $S[G]$ is open for every open $G\subseteq\beta\mathbf{N}\setminus\mathbf{N}$. (i) There are non-empty open sets G and H such that

$S[G] \cap S[H] = \emptyset$. [For otherwise $\bigcap_{U \in \mathscr{U}} S[U] \neq \emptyset$ for a base $\mathscr{U}$ of the topology, by 26C*b*.] (ii) S has $2^{\mathfrak{c}}$ equivalence classes. [Use the argument of 26C*b*. See KUCIA & SZYMAŃSKI 76, Theorems 6 and 7.]

(*f*) Let X be a partially ordered set of cardinal less than $\mathfrak{p}$. Then there is a function $f: X \to \mathscr{P}\mathbf{N}$ such that $f(x) \setminus f(y)$ is finite iff $x \leq y$. [Apply 26G to a suitable algebra of subsets of X. See KUNEN 80, Ex. 2.22.]

(*g*) In 26H we can add parts (vi) if X is zero-dimensional, so is $X \cup K$; (vii) if X is locally compact, so is $X \cup K$; (viii) if X is a Moore space, so is $X \cup K$; (ix) if X is Cech-complete, so is $X \cup K$; (x) if X is cometrizable, so is $X \cup K$; (xi) if X is copolish, so is $X \cup K$. [DOUWEN & PRZYMUSIŃSKI 80.]

26L Further results

(*a*) **Special types of ultrafilter on N.** Let $\mathscr{F}$ be a non-principal ultrafilter on $\mathbf{N}$. Say that $\mathscr{F}$ is **rare** if whenever $\langle I_n \rangle_{n \in \mathbf{N}}$ is a disjoint sequence in $[\mathbf{N}]^{<\omega}$ there is an $A \in \mathscr{F}$ such that $\#(A \cap I_n) \leq 1$ for every $n \in \mathbf{N}$. Say that $\mathscr{F}$ is **rapid** if for every $f \in \mathbf{N}^{\mathbf{N}}$ there is an $A \in \mathscr{F}$ such that $\#(A \cap f(n)) \leq n$ for every $n \in \mathbf{N}$. Say that $\mathscr{F}$ is **weakly Ramsey** if whenever $\mathscr{S}$ is a partition of $[\mathbf{N}]^2$ into three sets there are an $A \in \mathscr{F}$ and an $S \in \mathscr{S}$ such that $[A]^2 \cap S = \emptyset$. (Note: this conflicts with the terminology of COMFORT & NEGREPONTIS 74.) Write $\mathbf{N} \to (\mathscr{F}, k)^2$, where $k \geq 2$, if whenever $S \subseteq [\mathbf{N}]^2$ then either there is an $A \in \mathscr{F}$ such that $[A]^2 \subseteq S$ or there is a $B \in [\mathbf{N}]^k$ with $[B]^2 \cap S = \emptyset$. Observe that an ultrafilter is Ramsey iff it is a rare p-point ultrafilter [BOOTH 70]. Now [$\mathfrak{p} = \mathfrak{c}$] the following are true.

(i) There is a rapid $p(\mathfrak{c})$-point ultrafilter which is not rare. There is a $p(\mathfrak{c})$-point ultrafilter which is not rapid. [Cf. KUNEN 76.]

(ii) If $\kappa \leq \mathfrak{c}$ is a regular cardinal, there is a $p(\kappa)$-point ultrafilter which is not a $p(\kappa^+)$-point ultrafilter. [SZYMAŃSKI 77.]

(iii) If κ is regular and $\kappa^+ < \mathfrak{c}$, then the set of $p(\kappa)$-point ultrafilters which are not $p(\kappa^+)$-point ultrafilters is not homogeneous (under its topology inherited from $\beta\mathbf{N}$). [FRANKIEWICZ 77*a*.]

(iv) There is a Ramsey ultrafilter which is not a $p(\omega_2)$-point ultrafilter. [SOLOMON 77.]

(v) There is a weakly Ramsey ultrafilter which is not rare. There is a p-point ultrafilter $\mathscr{F}$ such that $\mathbf{N} \to (\mathscr{F}, k)^2$ for every $k \geq 2$ but $\mathscr{F}$ is not weakly Ramsey. If $k \geq 2$, there is a p-point ultrafilter $\mathscr{F}$ such that $\mathbf{N} \to (\mathscr{F}, k)^2$ but $\mathbf{N} \not\to (\mathscr{F}, k+1)^2$. If $k \geq 2$, there is a rare ultrafilter $\mathscr{F}$ such that $\mathbf{N} \to (\mathscr{F}, k)^2$ but $\mathbf{N} \not\to (\mathscr{F}, k+1)^2$. If $k \geq 3$, there is an ultrafilter $\mathscr{F}$ such that $\mathbf{N} \to (\mathscr{F}, k)^2$, $\mathbf{N} \not\to (\mathscr{F}, k+1)^2$, and $\mathscr{F}$ is neither rare nor a p-point ultrafilter. [BAUMGARTNER & TAYLOR 78.]

(*b*) **The Rudin–Keisler ordering** If $\mathscr{F}$ and $\mathscr{G}$ are ultrafilters on $\mathbf{N}$, say that $\mathscr{F} \leq_{\mathrm{RK}} \mathscr{G}$ if there is an $f:\mathbf{N}\to\mathbf{N}$ such that $\mathscr{F} = \{A: f^{-1}[A]\in\mathscr{G}\}$. Let $\mathfrak{F}$ be the set of equivalence classes in $\beta\mathbf{N}\setminus\mathbf{N}$ under the relation $\mathscr{F} \sim \mathscr{G}$ if $\mathscr{F} \leq_{\mathrm{RK}} \mathscr{G} \leq_{\mathrm{RK}} \mathscr{F}$. Let $\mathfrak{P} \subseteq \mathfrak{F}$ be the set of equivalence classes of p-point ultrafilters. Then $[\mathfrak{p} = \mathfrak{c}]$

(i) $\#(\mathfrak{P}) = 2^{\mathfrak{c}}$;

(ii) $\mathfrak{P}$ has no maximal element (for the induced partial order of $\mathfrak{F}$);

(iii) any increasing sequence in $\mathfrak{P}$ is bounded above in $\mathfrak{P}$;

(iv) there is an order-preserving embedding of $\mathbf{R}$ into $\mathfrak{P}$;

(v) there is a $p\in\mathfrak{P}$ such that $\{q: q\in\mathfrak{P}, q \leq p\}$ is not totally ordered.

[Blass 73.]

(*c*) (i) If $\mathfrak{A}$ is a Dedekind complete Boolean algebra and $\#(\mathfrak{A}) \leq \mathfrak{c}$, there is a surjective Boolean homomorphism from $\mathscr{P}\mathbf{N}/[\mathbf{N}]^{<\omega}$ onto $\mathfrak{A}$. (ii) If X is an extremally disconnected compact Hausdorff space and $w(X) \leq \mathfrak{c}$, then X can be embedded into $\beta\mathbf{N}\setminus\mathbf{N}$. (iii) $[\mathfrak{p} = \mathfrak{c}]$ If X is an extremally disconnected compact Hausdorff space and $w(X) \leq \mathfrak{c}$, then X can be embedded as a $p(\mathfrak{c})$-set in $\beta\mathbf{N}\setminus\mathbf{N}$. [Kunen 76.]

(*d*) $[\mathfrak{p} = \mathfrak{c}]$ Let X be a compact Hausdorff space and $\langle f_\xi \rangle_{\xi<\mathfrak{c}}$ a family of continuous open surjections from X onto $\beta\mathbf{N}$. Then there is an $x\in X$ such that, for every $\xi < \mathfrak{c}$, either $f_\xi(x)\in\mathbf{N}$ or $f_\xi(x)$ is a $p(\mathfrak{c})$-point ultrafilter. [Frankiewicz 81.]

(*e*) $[\mathfrak{p} = \mathfrak{c}]$ Let X be an extremally disconnected compact Hausdorff space with $w(X) = \mathfrak{c}$. Then there is a point in X which is not an accumulation point of any countable discrete subset of X. [Frankiewicz 81.]

(*f*) $[\mathfrak{p} = \mathfrak{c}]$ Let X be an extremally disconnected compact Hausdorff space of countable π-weight. Then there is a countable dense set $A \subseteq X$ such that no point of A is an accumulation point of any nowhere-dense subset of X of cardinal less than $\mathfrak{c}$. [Frankiewicz 81.]

(*g*) $[\mathfrak{p} = \mathfrak{c}]$ There is a non-empty countable set $A \subseteq \beta\mathbf{N}\setminus\mathbf{N}$, without isolated points, such that no point of A is an accumulation point of any discrete subset of $\beta\mathbf{N}\setminus\mathbf{N}$ with cardinal less than $\mathfrak{c}$. [See Kunen 76.]

(*h*) $[\mathfrak{p} = \mathfrak{c}]$ Suppose that $\omega < \kappa < \mathfrak{c}$ and that κ is regular. Let X be the compact Hausdorff space of uniform ultrafilters on κ. Then there is an $x\in X$ such that whenever $\mathscr{G}$ is a disjoint collection of open sets in $X\setminus\{x\}$ with $\#(\mathscr{G}) \leq \kappa$, and A is a set consisting of one point chosen from each member of $\mathscr{G}$, then $x\notin\bar{A}$. [Frankiewicz 81.]

(*i*) [$\mathfrak{p} > \omega_1$] Let X be the compact Hausdorff space of uniform ultrafilters on ω_1. Let $\langle F_n \rangle_{n \in \mathbf{N}}$ be a sequence of closed nowhere-dense subsets of X. If *either* each F_n is a G_δ set *or* $\langle F_n \rangle_{n \in \mathbf{N}}$ is disjoint, then $\bigcup_{n \in \mathbf{N}} F_n$ is nowhere dense. [SZYMAŃSKI 79.]

(*j*) [$\mathfrak{p} = \mathfrak{c}$] Let x be any point of $\beta\mathbf{N} \backslash \mathbf{N}$. (i) There is a dense $A \subseteq \beta\mathbf{N} \backslash \mathbf{N}$ such that $x \notin \bar{B}$ whenever $B \subseteq A$ and $\#(B) < \mathfrak{c}$. (iii) $(\beta\mathbf{N} \backslash \mathbf{N}) \backslash \{x\}$ is not normal. [MALYHIN & SAPIROVSKII 73.]

(*k*) [$\mathfrak{p} = \mathfrak{c}$] If $A \subseteq \beta\mathbf{N} \backslash \mathbf{N}$ is nowhere dense, there is a disjoint family $\mathscr{G}$ of open sets in $\beta\mathbf{N} \backslash \mathbf{N}$ such that $A \subseteq \bar{G} \backslash G$ for every $G \in \mathscr{G}$ and $\#(\mathscr{G}) = \mathfrak{c}$. [HECHLER 78.]

(*l*) There is a family $\mathscr{A}$ of open-and-closed subsets of $\beta\mathbf{N} \backslash \mathbf{N}$ such that $\#(\mathscr{A}) = \mathfrak{c}$ and whenever $\mathscr{B}$, $\mathscr{C}$ are disjoint subsets of $\mathscr{A}$ both of cardinal $< \mathfrak{p}$, then $\overline{\bigcup \mathscr{B}} \cap \overline{\bigcup \mathscr{C}} = \varnothing$.

(*m*) [$\mathfrak{p} = \mathfrak{c}$] There is a dense subset of $\beta\mathbf{N} \backslash \mathbf{N}$ homeomorphic to $\mathfrak{c}^{\mathfrak{c}}$ with its lexicographic ordering. [SIMON 78.]

(*n*) [$\mathfrak{p} = \mathfrak{c} > \omega_1$] There is a nowhere-dense closed $p(\omega_1)$-set in $\beta\mathbf{N} \backslash \mathbf{N}$ which is not a retract of $\beta\mathbf{N} \backslash \mathbf{N}$. [MILL 83, 2.8.1.]

(*o*) [$\mathfrak{p} > \omega_1$] If λ and κ are distinct infinite cardinals, then $\beta\lambda \backslash \lambda$ is not homeomorphic to $\beta\kappa \backslash \kappa$ [FRANKIEWICZ 77*b*, FRANKIEWICZ 78.]

(*p*) For other results associated with those above see MALYHIN 75*a*.

26M Problems

(*a*) [$\mathfrak{m} = \mathfrak{c} > \omega_1$] Is there a Ramsey $p(\mathfrak{c})$-point ultrafilter on $\mathbf{N}$ which is not $(\omega, \mathfrak{c})$-saturating?

(*b*) [$\mathfrak{m} = \mathfrak{c} > \omega_1$] Which Boolean algebras of cardinal $\mathfrak{c}$ can be embedded into $\mathscr{P}\mathbf{N}/[\mathbf{N}]^{<\omega}$? Which compact Hausdorff spaces of weight $\mathfrak{c}$ are continuous images of $\beta\mathbf{N} \backslash \mathbf{N}$?

(*c*) Do 26L*e*, 26L*k* and 26L*o* need special axioms?

(*d*) [$\mathfrak{m} = \mathfrak{c} > \omega_1$] Is there an infinite compact Hausdorff space X such that (i) there is no non-trivial convergent sequence in X, (ii) $\beta\mathbf{N}$ cannot be embedded into X?

Notes and comments

You will see that 26D*c*, 26K*a* and 26K*c* are nothing but reformulations of earlier results into terms of the topology of $\beta\mathbf{N} \backslash \mathbf{N}$. (One could say the same of 26C*a*.) I have taken the trouble to write them out because, while there is little or no new content to be found in this way, they awake different echoes. In 26K*b* I suggest a reverse translation to rediscover A.R.D. Mathias' concept of 'MH ideal'.

26C*b* is saying something new, though the idea is only a version of the Cech-Pospíšil theorem [A4J*b*]. We find ourselves saying that 'almost all' ultrafilters are Ramsey $P(\mathfrak{c})$-point ultrafilters [26*E*], even though there are familiar models with no Ramsey ultrafilters [Kunen 76, §5], and others with no *p*-point ultrafilters [Wimmers 82, Mills *a*]. I have written this out in terms of $(\omega, \mathfrak{p})$-saturating ultrafilters as well as $p(\mathfrak{p})$-point ultrafilters because the former seem to be closer to the natural limit of the argument of 26E*b*–*c*; but I am not sure there is a difference [26M*a*].

In 26F–G I turn to the Boolean algebra $\mathscr{P}$N/[N]$^{<\omega}$. 26G addresses the question (α) which Boolean algebras can be embedded in $\mathscr{P}$N/[N]$^{<\omega}$? On moving to the Stone spaces [see 12A*b*], this becomes (β) which zero-dimensional compact Hausdorff spaces are continuous images of βN\N? And a natural extension of this is (γ) which compact Hausdorff spaces are continuous images of βN\N?

Concerning (α) we have 26G, dealing with Boolean algebras of cardinal less than $\mathfrak{p}$, while the corresponding response to (γ) is 26I. I should mention here that it is theorem of ZFC, due to I.I. Parovičenko, that every compact Hausdorff space of weight less than or equal to ω_1 is a continuous image of βN\N (see Blaszczyk & Szymański 80 or Mill 83, 1.3.3); so that every Boolean algebra of cardinal less than or equal to ω_1 can be embedded in $\mathscr{P}$N\[N]$^{<\omega}$. Thus 26G and 26I have force only when $\mathfrak{p} > \omega_2$. At the same time, if $w(X) = \omega_1$, we can dispense with $\mathfrak{p} > \omega_1$ in 26H in everything except the claim that every point of X is the limit of a sequence in K.

These results leave several questions open. A number of authors have shown that Martin's axiom is not sufficient to ensure that every Boolean algebra of cardinal $\mathfrak{c}$ can be embedded into $\mathscr{P}$N/[N]$^{<\omega}$ (B3B*a*; note that CH is sufficient, by Parovičenko's theorem). However, we can still ask which algebras can be embedded [26M*b*]; in particular, I do not know whether the Lebesgue measure algebra can.

We can ask a similar question concerning the quotient algebras of $\mathscr{P}$N/[N]$^{<\omega}$ or the closed subsets of βN\N. For Dedekind complete algebras, or extremally disconnected compact Hausdorff spaces, this is answered by a theorem of B.A. Efimov [26L*c*(i)–(ii)]. K. Kunen's refinement [26L*c*(iii)] is what is needed to deduce 26L*g* from 26L*f*, and is the basis of 26L*n*. 26L*e*–*f* and 26L*h* are all corollaries of 26L*d*. See also 32Q*n*.

26H can be regarded as a kind of worked example on the concepts of general topology. 26K*g* is a supplement. Recall that 25M*k* gives another method of obtaining similar results. Differences between the two constructions seem to be that 26H preserves compactness and first-countability, while 25M*k* can be applied more readily to spaces of cardinal $\mathfrak{c}$ and has a better chance of preserving normality. (Observe that 25K shows that the construction of 26H applied to $\omega_1 \times \{0, 1\}$ can produce a non-normal space.)

In 26L*a* I list some of the special types of ultrafilter that can be constructed using Axiom P. Most of these are inductions of greater or lesser delicacy using 21E to enumerate the steps to be covered and 21A (possibly refined in the spirit of 21N*a*) to perform the inductive steps. 26L*b* also depends on efficient techniques for constructing special ultrafilters. Finally, I mention a problem of B.A. Efimov [26M*d*], which has been solved (affirmatively) under CH [FEDORČUK 75, TALAGRAND 80*d*]; it looks as if it may be adapted to Martin's axiom, but it is also possible that there is a solution in ZFC.

3

When $\mathfrak{m}_K > \omega_1$

In this chapter I deal with results which involve $\mathfrak{m}_K$ and in which it cannot (so far as I know) be replaced by $\mathfrak{p}$. Once again I give first place to combinatorial results (§31). I follow this with two sections on measure theory and functional analysis (§§32–3); then I give part of S. Shelah's solution to Whitehead's problem (§34) and A.D. Taylor's and R. Laver's theorems on the regularity of ideals of $\mathscr{P}\omega_1$ (§35).

31 Combinatorics

I begin with the fundamental theorem on extraction of sequences of directed sets from a partially ordered set satisfying Knaster's condition [31A], in a fairly general form, with its most commonly used corollaries [31B]. (Perhaps I should point out immediately that if $\mathfrak{m} > \omega_1$ then every ccc partially ordered set satisfies Knaster's condition [41A]; so that theorems which refer to $\mathfrak{m}_K$ and Knaster's condition can always be rewritten as theorems on ccc partially ordered sets, involving $\mathfrak{m}$). In 31C I describe a commonly-arising type of partially ordered set, the 'S-respecting' partial orders; I give lemmas for dealing with these [31C*b*–31F], and an application [31G]. 31H is a powerful general theorem on the piecing together of almost-consistent functions. 31J is one of K. Kunen's theorems on 'gaps' in $\mathscr{P}\mathbf{N}/[\mathbf{N}]^{<\omega}$. 31K concerns the cardinal $\mathfrak{m}_K$ itself.

31A **Theorem**

Let P be a partially ordered set satisfying Knaster's condition upwards. Let $\mathscr{Q}$, $\mathscr{R}$ and $\mathscr{S}$ be families of subsets of P such that

(i) $\#(\mathscr{Q}) < \mathfrak{m}_K$, $\#(\mathscr{R}) < \mathfrak{m}_K$, $\#(\mathscr{S}) < \mathfrak{m}_K$;

(ii) every member of $\mathscr{Q}$ is cofinal with P;

(iii) $S \cap \bigcap \mathscr{R}_0 \neq \varnothing$ for every $S \in \mathscr{S} \cup \{P\}$, finite $\mathscr{R}_0 \subseteq \mathscr{R}$.

Then there is a sequence $\langle P_i \rangle_{i \in \mathbf{N}}$ of subsets of P such that

(α) every P_i is upwards-directed, not empty;

(β) $P_i \cap Q \neq \varnothing$ for every $i \in \mathbf{N}$, $Q \in \mathscr{Q}$;

(γ) if $R \in \mathscr{R}$, $\{i : P_i \cap R = \emptyset\}$ is finite;

(δ) if $S \in \mathscr{S}$, $\{i : P_i \cap S \neq \emptyset\}$ is infinite.

Proof (a) Let $\mathscr{R}^*$ be $\{P \cap \bigcap \mathscr{R}_0 : \mathscr{R}_0 \subseteq \mathscr{R}$ finite$\}$. Then $P \in \mathscr{R}^*$, $\emptyset \notin \mathscr{R}^*$, $R \cap R' \in \mathscr{R}^*$ whenever R and R' belong to $\mathscr{R}^*$, and $R \cap S \neq \emptyset$ whenever $R \in \mathscr{R}^*$, $S \in \mathscr{S}$. For $R \in \mathscr{R}^*$ set $\hat{R} = \{q : \exists p \in R, q \geq p\}$.

Let X be the set of pairs

$$x = (\langle p_i^x \rangle_{i \leq n(x)}, R_x)$$

where $n(x) \in \mathbf{N}$, $p_i^x \in P$ for each $i \leq n(x)$, and $R_x \in \mathscr{R}^*$. Say that $x \leq y$ if

$$n(x) \leq n(y), p_i^x \leq p_i^y \text{ if } i \leq n(x), p_i^y \in \hat{R}_x \text{ if } n(x) < i \leq n(y), \text{and } R_y \subseteq R_x.$$

Then $x \leq y \leq z \Rightarrow x \leq z$ (because $q \geq p \in \hat{R}_x \Rightarrow q \in \hat{R}_x$), so $\leq$ is a partial order on X.

(b) X satisfies Knaster's condition upwards. **P** Let $Y \subseteq X$ be an uncountable set. Then there is an $m \in \mathbf{N}$ such that

$$Y' = \{y : y \in Y, n(y) = m\}$$

is uncountable. For $i \leq m$ define Y_i inductively so that Y_i is uncountable;

$$Y_0 \subseteq Y', Y_i \subseteq Y_{i-1} \quad \text{if} \quad i > 0;$$

and if $x, y \in Y_i$ then p_i^x, p_i^y have a common upper bound in P; using the fact that X satisfies Knaster's condition upwards. Now if $x, y \in Y_m$, we can find a finite sequence $\langle q_i \rangle_{i<m}$ such that $p_i^x \leq q_i$, $p_i^y \leq q_i$ for each $i \leq m$; in which case $(\langle q_i \rangle_{i \leq m}, R_x \cap R_y)$ is a common upper bound for x and y in X. Thus Y_m is upwards-linked in X. As Y is arbitrary, X satisfies Knaster's condition upwards. **Q**

(c) For $m \in \mathbf{N}$, $Q \in \mathscr{Q} \cup \{P\}$ set

$$T_{Qm} = \{x : x \in X, n(x) \geq m, p_m^x \in Q\}.$$

Then T_{Qm} is cofinal with X. **P** Let $x \in X$. If $m \leq n(x)$ set $y = x$. If $m > n(x)$, then take any $r \in R_x$ and define y by

$$n(y) = m, p_i^y = p_i^x \text{ if } i \leq n(x), p_i^y = r \text{ if } n(x) < i \leq m, R_y = R_x.$$

Then in either case we find that $x \leq y$ and $n(y) \geq m$. Now, because Q is cofinal with P, there is a $q \in Q$ such that $q \geq p_m^y$. So define z by

$$n(z) = n(y), p_i^z = p_i^y \text{ if } i \leq n(y) \text{ and } i \neq m, p_m^z = q, R_z = R_y.$$

Then $x \leq y \leq z \in T_{Qm}$. **Q**

(d) For $R \in \mathscr{R}$ set

$$U_R = \{x : x \in X, R_x \subseteq R\}.$$

Then U_R is cofinal with X. **P** Let $x\in X$. Then

$$x\leq(\langle p_i^x\rangle_{i\leq(x)},R\cap R_x)\in U_R. \quad \mathbf{Q}$$

(*e*) For $S\in\mathcal{S}$, $m\in\mathbf{N}$ set

$$V_{Sm}=\{x:x\in X,\exists i,m\leq i\leq n(x),p_i^x\in S\}.$$

Then V_{Sm} is cofinal with X. **P** Let $x\in X$. Then there is an $r\in R_x\cap S$. Define y by

$$n(y)=\max(m,n(x)+1),\quad p_i^y=p_i^x \text{ if } i\leq n(x),$$
$$p_i^y=r \text{ if } n(x)<i\leq n(y),\ R_y=R_x.$$

Then $x\leq y\in V_{Sm}$. **Q**

(*f*) Since $\mathcal{Q}$, $\mathcal{R}$ and $\mathcal{S}$ all have cardinals less than $\mathfrak{m}_K$, there is an upwards-directed $Z\subseteq X$ meeting every T_{Qm}, U_R and V_{Sm} for $Q\in\mathcal{Q}\cup\{P\}$, $R\in\mathcal{R}$, $S\in\mathcal{S}$ and $m\in\mathbf{N}$. Set

$$P_i=\{p:\exists z\in Z,n(z)\geq i,p\leq p_i^z\}.$$

(α) Because $Z\cap T_{Pi}\neq\emptyset$, $P_i\neq\emptyset$. If $p,q\in P_i$ there are $x,y\in Z$ such that $n(x)\geq i$, $n(y)\geq i$, $p\leq p_i^x$ and $q\leq p_i^y$. Now x and y have a common upper bound $z\in Z$; in which case $n(z)\geq i$ and $p_i^z\in P_i$ is a common upper bound of p_i^x, p_i^y, p and q. As p and q are arbitrary, P_i is upwards-directed.

(β) If $Q\in\mathcal{Q}$ and $i\in\mathbf{N}$, then Z meets T_{Qi}; say $z\in Z\cap T_{Qi}$. In this case $n(z)\geq i$ and $p_i^z\in P_i\cap Q$. So $P_i\cap Q$ is not empty.

(γ) If $R\in\mathcal{R}$ then Z meets U_R; say $z\in Z\cap U_R$. If $i>n(z)$, then we know that there is an $x\in T_{Pi}\cap Z$. Let y be a common upper bound of x and z in Z. Then $n(y)\geq n(x)\geq i>n(z)$, so $p_i^y\in\hat{R}_z$. Let $r\in R_z$ be such that $r\leq p_i^y$; then $r\in P_i\cap R_z\subseteq P_i\cap R$.

Thus P_i meets R for every $i>n(z)$, and $\{i:P_i\cap R\neq\emptyset\}$ is finite.

(δ) If $S\in\mathcal{S}$, $m\in\mathbf{N}$ the Z meets V_{Sm}; say $z\in Z\cap V_{Sm}$. Then there is an i such that $m\leq i\leq n(z)$ and $p_i^z\in S$. In this case $P_i\cap S\neq\emptyset$. As m is arbitrary, $\{i:P_i\cap S\neq\emptyset\}$ is infinite.

So the sequence $\langle P_i\rangle_{i\in\mathbf{N}}$ fulfils all the conditions imposed.

31B **Corollary**

Let P be a non-empty partially ordered set satisfying Knaster's condition upwards.

(*a*) If $\mathcal{Q}$ is a family of cofinal subsets of P with $\#(\mathcal{Q})<\mathfrak{m}_K$, and $A\subseteq P$ has cardinal less than $\mathfrak{m}_K$, then there is a sequence $\langle P_i\rangle_{i\in\mathbf{N}}$ of upwards-directed subsets of P, all meeting every member of $\mathcal{Q}$, such that $A\subseteq\bigcup_{i\in\mathbf{N}}P_i$.

(*b*) If $\#(P)<\mathfrak{m}_K$ then P is σ-centered upwards.

Proof (*a*) Take $\mathscr{R} = \emptyset$, $\mathscr{S} = \{\{p\}: p \in A\}$ in 31A.

(*b*) Take $\mathscr{Q} = \emptyset$, $A = P$ in (*a*).

31C **S-respecting orders**

(*a*) Let X and Y be sets, and $S \subseteq X \times Y$ a relation. Write $P = [X]^{<\omega} \times [Y]^{<\omega}$ and order P by saying

$(I, J) \leq (K, L)$ if $I \subseteq K, J \subseteq L$ and $S[I] \cap L \subseteq J$.

It is easy to check that this is a partial order on P; I shall call it the **S-respecting** partial order on P. Note that it is not symmetric between X and Y.

(*b*) It will be convenient to have code-names for some elementary facts.

(i) If $x \in X$, then

$Q_x = \{(I, J): (I, J) \in P, x \in I\}$

is cofinal with P, because $(I, J) \leq (I \cup \{x\}, J)$ for any $(I, J) \in P$.

(ii) *If* $R \subseteq P$ is upwards-linked, set

$W = \bigcup \{J: (I, J) \in R\} \subseteq Y$.

Then $W \cap S[I] \subseteq J$ whenever $(I, J) \in R$. **P** If $y \in W \cap S[I]$ there is an $(I', J') \in R$ such that $y \in J'$; now there is a $(K, L) \in P$ which is a common upper bound of (I, J) and (I', J'). As $J' \subseteq L$, $y \in L \cap S[I] \subseteq J$. **Q**

(iii) If $S[X]$ is countable then P is σ-centered upwards, because $\{(I, J): J \cap S[X] = K\}$ is upwards-centered in P for any finite $K \subseteq S[X]$.

(iv) For a set $R \subseteq P$, the following are equivalent: (α) R is upwards-linked; (β) R is upwards-centered; (γ) there is an upwards-directed $R' \supseteq R$; (δ) $S[I] \cap L \subseteq J$ whenever (I, J) and (K, L) belong to R. **P** (α) $\Rightarrow$ (δ) $\Rightarrow$ (γ). **Q**

31D **Lemma**

Let X and Y be sets and $S \subseteq X \times Y$ a relation. Let P be $[X]^{<\omega} \times [Y]^{<\omega}$ with the S-respecting partial order.

(*a*) If P is upwards-ccc then the vertical sections of S are countable.

(*b*) If vertical sections of S are countable then for any family $\langle (I_\xi, J_\xi) \rangle_{\xi < \omega_1}$ in P there is an uncountable $A \subseteq \omega_1$ such that $\langle I_\xi \rangle_{\xi \in A}$ and $\langle J_\xi \rangle_{\xi \in A}$ are constant-size Δ-systems and $(I_\eta, J_\eta) \leq (I_\eta \cup I_\xi, J_\eta \cup J_\xi)$ whenever $\xi \in A$ and $\eta \in A \cap \xi$.

(*c*) P satisfies Knaster's condition upwards iff (α) vertical sections of S are countable, (β) whenever $\langle x_\xi\rangle_{\xi<\omega_1}$ and $\langle y_\xi\rangle_{\xi<\omega_1}$ are families of distinct elements of X, Y respectively, there is an uncountable $B\subseteq\omega_1$ such that $(x_\xi, y_\eta)\notin S$ whenever $\xi\in B$ and $\eta\in B\cap\xi$.

Proof (*a*) If $x\in X$ then $\{(\{x\},\{y\}):(x,y)\in S\}$ is an up-antichain in P; so if P is upwards-ccc then $S[\{x\}]$ must be countable.

(*b*) By the Δ-system lemma [12H] there is an uncountable $C\subseteq\omega_1$ such that $\langle I_\xi\rangle_{\xi\in C}$ and $\langle J_\xi\rangle_{\xi\in C}$ are constant-size Δ-systems with roots I, J respectively. If M is a countable subset of X, then $S[M]$ is countable (being a countable union of vertical sections of S), and $\{\xi:S[M]\cap (J_\xi\setminus J)\neq\emptyset\}$ is countable. So there is an uncountable $A\subseteq C$ such that

$$S[\bigcup_{\eta\in A\cap\xi}I_\eta]\cap(J_\xi\setminus J)=\emptyset\,\forall\xi\in A.$$

Now if $\xi\in A$ and $\eta\in A\cap\xi$ we have

$$S[I_\eta]\cap(J_\eta\cup J_\xi)\subseteq J_\eta\cup(S[I_\eta]\cap(J_\xi\setminus J))=J_\eta,$$

so that $(I_\eta, J_\eta)\leq(I_\eta\cup I_\xi, J_\eta\cup J_\xi)$.

(*c*) (i) If P satisfies Knaster's condition upwards, then the vertical sections of S are countable, by (*a*). Moreover, if $\langle x_\xi\rangle_{\xi<\omega_1}$ and $\langle y_\xi\rangle_{\xi<\omega_1}$ are families of distinct elements of X, Y respectively, there is an uncountable $B\subseteq\omega_1$ such that $\{(\{x_\xi\},\{y_\xi\}):\xi\in B\}$ is upwards-linked in P; now by 31C*b*(ii), $S[\{x_\xi\}]\cap\{y_\eta:\eta\in B\}\subseteq\{y_\xi\}$ for all $\xi\in B$ i.e. $(x_\xi, y_\eta)\notin S$ whenever ξ, η are distinct elements of B.

(ii) If S satisfies (α) and (β), and $\langle(I_\xi, J_\xi)\rangle_{\xi<\omega_1}$ is any family in P, then by (*b*) there is an uncountable $A\subseteq\omega_1$ such that $\langle I_\xi\rangle_{\xi\in A}$ and $\langle J_\xi\rangle_{\xi\in A}$ are constant-size Δ-systems and $(I_\eta, J_\eta)\leq(I_\eta\cup I_\xi, J_\eta\cup J_\xi)$ whenever $\xi\in A$ and $\eta\in A\cap\xi$. Set $\zeta=\min A$. Let I, J be the roots of $\langle I_\xi\rangle_{\xi\in A}$, $\langle J_\xi\rangle_{\xi\in A}$ respectively, and write m and n for the common values of $\#(I_\xi\setminus I)$ and $\#(J_\xi\setminus J)$ for $\xi\in A$. For each $\xi\in A$ let $\langle x_i^\xi\rangle_{i<m}$ and $\langle y_j^\xi\rangle_{j<n}$ enumerate $I_\xi\setminus I$ and $J_\xi\setminus J$. Applying (β) mn times, there is an uncountable $E\subseteq A\setminus\{\zeta\}$ such that $(x_i^\xi, y_j^\eta)\notin S$ whenever $\xi\in E$, $\eta\in E\cap\xi$, $i<m$ and $j<n$. Now if $\xi\in E$ and $\eta\in E\cap\xi$, $S[I_\xi\setminus I]\cap(J_\eta\setminus J)=\emptyset$; but also $S[I]\cap(J_\eta\setminus J)\subseteq S[I_\zeta]\cap(J_\eta\setminus J)=\emptyset$ because $(I_\zeta, J_\zeta)\leq(I_\zeta\cup I_\eta, J_\zeta\cup J_\eta)$. So $S[I_\xi]\cap(J_\eta\setminus J)=\emptyset$, and $(I_\xi, J_\xi)\leq(I_\eta\cup I_\xi, J_\eta\cup J_\xi)$. Thus $\{(I_\xi, J_\xi):\xi\in E\}$ is upwards-linked in P. As $\langle(I_\xi, J_\xi)\rangle_{\xi<\omega_1}$ is arbitrary, P satisfies Knaster's condition upwards.

31E **Lemma**

Let X and Y be sets and $S\subseteq X\times Y$. Let Z be a subset of Y and $f:Z\to X$ a function; take $\mathscr{B}\subseteq\mathscr{P}Y$. Suppose that

(i) $[X]^{<\omega}\times[Y]^{<\omega}$ satisfies Knaster's condition upwards when given the S-respecting partial order;

(ii) $B\setminus S[I] \neq \emptyset$ for $B\in\mathscr{B}$, $I\in[X]^{<\omega}$;
(iii) $\#(X) < \mathfrak{m}_K$, $\#(\mathscr{B}) < \mathfrak{m}_K$ and $\#(Z) < \mathfrak{m}_K$.

Then there are sequences $\langle Y_n\rangle_{n\in\mathbf{N}}$, $\langle Z_n\rangle_{n\in\mathbf{N}}$ of sets such that the following are true.

(*a*) $Y = \bigcup_{n\in\mathbf{N}} Y_n$; $Y_n \cap S[\{x\}]$ is finite for every $x\in X$, $n\in\mathbf{N}$; and $Y_n\cap B \neq \emptyset$ for every $B\in\mathscr{B}$, $n\in\mathbf{N}$.

(*b*) $Z = \bigcup_{n\in\mathbf{N}} Z_n$; $Z_n\cap S[\{x\}]$ is finite for every $x\in X$, $n\in\mathbf{N}$; and $(f(z), w)\notin S$ whenever $n\in\mathbf{N}$ and w, z are distinct members of Z_n.

Proof Write $P = [X]^{<\omega}\times[Y]^{<\omega}$ with the S-respecting partial order. For $x\in X$ set

$$Q_x = \{(I, J) : (I, J)\in P, x\in I\},$$

so that Q_x is cofinal with P [31C*b*(i)]. For $B\in\mathscr{B}$ set

$$Q'_B = \{(I, J) : (I, J)\in P, J\cap B \neq \emptyset\}.$$

Then Q'_B is cofinal with P. **P** If $(I, J)\in P$ then, by condition (ii), there is a $y\in B\setminus S[I]$, and now $(I, J) \leq (I, J\cup\{y\})\in Q'_B$. **Q**

Note that vertical sections of S are countable [31D*a*], so that $\#(S[X]) \leq \max(\omega, \#(X)) < \mathfrak{m}_K$. By 31B*a* there is a sequence $\langle P_n\rangle_{n\in\mathbf{N}}$ of upwards-directed subsets of P such that

$$\bigcup_{n\in\mathbf{N}} P_n \supseteq \{(\emptyset, \{y\}) : y\in S[X]\}\cup\{(\{f(z)\}, \{z\}) : z\in Z\}$$

and every P_n meets every Q_x and every Q'_B. Set

$$W_n = \bigcup\{J : (I, J)\in P_n\},\ Y_n = W_n\cup(Y\setminus S[X]),$$

$$Z_n = \{z : z\in Z, (\{f(z)\}, \{z\})\in P_n\},$$

so that $\bigcup_{n\in\mathbf{N}} W_n \supseteq S[X]$, $\bigcup_{n\in\mathbf{N}} Y_n = Y$ and $\bigcup_{n\in\mathbf{N}} Z_n = Z$, while $Z_n \subseteq W_n \subseteq Y_n$ for each $n\in\mathbf{N}$. If $x\in X$ and $n\in\mathbf{N}$ there is an $(I, J)\in Q_x\cap P_n$, and now

$$Z_n\cap S[\{x\}] \subseteq Y_n\cap S[\{x\}] = W_n\cap S[\{x\}] \subseteq W_n\cap S[I] \subseteq J$$

[31C*b*(ii)], so that $Z_n\cap S[\{x\}]$ and $Y_n\cap S[\{x\}]$ are finite. If $B\in\mathscr{B}$ and $n\in\mathbf{N}$ there is an $(I, J)\in Q'_B\cap P_n$, and now

$$Y_n\cap B \supseteq W_n\cap B \supseteq J\cap B \neq \emptyset.$$

If $n\in\mathbf{N}$ and $z\in Z_n$ then

$$Z_n\cap S[\{f(z)\}] \subseteq W_n\cap S[\{f(z)\}] \subseteq \{z\}$$

by 31C*b*(ii) again, i.e. $(f(z), w)\notin S$ for any $w\in Z_n\setminus\{z\}$.

31F **Lemma**

Let Y be a well-ordered set, X any set, and $S \subseteq X \times Y$ a relation such that

$$\alpha = \sup_{x \in X} \operatorname{otp}(S[\{x\}]) < \omega_1.$$

Then $[X]^{<\omega} \times [Y]^{<\omega}$, with the S-respecting partial order, satisfies Knaster's condition upwards.

Proof I use 31D and A2L. Clearly the vertical sections of S are countable. If $\langle x_\xi \rangle_{\xi < \omega_1}$ and $\langle y_\xi \rangle_{\xi < \omega_1}$ are families of distinct elements of X and Y respectively, then set

$$C = \{\xi : \xi < \omega_1, y_\xi \leq y_\zeta \, \forall \zeta \in \omega_1 \setminus \xi\}.$$

Observe that C is uncountable and that $\xi \mapsto y_\xi : C \to Y$ is strictly increasing. Set

$$T = \{\{\eta, \xi\} : \eta < \xi < \omega_1, (x_\xi, y_\eta) \in S\}.$$

For $\xi \in C$,

$$\operatorname{otp}(\{\eta : \eta \in C \cap \xi, \{\eta, \xi\} \in T\} \leq \operatorname{otp}(S[\{x_\xi\}]) \leq \alpha.$$

So by A2L there is an uncountable $A \subseteq C$ such that $[A]^2 \cap T = \emptyset$ i.e. $(x_\xi, y_\eta) \notin S$ whenever $\xi \in A$ and $\eta \in A \cap \xi$. By 31D*c*, $[X]^{<\omega} \times [Y]^{<\omega}$ satisfies Knaster's condition upwards.

31G **Proposition**

Let $\beta < \mathfrak{m}_K$ be an ordinal and $S \subseteq [\beta]^2$ a set such that

$$\sup_{\xi < \beta} \operatorname{otp}(\{\eta : \eta < \xi, \{\eta, \xi\} \in S\} < \omega_1.$$

Then β is expressible as $\bigcup_{n \in \mathbf{N}} Z_n$ where $[Z_n]^2 \cap S = \emptyset$ for every $n \in \mathbf{N}$.

Proof Set $S' = \{(\xi, \eta) : \eta < \xi < \beta, \quad \{\eta, \xi\} \in S\} \subseteq \beta^2$. Then $\sup_{\xi < \beta} \operatorname{otp}(S'[\{\xi\}]) < \omega_1$. So $[\beta]^{<\omega} \times [\beta]^{<\omega}$, with the S'-respecting partial order, satisfies Knaster's condition upwards [31F]. In 31E set $X = Y = Z = \beta$, $\mathscr{B} = \emptyset$ and $f(\xi) = \xi$ for $\xi < \beta$, and see that β is expressible as $\bigcup_{n \in \mathbf{N}} Z_n$ where

$$Z_n \cap S'[\{\xi\}] \subseteq \{\xi\} \, \forall n \in \mathbf{N}, \xi \in Z_n.$$

But this says just that $[Z_n]^2 \cap S = \emptyset$ for every $n \in \mathbf{N}$.

31H **Theorem**

Suppose $\kappa < \mathfrak{m}_K$. Let X be a set, and $\langle Y_\xi \rangle_{\xi < \kappa}$ a family of countable subsets of X such that $\{\xi : Y_\xi \cap Y \text{ infinite}\}$ is countable for every countable

$Y \subseteq X$. Suppose we are given, for each $\xi < \kappa$, a function $f_\xi : Y_\xi \to \mathbf{N}$, in such a way that

$$\{x : x \in Y_\xi \cap Y_\eta, f_\xi(x) \neq f_\eta(x)\}$$

is finite for all $\xi, \eta < \kappa$. Then there is a function $f : X \to \mathbf{N}$ such that

$$\{x : x \in Y_\xi, f(x) \neq f_\xi(x)\}$$

is finite for every $\xi < \kappa$.

Proof (*a*) For $I \subseteq \kappa$ set $Z(I) = \bigcup_{\xi \in I} Y_\xi$. If $\langle I_\xi \rangle_{\xi < \omega_1}$ is any disjoint family of finite subsets of κ, there is an uncountable set $D \subseteq \omega_1$ and a finite $W \subseteq X$ such that

$$Z(I_\xi) \cap Z(I_\eta) = W$$

whenever ξ, η are distinct members of D.

P If $Y \subseteq X$ is countable, then $\{\xi : Z(I_\xi) \cap Y \text{ infinite}\}$ is countable, because the I_ξ are disjoint. So we can find an uncountable set $A \subseteq \omega_1$ such that

$$V_\xi = Z(I_\xi) \cap \bigcup_{\eta \in A, \eta < \xi} Z(I_\eta)$$

if finite whenever $\xi \in A$. Now there is an uncountable $B \subseteq A$ such that $\langle V_\xi \rangle_{\xi \in B}$ is a Δ-system with root V say. If $\xi, \eta, \zeta \in B$ and $\zeta < \eta < \xi$, then

$$Z(I_\xi) \cap Z(I_\eta) \cap Z(I_\zeta) \subseteq V_\xi \cap V_\eta = V.$$

So if $\xi \in B$, $\langle Z(I_\eta) \cap Z(I_\xi) \setminus V \rangle_{\eta < \xi, \eta \in B}$ is a disjoint family of subsets of $V_\xi \setminus V$, and

$$\{\eta : \eta \in B, \eta < \xi, Z(I_\eta) \cap Z(I_\xi) \nsubseteq V\}$$

must be finite. By A2K or A2L, there is an uncountable $C \subseteq B$ such that $Z(I_\eta) \cap Z(I_\xi) \subseteq V$ whenever η, ξ are distinct members of C. Now there is a $W \subseteq V$ such that

$$D = \{\xi : \xi \in C, Z(I_\xi) \cap V = W\}$$

is uncountable, and this pair D, W will serve. **Q**

(*b*) Let P be the set of all pairs (I, g) where $I \subseteq \kappa$ is finite, $g : Z(I) \to \mathbf{N}$ is a function, and

$$\{x : x \in Y_\xi, g(x) \neq f_\xi(x)\}$$

is finite for every $\xi \in I$. Order P by saying that $(I, g) \leq (J, h)$ if $I \subseteq J$ and h extends g. Observe that if $I \subseteq \kappa$ is finite, then $\{g : (I, g) \in P\}$ is countable (because each Y_ξ is countable).

(*c*) P satisfies Knaster's condition upwards. **P** Let $\langle (I_\xi, g_\xi)\rangle_{\xi<\omega_1}$ be a family in P. Let $A \subseteq \omega_1$ be an uncountable set such that $\langle I_\xi \rangle_{\xi\in A}$ is a Δ-system with root I say. By (*a*), there is an uncountable $D \subseteq A$ and a finite $W \subseteq X$ such that $Z(I_\xi \setminus I) \cap Z(I_\eta \setminus I) = W$ whenever ξ, η are distinct members of D. In this case $Z(I_\xi) \cap Z(I_\eta) = Z(I) \cup W$ whenever $\xi, \eta \in D$ are distinct. Now

$$\{g_\xi \restriction W : \xi \in D\} \subseteq \mathbf{N}^W,$$
$$\{g_\xi \restriction Z(I) : \xi \in D\} \subseteq \{g : (I, g) \in P\}$$

are both countable. So there is an uncountable $E \subseteq D$ such that

$$g_\xi \restriction Z(I) \cup W = g_\eta \restriction Z(I) \cup W$$

whenever $\xi, \eta \in E$. In this case, if $\xi, \eta \in E$, then g_ξ and g_η have a common extension $g : Z(I_\xi \cup I_\eta) \to \mathbf{N}$, and $(I_\xi \cup I_\eta, g) \in P$. Thus $\{(I_\xi, g_\xi) : \xi \in E\}$ is upwards-linked. As $\langle (I_\xi, g_\xi)\rangle_{\xi<\omega_1}$ is arbitrary, P satisfies Knaster's condition upwards. **Q**

(*d*) For $\zeta < \kappa$ set $Q_\zeta = \{(I, g) : (I, g) \in P, \zeta \in I\}$. Then Q_ζ is cofinal with P. **P** If $(I, g) \in P$, define $h : Z(I) \cup Y_\zeta \to \mathbf{N}$ by

$$h(x) = g(x) \text{ if } x \in Z(I), \quad h(x) = f(x) \text{ if } x \in Y_\zeta \setminus Z(I).$$

Since

$$\{x : x \in Z(I) \cap Y_\zeta, g(x) \neq f_\zeta(x)\} \subseteq \bigcup_{\xi\in I} \{x : x \in Y_\xi \cap Y_\zeta, f_\xi(x) \neq f_\zeta(x)\}$$
$$\cup \bigcup_{\xi\in I} \{x : x \in Y_\xi, g(x) \neq f_\xi(x)\}$$

is finite, $(I \cup \{\zeta\}, h) \in P$ and $(I, g) \leq (I \cup \{\zeta\}, h) \in Q_\zeta$. **Q**

(*e*) Because $\kappa < \mathfrak{m}_K$, there is an upwards-directed set $R \subseteq P$ meeting every Q_ζ. Let $f : X \to \mathbf{N}$ be any function extending g for every $(I, g) \in R$. If $\zeta < \kappa$ there is an $(I, g) \in R \cap Q_\zeta$; in which case

$$\{x : x \in Y_\zeta, f(x) \neq f_\zeta(x)\} = \{x : x \in Y_\zeta, g(x) \neq f_\zeta(x)\}$$

is finite. So this f has the property we seek.

31I **Proposition**

Let $\mathscr{A}$ and $\mathscr{B}$ be two families of subsets of $\mathbf{N}$ such that

(i) $\#(\mathscr{A}) < \mathfrak{m}_K$, $\#(\mathscr{B}) < \mathfrak{m}_K$;

(ii) if $\mathscr{C} \subseteq \mathscr{A}$, $\mathscr{D} \subseteq \mathscr{B}$, $\#(\mathscr{C}) \leq \omega_1$ and $\#(\mathscr{D}) \leq \omega_1$, there is an $H \subseteq \mathbf{N}$ such that $C \cap H$ and $D \setminus H$ are finite for all $C \in \mathscr{C}$ and $D \in \mathscr{D}$.

Then there is an $H \subseteq \mathbf{N}$ such that $A \cap H$ and $B \setminus H$ are finite for all $A \in \mathscr{A}$, $B \in \mathscr{B}$.

Proof Let $\mathscr{A}^*$, $\mathscr{B}^*$ be the ideals of $\mathscr{P}\mathbf{N}$ generated by $\mathscr{A}$, $\mathscr{B}$ respectively; then the pair $\mathscr{A}^*$, $\mathscr{B}^*$ also satisfies the condition (ii) above.

Set $P = \{(A, B): A \in \mathscr{A}^*, B \in \mathscr{B}^*, A \cap B = \emptyset\}$, and say that $(A, B) \leq (C, D)$ if $A \subseteq C$ and $B \subseteq D$. Then P satisfies Knaster's condition upwards. **P** Let $R \subseteq P$ have cardinal ω_1. By (ii), applied to $\{A : (A, B) \in R\}$ and $\{B : (A, B) \in R\}$, there is an $H \subseteq \mathbf{N}$ such that $A \cap H$ and $B \setminus H$ are finite for every $(A, B) \in R$. Let K, L be such that

$$R_1 = \{(A, B) : (A, B) \in R, A \cap H = K, B \setminus H = L\}$$

is uncountable; these exist because $[\mathbf{N}]^{<\omega}$ is countable. If (A, B) and $(C, D) \in R_1$, then

$$(A \cup C) \cap (B \cup D) \subseteq ((A \cup C) \cap H) \cup ((B \cup D) \setminus H) = K \cup L;$$

as $\quad ((A \cup C) \cap L) \cup ((B \cup D) \cap K) \subseteq (A \cap B) \cup (C \cap D) = \emptyset,$
$(A \cup C) \cap (B \cup D) = \emptyset$ and $(A \cup C,\ B \cup D)$ is a common upper bound of (A, B) and (C, D) in P. So R_1 is upwards-linked in P. As R is arbitrary, P satisfies Knaster's condition upwards. **Q**

For $A \in \mathscr{A}$, $B \in \mathscr{B}$ set

$$Q_{AB} = \{(C, D) : (C, D) \in P, A \setminus C \text{ is finite}, B \setminus D \text{ is finite}\}.$$

Then Q_{AB} is cofinal with P. **P** If $(C, D) \in P$ then $J = (A \cup C) \cap (B \cup D)$ is finite (applying (ii) to $\{A, C\}$ and $\{B, D\}$); now $(C, D) \leq (C \cup (A \setminus J), B \cup (D \setminus J)) \in Q_{AB}$. **Q**

As $\mathscr{A}$ and $\mathscr{B}$ have cardinals less than $\mathfrak{m}_K$, there is an upwards-directed $R \subseteq P$ meeting every Q_{AB}. Set $H = \bigcup\{B : (A, B) \in R\}$. Observe that as R is upwards-directed, $A \cap H = \emptyset$ for every $(A, B) \in R$. Now if $A \in \mathscr{A}$ and $B \in \mathscr{B}$, R meets Q_{AB} in (C, D), say; so

$$A \cap H \subseteq (A \setminus C) \cup (C \cap H) = A \setminus C,$$
$$B \setminus H \subseteq (B \setminus D) \cup (D \setminus H) = B \setminus D$$

are both finite, and this H will serve.

31J Corollary

Let α, β be ordinals less than $\mathfrak{m}_K$. Let $\langle A_\xi \rangle_{\xi < \alpha}$ and $\langle B_\eta \rangle_{\eta < \beta}$ be indexed families of subsets of $\mathbf{N}$ such that

$A_\eta \setminus A_\xi$ is finite whenever $\eta \leq \xi < \alpha$,
$B_\eta \setminus B_\xi$ is finite whenever $\eta \leq \xi < \beta$,
$A_\xi \cap B_\eta$ is finite whenever $\xi < \alpha$, $\eta < \beta$,

and suppose that α, β are not both limit ordinals of cofinality ω_1. Then there is an $H \subseteq \mathbf{N}$ such that $A_\xi \cap H$ and $B_\eta \setminus H$ are finite for all $\xi < \alpha$, $\eta < \beta$.

Proof (*a*) If $\alpha=0$ take $H=\mathbf{N}$; if $\beta=0$ take $H=\emptyset$. If $\alpha=\zeta+1$, take $H=\mathbf{N}\setminus A_\zeta$; if $\beta=\theta+1$, take $H=B_\theta$.

(*b*) If $\mathrm{cf}(\alpha)=\omega$, let $\langle\zeta(n)\rangle_{n\in\mathbf{N}}$ be an increasing sequence in α with supremum α. Apply 21A with $\mathscr{A}=\{A_{\zeta(n)}:n\in\mathbf{N}\}$ and $\mathscr{C}=\{B_\eta:\eta<\beta\}$ to find an $I\subseteq N$ such that $A_{\zeta(n)}\setminus I$ and $B_\eta\cap I$ are finite for every $n\in\mathbf{N}$, $\eta<\beta$; now take $H=\mathbf{N}\setminus I$.

The same trick works if $\mathrm{cf}(\beta)=\omega$.

(*c*) If $\mathrm{cf}(\alpha)\geq\omega_1$ and $\mathrm{cf}(\beta)\geq\omega_1$, apply 31I to

$$\mathscr{A}=\{A_\xi:\xi<\alpha\},\ \mathscr{B}=\{B_\eta:\eta<\beta\}.$$

If $\mathscr{C}\subseteq\mathscr{A}$, $\mathscr{D}\subseteq\mathscr{B}$, and $\#(\mathscr{C}\cup\mathscr{D})\leq\omega_1$, express them as $\mathscr{C}=\{A_\xi:\xi\in C\}$, $\mathscr{D}=\{B_\eta:\eta\in D\}$ where $\#(C\cup D)\leq\omega_1$. Set $\zeta=\sup C$, $\theta=\sup D$. If $\mathrm{cf}(\alpha)>\omega_1$, then $\zeta<\alpha$ and we can take $H=\mathbf{N}\setminus A_\zeta$; if $\mathrm{cf}(\beta)>\omega_1$, then $\theta<\beta$ and we can take $H=B_\theta$. Thus in either case the condition (ii) of 31I is satisfied. Now the H provided by 31I is the H we seek.

31K **Proposition**

$\mathrm{cf}(\mathfrak{m}_K)>\omega$.

Proof Let $\langle\kappa_n\rangle_{n\in\mathbf{N}}$ be an increasing sequence of infinite cardinals less than $\mathfrak{m}_K$, with supremum κ. Let P be a non-empty partially ordered set satisfying Knaster's condition upwards, and $\mathscr{Q}$ a family of cofinal subsets of P with $\#(\mathscr{Q})\leq\kappa$. Express $\mathscr{Q}$ as $\bigcup_{n\in\mathbf{N}}\mathscr{Q}_n$ where $\#(\mathscr{Q}_n)\leq\kappa_n$ for each $n\in\mathbf{N}$. Choose inductively, for $n\in\mathbf{N}$, sets A_n, B_n, $R_n^i\subseteq P$ as follows. Start by taking any $p_0\in P$, and set $B_0=\{p_0\}$. Given that $\#(B_n)\leq\kappa_n$, let $A_n\subseteq P$ be such that $\#(A_n)\leq\kappa_n$ and $\forall p\in B_n$, $Q\in\bigcup_{m\leq n}\mathscr{Q}_m$ $\exists q\in A_n\cap Q$ such that $p\leq q$. By 31B*a* there is a sequence $\langle P_n^i\rangle_{i\in\mathbf{N}}$ of upwards-directed subsets of P covering A_n. For each $i\in\mathbf{N}$ let $R_n^i\subseteq P_n^i$ be an upwards-directed set such that $A_n\cap P_n^i\subseteq R_n^i$ and $\#(R_n^i)\leq\kappa_n$. Set $B_{n+1}=B_n\cup\bigcup_{i\in\mathbf{N}}R_n^i$. Continue.

Set $P'=\bigcup_{i,n\in\mathbf{N}}R_n^i$. Then P' is σ-centered upwards. Also, $P'=\bigcup_{n\in\mathbf{N}}B_n$ and $P'\supseteq\bigcup_{n\in\mathbf{N}}A_n$; so $Q\cap P'$ is cofinal with P' for every $Q\in\mathscr{Q}$. Now we know by 21K that $\kappa<\mathfrak{p}$. So there is an upwards-directed $R\subseteq P'$ meeting every $Q\in\mathscr{Q}$.

As P and $\mathscr{Q}$ are arbitrary, $\kappa<\mathfrak{m}_K$. As $\langle\kappa_n\rangle_{n\in\mathbf{N}}$ is arbitrary, $\mathrm{cf}(\mathfrak{m}_K)>\omega$.

31L **Sources**

Malyhin & Sapirovskii 73 for 31B*a*. Hajnal & Máté 75 for a special case of 31G. Arhangel'skii 76 for 31A. Levy 79 for 31G. Kunen *b* and Baumgartner 83 for 31I–J. Baumgartner & Taylor 82*a* for a version of 31F (attributed to F. Galvin and A. Hajnal).

31M Exercises

(*a*) Let $\langle P_\iota\rangle_{\iota\in I}$ be a family of partially ordered sets. Write $P=\bigcup\{\prod_{\iota\in J}P_\iota : J\in[I]^{<\omega}\}$ and order P by writing

$$f\le g \text{ if } \mathrm{dom}(f)\subseteq\mathrm{dom}(g) \text{ and } f(\iota)\le g(\iota)\forall\iota\in\mathrm{dom}(f)$$

If every P_ι satisfies Knaster's condition upwards, so does P.

(*b*) Let X and Y be sets and $S\subseteq X\times Y$ a relation. Let P be $[X]^{<\omega}\times[Y]^{<\omega}$ with the S-respecting partial order and let Q be $\{(I,J): (I,J)\in P, S[I]\cap J=\varnothing\}$. (i) For (I,J) and $(K,L)\in Q$, $(I,J)\le(K,L)$ iff $I\subseteq K$ and $J\subseteq L$. (ii) A subset of Q is bounded above in Q iff it is bounded above in P. (iii) If P satisfies Knaster's condition upwards, so does Q. (iv) P is upwards-ccc iff Q is upwards-ccc and S has countable vertical sections.

(*c*) Let X be a set and $S\subseteq X^2$ a relation. Let P be $[X]^{<\omega}\times[X]^{<\omega}$ with the S-respecting partial order and let R be

$$\{I : I\in[X]^{<\omega}, S[\{x\}]\cap I\subseteq\{x\}\ \forall x\in I\}.$$

(i) For I and $K\in R$, $I\subseteq K$ iff $(I,I)\le(K,K)$ in P. (ii) A set $U\subseteq R$ is bounded above in R iff $\{(I,I): I\in U\}$ is bounded above in P. (iii) If P satisfies Knaster's condition upwards, so does R. (iv) If P is upwards-ccc, so is R. (v) If R satisfies Knaster's condition upwards and S has countable vertical sections, then P is upwards-ccc.

(*d*) Let X be a set and $\mathcal{A}$ a σ-point-finite family of subsets of X. Set $S=\{(x,A): x\in A\in\mathcal{A}\}$. Then $[X]^{<\omega}\times[\mathcal{A}]^{<\omega}$, with the S-respecting partial order, satisfies Knaster's condition upwards. [Use 31F with a suitable well-ordering of $\mathcal{A}$. Compare FLEISSNER & REED 78, 4.1.]

(*e*) [$\mathfrak{m}_K>\omega_1$] Let $\langle\theta(\zeta,n)\rangle_{n\in\mathbf{N},\zeta\in\Omega}$ be a ladder system on ω_1 [definition: A2G]. (i) There is a stationary $C\subseteq\omega_1$ such that $\{n:\theta(\zeta,n)\in C\}$ is finite for every $\zeta\in\Omega$. [Use 31G.] (ii) If $\langle f_\zeta\rangle_{\zeta\in\Omega}$ is a family in $\mathbf{N}^{\mathbf{N}}$, there is a function $f:\omega_1\to\mathbf{N}$ such that $\{n: f(\theta(\zeta,n))\ne f_\zeta(n)\}$ is finite for every $\zeta\in\Omega$. [Hint: in 31H, set $Y_\zeta=\{\theta(\zeta,n): n\in\mathbf{N}\}$. See DEVLIN & SHELAH 78, 5.2.] (iii) If $A\subseteq\Omega$ there is a $G\subseteq\omega_1$ such that $\{n: n\in\mathbf{N}, \theta(\zeta,n)\in G\}$ is cofinite for $\zeta\in A$, finite for $\zeta\in\Omega\setminus A$. [Use (ii).]

(*f*) [$\mathfrak{m}_K>\omega_1$] Let $\langle\theta(\zeta,n)\rangle_{n\in\mathbf{N},\zeta\in\Omega}$ be a ladder system on ω_1. Give ω_1 the finest topology $\mathfrak{T}$ for which $\zeta=\lim_{n\to\infty}\theta(\zeta,n)$ for every $\zeta\in\Omega$. (i) $\mathfrak{T}$ is Hausdorff and zero-dimensional. (ii) There is a sequence of closed discrete sets covering ω_1. [Hint: (*e*) (i) above.] It follows that ω_1 is a Q-space. (iii) ω_1 is hereditarily normal. [Hint: by 31Me(iii), if $E\subseteq\omega_1$, there is a $G_0\subseteq\omega_1$ such that $E\subseteq G_0\subseteq\bar{G}_0\subseteq\bar{E}\cup G_0$ and, for every $\zeta\in E\cap\Omega$, $\{i:\theta(\zeta,i)\notin G_0\}$ is finite.] Consequently ω_1 is countably paracompact. (iv) ω_1 is not collectionwise Hausdorff. [Hint: there is a closed discrete stationary set.] (v) ω_1 is not metalindelöf [compare 25G prf *a*], therefore not paracompact or metrizable. (vi) There is a separable metrizable topology $\mathfrak{S} \subseteq \mathfrak{T}$. [Use (ii) and (iii).]

(*g*) **De Caux's space** Let $\langle \theta(\zeta, n) \rangle_{n \in \mathbf{N}, \zeta \in \Omega}$ be a ladder system on ω_1. Set $X = \mathbf{N} \times \omega_1$ and give X the finest topology such that $(m+1, \zeta) = \lim_{n \to \infty}(m, \theta(\zeta, n))$ for every $\zeta \in \Omega$, $m \in \mathbf{N}$. [$\mathfrak{m}_K > \omega_1$] X is expressible as the union of a sequence of closed discrete sets and is hereditarily normal. [FLEISSNER 80.]

(*h*) Let $\langle Y_\xi \rangle_{\xi < \omega_1}$ be an almost-disjoint family of infinite subsets of $\mathbf{N}$. For each $\xi < \omega_1$ define $f_\xi : Y_\xi \to \mathbf{N}$ by writing

$$f_\xi(n) = \min\{i : i \in Y_\xi, i > n\}.$$

Then there is no $g : \mathbf{N} \to \mathbf{N}$ such that $\{n : n \in Y_\xi, g(n) \neq f_\xi(n)\}$ is finite for every $\xi < \omega_1$.

(*i*) [$\mathfrak{m}_K > \omega_1$] Let $(S, \mathfrak{T})$ be a topological space and $Z = [S]^{<\omega}$ with its Pixley–Roy topology [25H]. Then Z satisfies Knaster's condition iff whenever $T \subseteq S$, $\mathscr{G} \subseteq \mathfrak{T}$ both have cardinals less than or equal to ω_1 there is a second-countable topology $\mathfrak{S}$ on T including $\{G \cap T : G \in \mathscr{G}\}$. [Hint: set $P = \{(I, G) : G \in \mathscr{G}, I \in [G]^{<\omega}\}$. Say $(I, G) \leq (J, H)$ if $I \subseteq J$ and $H \subseteq G$. If Z satisfies Knaster's condition apply 31B*a* to $A = \{(\{s\}, G) : G \in \mathscr{G}, s \in G \cap T\}$. See HAJNAL & JUHÁSZ 82.]

(*j*) Let $\mathscr{I}$ be the ideal of $\mathscr{P}\omega_1$ generated by the sets of order type ω. (i) If $\mathscr{I}_0 \subseteq \mathscr{I}$ and $\#(\mathscr{I}_0) < \mathfrak{m}_K$ there is a countable cover $\mathscr{E}$ of ω_1 such that $E \cap I$ is finite for every $E \in \mathscr{E}$ and $I \in \mathscr{I}_0$. [Hint: $\sup_{I \in \mathscr{I}} \operatorname{otp}(I) < \omega_1$.] (ii) If $\mathscr{I}_0 \subseteq \mathscr{I}$ and $\#(\mathscr{I}_0) < \mathfrak{m}_K$ there is a second-countable topology $\mathfrak{S}$ on ω_1 such that every member of $\mathscr{I}_0$ is $\mathfrak{S}$-closed. [Hint: take $\mathfrak{S} \supseteq \mathscr{E}$.] (iii) [$\mathfrak{m}_K > \omega_1$] Let $\mathscr{D} \subseteq \mathscr{P}\omega_1$ be countable and let $\mathfrak{T}$ be the topology on ω_1 generated by $\{D \setminus I : D \in \mathscr{D}, I \in \mathscr{I}\}$. Let $X \subseteq \mathscr{P}\omega_1$ be a set including $[\omega_1]^{<\omega}$ and give X the Pixley–Roy topology derived from $\mathfrak{T}$. Then X satisfies Knaster's condition. [Use 31M*i*. See HAJNAL & JUHÁSZ 82.]

(*k*) Let $\mathfrak{A}$ be the Boolean algebra $\mathscr{P}\mathbf{N} \setminus [\mathbf{N}]^{<\omega}$. Let $A, B \subseteq \mathfrak{A}$ be such that $\#(A \cup B) < \mathfrak{m}_K$ and whenever $C \subseteq A$, $D \subseteq B$ and $\#(C \cup D) \leq \omega_1$ there is an $h \in \mathfrak{A}$ such that $c \subseteq h \subseteq d$ for every $c \in C$, $d \in D$. Then there is an $h \in \mathfrak{A}$ such that $a \subseteq h \subseteq b$ for every $a \in A$, $b \in B$.

(*l*) Let $\mathscr{B} \subseteq \mathscr{P}(\mathbf{N} \times \mathbf{N})$, $\#(\mathscr{B}) < \mathfrak{m}_K$. Suppose that

$$H(n) = \{i : i \in \mathbf{N}, \exists \mathscr{B}_0 \in [\mathscr{B}]^{\leq n}, i \notin \pi_1[\bigcap \mathscr{B}_0]\}$$

is finite for every $n \in \mathbf{N}$. Then there is an $f:\mathbf{N} \to \mathbf{N}$ such that $\{n : n \in \mathbf{N}, (n, f(n)) \notin B\}$ is finite for every $B \in \mathscr{B}$. [Let P be

$$\{(g, \mathscr{B}_0) : \exists n \in \mathbf{N}, \mathscr{B}_0 \in [\mathscr{B}]^{\leq n}, g \in \mathbf{N}^{H(n) \cup n}\}.$$

Say $(g, \mathscr{B}_0) \leq (h, \mathscr{B}_1)$ if $\mathscr{B}_0 \subseteq \mathscr{B}_1$, h extends g, and $(i, h(i)) \in \bigcap \mathscr{B}_0$ if $i \in \mathrm{dom}(h) \setminus \mathrm{dom}(g)$.]

(*m*) $\mathfrak{p} = \mathfrak{m}_K$ iff every partially ordered set of cardinal less than $\mathfrak{p}$ which satisfies Knaster's condition upwards is σ-centered upwards.

31N Further results

(*a*) If $\mathfrak{m}_K > \omega_2$ then the proper forcing axiom is false. [Baumgartner 83.]

(*b*) Suppose that $\kappa < \mathfrak{m}_K$, $n \geq 1$ and $\mathscr{S} \subseteq [\kappa]^{2n+1}$. Then either there is an $f:\kappa \to \mathbf{N}$ not constant on any member of $\mathscr{S}$ or there is an $A \in [\kappa]^{n+1}$ such that $\{S : A \subseteq S \in \mathscr{S}\}$ is infinite. [Erdös, Galvin & Hajnal 75.]

(*c*) $[\mathfrak{m}_K > \omega_2]$ Let X be the compact Hausdorff space of uniform ultrafilters on ω_1. Then X cannot be embedded as a $p(\omega_3)$-set in $\beta\mathbf{N} \setminus \mathbf{N}$. [Balcar, Frankiewicz & Mills 80.] [Contrast 26Lc(iii).]

31O Problem

In 31M*f*, under what circumstances is ω_1^2 normal?

Notes and comments

Unlike $\mathfrak{p} = \mathfrak{c}$ and MA, the axiom $\mathfrak{m}_K = \mathfrak{c}$ has not attracted a great deal of attention in the past. I have chosen to give it prominence in this book for three reasons. First, it provides a criterion for dividing the material I wish to present into chapters of tolerably balanced lengths. Second, the division is not an unnatural one; there is to my mind a real difference in the character of the results which involve $\mathfrak{m}_K$ which separates them from those which need the full strength of $\mathfrak{m}$. And third, it may one day turn out to be significant that all the results of this chapter are consistent with (for instance) the existence of a Souslin tree.

31A is our first encounter with one of the most powerful arguments in the theory. It will be used more often in Chapter 4 than in Chapter 3; but this is only because the hypothesis 'let X be a ccc topological space' trips more readily from the tongue than the hypothesis 'let X be a topological space satisfying Knaster's condition'. In practice, if we set up a partially ordered set, and if we can prove that it's ccc, then more often that not we can prove that it satisfies Knaster's condition. In view of 41A, this is bound to be so.

The partially ordered set X of 31A bears an obvious family relationship to those which we have seen in 11*Cc* and 21A, but of course the components p_i^x are essentially more complicated than their opposite numbers in the other theorems. The essence of the proof is in part (*b*), which in effect says that the product of finitely many partially ordered sets satisfying Knaster's condition again satisfies Knaster's condition; compare 12M*a*. I think in fact that this is the real point of Knaster's condition. The idea of using products in this way appears to derive ultimately from HAJNAL & JUHASZ 71. An occasionally useful extension (compare 12M*a* again) is in 31M*a*.

In many applications, of course, one of the simpler forms in 31B is all that we need. Note 31B*b*, which gives a condition for $\mathfrak{p}$ to be equal to $\mathfrak{m}_K$ [31M*m*]. Similar conditions for $\mathfrak{p} = \mathfrak{m}$ and $\mathfrak{m}_K = \mathfrak{m}$ will be discussed at greater length in §41.

31C–31G are devoted to what I call 'S-respecting partial orders'. Readers familiar with combinatorial applications of Martin's axiom will recognize here a standard method, which is very close to that of 11*Cc*. The three pages spent on 31C–31F actually amount to a rather ponderous exposition of the ideas needed for 31G. I have taken the trouble to spell the lemmas out at length because the machinery developed here can be used in several ways (see §42), and I think it likely that there are many further applications.

There are other partially ordered sets similarly associated with a relation $S \subseteq X \times Y$; I describe two in 31M*b*–*c*; see also 41H. It is clear that S-respecting partial orders are useful to us only for relations S with countable vertical sections [31D*a*]; this is not obviously the case in 31M*b*–*c*, but as it happens I do not know of any applications of the latter in which S does not have countable sections. 31M*b*(iv) and 31M*c*(v) skirt round some curious technical problems (remember that in the contexts most interesting to us, 'Knaster's condition' and 'ccc' collapse together). 31G seems to be a folk-lore result; its precursor in HAJNAL & MÁTÉ 75 relied on different ideas.

31H is a descendant of a result of S. Shelah [31M*e*(ii)]. Note that we seem to need five conditions: (i) each of the given functions f_ξ must have countable domain; (ii) there must not be too many of them; (iii) the domains must not overlap too much; (iv) the common codomain must be countable; (v) for any two of the given functions, the set on which they disagree must be finite. Of these, only (v) is obviously necessary for the conclusion. I am sure that none of the other four can be dispensed with altogether, but perhaps there is scope for variation in them. (I give 31M*h* to indicate one of the obstacles. See also SHELAH 82, §II.4.) The partially ordered set in the proof of 31H is the obvious one; the proof that it satisfies Knaster's condition is a pleasing combination of two of the most important methods, the Δ-system lemma and $\omega_1 \to (\omega_1, \omega+1)^2$.

31M*e*(ii)–(iii) and 31M*f*–*g* are corollaries of 31H. For a similar situation which does not seem amenable to the same methods, see 31O. 31M*i*–*j* provide examples of Pixley–Roy spaces which are ccc without being σ-centered; by 43O*i*, some of these will satisfy the hypotheses of Theorem 43E.

31I–J shows how special is the role of ω_1 in Hausdorff's gap [21L]. These results have simple translations in terms of the algebra $\mathscr{P}\mathbf{N}/[\mathbf{N}]^{<\omega}$ [31M*k*]. For a discussion of the problems which arise in 31J if α or β is $\mathfrak{c}$, see B3A. 31N*c* is a consequence of 31J and 21N*l*; 31N*a* also depends on 31J. 31K is the result for $\mathfrak{m}_K$ corresponding to 21K for $\mathfrak{p}$. The situation here is different as $\mathfrak{m}_K$ need not be regular [B2A].

31M*l* is not connected to the work of this section; I include it because it may be relevant to Problem 26M*a*.

32 Measure theory I

The main work of this section falls into two unequal parts. The first [32B–C] concerns Radon measure spaces. In these, we find that no non-negligible measurable set can be covered by fewer than $\mathfrak{m}_K$ negligible sets; the argument is essentially the same as that of (i)$\Rightarrow$(iii) in Theorem 13A. In the second part [32D–32I] I explore the consequences of the argument devised by D.A. Martin and R.M. Solovay to show that the union of fewer than $\mathfrak{m}$ Lebesgue negligible sets is Lebesgue negligible. The framework I choose for this discussion is the theory of 'quasi-Radon' measure spaces; see 32D for an advertisement, and §A7 for a fuller exposition of this concept. I suggest, however, that anyone who is not particularly interested in general measure theory should start by taking every measure space in this part to be Lebesgue measure on **R**; even in this case, I think he will find the results surprising. All the results are presented as corollaries of the fundamental theorem 32E; the standard applications are spelt out in 32G. One of the most useful is 32G*a*, which I have expressed in terms of 'properly based' measure spaces; see A6L and A7B*l* for the definition and some typical properties of these spaces. I end the section with a group of results [32J–32N] from the abstract theory of topological measure spaces.

32A Lemma

Let $\mathfrak{A}$ be a Boolean algebra, $\nu:\mathfrak{A}\to\mathbf{R}$ a non-negative additive functional. Then $\mathfrak{A}^+ = \{a:\nu a>0\}$ satisfies Knaster's condition downwards.

Proof If $R \subseteq \mathfrak{A}^+$ is uncountable, there must be an $s > 0$ such that $R_1 = \{a : a \in R, \nu a \geq s\}$ is uncountable; let $\langle a_\xi \rangle_{\xi < \omega_1}$ enumerate a family in R_1. Set $S = \{\{\xi, \eta\} : \xi < \eta < \omega_1, a_\xi \cap a_\eta \neq 0\}$. **?** If $\langle \zeta(i) \rangle_{i \in \mathbf{N}}$ is a strictly increasing sequence in ω_1 such that $[\{\zeta(i) : i \in \mathbf{N}\}]^2 \cap S = \emptyset$, then $\langle a_{\zeta(i)} \rangle_{i \in \mathbf{N}}$ is an infinite disjoint sequence in $\mathfrak{A}$; but now

$$\nu 1 \geq \sum_{i \leq n} \nu a_{\zeta(i)} \geq (n+1)s \; \forall n \in \mathbf{N}$$

which is impossible. **X** By A2K or A2L, there must be an uncountable $A \subseteq \omega_1$ such that $[A]^2 \subseteq S$, and now $\{a_\xi : \xi \in A\}$ is an uncountable downwards-linked subset of R.

32B **Theorem**

Let $(X, \mathfrak{T}, \Sigma, \mu)$ be a Radon measure space. Suppose that $E_0 \in \Sigma$ and that $E_0 \subseteq \bigcup \mathscr{E}$, where $\mathscr{E}$ is a family of negligible sets with $\#(\mathscr{E}) < \mathfrak{m}_K$. Then $\mu E_0 = 0$.

Proof **?** If $\mu E_0 > 0$ there is a compact $F \subseteq E_0$ with $\mu F > 0$. Let $\mathscr{K}$ be $\{K : K \subseteq F$ is compact, $\mu K > 0\}$. By 32A, $\mathscr{K}$ satisfies Knaster's condition downwards. (Formally speaking, we must apply 32A to $\mathfrak{A} = \Sigma$ and take $\nu H = \mu(F \cap H)$ for $H \in \Sigma$; and observe that a subset of $\mathscr{K}$ is downwards-linked in $\mathscr{K}$ iff it is downwards-linked in $\mathfrak{A}^+$.) For each $E \in \mathscr{E}$, set

$$\mathscr{Q}_E = \{K : K \in \mathscr{K}, E \cap K = \emptyset\}.$$

Then $\mathscr{Q}_E$ is coinitial with $\mathscr{K}$. **P** If $K \in \mathscr{K}$, then $\mu(K \setminus E) = \mu K > 0$, so there is a compact $L \subseteq K \setminus E$ such that $\mu L > 0$; now $K \supseteq L \in \mathscr{Q}_E$. **Q** As $\#(\mathscr{E}) < \mathfrak{m}_K$, there is a downwards-directed $\mathscr{R} \subseteq \mathscr{K}$ meeting every $\mathscr{Q}_E$. $\mathscr{R}$ is a family of compact sets with the finite intersection property, so $\bigcap \mathscr{R} \neq \emptyset$; but $\bigcap \mathscr{R} \subseteq F \setminus \bigcup \mathscr{E}$, so $E_0 \nsubseteq \bigcup \mathscr{E}$. **X**

32C **Corollary**

(*a*) Let $(X, \mathfrak{T}, \Sigma, \mu)$ be a Radon measure space, and $\mathscr{E} \subseteq \Sigma$ a set of cardinal $< \mathfrak{m}_K$. Suppose that $\bigcup \mathscr{E} \in \Sigma$. Then $\mu(\bigcup \mathscr{E}) \leq \sum_{E \in \mathscr{E}} \mu E$. If $\mathscr{E}$ is disjoint, $\mu(\bigcup \mathscr{E}) = \sum_{E \in \mathscr{E}} \mu E$; if $\mathscr{E}$ is upwards-directed, then $\mu(\bigcup \mathscr{E}) = \sup_{E \in \mathscr{E}} \mu E$.

(*b*) $[\mathfrak{m}_K = \mathfrak{c}]$ (i) If $(X, \mathfrak{T}, \Sigma, \mu)$ is a Radon measure space and $d(L^1(X)) \leq \mathfrak{c}$, then (X, Σ, μ) is properly $\mathfrak{c}$-based. (ii) If moreover μ is atomless, then (X, Σ, μ) has a $\mathfrak{c}$-Sierpiński set.

Proof (*a*) If $F \subseteq \bigcup \mathscr{E}$ is compact, there is a countable $\mathscr{E}_1 \subseteq \mathscr{E}$ such that $\mu(F \cap E \setminus \bigcup \mathscr{E}_1) = 0$ for every $E \in \mathscr{E}$ [A6G*a*]. By 32B, $\mu(F \setminus \bigcup \mathscr{E}_1) = 0$, and

$$\mu F \leq \mu(\bigcup \mathscr{E}_1) = \sup \{\mu(\bigcup \mathscr{E}_0) : \mathscr{E}_0 \in [\mathscr{E}_1]^{<\omega}\}.$$

Accordingly

$$\mu(\bigcup\mathscr{E}) = \sup\{\mu F : F \text{ compact}, F \subseteq \bigcup\mathscr{E}\}$$
$$\leq \sup\{\mu(\bigcup\mathscr{E}_0) : \mathscr{E}_0 \in [\mathscr{E}]^{<\omega}\}.$$

The rest of (*a*) follows at once, as in 22H*a*.

(*b*) (i) Let $\mathscr{K}$ be the family of all non-empty supporting [A7B*g*] compact sets in X. By 32B, no member of $\mathscr{K}$ can be covered by fewer than $\mathfrak{c}$ negligible sets; by A7B*g*, every non-negligible compact set includes a member of $\mathscr{K}$, so every non-negligible measurable set includes a member of $\mathscr{K}$. If $K, L \in \mathscr{K}$ and $K \nsubseteq L$, then $\mu(K \setminus L) > 0$ because K is supporting. So the map $K \mapsto (\chi K)^{\cdot} : \mathscr{K} \to L^1(X)$ is injective, and $\#(\mathscr{K}) \leq \#(L^1(X)) \leq \mathfrak{c}$. Thus (X, Σ, μ) is properly $\mathfrak{c}$-based. (ii) now follows from A7B*k*–*l*.

32D Quasi-Radon measures with separable $\mathbf{L}^1$ spaces

For most of the rest of this section we shall be considering spaces of this kind. To persuade you that they are worth the trouble I give here some examples:

(*a*) Lebesgue measure on $\mathbf{R}^n$;
(*b*) any Radon measure on any analytic space;
(*c*) the completion of any locally finite Borel measure on any analytic space (of course this is the same thing as (*b*));
(*d*) the completion of any σ-finite measure defined on a countably-generated σ-algebra (you have to pick a suitable topology, which can be separable and metrizable);
(*e*) the usual measure on the split interval;
(*f*) the hyperstonian space of the unit interval;
(*g*) any finite product of spaces of the types above;
(*h*) any countable product of probability spaces of the types above;
(*i*) any subspace of any space of the types above.

For details, see §A7; alternatively, you can check that the proof of Theorem 32E works for whichever of the examples above you find sufficiently interesting.

32E Theorem

Let $(X, \mathfrak{T}, \Sigma, \mu)$ be a quasi-Radon measure space such that $L^1(X)$ is separable, and $\mathscr{E}$ a collection of measurable sets of finite measure with $\#(\mathscr{E}) < \mathfrak{m}_K$. Let $s > 0$. Then there is a measurable set $V \subseteq X$ such that (i) $\mu(X \setminus V) \leq s$, (ii) $E \cap V$ is closed for every $E \in \mathscr{E}$, (iii) $\{E \cap V : E \in \mathscr{E}\}$ is countable. If X can be covered by a sequence of open sets of finite measure, then V can be taken to be closed.

Proof (*a*) Let $\langle U_n \rangle_{n\in\mathbf{N}}$ be a sequence of open sets of finite measure such that $\mu(X\setminus\bigcup_{n\in\mathbf{N}}U_n)=0$ [A7B*c*]; set $U=\bigcup_{n\in\mathbf{N}}U_n$. Then

$$\mu(E\cap U)=\inf\{\mu G: G \text{ open}, G\supseteq E\cap U\}$$

for every $E\in\Sigma$ [A7B*d*].

Let P be the set of all finite sequences

$$p=\langle (F_i^p, G_i^p)\rangle_{i\leq n(p)}$$

where $n(p)\in\mathbf{N}$, and for $i\leq n(p)$ F_i^p is closed, G_i^p is open and of finite measure, $F_i^p\subseteq G_i^p\subseteq U$, and $\sum_{i\leq n(p)}\mu(G_i^p\setminus F_i^p)<s$. Say $p\leq q$ if $n(p)\leq n(q)$ and $F_i^p\supseteq F_i^q$, $G_i^p\subseteq G_i^q$ for $i\leq n(p)$. Clearly P is partially ordered.

(*b*) P satisfies Knaster's condition upwards. **P** Let $R\subseteq P$ be uncountable. Then there are $m\in\mathbf{N}$ and $t>0$ such that

$$R_1=\{p:p\in R, n(p)=m, \textstyle\sum_{i\leq m}\mu(G_i^p\setminus F_i^p)<s-t\}$$

is uncountable. As $L^1(X)$ is separable, and all the F_i^p, G_i^p are of finite measure, there are $G_0,\ldots,G_m$, $F_0,\ldots,F_m\in\Sigma$ such that

$$R_2=\left\{p:p\in R_1, \mu(G_i^p\triangle G_i)\leq\frac{t}{4(m+1)},\right.$$
$$\left.\mu(F_i^p\triangle F_i)\leq\frac{t}{4(m+1)}\ \forall i\leq m\right\}$$

is uncountable. Now, if $p,q\in R_2$, a simple calculation shows that

$$\langle (F_i^p\cap F_i^q, G_i^p\cup G_i^q)\rangle_{i\leq m}$$

is a common upper bound for p and q in P, so that R_2 is upwards-linked. **Q**

(*c*) For $E\in\mathscr{E}$ set

$$Q_E=\{p:p\in P, \exists i\leq n(p), F_i^p\subseteq E\cap U\subseteq G_i^p\}.$$

Then Q_E is cofinal with P. **P** Let $p\in P$. Then $t=s-\sum_{i\leq n(p)}\mu(G_i^p\setminus F_i^p)>0$. Let G be an open set such that $E\cap U\subseteq G\subseteq U$ and $\mu G-\mu(E\cap U)<t$ (recalling that $\mu(E\cap U)\leq\mu E<\infty$); let $F\subseteq E\cap U$ be a closed set such that $\mu G-\mu F<t$. Define q by saying that $n(q)=n(p)+1$, $F_i^q=F_i^p$ and $G_i^q=G_i^p$ for $i\leq n(p)$, while $G_{n(p)+1}^q=G$, $F_{n(p)+1}^q=F$; then $p\leq q\in Q_E$. **Q**

(*d*) As $\#(\mathscr{E})<\mathfrak{m}_K$, there is an upwards-directed $R\subseteq P$ meeting every Q_E. Now if for $p \in P$ we set

$$H_p = \bigcup_{i \leq n(p)} G_i^p \setminus F_i^p,$$

we see that each H_p is an open set of measure less than s and that $H_p \subseteq H_q$ if $p \leq q$. So $H = \bigcup_{p \in R} H_p$ is open and $\mu H = \sup_{p \in R} \mu H_p \leq s$, because μ is τ-additive. Set $V = U \setminus H$; then $\mu(X \setminus V) = \mu H \leq s$.

(*e*) If $p, q \in R$ and $i \leq \min(n(p), n(q))$, then $F_i^p \cap V = F_i^q \cap V$. **P** p and q have a common upper bound $r \in R$; now

$$F_i^r \subseteq F_i^p \subseteq G_i^p \subseteq G_i^r$$

and $G_i^r \setminus F_i^r \subseteq H_r \subseteq H \subseteq X \setminus V$, so that $F_i^r \cap V = F_i^p \cap V$. Similarly $F_i^r \cap V = F_i^q \cap V$, so that $F_i^p \cap V = F_i^q \cap V$. **Q** It follows that

$$\mathscr{D} = \{F_i^p \cap V : p \in R, i \leq n(p)\}$$

is countable. Note also that each member of $\mathscr{D}$ is of the form $F_i^p \setminus H$, so is closed.

(*f*) We have already seen that V satisfies (i). If $E \in \mathscr{E}$, there is a $p \in R \cap Q_E$ and an $i \leq n(p)$ such that $F_i^p \subseteq E \cap U \subseteq G_i^p$; now $V \cap (G_i^p \setminus F_i^p) = \emptyset$, so $E \cap V = E \cap U \cap V = F_i^p \cap V \in \mathscr{D}$. From (*e*) above, this proves that V satisfies (ii) and (iii). Finally, if X can be covered by a sequence of open sets of finite measure, then we can suppose that they all belong to $\mathscr{E}$, in which case V will be closed.

32F Corollary

Let $(X, \mathfrak{T}, \Sigma, \mu)$ be a quasi-Radon measure space such that $L^1(X)$ is separable. Then μ is $\mathfrak{m}_K$-additive.

Proof Let $\mathscr{E}$ be a collection of negligible sets such that $\#(\mathscr{E}) < \mathfrak{m}_K$. Let $s > 0$. Take V as in Theorem 32E. Then $\bigcup_{E \in \mathscr{E}} E \cap V$ is a countable union of negligible sets, so is negligible, and

$$\bigcup \mathscr{E} \subseteq (X \setminus V) \cup \bigcup_{E \in \mathscr{E}} E \cap V$$

has outer measure less than or equal to s. As s is arbitrary, $\mu^*(\bigcup \mathscr{E}) = 0$. As μ is complete and locally determined, this is enough to prove that it is $\mathfrak{m}_K$-additive [A6O*c*].

32G Corollaries

(*a*) [$\mathfrak{m}_K = \mathfrak{c}$] (i) If $(X, \mathfrak{T}, \Sigma, \mu)$ is a quasi-Radon measure space and $L^1(X)$ is separable, then (X, Σ, μ) is properly $\mathfrak{c}$-based. (ii) If, moreover, $\mu\{x\} = 0$ for every $x \in X$, then (X, Σ, μ) has a $\mathfrak{c}$-Sierpiński subset.

(*b*) [$\mathfrak{m}_K > \omega_1$] Any PCA or CPCA set in a Polish space is universally measurable.

(*c*) Let (X, Σ, μ) be a σ-finite measure space in which Σ is countably generated (as σ-algebra), and $(X, \hat{\Sigma}, \hat{\mu})$ its completion [A6D]. (i) The union of fewer than $\mathfrak{m}_K$ μ-negligible sets is μ-negligible. (ii) $\hat{\mu}$ is $\mathfrak{m}_K$-additive. (iii) [$\mathfrak{m}_K = \mathfrak{c}$] (X, Σ, μ) and $(X, \hat{\Sigma}, \hat{\mu})$ are properly $\mathfrak{c}$-based. (iv) [$\mathfrak{m}_K = \mathfrak{c}$] If singleton sets in X are negligible, then there is a $\mathfrak{c}$-Sierpiński set for (X, Σ, μ) and $(X, \hat{\Sigma}, \hat{\mu})$.

Proof (*a*) Take $\mathscr{K}$ to be the family of non-empty supporting closed sets of finite measure, and argue as in 32C*b*.

(*b*) We need only recall that any PCA set is expressible as the union of ω_1 Borel sets [A5G*c*], and use 32F.

(*c*) Give X a topology $\mathfrak{T}$ for which $(X, \mathfrak{T}, \hat{\Sigma}, \hat{\mu})$ is a quasi-Radon measure space [A7B*f*]. Then $L^1(X, \hat{\mu}) \cong L^1(X, \mu)$ is separable, so the results concerning $(X, \hat{\Sigma}, \hat{\mu})$ are immediate from 32F and (*a*). Because (X, Σ, μ) has the same negligible sets and the same Sierpiński sets as $(X, \hat{\Sigma}, \hat{\mu})$ [A3E*d*], and is properly $\mathfrak{c}$-based iff $(X, \hat{\Sigma}, \hat{\mu})$ is [A6L*b*], the results for (X, Σ, μ) also follow.

32H **Corollary**

Let $(X, \mathfrak{T}, \Sigma, \mu)$ be a quasi-Radon measure space such that $L^1(X)$ is separable.

(*a*) If $\mathscr{E} \subseteq \Sigma$ and $\#(\mathscr{E}) < \mathfrak{m}_K$, there is a countable $\mathscr{D} \subseteq \Sigma$ such that $\mu E = \sup\{\mu D : D \in \mathscr{D}, D \subseteq E\}$ for every $E \in \mathscr{E}$.

(*b*) If, moreover, every member of $\mathscr{E}$ has finite measure, then we can choose $\mathscr{D}$ such that $\mu E = \inf\{\mu D : D \in \mathscr{D}, D \supseteq E\}$ for every $E \in \mathscr{E}$.

(*c*) If $s \in]0, \infty]$, then $\{E : E \in \Sigma, \mu E < s\}$ has the $(<\mathfrak{m}_K, \omega)$-covering property.

Proof I deal with (*b*) first. In this case, use 32E to choose for each $n \in \mathbf{N}$ a set $V_n \in \Sigma$ such that $\mu(X \setminus V_n) \leq 2^{-n}$ and $\{E \cap V_n : E \in \mathscr{E}\}$ is countable. Set

$$\mathscr{D} = \{E \cap V_n : n \in \mathbf{N}, E \in \mathscr{E}\} \cup \{(X \setminus V_n) \cup (E \cap V_n) : n \in \mathbf{N}, E \in \mathscr{E}\}.$$

Then $\mathscr{D}$ is countable. If $E \in \mathscr{E}$, then for any $n \in \mathbf{N}$

$$E \cap V_n \subseteq E \subseteq (E \cap V_n) \cup (X \setminus V_n)$$

so that $\mu E = \sup\{\mu D : D \in \mathscr{D}, D \subseteq E\} = \inf\{\mu D : D \in \mathscr{D}, D \supseteq E\}$.

To deduce (*a*), let $\langle X_n \rangle_{n\in\mathbb{N}}$ be an increasing sequence of measurable sets of finite measure covering X, and apply (*b*) to $\{E\cap X_n : E\in\mathscr{E}, n\in\mathbb{N}\}$; then for any $E\in\mathscr{E}$,

$$\mu E = \sup_{n\in\mathbb{N}}\mu(E\cap X_n) = \sup\{\mu D : D\in\mathscr{D}, D\subseteq E\}.$$

Finally, for (*c*), apply (*b*) to any collection $\mathscr{E}$ of sets of measure less than s with cardinal less than $\mathfrak{m}_K$, and consider $\{D : D\in\mathscr{D}, \mu D < s\}$.

32I Corollary

Let $(X, \mathfrak{T}, \Sigma, \mu)$ be a quasi-Radon measure space such that $L^1(X)$ is separable, and $\mathscr{E}$ a non-empty downwards-directed family of non-negligible measurable sets with $\#(\mathscr{E}) < \mathfrak{m}_K$. Then there is a decreasing sequence $\langle F_n \rangle_{n\in\mathbb{N}}$ of non-negligible measurable sets such that every member of $\mathscr{E}$ includes some F_n.

Proof Take $\mathscr{D}$ as in 32H*a*. Let $\langle D_n \rangle_{n\in\mathbb{N}}$ be a sequence in $\mathscr{D}$ such that each element of $\mathscr{D}$ appears infinitely often in the sequence. For $E\in\mathscr{E}$, set

$$A_E = \{n : n\in\mathbb{N}, D_n\subseteq E, \mu D_n > 0\}.$$

Then if $\mathscr{E}_0\subseteq\mathscr{E}$ is finite, $\bigcap\{A_E : E\in\mathscr{E}_0\}$ is infinite. **P** As $\mathscr{E}$ is downwards-directed, there is an $F\in\mathscr{E}$ such that $F\subseteq\bigcap\mathscr{E}_0$. Now $0<\mu F = \sup\{\mu D : D\in\mathscr{D}, D\subseteq F\}$ so there is a $D\in\mathscr{D}$ such that $D\subseteq F$ and $\mu D > 0$. So $\bigcap\{A_E : E\in\mathscr{E}_0\}\supseteq\{n : D_n = D\}$ is infinite. **Q** As $\#(\mathscr{E}) < \mathfrak{m}_K \leq \mathfrak{p}$, there is an infinite $I\subseteq\mathbb{N}$ such that $I\setminus A_E$ is finite for each $E\in\mathscr{E}$. Set

$$F_n = \bigcup_{i\in I, i\geq n} D_i \ \forall n\in\mathbb{N},$$

and see that $\langle F_n \rangle_{n\in\mathbb{N}}$ has the properties required.

32J Corollary

Let $(X, \mathfrak{T}, \Sigma, \mu)$ be a quasi-Radon measure space such that $L^1(X)$ is separable and $w(X) < \mathfrak{m}_K$. Then

(i) for any $s>0$ there is a second-countable measurable set $V\subseteq X$ such that $\mu(X\setminus V)\leq s$;
(ii) μ is inner regular for the second-countable subsets of X;
(iii) if $\mathfrak{T}$ is regular and Hausdorff, then μ is inner regular for the metrizable subsets of X.

Proof Let Y be the union of the open sets of finite measure in X, so that $\mu(X\setminus Y) = 0$. Let $\mathscr{U}$ be a base for the subspace topology of Y consisting of sets of finite measure with $\#(\mathscr{U}) < \mathfrak{m}_K$. By 32E, there is for any $s>0$ a measurable $V\subseteq X$ with $\mu(X\setminus V)\leq s$ and $\{V\cap U : U\in\mathscr{U}\}$ countable. Now $V\cap Y$ is second-countable and $\mu(X\setminus(V\cap Y))\leq s$.

If $E \in \mathscr{E}$ and $s < \mu E$, then by (i) there is a measurable second-countable V such that $s + \mu(X \setminus V) \leq \mu E$; now $E \cap V$ is second-countable and $\mu(E \cap V) \geq s$. This shows that μ is inner regular for the second-countable sets. Part (iii) follows at once.

32K **Corollary**

Let $(X, \mathfrak{T}, \Sigma, \mu)$ be a quasi-Radon measure space such that $L^1(X)$ is separable, and Y a topological space of weight less than $\mathfrak{m}_K$. If $f: X \to Y$ is a function, then the following are equivalent:

(i) f is measurable;
(ii) f is almost continuous;
(iii) there is a base $\mathscr{U}$ for the topology of Y such that $f^{-1}[U]$ is measurable for each $U \in \mathscr{U}$.

Proof (ii)$\Rightarrow$(i) is A7G*b*; (i)$\Rightarrow$(iii) is elementary. For (iii)$\Rightarrow$(ii), begin by taking a base $\mathscr{V}$ for the topology of Y with $\mathscr{V} \subseteq \mathscr{U}$ and $\#(\mathscr{V}) < \mathfrak{m}_K$ [A4B*j*]. Now suppose that $E \in \Sigma$ and that $s < \mu E$. Take $E_0 \in \Sigma$ such that $E_0 \subseteq E$ and $s < \mu E_0 < \infty$. Applying 32E to $\{E_0 \setminus f^{-1}[V]: V \in \mathscr{V}\}$, we can find a $W \subseteq X$ such that $\mu(X \setminus W) \leq \mu E_0 - s$ and $W \cap E_0 \setminus f^{-1}[V]$ is closed for every $V \in \mathscr{V}$. Now $W \cap E_0 \subseteq E$, $\mu(W \cap E_0) \geq s$ and $f \restriction W \cap E_0$ is continuous. As E and s are arbitrary, f is almost continuous.

32L **Proposition**

Let $(X, \mathfrak{T}, \Sigma, \mu)$ be a first-countable compact Radon measure space with $w(X) < \min(\mathfrak{m}_K, \omega_\omega)$. Then μ is inner regular for the compact metrizable G_δ sets.

Proof Write $\kappa = w(X)$. Let $\langle f_\xi \rangle_{\xi < \kappa}$ be a family of continuous functions from X to $[0,1]$ which separates the points of X. Let $\mathscr{C}$ be a family of countable subsets of κ such that every countable subset of κ is included in some member of $\mathscr{C}$, and $\#(\mathscr{C}) \leq \kappa$ [A2I]. (This is where I need the hypothesis that $\kappa < \omega_\omega$.) For $C \in \mathscr{C}$ set

$$g_C(x) = \langle f_\xi(x) \rangle_{\xi \in C} \in [0,1]^C \; \forall x \in X,$$

and

$$E_C = \{x : x \in X, g_C(x) = g_C(y) \Rightarrow x = y\}.$$

Then E_C is measurable. **P** $E_C = g_C^{-1}[g_C[E_C]]$. At the same time,

$$X \setminus E_C = \{x : \exists y \in X, k \in \mathbf{N}, \xi < \kappa \text{ such that } g_C(y) = g_C(x) \text{ and } |f_\xi(y) - f_\xi(x)| \geq 2^{-k}\}$$

is the union of κ compact sets. So $g_C[X] \setminus g_C[E_C] = g_C[X \setminus E_C]$ is the union of κ compact sets. But the Radon measure μg_C^{-1} on $[0,1]^C$ [A7Gc] has a separable L^1 space, because $[0,1]^C$ is metrizable; so $g_C[E_C]$ is measurable for μg_C^{-1}, by 32F, and $E_C = g_C^{-1}[g_C[E_C]]$ is measurable for μ. **Q**

As X is first-countable, $X = \bigcup_{C \in \mathscr{C}} E_C$. By 32C*a*, $\mu X = \sup_{C \in \mathscr{C}} \mu E_C$ (because $C \mapsto E_C$ is increasing, so that $\{E_C : C \in \mathscr{C}\}$ is directed upwards). But if $K \subseteq E_C$ is any compact set, $K = g_C^{-1}[g_C[K]]$ is G_δ and $K \cong g_C[K]$ is metrizable. So

$\mu X = \sup\{\mu K : K \subseteq X$ is a compact metrizable G_δ set$\}$, and it follows easily that μ is inner regular for the compact metrizable G_δ sets.

32M Corollary

(*a*) If X is a first-countable σ-finite Radon measure space and $w(X) < \min(\mathfrak{m}_K, \omega_\omega)$ then $L^1(X)$ is separable.

(*b*) A first-countable compact Hausdorff space of weight less than $\min(\mathfrak{m}_K, \omega_\omega)$ is a Radon space.

Proof (a) There is a sequence $\langle K_n \rangle_{n \in \mathbf{N}}$ of compact metrizable subsets of X such that $\mu(X \setminus \bigcup_{n \in \mathbf{N}} K_n) = 0$. Now $L^1(K_n)$ is separable for each $n \in \mathbf{N}$ so $L^1(X)$ is separable.

(*b*) For every Radon measure on X is completion regular [A7I*b*].

32N Theorem

$[\mathfrak{m}_K > \omega_1]$ Let $(X, \mathfrak{T}, \Sigma, \mu)$ be a hereditarily Lindelöf compact Radon measure space. Then there is a metrizable space Z and a continuous function $f: X \to Z$ such that $\mu(f^{-1}[f[E]])$ exists and is equal to μE for every $E \in \Sigma$. Consequently $L^1(X)$ is separable.

Proof ? Suppose the result is false. Then I can define continuous functions $f_\xi : X \to \mathbf{R}^{\mathbf{N}}$ inductively, for $\xi < \omega_1$, as follows. Given $\langle f_\eta \rangle_{\eta < \xi}$, where $\xi < \omega_1$, let $g_\xi : X \to (\mathbf{R}^{\mathbf{N}})^\xi$ be given by $g_\xi(x) = \langle f_\eta(x) \rangle_{\eta < \xi}$. As $(\mathbf{R}^{\mathbf{N}})^\xi$ is metrizable, the counter-hypothesis declares that there is a measurable set $E_\xi \subseteq X$ such that

$$\mu^*(g_\xi^{-1}[g_\xi[E_\xi]]) > \mu E_\xi.$$

Let E'_ξ be a G_δ set including E_ξ and of the same measure, and let $f_\xi : X \to \mathbf{R}^{\mathbf{N}}$ be a continuous function such that $f_\xi^{-1}[f_\xi[E'_\xi]] = E'_\xi$; this exists because every open set in X is a cozero set. Set $g(x) = \langle f_\xi(x) \rangle_{\xi < \omega_1}$ for each $x \in X$.

Consider $Z = g[X] \subseteq (\mathbf{R}^{\mathbf{N}})^{\omega_1}$, and the Radon measure μg^{-1} on Z [A7G*c*]. Because Z is a continuous image of a hereditarily Lindelöf space, it is hereditarily Lindelöf; in particular, it is first-countable. Also $w(Z) \leq \omega_1$. So by 32L μg^{-1} is inner regular for the metrizable G_δ sets. Let $s > 0$ be such that

$$C = \{\xi : \mu^*(g_\xi^{-1}[g_\xi[E_\xi]]) > \mu E_\xi + s\}$$

is uncountable, and let $K \subseteq Z$ be a compact metrizable G_δ set such that $\mu g^{-1}(Z \backslash K) \leq s$. Because K is G_δ and second-countable, there is a $\zeta < \omega_1$ such that $\pi_\zeta^{-1}[\pi_\zeta[K]] = K$ and π_ζ is injective on K, where $\pi_\zeta : Z \to (\mathbf{R}^{\mathbf{N}})^\zeta$ is the natural projection. Let $\xi \in C$ be greater than or equal to ζ. As $\mu g^{-1}[K] \geq \mu X - s$, there is an x belonging to

$$g^{-1}[K] \cap g_\xi^{-1}[g_\xi[E_\xi]] \backslash E'_\xi,$$

and there is a $y \in E_\xi \subseteq E'_\xi$ such that $g_\xi(y) = g_\xi(x)$. So

$$\pi_\zeta(g(y)) = g_\zeta(y) = g_\zeta(x) = \pi_\zeta(g(x))$$

and $g(y) = g(x)$ because $\pi_\zeta^{-1}[\pi_\zeta[K]] = K$ and π_ζ is injective on K. But $g^{-1}[g[E'_\xi]] = E'_\xi$ because $f_\xi^{-1}[f_\xi[E'_\xi]] = E'_\xi$, so we find that x belongs to E'_ξ; and we chose $x \notin E'_\xi$. **X**

It follows at once that $L^1(X) \cong L^1(Z)$ is separable.

32O Sources

Horn & Tarski 48 for 32A. Martin & Solovay 70 for 32G*b* and (in essence) 32F. Kunen 76 for 32I. Talagrand 80*a* for 32G*c* (in part). Fremlin 81 for 32K.

32P Exercises

(*a*) Let P be a partially ordered set expressible as $\bigcup_{n \in \mathbf{N}} P_n$, where any up-antichain in P has finite intersection with every P_n. Show that P satisfies Knaster's condition upwards.

(*b*) In $\{0,1\}^{\omega_1}$ set $F_\xi = \{x : x(\eta) = 0 \ \forall \eta \geq \xi\}$. Considering $\bigcup_{\xi < \omega_1} F_\xi$, show that (i) there is a Radon probability space in which the union of ω_1 closed negligible sets need not be negligible, (ii) there is a quasi-Radon probability space which is the union of ω_1 closed negligible sets, (iii) there is a Radon probability space with a separable L^1 space in which the family of negligible compact sets does not have the (ω_1, ω)-covering property. [R.G. Haydon, oral communication.]

(*c*) Let $(X, \mathfrak{T}, \Sigma, \mu)$ be a quasi-Radon measure space such that $L^1(X)$ is separable. Show that, for any $s > 0$, $\{G : G \in \mathfrak{T}, \mu G < s\}$ satisfies Knaster's condition upwards. Use this to prove 32F. [Take $\mu X < \infty$ first.]

(*d*) $[\mathfrak{m}_K = \mathfrak{c}]$ If $E \subseteq [0,1]^2$ has planar Lebesgue measure 1, there is an $A \subseteq [0,1]$ such that A has linear outer Lebesgue measure 1 and $(s,t) \in E$ whenever s, t are distinct members of A. [See A6L*c*.]

(*e*) Let Z be the hyperstonian space of the unit interval. Then (i) the union of fewer than $\mathfrak{m}_K$ nowhere dense sets in Z is nowhere dense, (ii) if $\mathscr{E}$ is a family of open-and-closed sets in Z with the finite intersection property, and $\#(\mathscr{E}) < \mathfrak{m}_K$, there is a non-empty zero set $F \subseteq \bigcap \mathscr{E}$. [Use 32I. See Frankiewicz 81.]

(*f*) $[\mathfrak{m}_K = \mathfrak{c}]$ (i) There is a $\mathfrak{c}$-Sierpiński subset of $\mathbf{R}$ which is an additive subgroup of $\mathbf{R}$. [Use the method of A3F*a*, but requiring

$$x_\xi \in \mathbf{R} \setminus \bigcup \{E_\eta + Y_\xi : \eta < \xi, E_\eta \in \mathscr{I}\},$$

where Y_ξ is the subgroup generated by $\{x_\eta : \eta < \xi\}$. Compare 22N*f*.] (ii) If $A \subseteq \mathbf{R}$ is any $\mathfrak{c}$-Sierpiński set then every relatively Borel subset of A is relatively G_δ but not every relatively Borel subset of A is relatively open. [Use 23B. See Miller 79, Theorem 17; and 23N*f*.]

(*g*) $[\mathfrak{m}_K = \mathfrak{c}]$ Let Σ be the algebra of subsets of $\mathbf{R}$ which are universally measurable and have the strong Baire property, and let μ be the restriction of Lebesgue measure to Σ. Then $(\mathbf{R}, \Sigma, \mu)$ has a multiplicative lifting. [Compare Mauldin 78, Theorem 5.2.]

(*h*) Let $(X, \mathfrak{T}, \Sigma, \mu)$ be a quasi-Radon measure space such that $L^1(X)$ is separable, and $\mathscr{E} \subseteq \Sigma$ a non-empty family of non-negligible sets with $\#(\mathscr{E}) < \mathfrak{m}_K$. Then $\mathscr{E}$ is expressible as $\bigcup_{n \in \mathbf{N}} \mathscr{E}_n$, where $\mu(\bigcap \mathscr{E}_n) > 0$ for each $n \in \mathbf{N}$. [Kunen 80, Ex. 2.27.]

(*i*) Let (X, μ) be a σ-finite quasi-Radon measure space such that X is the support of μ. Suppose *either* (α) that μ is a Radon measure *or* (β) that $L^1(X)$ is separable. (i) If $\mathscr{G}$ is a family of non-empty open sets in X and $\#(\mathscr{G}) < \mathfrak{m}_K$, there is a countable subset of X which meets every member of $\mathscr{G}$. [For (α), first take X compact and use the method of 43E. For (β), use 32H*a*.] (ii) If $\omega < \mathrm{cf}(\kappa) \le \kappa < \mathfrak{m}_K$ then κ is a caliber of X. [See Tall 78, Theorem 13(11).] (iii) $[\mathfrak{m}_K > \omega_1]$ If X is metalindelöf and μ is locally finite, there is a sequence of open sets of finite measure covering X. [See Gardner & Pfeffer 80, 3.6.]

(*j*) Let (X, Σ, μ) be a complete σ-finite measure space such that $L^1(X)$ is separable and μ is $\mathfrak{m}_K$-additive. (i) If $\mathscr{E}$ is a collection of measurable sets of finite measure and $\#(\mathscr{E}) < \mathfrak{m}_K$, and if $s > 0$, there is a $V \in \Sigma$ such that $\mu(X \setminus V) \le s$ and $\{E \cap V : E \in \mathscr{E}\}$ is countable. (ii) (X, Σ, μ) has the properties of 32H–I.

(*k*) Let $(X, \mathfrak{T}, \Sigma, \mu)$ be a quasi-Radon measure space such that

$L^1(X)$ is separable, and let $\langle Y_\iota \rangle_{\iota\in I}$ be a family of topological spaces with $\#(I) < \mathfrak{m}_K$. Let $f: X \to \prod_{\iota\in I} Y_\iota$ be a function. Then f is almost continuous iff $\pi_\iota f: X \to Y_\iota$ is almost continuous for every $\iota \in I$, where π_ι is the ι-coordinate map. [Hint: Suppose $\mu X < \infty$. Given $s > 0$, show that $\{G: \pi_\iota f \restriction X \setminus G$ is continuous$\}$ is cofinal with $\{G: G \in \mathfrak{T}, \mu G < s\}$. See FREMLIN 81.]

(*l*) Let X be a compact Hausdorff space of weight $< \mathfrak{m}_K$, and μ a completion regular Radon measure on X such that $L^1(X)$ is separable. Then $\chi(x, X) \leq \omega$ for almost all $x \in X$. [Use 32J. Compare 22N*d*.]

(*m*) (i) Let X be a quasi-Radon measure space such that $L^1(X)$ is separable, and E a Banach space such that $d(E') < \mathfrak{m}_K$. If $\varphi: X \to E$ is a bounded scalarly measurable function it is almost continuous for the weak topology on E; consequently there is a separable subspace F of E such that $\varphi^{-1}[E \setminus F]$ is negligible, and φ is almost continuous for the norm topology of E. [Use 32K and Tortrat's theorem, A8F.] (ii) $[\mathfrak{m}_K > \omega_1]$ Let $\varphi: [0,1] \to C([0, \omega_1])$ be bounded and scalarly measurable. Then φ is Bochner integrable. [EDGAR *a*.]

(*n*) $[\mathfrak{m}_K > \omega_1]$ Let X be a hereditarily Lindelöf compact Hausdorff space. Then the positive cone of $C(X)'$ is first-countable for the vague topology $\mathfrak{T}_s(C(X)', C(X))$. [Hint: use 32N.]

32Q **Further results**

(*a*) $[\mathfrak{m}_K = \mathfrak{c}]$ Let $\langle I(n) \rangle_{n\in\mathbf{N}}$ be a disjoint sequence in $[\mathbf{N}]^{<\omega}$ such that $\sum_{n\in\mathbf{N}} s^{\#(I(n))} < \infty$ for every $s \in [0, 1[$. Then there is a non-Haar-measurable filter $\mathscr{F}$ on $\mathbf{N}$ such that $\{n: A \cap I_n = \emptyset\}$ is finite for every $A \in \mathscr{F}$. [TALAGRAND 80*a*.]

(*b*) $[\mathfrak{m}_K = \mathfrak{c}]$ There is a non-Haar-measurable filter $\mathscr{F}$ on $\mathbf{N}$ with the strong Baire property. [TALAGRAND 83.]

(*c*) $[\mathfrak{m}_K = \mathfrak{c}]$ Any infinite-dimensional separable Banach space has a meagre dense hyperplane. [ARIAS DE REYNA 80.]

(*d*) $[\mathfrak{m}_K = \mathfrak{c}]$ Let X be a compact Hausdorff topological group and μ its left-invariant Haar measure. For $f \in \mathscr{L}^\infty(X)$, $x \in X$ set $(L_x f)(y) = f(xy)$. Then for $f \in \mathscr{L}^\infty$ the following are equivalent: (i) there is a $g \in \mathscr{L}^\infty$ such that $g = f$ almost everywhere and g is continuous at almost every point; (ii) for every multiplicative homomorphism $\varphi: L^\infty(X) \to \mathbf{R}$ the map $x \mapsto \varphi((L_x f)^\cdot)$ is measurable. [TALAGRAND 82, TALAGRAND *a*.]

(*e*) $[\mathfrak{m}_K = \mathfrak{c}]$ Let (X, Σ, μ) be a probability space, $K \subseteq \mathbf{R}^X$ a set of measurable functions which is compact for the topology of $\mathbf{R}^X$ but is not compact for the topology of convergence in measure. Then (i) there are $s < t$ in $\mathbf{R}$ and a $Y \subseteq X$ such that $\#(Y) = \mathfrak{c}$ and, for every $A \subseteq Y$, there is an $f \in K$ such that $f(x) \leq s$ for $x \in A$ and $f(x) \geq t$ for $x \in Y \setminus A$; (ii) $\beta\mathbf{N}$ can be embedded into K; (iii) Σ is not the completion of a σ-algebra generated by $\mathfrak{c}$ or fewer sets. [TALAGRAND 80*a*, TALAGRAND *a*.]

(*f*) [$\mathfrak{m}_K = \mathfrak{c}$] Let E be a Banach space. Then *either* $\ell^\infty(\mathbf{N})$ is isomorphic to a quotient of E *or* whenever (X, Σ, μ) is a probability space and $\varphi: X \to E$ is a bounded Pettis integrable function, the indefinite Pettis integral of φ has relatively compact range in E. [TALAGRAND 80*b*.]

(*g*) [$\mathfrak{m}_K = \mathfrak{c}$] There is a set $A \subseteq \mathbf{R}$ such that $\#(A) = \mathfrak{c}$, $\#(A \cap G) < \mathfrak{c}$ for every open set G including $\mathbf{Q}$, and $A + E$ is (Lebesgue) negligible for every negligible $E \subseteq \mathbf{R}$. [ERDÖS, KUNEN & MAULDIN 81.]

(*h*) [$\mathfrak{m}_K > \omega_2$] Suppose there is a two-valued-measurable cardinal. Then every PCPCA set in a Polish space is universally measurable. [MARTIN & SOLOVAY 70.]

(*i*) [$\mathfrak{m}_K > \omega_1$] Let X and Y be Polish spaces and $A \subseteq X \times Y$ an analytic set such that all vertical sections of A are Borel. Let μ be a Radon measure on X. Then there is a μ-negligible set $E \subseteq X$ such that $A \setminus (E \times Y)$ is Borel. [MAULDIN *a*.]

(*j*) Let X be a topological space, μ_L Lebesgue measure on $[0,1]$. Let us say that X has the 'measure extension property' if whenever $f: X \to [0,1]$ is Borel measurable and $\mu_L^*(f[X]) = 1$, there is a topological measure on X for which f is inverse-measure-preserving. [$\mathfrak{m}_K > \omega_1$] Every PCA set in a Polish space has the measure extension property. [MAITRA, RAO & RAO 79.]

(*k*) Let X be a Radon measure space, Y a complete metric space, and $R \subseteq X \times Y$ a set such that all vertical sections of R are closed and separable, and $R^{-1}[G]$ is measurable for every open $G \subseteq Y$. Assume *either* $\mathfrak{m}_K = \mathfrak{c}$ *or* $\mathfrak{m}_K > \omega_1$ and that there is no proper ω_1-saturated σ-ideal of $\mathscr{P}\mathfrak{c}$ containing all singletons [B2D]. Then R has an almost continuous selector. [FREMLIN 82. Compare 43Q*b*.]

(*l*) (i) If X is a compact Hausdorff space, $\mathscr{E}$ a non-empty family of universally measurable subsets of X with $\#(\mathscr{E}) < \mathfrak{m}_K$, then $\bigcap \mathscr{E}$ is measure-compact. [FREMLIN 77*b*.] (ii) $\mathbf{R}^\kappa \times Y$ is measure-compact for any separable metric space Y and any $\kappa < \mathfrak{m}_K$. [KOUMOULLIS 83.]

(*m*) [$\mathfrak{m}_K > \omega_1$] If $\kappa \geq \omega_1$, there is no completion regular Radon probability measure on $\{0,1\}^\kappa$ with a separable L^1 space. [CHOKSI & FREMLIN 79.]

(*n*) [$\mathfrak{m}_K = \mathfrak{c}$] Let X be *either* the hyperstonian space of the unit interval *or* $\beta\mathbf{N}\setminus\mathbf{N}$. Then (i) there is a non-*p*-point $x \in X$ which is not in $\bar{B}$ for any $B \in [X\setminus\{x\}]^{<\mathfrak{c}}$ (ii) there is a non-empty countable $A \subseteq X$, without isolated points, such that $A \cap \bar{B} = \emptyset$ whenever $B \subseteq X\setminus A$ is a discrete set of cardinal less than $\mathfrak{c}$. [Kunen 76; see also Frankiewicz 81.]

32R **Problems**

(*a*) Give ω_1 its order topology, and let $f:[0,1] \to \omega_1$ be measurable. Does f have to be almost continuous? [Fremlin 81.]

(*b*) Are 32L and 32M true if $\omega_\omega \leq w(X) < \mathfrak{m}_K$?

(*c*) [$\mathfrak{m}_K > \omega_1$] Let X be a first-countable compact Radon measure space. Is $L^1(X)$ necessarily separable?

(*d*) Let $\mathscr{E}$ be the ideal of sets $E \subseteq \mathbf{R}^2$ such that for every $\varepsilon > 0$ there is an open set $G \supseteq E$ such that all the vertical sections of G have linear Lebesgue measure less than or equal to ε. Is $\mathscr{E}$ always $\mathfrak{m}$-additive? [G. Mokobodzki.] *Added in proof*: No (J. Pawlikowski).

(*e*) [$\mathfrak{m} = \mathfrak{c}$] Does the restriction of Lebesgue measure to the Borel sets of $\mathbf{R}$ have a multiplicative lifting? [Mauldin 78.]

(*f*) Do the results of 32P*d*, 32Q*a*–*e* and 32Q*k* require special axioms? If in 32P*d* we take E to be open, $\mathfrak{p} = \mathfrak{c}$ will be enough; will it be necessary?

Notes and comments

The simple fact that the support of a finite Radon measure has the ccc gives us reason to suppose that Martin's axiom will have an effect in the theory of Radon measures; and 32B–C are really no more than working out the details. This does not, however, take us very far (though 32Q*a*–*d*, 32Q*k* and 32Q*l*(i) can be obtained from 32C alone); the problem being that (as 32P*b* shows) the union of 'few' measurable sets can be non-measurable, even though it cannot be a non-negligible measurable set.

In §22, of course, we have already seen that the corresponding problem for nowhere-dense sets changes character completely in second-countable spaces. The same thing happens for measure, for different reasons; the union of fewer than $\mathfrak{p}$ meagre sets in $\mathbf{R}$ is meagre, and the union of fewer than $\mathfrak{m}_K$ negligible sets in $\mathbf{R}$ is negligible. As far as I know, there is only one standard proof of this, which goes back to Martin & Solovay 70. It relies on finding a family $\mathfrak{T}$ of sets such that (i) if $s > 0$, then $\{G : G \in \mathfrak{T}, \mu G < s\}$ is upwards-ccc, (ii) if E is negligible, then

$\inf\{\mu G: G \in \mathfrak{T}, G \supseteq E\} = 0$, (iii) if $\mathscr{G} \subseteq \mathfrak{T}$ and $\mathscr{G}$ is upwards-directed, then $\mu(\bigcup \mathscr{G}) = \sup_{G \in \mathscr{G}} \mu G$. Of these, (i) requires L^1 to be separable, while (ii) and (iii) describe the role of the open sets in a quasi-Radon measure space. Accordingly I have expressed the results here in terms of the theory of quasi-Radon measure spaces with separable L^1 spaces.

I have not in fact used the 'standard' proof here (it is outlined in 32P*c*), since it seems to save time to go straight to 32E. This is not only striking in itself but seems to encapsulate the whole of this part of the theory; we do not need another partially ordered set in the whole section (except perhaps in 32P*k*). In fact, however, 32E can be derived from 32F; see the notes to §33, and 32P*j*. The proof I give of 32E uses ideas from K. Kunen's proof of 32I, and is clearly associated with the method of 31A.

32G lists the most frequently used consequences of 32F. 32C*b*, 32G*a* and 32G*c*(iii) give us three ways in which Martin's axiom provides us with properly based measure spaces. A6L and A7B*l* give some constructions involving properly based measures; others are used in 32Q*a*–*b*, 32Q*d* and 32Q*k*. (I should remark that I do not know that special axioms are necessary for any of these.) 32G*c* spells out the standard method of applying results on quasi-Radon measure spaces to measures on countably-generated σ-algebras.

32H*a* is a curious fact; I use it here on the way to 32H*c*, 32I and 32P*h*. 32I has been applied by K. Kunen and R. Frankiewicz to 32P*e*(ii) and 32Q*n*. Another result of this kind is in 33D.

32J–N should be regarded as a progress report on my current research. I shall be surprised if they are really the best possible results; see, in particular, 32R*b*–*c*. Concerning 32R*b*, it seems to be likely that $[\omega_\omega]^\omega$ need not always have a cofinal subset of cardinal $\omega_{\omega+1}$. 32N shows that Martin's axiom has effects on the measure theory of hereditarily Lindelöf spaces as well as on their topological properties, discussed in §44. The standard example of the phenomenon of 32N is the split interval A7J] with the vertical projection onto the ordinary interval. Note that in KUNEN 81 there is an example, depending on CH, of a hereditarily Lindelöf compact Radon measure space with a non-separable L^1 space [see 44J*c*.].

Among the exercises, 32P*a* is a trivial generalization of 32A. I have dressed 32P*g* up as interestingly as I can; but note that it would evaporate if 32R*e* could be answered positively. (Subject to the continuum hypothesis, there is certainly a multiplicative lifting of the restriction of Lebesgue measure to the Borel sets. SHELAH *b* describes a model in which there is no such lifting.) 32P*k* is an easy corollary of 33D below, but also has a direct proof, as indicated.

32Q*e* uses 32G*c*; 32Q*f* uses 32Q*e*(i). 32Q*h* corresponds to 23N*b*. 32Q*i*–*j* are associated with 32G*b*. 32Q*l*(ii) depends on 32F, and 32Q*m* can be thought of as a corollary of 32P*l*. Finally, 32F provides us with a positive answer to 32R*a* when $\mathfrak{m}_K > \omega_1$; a more interesting argument gives the same answer when Lebesgue measure is properly ω_1-based [FREMLIN 81]; I have no idea whether there is an answer in ZFC.

33 Measure theory II

In this section I collect the overflow from §32, consisting of results which stand a little apart from the main body. I begin with a theorem on sets of strong measure 0 [33B] which is what one would expect; I continue with similar results on the domination of summable sequences [33C] and a kind of $(<\mathfrak{m}_K, \omega)$-covering property for null sequences of measurable sets [33D]. The second part of the section is devoted to S.A. Argyros' results on the embedding of $\ell^1(\omega_1)$ into Banach spaces.

33A Definition

Let (X, ρ) be a non-empty metric space. A set $A \subseteq X$ is of **strong measure 0** if for any sequence $\langle s_n \rangle_{n\in\mathbf{N}}$ of strictly positive real numbers there is a sequence $\langle x_n \rangle_{n\in\mathbf{N}}$ in X such that $A \subseteq \bigcup_{n\in\mathbf{N}} V(x_n, s_n)$, where

$$V(x, s) = \{y : \rho(y, x) \leq s\}.$$

If $X = \varnothing$ then I will still allow $\varnothing$ as a set of strong measure 0.

33B Theorem

In a separable metric space, the union of fewer than $\mathfrak{m}_K$ sets of strong measure 0 has strong measure 0.

Proof (*a*) Let (X, ρ) be a separable metric space. If $X = \varnothing$ there is nothing to prove; take $X \neq \varnothing$. Let $\mathscr{E}$ be a family of sets of strong measure 0 in X with $\#(\mathscr{E}) < \mathfrak{m}_K$. Fix on a sequence $\langle s_n \rangle_{n\in\mathbf{N}}$ of strictly positive real numbers and set $s'_n = \min_{i \leq n} s_i$, so that $\langle s'_n \rangle_{n\in\mathbf{N}}$ is a decreasing sequence of strictly positive numbers. Let $D \subseteq X$ be a countable dense set.

(*b*) Let P be the set of sequences $\langle I_k \rangle_{k\in\mathbf{N}}$ in $[D]^{<\omega}$ such that $\lim_{n\to\infty}(1/n)\,\#(\bigcup_{k<n} I_k) = 0$ and $\#(\bigcup_{k<n} I_k) \leq n$ for every $n\in\mathbf{N}$. Say $\langle I_k \rangle_{k\in\mathbf{N}} \leq \langle J_k \rangle_{k\in\mathbf{N}}$ if $I_k \subseteq J_k$ for every $k\in\mathbf{N}$. Then P satisfies Knaster's condition upwards. **P** Let $R \subseteq P$ be uncountable. Then there is an $m\in\mathbf{N}$ such that

$$R_1 = \{\langle I_k \rangle_{k\in\mathbf{N}} : \langle I_k \rangle_{k\in\mathbf{N}} \in R, (1/n)\,\#(\textstyle\bigcup_{k<n} I_k) \leq \tfrac{1}{2}\ \forall n \geq m\}$$

is uncountable. As $[D]^{<\omega}$ is countable (this is where we use the separability of X), there is a finite sequence $\langle K_k \rangle_{k<m}$ such that

$$R_2 = \{\langle I_k \rangle_{k\in\mathbf{N}} : \langle I_k \rangle_{k\in\mathbf{N}} \in R_1, I_k = K_k \, \forall k < m\}$$

is uncountable. Now if $\langle I_k \rangle_{k\in\mathbf{N}}$ and $\langle J_k \rangle_{k\in\mathbf{N}}$ are any two members of R_2, $\langle I_k \cup J_k \rangle_{k\in\mathbf{N}} \in P$ and is a common upper bound of $\langle I_k \rangle_{k\in\mathbf{N}}$ and $\langle J_k \rangle_{k\in\mathbf{N}}$. Thus R_2 is upwards-linked; as R is arbitrary, P satisfies Knaster's condition upwards. **Q**

(*c*) For $E \in \mathscr{E}$ set

$$Q_E = \{\langle I_k \rangle_{k\in\mathbf{N}} : \langle I_k \rangle_{k\in\mathbf{N}} \in P, E \subseteq \bigcup \{V(x, s'_k) : k \in \mathbf{N}, x \in I_k\}\}.$$

Then Q_E is cofinal with P. **P** Let $\langle I_k \rangle_{k\in\mathbf{N}} \in P$. Let $m \in \mathbf{N}$ be such that $\#(\bigcup_{k<n} I_k) \leq \frac{1}{2}n$ for every $n \geq m$. Write $h(i) = 2 + m + i^2$ for $i \in \mathbf{N}$. Because E is of strong measure 0, there is a sequence $\langle y_i \rangle_{i\in\mathbf{N}}$ in X such that $E \subseteq \bigcup_{i\in\mathbf{N}} V(y_i, \frac{1}{2}s'_{h(i)})$. Choose $x_i \in D$ such that $\rho(x_i, y_i) \leq \frac{1}{2}s'_{h(i)}$, and set $J_{h(i)} = I_{h(i)} \cup \{x_i\}$ for each $i \in \mathbf{N}$; while for $k \in \mathbf{N} \setminus h[\mathbf{N}]$ set $J_k = I_k$. Then (because I chose h carefully) $\langle J_k \rangle_{k\in\mathbf{N}} \in P$ and $\langle I_k \rangle_{k\in\mathbf{N}} \leq \langle J_k \rangle_{k\in\mathbf{N}} \in Q_E$, since $E \subseteq \bigcup_{i\in\mathbf{N}} V(x_i, s'_{h(i)})$. **Q**

(*d*) Let $R \subseteq P$ be an upwards-directed set meeting every Q_E. Set

$$C_n = \bigcup \{I_n : \langle I_k \rangle_{k\in\mathbf{N}} \in R\}$$

for each $n \in \mathbf{N}$. Then we have

$$\#(\bigcup_{k<n} C_k) \leq n \, \forall n \in \mathbf{N}$$

because R is upwards-directed. So we can find a sequence $\langle x_i \rangle_{i\in\mathbf{N}}$ in D such that $C_n \subseteq \{x_i : i \leq n\}$ for every $n \in \mathbf{N}$. Now $\bigcup \mathscr{E} \subseteq \bigcup_{i\in\mathbf{N}} V(x_i, s_i)$. **P** If $y \in E \in \mathscr{E}$ there is an $\langle I_k \rangle_{k\in\mathbf{N}}$ in $R \cap Q_E$. Now there are a $k \in \mathbf{N}$ and an $x \in I_k$ such that $y \in V(x, s'_k)$. In this case $x \in C_k$ so $x = x_i$ for some $i \leq k$ and $y \in V(x_i, s'_k) \subseteq V(x_i, s_i)$. **Q**

(*e*) As $\langle s_n \rangle_{n\in\mathbf{N}}$ is arbitrary, $\bigcup \mathscr{E}$ is of strong measure 0, as required.

33C Proposition

Let A be a family of absolutely summable sequences in $\mathbf{R}$ with $\#(A) < \mathfrak{m}_K$. Then there is a strictly positive summable sequence h with $\lim_{n\to\infty}[f(n)/h(n)] = 0$ for every $f \in A$.

Proof Let P be the set

$$\{g : g \in \ell^1(\mathbf{N}), g(n) > 0 \, \forall n \in \mathbf{N}, \sum_{n\in\mathbf{N}} g(n) < 1\}.$$

Then P satisfies Knaster's condition upwards. **P** If $R \subseteq P$ is uncountable, there is an $\varepsilon > 0$ such that

$$R_1 = \{g : g \in R, \sum_{n \in \mathbb{N}} g(n) < 1 - \varepsilon\}$$

is uncountable. As $\ell^1(\mathbb{N})$ is separable, there is an uncountable $R_2 \subseteq R_1$ with $\| \|_1$-diameter $\leq \varepsilon$; now R_2 is upwards-linked in P. **Q**

For $f \in A$, set

$$Q_f = \{g : g \in P, \lim_{n \to \infty} [f(n)/g(n)] = 0\}.$$

Then Q_f is cofinal with P. As $\#(A) < \mathfrak{m}_K$, there is an upwards-directed $R \subseteq P$ meeting every Q_f. Set

$$h(n) = \sup \{g(n) : g \in R\};$$

then (because R is upwards-directed) $\sum_{n \in \mathbb{N}} h(n) \leq 1$, and h is summable; while if $f \in A$, then $\lim_{n \to \infty} [f(n)/h(n)] = 0$ because R meets Q_f.

33D **Proposition**

Let $(X, \mathfrak{T}, \Sigma, \mu)$ be a quasi-Radon measure space with a separable L^1 space, and $\kappa < \mathfrak{m}_K$. Let $\langle E_i^\xi \rangle_{i \in \mathbb{N}, \xi < \kappa}$ be a family in Σ such that $\lim_{i \to \infty} \mu E_i^\xi = 0$ for every $\xi < \kappa$. Then there is a sequence $\langle F_i \rangle_{i \in \mathbb{N}}$ in Σ such that $\lim_{i \to \infty} \mu F_i = 0$ and $\{i : E_i^\xi \not\subseteq F_i\}$ is finite for every $\xi < \kappa$. If $\langle E_i^\xi \rangle_{i \in \mathbb{N}}$ is a decreasing sequence for every ξ, then we can take $\langle F_i \rangle_{i \in \mathbb{N}}$ to be decreasing also.

Proof (*a*) There is a decreasing sequence $\langle s_i \rangle_{i \in \mathbb{N}}$ of strictly positive real numbers such that $\lim_{n \to \infty} \mu E_i^\xi / s_i = 0$ for every $\xi < \kappa$. **P** Apply 14B to $\langle f_\xi \rangle_{\xi < \kappa}$ where $f_\xi : \mathbb{N} \to \mathbb{N}$ is such that $\mu E_i^\xi \leq 2^{-n}$ for $i \geq f_\xi(n)$. **Q** Let P be the set of all sequences $\langle E_i \rangle_{i \in \mathbb{N}}$ in Σ such that $\mu E_i < s_i$ for every $i \in \mathbb{N}$ and $\lim_{i \to \infty} \mu E_i / s_i = 0$. Give P the natural order

$$\langle E_i \rangle_{i \in \mathbb{N}} \leq \langle F_i \rangle_{i \in \mathbb{N}} \text{ if } E_i \subseteq F_i \text{ for every } i \in \mathbb{N}.$$

(*b*) P satisfies Knaster's condition upwards. **P** Let $R \subseteq P$ be uncountable. Then there are $m \in \mathbb{N}$, $\varepsilon > 0$ such that

$$R_1 = \{\langle E_i \rangle_{i \in \mathbb{N}} : \langle E_i \rangle_{i \in \mathbb{N}} \in R, \mu E_i < \tfrac{1}{2} s_i \, \forall i > m, \mu E_i < s_i - \varepsilon \, \forall i \leq m\}$$

is uncountable. As $L^1(X)$ is separable there is an uncountable $R_2 \subseteq R_1$ such that

$$\mu(E_i \triangle F_i) \leq \varepsilon \text{ whenever } i \leq m \text{ and } \langle E_i \rangle_{i \in \mathbb{N}}, \langle F_i \rangle_{i \in \mathbb{N}} \in R_2.$$

Now R_2 is upwards-linked in P. **Q**

(*c*) For $\xi < \kappa$ set

$Q_\xi = \{\langle E_i \rangle_{i \in \mathbb{N}} : \langle E_i \rangle_{i \in \mathbb{N}} \in P, \ \exists n \in \mathbb{N}, E_i^\xi \subseteq E_i \, \forall i \geq n\}$. Then Q_ξ is cofinal with P. **P** If $\langle E_i\rangle_{i\in\mathbf{N}}\in P$, let $m\in\mathbf{N}$ be such that $\mu(E_i\cup E_i^\xi)<s_i$ for $i\ge m$; then set $F_i=E_i$ for $i<m$, $E_i\cup E_i^\xi$ for $i\ge m$, and see that $\langle E_i\rangle_{i\in\mathbf{N}}\le\langle F_i\rangle_{i\in\mathbf{N}}\in Q_\xi$. **Q**

(*d*) As $\kappa<\mathfrak{m}_K$ there is an upwards-directed $R\subseteq P$ meeting every Q_ξ. Set

$$F_j=\bigcup\{E_j:\langle E_i\rangle_{i\in\mathbf{N}}\in R\}$$

and observe that by 32F $\mu F_j\le s_j$ for every $j\in\mathbf{N}$, so that $\lim_{j\to\infty}\mu F_j=0$. Because R meets Q_ξ, $\{i:E_i^\xi\not\subseteq F_i\}$ is finite for every $\xi<\kappa$.

(*e*) If every $\langle E_i^\xi\rangle_{i\in\mathbf{N}}$ is decreasing, let P^* be the set of decreasing sequences in P. Because a subset of P^* is bounded above in P^* iff it is bounded above in P, P^* also satisfies Knaster's condition upwards. Observe next that $Q_\xi\cap P^*$ is cofinal with P^* for each ξ. **P** Let $\langle E_i\rangle_{i\in\mathbf{N}}\in P^*$. Let $m\in\mathbf{N}$ be such that $\mu(E_i\cup E_i^\xi)<s_i$ for $i\ge m$. Let $k\ge m$ be such that

$$\mu E_k^\xi<s_i-\mu E_i\ \forall i<m.$$

Set

$$F_i=E_i\cup E_k^\xi \text{ for } i\le k,$$
$$=E_i\cup E_i^\xi \text{ for } i>k.$$

Then $\langle F_i\rangle_{i\in\mathbf{N}}$ is decreasing. If $i<m$, then $\mu F_i\le\mu E_i+\mu E_k^\xi<s_i$; if $m\le i\le k$, then $\mu F_i=\mu(E_i\cup E_k^\xi)\le\mu(E_i\cup E_i^\xi)<s_i$; if $i>k$, then $\mu F_i=\mu(E_i\cup E_i^\xi)<s_i$. So $\langle F_i\rangle_{i\in\mathbf{N}}\in P^*$ and $\langle E_i\rangle_{i\in\mathbf{N}}\le\langle F_i\rangle_{i\in\mathbf{N}}\in Q_\xi\cap P^*$. **Q**

Accordingly we can find an upwards-directed $R^*\subseteq P^*$ meeting every Q_ξ, and setting

$$F_j^*=\bigcup\{E_j:\langle E_i\rangle_{i\in\mathbf{N}}\in R^*\}$$

we get the decreasing sequence we want.

33E **Theorem**

[$\mathfrak{m}_K>\omega_1$] Let X be a probability space, and let $V\subseteq L^\infty(X)$ be such that V is not $\|\ \|_1$-separable. Then there is a family $\langle v_\xi\rangle_{\xi<\omega_1}$ in V and an $\varepsilon>0$ such that

$$\|\textstyle\sum_{\xi\in I}s_\xi v_\xi\|_\infty\ge\varepsilon\sum_{\xi\in I}|s_\xi|$$

whenever $I\subseteq\omega_1$ is finite and $\langle s_\xi\rangle_{\xi\in I}\in\mathbf{R}^I$.

Proof (*a*) By Maharam's theorem [A6F], X can be expressed as a countable union of homogeneous parts X_n. V is included in the closure of the set

$$\{\textstyle\sum_{i\le n}u\times\chi X_i:u\in V,n\in\mathbf{N}\}$$

so there must be some n such that $\{u \times \chi X_n : u \in V\}$ is non-separable. Accordingly there is no loss of generality in supposing that X is homogeneous. But in this case we can suppose that $X = \{0,1\}^\lambda$ for some cardinal λ; and this is what I shall do for the rest of the proof.

(*b*) For $A \subseteq \lambda$ let $E_A \subseteq L^1$ be the set of equivalence classes of integrable functions which depend only on coordinates in A, and let $\pi_A : L^1 \to E_A$ be the corresponding conditional-expectation projection [A6N*b*]. Note that

$$\pi_A \pi_B = \pi_{A \cap B} \text{ for all } A, B \subseteq \lambda,$$
$$E_A = \{u : u \in L^1, \pi_I(u) \text{ is constant } \forall I \in [\omega_1 \backslash A]^{<\omega}\}.$$

(*c*) Suppose that $U \subseteq L^\infty$ is finite, $J \subseteq \lambda$ is finite, $m \in \mathbf{N}$ and $\delta > 0$. Then there is a finite $K \subseteq \lambda$ such that

$$\|\pi_{J \cup M}(u) - \pi_J(u)\|_\infty \leq \delta \ \forall u \in U, M \in [\lambda \backslash K]^{\leq m}.$$

P As $\bigcup\{E_K : K \in [\lambda]^{<\omega}\}$ is dense in L^1, there is a finite $K \subseteq \lambda$ such that $K \supseteq J$ and $\|\pi_K(u) - u\|_1 \leq 2^{-m-\#(J)}\delta$ for every $u \in U$. Now if $M \in [\lambda \backslash K]^{\leq m}$, $\pi_J = \pi_{J \cup M} \pi_K$, so that, for $u \in U$,

$$\begin{aligned} \|\pi_{J \cup M}(u) - \pi_J(u)\|_\infty &\leq 2^{\#(J \cup M)} \|\pi_{J \cup M}(u) - \pi_J(u)\|_1 \\ &\leq 2^{m+\#(J)} \|\pi_{J \cup M}(u - \pi_K(u))\|_1 \\ &\leq 2^{m+\#(J)} \|\pi_K(u) - u\|_1 \leq \delta. \ \mathbf{Q} \end{aligned}$$

(*d*) There is a disjoint family $\langle I(\xi)\rangle_{\xi<\omega_1}$ of finite sets in λ and a family $\langle v_\xi \rangle_{\xi<\omega_1}$ in V such that $\pi_{I(\xi)} v_\xi$ is non-constant for every $\xi < \omega_1$ **P** Construct the v_ξ, $I(\xi)$ inductively, as follows. Given $\langle I(\eta)\rangle_{\eta<\xi}$, set $A = \bigcup_{\eta<\xi} I(\eta)$. Then E_A is separable so there is a $v_\xi \in V \backslash E_A$. Now there is a finite $I(\xi) \subseteq \lambda \backslash A$ such that $\pi_{I(\xi)} v_\xi$ is non-constant. **Q**

(*e*) Each $\pi_{I(\xi)} v_\xi$ is expressible (uniquely) as $f_\xi^{\cdot}$ where $f_\xi : X \to \mathbf{R}$ depends only on coordinates in $I(\xi)$. Let Y_ξ be the open-and-closed set on which f_ξ takes its minimum value; let Z_ξ be the open-and-closed set on which f_ξ takes its maximum value; and let r_ξ, s_ξ be rational numbers such that

$$f_\xi(y) < r_\xi < s_\xi < f_\xi(z) \ \forall y \in Y_\xi, z \in Z_\xi.$$

Let H_ξ be

$$\{h^{\cdot} : h \in \mathscr{L}^\infty(X), \exists \delta > 0, h(y) \leq r_\xi - \delta \ \forall y \in Y_\xi,$$
$$h(z) \geq s_\xi + \delta \ \forall z \in Z_\xi\},$$

so that H_ξ is a $\| \ \|_\infty$-open set in L^∞, containing $\pi_{I(\xi)} v_\xi$, for every $\xi < \omega_1$.

Let $r, s \in \mathbf{Q}$ be such that

$$C = \{\xi : \xi < \omega_1, r_\xi = r, s_\xi = s\}$$

is uncountable. Set $\varepsilon = \frac{1}{2}(s - r)$.

(*f*) For $J \subseteq C$ set $M(J) = \bigcup_{\xi \in J} I(\xi)$. Let P be

$$\{J : J \in [C]^{<\omega}, \pi_{M(J)} v_\xi \in H_\xi \ \forall \xi \in J\}.$$

Then $\{\xi\} \in P$ for every $\xi \in C$ so P is uncountable.

(*g*) P satisfies Knaster's condition upwards. **P** Let $R \subseteq P$ be uncountable. Then there are $m \in \mathbf{N}$, $\delta > 0$ such that

$$R_1 = \{J : J \in R, \#(M(J)) \le m, u \in H_\xi \text{ whenever } \xi \in J \text{ and } \|u - \pi_{M(J)}(v_\xi)\|_\infty \le \delta\}$$

is uncountable. For each $J \in R_1$ let $K(J)$ be a finite subset of λ such that

$$\|\pi_{M(J) \cup M}(v_\xi) - \pi_{M(J)}(v_\xi)\|_\infty \le \delta \ \forall \xi \in J,\ M \in [\lambda \backslash K(J)]^{\le m};$$

such a set exists by (*c*) above.

Let $R_2 \subseteq R_1$ be an uncountable Δ-system; let J^* be its root. Then (because the $I(\xi)$ are disjoint)

$$M(J) \cap M(J') = M(J^*)$$

for any distinct $J, J' \in R_2$. So we can choose a family $\langle J_\xi \rangle_{\xi < \omega_1}$ in R_2 such that $M(J_\eta) \backslash M(J^*)$ is disjoint from $K(J_\zeta)$ whenever $\zeta < \eta$.

? Suppose, if possible, that there is a $\zeta < \omega_1$ such that

$$\{\eta : \eta < \zeta, J_\eta \cup J_\zeta \notin P\}$$

is infinite. As the $M(J_\eta) \backslash M(J^*)$ are disjoint for $\eta < \zeta$, there is an $\eta < \zeta$ such that $K(J_\zeta) \cap M(J_\eta) \backslash M(J^*) = \emptyset$. But now, writing $J = J_\eta \cup J_\zeta$,

$$M(J) \backslash M(J_\eta) = M(J_\zeta) \backslash M(J^*) \in [\lambda \backslash K(J_\eta)]^{\le m},$$
$$M(J) \backslash M(J_\zeta) \in [\lambda \backslash K(J_\zeta)]^{\le m}.$$

So

$$\|\pi_{M(J)}(v_\xi) - \pi_{M(J_\eta)}(v_\xi)\|_\infty \le \delta \ \forall \xi \in J_\eta,$$

and $\pi_{M(J)}(v_\xi) \in H_\xi$ for every $\xi \in J_\eta$, because $J_\eta \in R_1$. Similarly $\pi_{M(J)}(v_\xi) \in H_\xi$ for every $\xi \in J_\zeta$, and $J \in P$. **X**

From A2K or A2L it follows that, for some uncountable $A \subseteq \omega_1$, $J_\eta \cup J_\zeta \in P$ for every η, $\zeta \in A$; and now $\{J_\eta : \eta \in A\}$ is an uncountable upwards-linked subset of R. As R is arbitrary, P satisfies Knaster's condition upwards. **Q** (*h*) By 31B there is an uncountable upwards-directed $R \subseteq P$. Set $A = \bigcup R \subseteq \omega_1$. Then A is uncountable.

If K is a finite subset of A and $\langle t_\xi \rangle_{\xi \in K} \in \mathbf{R}^K$, there is a $J \in R$ such that $K \subseteq J$. Now $\pi_{M(J)}(v_\xi) \in H_\xi$ for every $\xi \in K$. Let $g_\xi : X \to \mathbf{R}$ be the (unique) continuous function representing $\pi_{M(J)}(v_\xi)$. As $\pi_{M(J)}(v_\xi) \in H_\xi$ we have

$$g_\xi(y) \leq r \, \forall y \in Y_\xi, \quad g_\xi(z) \geq s \, \forall z \in Z_\xi$$

for each $\xi \in K$. But observe now that because the $I(\xi)$ are disjoint there are $y, z \in X$ such that

$$y \in Y_\xi \text{ if } t_\xi \geq 0, \; Z_\xi \text{ if } t_\xi < 0,$$
$$z \in Z_\xi \text{ if } t_\xi \geq 0, \; Y_\xi \text{ if } t_\xi < 0.$$

So we get

$$t_\xi(g_\xi(z) - g_\xi(y)) \geq (s - r)|t_\xi| = 2\varepsilon |t_\xi|$$

for each $\xi \in K$, and

$$\begin{aligned}\|\textstyle\sum_{\xi \in K} t_\xi v_\xi\|_\infty &\geq \|\pi_{M(J)} \textstyle\sum_{\xi \in K} t_\xi v_\xi\|_\infty = \|\textstyle\sum_{\xi \in K} t_\xi g_\xi\|_\infty \\ &\geq \tfrac{1}{2}(\textstyle\sum_{\xi \in K} t_\xi g_\xi(z) - \textstyle\sum_{\xi \in K} t_\xi g_\xi(y)) \geq \varepsilon \textstyle\sum_{\xi \in K} |t_\xi|.\end{aligned}$$

As $\langle t_\xi \rangle_{\xi \in K}$ is arbitrary, $\langle v_\xi \rangle_{\xi \in A}$ is a family of the required type, since $\#(A) = \omega_1$.

33F **Corollary**

[$\mathfrak{m}_K > \omega_1$] Let X be a probability space, E a Banach space, and $T : L^1(X) \to E'$ a continuous linear operator. Then *either* $\ell^1(\omega_1)$ is isomorphically embedded in E *or* there is a separable complemented subspace $F \subseteq L^1(X)$, with associated projection $\pi : L^1(X) \to F$, such that $T = T\pi$. In particular, $\ell^1(\omega_1)$ is embedded in E if either $L^1(X)$ is non-separable and T is injective, or $T[L^1(X)]$ is non-separable.

Proof Consider $V = T'[E] \subseteq L^1(X)' \cong L^\infty(X)$.

(*a*) If V is $\| \; \|_1$-separable, then the $\| \; \|_1$-closed Riesz subspace F of L^1 generated by V is also separable. Let $\pi : L^1 \to F$ be the conditional-expectation projection [A8D*b*]. For $u \in L^1$,

$$\pi u = 0 \Rightarrow \int u \times T'e = 0 \, \forall e \in E \Rightarrow Tu = 0.$$

So $T = T\pi$.

In this case, of course, $T[L^1]$ will be separable; and if L^1 is not separable, T cannot be injective.

(*b*) If V is not $\| \; \|_1$-separable, then by 33E there is a family

$\langle v_\xi \rangle_{\xi<\omega_1}$ in V and an $\varepsilon > 0$ such that

$$\|\sum_{\xi\in I} t_\xi v_\xi\|_\infty \geq \varepsilon \sum_{\xi\in I}|t_\xi| \; \forall I\in[\omega_1]^{<\omega}, \langle t_\xi\rangle_{\xi\in I}\in \mathbf{R}^I.$$

Choose $e_\xi \in E$ such that $T'e_\xi = v_\xi$ for $\xi < \omega_1$. There is an $m\in\mathbf{N}$ such that $A = \{\xi : \|e_\xi\| \leq m\}$ is uncountable. Now if $I\in[A]^{<\omega}$ and $\langle t_\xi\rangle_{\xi\in I}\in\mathbf{R}^I$,

$$\begin{aligned}\varepsilon\sum_{\xi\in I}|t_\xi| \leq \|\sum_{\xi\in I}t_\xi v_\xi\|_\infty &\leq \|T\|\,\|\sum_{\xi\in I}t_\xi e_\xi\| \\ &\leq m\|T\|\sum_{\xi\in I}|t_\xi|.\end{aligned}$$

So we have an isomorphic embedding of $\ell^1(A)$ into E given by

$$\langle t_\xi\rangle_{\xi\in A} \mapsto \sum_{\xi\in A} t_\xi e_\xi.$$

33G **Remark**

This corollary deals with a particular case of the following problem, due to A. Pełczyński: If E is a Banach space and $L^1(\{0,1\}^\kappa)$ embeds isomorphically into E', does $\ell^1(\kappa)$ embed isomorphically into E? For $\kappa = \omega$, yes (Pełczyński); for $\kappa \geq \omega_2$, yes (S.A. Argyros); for $\kappa = \omega_1$, undecidable (R.G. Haydon and Argyros). See ARGYROS 82.

33H **Sources**

CARLSON 80 for 33B. ARGYROS 82 and ARGYROS & ZACHARIADES *a* for 33E-G. F. Galvin gave me the proof of 33B.

33I **Exercises**

(*a*) If X is a separable metric space, the union of fewer than $\mathfrak{m}_K$ sets with Rothberger's property in X [22N*q*] has Rothberger's property.

(*b*) [$\mathfrak{m}_K = \mathfrak{c}$] $\ell^1(\mathbf{N})$ has an up-scale of rank $\mathfrak{c}$ [22K].

(*c*) The results of 33D are valid for any semi-finite measure space with an $\mathfrak{m}_K$-additive measure and a separable L^1 space.

(*d*) Let X be a set, $\langle A_n\rangle_{n\in\mathbf{N}}$ a sequence of subsets of X, $\langle s_n\rangle_{n\in\mathbf{N}}$ a sequence of non-negative real numbers. For $Y \subseteq X$ set

$$\Lambda^*(Y) = \sup_{n\in\mathbf{N}}\inf\{\sum_{i\in I}s_i : I\subseteq \mathbf{N}\setminus n,\, Y\subseteq \bigcup_{i\in I}A_i\},$$

taking $\inf\emptyset = \infty$. If $\mathscr{Y}$ is a family of subsets of X, $\Lambda^*(Y) = 0$ for every $Y\in\mathscr{Y}$, and $\#(\mathscr{Y}) < \mathfrak{m}_K$, then $\Lambda^*(\bigcup\mathscr{Y}) = 0$. [Show that $\{I : I\subseteq\mathbf{N}, \sum_{i\in I}s_i < s\}$ satisfies Knaster's condition upwards for every $s > 0$. See OSTASZEWSKI 74, Theorem 2.]

(*e*) Let $\langle s_n\rangle_{n\in\mathbf{N}}$ be any sequence of non-negative real numbers. Set $\mathscr{I} = \{I : I\subseteq\mathbf{N}, \sum_{n\in I}s_n < \infty\}$. If $\mathscr{C}\subseteq\mathscr{I}$ and $\#(\mathscr{C}) < \mathfrak{m}_K$ there is an $I\in\mathscr{I}$ such that $C\setminus I$ is finite for every $C\in\mathscr{C}$.

(*f*) Let $\mathscr{Z}$ be the set of subsets of $\mathbf{N}$ with zero asymptotic density. If $\mathscr{C} \subseteq \mathscr{Z}$ and $\#(\mathscr{C}) < \mathfrak{m}_K$ there is an $I \in \mathscr{Z}$ such that $C \setminus I$ is finite for every $C \in \mathscr{C}$. [(α) If $s > 0$, then

$$\{A : A \in \mathscr{Z},\ \#(A \cap k) \leq sk\ \forall k \in \mathbf{N}\}$$

satisfies Knaster's condition upwards. (β) For each $n \in \mathbf{N}$ there is an $A_n \subseteq \mathbf{N}$ such that $\#(A_n \cap k) \leq 2^{-n}k$ for every $k \in \mathbf{N}$ and $C \setminus A_n$ is finite for each $C \in \mathscr{C}$. (γ) Now use 21A. See FRANKIEWICZ & GUTEK 81.]

(*g*) Express (*e*) and (*f*) in terms of (*i*) $p(\mathfrak{m}_K)$-sets in $\beta\mathbf{N} \setminus \mathbf{N}$ (ii) the $(<\mathfrak{m}_K, \omega)$-covering property.

(*h*) Let X be a Polish space with a lattice structure for which the lattice operations $\vee$ and $\wedge$ are continuous. Suppose that $\mathscr{A}$ is a non-empty family of subsets of X with $\#(\mathscr{A}) < \mathfrak{m}_K$, and that $x \in \bigcap_{A \in \mathscr{A}} \bar{A}$. Then there are monotonic sequences $\langle y_n \rangle_{n \in \mathbf{N}}$ in X such that $x = \lim_{n \to \infty} y_n = \lim_{n \to \infty} z_n$ and $A \cap [y_n, z_n] \neq \emptyset$ for every $n \in \mathbf{N}$ and $A \in \mathscr{A}$. [Set $P = \{(y, z, n) : y \leq x \leq z\} \subseteq X \times X \times \mathbf{N}$, ordered by saying that $(y, z, n) \leq (u, v, m)$ if $u \leq y$, $z \leq v$, $n \leq m$ and $\rho(u, y) + \rho(v, z) \leq 2^{-n} - 2^{-m}$. Show that P satisfies Knaster's condition upwards and that if R is an upwards-directed subset of P then $\inf\{y : (y, z, n) \in R\}$, $\sup\{z : (y, z, n) \in R\}$ exist in X. Compare HOFFMANN-JØRGENSEN 78, Theorem 5.4.]

33J Further results

(*a*) $[\mathfrak{m}_K > \omega_1]$ Let E be a Banach space, X a measure space. Suppose there are continuous linear maps

$$T : L^1(\{0,1\}^{\omega_1}) \to E', \quad S : E' \to L^1(X)$$

such that $ST : L^1(\{0,1\}^{\omega_1}) \to L^1(X)$ is an isomorphic embedding. Then the ℓ^1-sum

$$(\bigoplus_{n \in \mathbf{N}, \xi < \omega_1} F_{n\xi})_1$$

embeds isomorphically into E, where $F_{n\xi} = \ell^\infty(n)$ for each $n \in \mathbf{N}$, $\xi < \omega_1$. [ARGYROS 82.]

(*b*) $[\mathfrak{m}_K > \omega_1]$ Let X be a measure space and suppose that $L^1(X)$ is isomorphic to the dual of some Banach space. Then $L^1(X)$ is isomorphic to an ℓ^1-sum of spaces of the form $C(\{0,1\}^\kappa)'$. [HAYDON 80.]

(*c*) Let $\mathfrak{A}$ be a Boolean algebra of cardinal less than or equal to $\mathfrak{m}_K$ and ν a non-negative additive functional on $\mathfrak{A}$ such that $\nu 1 = 1$. Let $\mathscr{Z} \subseteq \mathscr{P}\mathbf{N}$ be the ideal of sets of asymptotic density 0. Then there is a

Boolean homomorphism $\theta:\mathfrak{A} \to \mathscr{P}\mathbf{N}/\mathscr{Z}$ such that

$$\nu a = \lim_{n\to\infty}(1/n)\,\#(A\cap n)$$

whenever $a\in\mathfrak{A}$, $A\subseteq \mathbf{N}$ and $A^{\cdot} = \theta a$ in $\mathscr{P}\mathbf{N}/\mathscr{Z}$.

33K **Problems**

(*a*) Is 33J*c* true for $\#(\mathfrak{A}) \le \mathfrak{c}$ without special axioms?

(*b*) [$\mathfrak{m}_K > \omega_1$] Let X be a compact Radon measure space such that $L^1(X)$ is not separable. Is there necessarily a continuous surjection from X onto $[0,1]^{\omega_1}$? [R.G. Haydon.]

Notes and Comments

In view of the similar results in §§22 and 32, 33B–D are all unsurprising. I have phrased 33D in terms of quasi-Radon measures in order to link it with the right bits of §32; but of course no topology enters the argument; the conditions we really use are those of 33I*c*. Recently (1982–3) T.C. Bartoszyński and other Polish and French mathematicians have shown that throughout 32E, 32G–H, 32J–K and 33B–D the cardinal $\mathfrak{m}_K$ can be replaced by the additivity of Lebesgue measure (the largest cardinal κ for which Lebesgue measure is κ-additive); that is, they are all consequences of 32F.

With 33E we find ourselves working much harder. I have stated 33E as a 'theorem' and 33F as a 'corollary', but of course they could just as well be called 'lemma' and 'theorem'; most of the ideas come from Argyros' solution of Pełczyński's problem [33G], and I have no other consequences of 33E to present. (33J*a* uses a similar, but more complex argument.) Note that the temptation to generalise 33E to cardinals $\kappa > \omega_1$ should be resisted. Replacing the hypothesis 'V is not $\|\ \|_1$-separable' by '$d(V) \ge \kappa$', where $d(V)$ is the $\|\ \|_1$-density of V, we find that the arguments go through very nicely if $\mathrm{cf}(\kappa) > \omega$ and $\kappa < \mathfrak{m}_K$, and we get a family $\langle v_\xi\rangle_{\xi<\kappa}$. But actually such a family can be constructed, using ZFC alone, for *any* $\kappa \ge \omega_2$ of uncountable cofinality, by arguments in Argyros 82. It is therefore particularly important to check that ω_1 really is exceptional, as is shown by Haydon 78, using CH.

33I*d*–*g* are variations on the theme of 33B–D; 33I*h* seems to have some of the same qualities. 33J*c* may be related to 26M*b*; I do not know whether $\mathfrak{m} = \mathfrak{c}$ is enough to ensure that $\mathscr{P}\mathbf{N}/\mathscr{Z}$ can be embedded into $\mathscr{P}\mathbf{N}/[\mathbf{N}]^{<\omega}$. It seems possible that 33K*b* is connected with 33E. An affirmative answer here would also settle 32R*c*, since $[0,1]^{\omega_1}$ is not a continuous image of any first-countable compact Hausdorff space.

34 Whitehead's problem

I give P.C. Eklof's version of half of S. Shelah's solution of J.H.C. Whitehead's problem: if Martin's axiom is true and the continuum hypothesis is false, there is a non-free Whitehead group.

34A The problem

Let G and M be groups, $g:M\to G$ a surjective homomorphism. We say that g **splits** if there is a homomorphism $f:G\to M$ such that gf is the identity on G. We say that an abelian group G is **free** if it is isomorphic to a direct sum of copies of $\mathbf{Z}$.

It is easy to see that, if G is a free abelian group, then every surjection from an abelian group onto G splits [34B*d*(ii)]; and in fact this characterizes the free abelian groups [Fuchs 70, Theorem 14.6]. But, if we restrict M and g, the question changes character. A **Whitehead group** is an abelian group G such that, whenever M is an abelian group and $g:M\to G$ is a surjective homomorphism with kernel isomorphic to $\mathbf{Z}$, then g splits. **Whitehead's problem** is: is every Whitehead group free?

Countable Whitehead groups are indeed free [Fuchs 73, Theorem 99.1; Eklof 76, Theorem 4.1]. But for uncountable G the question is undecidable. Here I shall present only half the answer: there is a non-free abelian group G, of cardinal ω_1, which (if $\mathfrak{m}_K > \omega_1$) has the property that, if M is an abelian group and $g:M\to G$ is a surjective homomorphism with countable kernel, then g splits. It is quite easy to write down a formula for G [34C], and the proof that it is not free is not hard [34E]; the proof of its splitting property, however, is delicate [34D, 34F–J].

34B Abelian groups

We need the following definitions and facts.

(*a*) If G is an abelian group, a **basis** for G is a set $B\subseteq G$ such that each element of G is uniquely expressible as $\sum_{x\in B} n_x x$, where $n_x\in\mathbf{Z}$ for each $x\in B$ and $\{x:n_x\neq 0\}$ is finite. By definition, G is free iff it has a basis. If G is free, then every basis of G has the same cardinality [Fuchs 70, Proposition 14.1]; call this cardinal the **rank** $r(G)$ of G.

(*b*) If G is a free abelian group and H is a subgroup of G, then H is free and $r(H)\leq r(G)$ [Fuchs 70, Theorem 14.5].

(*c*) If G is any abelian group, a subgroup H of G is **pure** if whenever

$x \in G$ and $n \in \mathbf{Z}$ and $nx \in H$, there is a $y \in H$ such that $ny = nx$. G is **torsion-free** if every element of G, other than 0, has infinite order. In this case, a subgroup H of G is pure iff whenever $x \in G$, $n \in \mathbf{Z} \setminus \{0\}$ and $nx \in H$ then $x \in H$. If G is torsion-free, and H is any subgroup of G, then

$$K = \{x : \exists n \in \mathbf{Z} \setminus \{0\}, nx \in H\}$$

is the smallest pure subgroup of G including H; we call it the **pure closure** of H in G.

(*d*) If G is an abelian group and H is a subgroup of G such that G/H is free, then there is a free subgroup K of G such that $G = H \oplus K$ [Fuchs 70, Theorem 14.4]. Consequently (i) if H is free, so is G, and every basis of H is a subset of a basis of G; (ii) if M is an abelian group and $g : M \to G$ is a surjective homomorphism, and if there is a homomorphism $f_0 : H \to M$ such that $g(f_0(x)) = x$ for every $x \in H$, then there is a homomorphism $f : G \to M$, extending f_0, such that $g(f(x)) = x$ for every $x \in G$; (iii) if L is a subgroup of H such that H/L is free, then G/L is free.

(*e*) A finitely generated torsion-free abelian group is free. [Use Fuchs 70, Theorem 15.5] Consequently, if G is a finitely generated torsion-free abelian group and H is a pure subgroup of G, then G/H is free; and if $H \neq G$ then $r(H) < r(G)$.

34C Construction

The terminology introduced here will remain in force down to 34J below.

Let $\langle \theta(\zeta, n) \rangle_{n \in \mathbf{N}, \zeta \in \Omega}$ be a ladder system on ω_1 [definition: A2G]. Let G be the abelian group with generators x_ξ (for $\xi < \omega_1$) and z_ζ^n (for $\zeta \in \Omega$ and $n \in \mathbf{N}$), and relations

$$z_\zeta^n = x_{\theta(\zeta,n)} + 2z_\zeta^{n+1} \; \forall n \in \mathbf{N}, \zeta \in \Omega.$$

For $\xi < \omega_1$ let G_ξ be the subgroup of G generated by

$$\{x_\eta : \eta < \xi\} \cup \{z_\zeta^n : \zeta \in \Omega, \zeta \le \xi, n \in \mathbf{N}\}.$$

34D Lemma

G_ζ/G_ξ is free whenever $\xi \le \zeta < \omega_1$. In particular, every G_ξ is free, and G is torsion-free.

Proof Induce on ζ. (*a*) For $\zeta = 0$ this is trivial.

(*b*) If $\zeta = \eta + 1$, then $G_\zeta = G_\eta \oplus \langle x_\eta \rangle$. We have $G_\zeta/G_\zeta = \{0\}$, $G_\zeta/G_\eta \cong \mathbf{Z}$, and for $\xi < \eta$, $G_\zeta/G_\xi \cong (G_\eta/G_\xi) \oplus \mathbf{Z}$ is free, by the inductive hypothesis.

(c) If $\zeta \in \Omega$, we have

$$G_{\theta(\zeta,0)} \cong G_{\theta(\zeta,0)}/G_0 \text{ free;}$$
$$G_{\theta(\zeta,n)+1} = G_{\theta(\zeta,n)} \oplus \langle x_{\theta(\zeta,n)} \rangle;$$
$$G_{\theta(\zeta,n+1)}/G_{\theta(\zeta,n)+1} \text{ free,}$$

for every $n \in \mathbf{N}$, by the inductive hypothesis. So we can choose inductively sets $B_n \subseteq G_{\theta(\zeta,n)}$, disjoint from each other and from $\{x_{\theta(\zeta,i)} : i \in \mathbf{N}\}$, such that

$$\bigcup_{i \leq n} B_i \cup \{x_{\theta(\zeta,i)} : i < n\}$$

is a basis of $G_{\theta(\zeta,n)}$ for each $n \in \mathbf{N}$.

For any $n \in \mathbf{N}$, consider

$$C_n = \bigcup_{i \in \mathbf{N}} B_i \cup \{x_{\theta(\zeta,i)} : i < n\} \cup \{z_\zeta^i : i \geq n\}.$$

Then $C_n \subseteq G_\zeta$ and C_n includes the constructed basis of $G_{\theta(\zeta,n)}$. The point is that C_n is a basis of G_ζ. **P** We have

$$x_{\theta(\zeta,i)} + 2z_\zeta^{i+1} = z_\zeta^i \ \forall i \in \mathbf{N},$$

so that $\langle C_n \rangle$ contains $x_{\theta(\zeta,i)}$ for $i \geq n$ and also z_ζ^i for $i < n$, and is the whole of G_ζ. Next, let us consider a finite combination of members of C_n. This must be of the form

$$w = \sum_{j \leq r} m_j y_j + \sum_{i < n} p_i x_{\theta(\zeta,i)} + \sum_{n \leq i \leq s} q_i z_\zeta^i,$$

where $r, s \in \mathbf{N}$, $s \geq n$; m_j, p_i, q_i all belong to $\mathbf{Z}$; and the y_j all belong to $\bigcup_{i \in \mathbf{N}} B_i$. Observe that $\sum_{n \leq i \leq s} q_i z_\zeta^i$ can be expressed as $q z_\zeta^s + \sum_{n \leq i < s} p_i x_{\theta(\zeta,i)}$ where q and the new p_i also belong to $\mathbf{Z}$. Now suppose that $w = 0$. Since there is a homomorphism $h: G \to \mathbf{Q}$ defined by

$$h(z_\zeta^i) = 2^{-i} \ \forall i \in \mathbf{N}, \quad h(x_\xi) = 0 \ \forall \xi < \omega_1,$$
$$h(z_\xi^i) = 0 \ \forall \xi \in \Omega \setminus \{\zeta\}, i \in \mathbf{N},$$

we see that

$$h(x) = 0 \ \forall x \in \bigcup_{\xi < \zeta} G_\xi,$$

so that $h(y_j) = 0$ for each $j \leq r$ and

$$0 = h(w) = q h(z_\zeta^s) = 2^{-s} q,$$

so that $q = 0$. Thus

$$\sum_{j \leq r} m_j y_j + \sum_{i < s} p_i x_{\theta(\zeta,i)} = 0.$$

But there is a $k \geq s$ such that $y_j \in \bigcup_{i \leq k} B_i$ for every $j \leq r$, and now $\bigcup_{i \leq k} B_i \cup \{x_{\theta(\zeta,i)} : i < k\}$ is a basis of $G_{\theta(\zeta,k)}$, by construction. It follows that $m_j = p_i = 0$ for $j \leq r$ and $i < s$.

So we have $\sum_{n \le i \le s} q_i z_\zeta^i = q z_\zeta^s + \sum_{n \le i < s} p_i x_{\theta(\zeta,i)} = 0$. Now, for any $j \in \mathbf{N}$, there is a homomorphism $h_j : G \to \mathbf{Q}$ defined as follows:

$$h_j(z_\zeta^j) = 1,\ h_j(z_\zeta^i) = 0\ \forall i \ne j;$$
$$h_j(x_{\theta(\zeta,i)}) = h_j(z_\zeta^i) - 2h_j(z_\zeta^{i+1})\ \forall i \in \mathbf{N},$$
$$h_j(x_\xi) = 0 \text{ for all other } \xi < \omega_1;$$
$$h_j(z_\xi^0) = 0,\ h_j(z_\xi^{i+1}) = \tfrac{1}{2}(h_j(z_\xi^i) - h_j(x_{\theta(\xi,i)}))$$
$$\text{if } i \in \mathbf{N},\ \xi \in \Omega \setminus \{\zeta\}.$$

In this case we have, for $n \le j \le s$,

$$q_j = h_j(\textstyle\sum_{n \le i \le s} q_i z_\zeta^i) = 0.$$

This shows at last that the form

$$w = \textstyle\sum_{j \le r} m_j y_j + \sum_{i < n} p_i x_{\theta(\zeta,i)} + \sum_{n \le i \le s} q_i z_\zeta^i$$

is the trivial form, with all the coefficients zero. This is what we need in order to see that C_n is indeed a basis of G_ζ. **Q**

Thus $G_{\theta(\zeta,n)}$ is a direct summand of G_ζ for every $n \in \mathbf{N}$, and the quotients $G_\zeta / G_{\theta(\zeta,n)}$ are free. If $\xi < \zeta$, there is an $n \in \mathbf{N}$ such that $\xi \le \theta(\zeta, n)$; by the inductive hypothesis, $G_{\theta(\zeta,n)}/G_\xi$ is free; so by 34B*d*(iii), G_ζ/G_ξ is free.

Thus the induction continues. Now $G_\xi = G_\xi/G_0$ is free for every $\xi < \omega_1$. As every element of G belongs to a free subgroup of G, G must be torsion-free.

34E Lemma

G is not free.

Proof ? Suppose, if possible, that G had a basis B. Choose inductively, for $n \in \mathbf{N}$, countable sets $B_n \subseteq B$ and ordinals $\zeta(n)$, as follows. $\zeta(0) = 0$. Given $\zeta(n)$, $G_{\zeta(n)}$ is countable, so there is a countable set $B_n \subseteq B$ such that $G_{\zeta(n)} \subseteq \langle B_n \rangle$. Now $\langle G_\xi \rangle_{\xi < \omega_1}$ is an increasing family with union G, so there is a $\zeta(n+1) > \zeta(n)$ such that $B_n \subseteq G_{\zeta(n)}$. Continue.

Set $\zeta = \sup_{n \in \mathbf{N}} \zeta(n)$; then $\zeta \in \Omega$,

$$\bigcup_{n \in \mathbf{N}} B_n \subseteq \bigcup_{n \in \mathbf{N}} G_{\zeta(n+1)} \subseteq \bigcup_{n \in \mathbf{N}} \langle B_{n+1} \rangle \subseteq \langle \bigcup_{n \in \mathbf{N}} B_n \rangle.$$

So writing $B' = \bigcup_{n \in \mathbf{N}} B_n$, $G'_\zeta = \bigcup_{n \in \mathbf{N}} G_{\zeta(n)} = \bigcup_{\xi < \zeta} G_\xi$, we have $G'_\zeta = \langle B' \rangle$ and G'_ζ is a direct summand of G. Thus G/G'_ζ is free. But consider the equivalence classes $(z_\zeta^n)^\cdot$ of z_ζ^n in G/G'_ζ. We have

$$(z_\zeta^n)^\cdot = 2(z_\zeta^{n+1})^\cdot\ \forall n \in \mathbf{N},$$

so that $(z_\zeta^0)^\cdot$ is divisible by all powers of 2; which in a free group is impossible. **X** Thus G has no basis and is not free.

34F **Lemma**

Let $H \subseteq G$ be a finitely generated pure subgroup and $A \subseteq G$ a finite set. Then H is free and there is a finitely generated pure subgroup K of G such that $H \cup A \subseteq K$ and K/H is free.

Proof As H is countable, there is a $\zeta < \omega_1$ such that $H \cup A \subseteq G_\zeta$. By 34D, G_ζ is free; let B be a basis for G_ζ. As H is finitely generated and A is finite, there is a finite $B' \subseteq B$ such that $K = \langle B' \rangle \supseteq H \cup A$. K is finitely generated. To see that K is pure, note that if $x \in G$ and $nx \in K$, where $n \in \mathbf{Z} \setminus \{0\}$, there is a $\xi \geq \zeta$ such that $x \in G_\xi$; as G_ξ/G_ζ is free, $x \in G_\zeta$; as G_ζ/K is free, $x \in K$. As K is free and finitely generated, and H is pure in K (because H is pure in G), K/H is free [34B*e*]. At the same time, H is free by 34B*b*.

34G **The partially ordered set**

From here down to 34J, let M be an abelian group and $g: M \to G$ a surjective homomorphism with countable kernel. Let P be the set of all 'partial splittings' of g, i.e. functions f such that

dom(f) is a finitely generated pure subgroup of G;

f:dom(f)$\to M$ is a homomorphism;

$gf(x) = x \; \forall x \in \text{dom}(f)$.

34H **Lemma**

Let $R \subseteq P$ be uncountable. Then there is a free pure subgroup H of G such that $\{f : f \in R, \text{dom}(f) \subseteq H\}$ is uncountable.

Proof Note from 34F that dom(f) is free for each $f \in R$. Let $m_1 \in \mathbf{N}$ be such that

$$R_1 = \{f : f \in R, r(\text{dom}(f)) = m_1\}$$

[34B*a*] is uncountable. Let $\mathscr{K}$ be

$$\{K : K \text{ is a pure subgroup of } G, \{f : f \in R_1, K \subseteq \text{dom}(f)\} \text{ is uncountable}\}.$$

Then $\{0\} \in \mathscr{K}$ and $r(K) \leq m_1$ for every $K \in \mathscr{K}$, by 34B*b*. Set $m_0 = \sup_{K \in \mathscr{K}} r(K)$ and take $K_0 \in \mathscr{K}$ with $r(K_0) = m_0$. Set

$$R_2 = \{f : f \in R_1, K_0 \subseteq \text{dom}(f)\}$$

so that R_2 is uncountable.

Now choose inductively countable pure subgroups K_ξ of G and functions

$f_\xi \in R_2$, for $\xi < \omega_1$, as follows. We already have K_0. Given that $K_\xi \supseteq K_0$ is countable, let ζ be such that $K_\xi \subseteq G_\zeta$. Then

$$\{f : f \in R_2, G_\zeta \cap \mathrm{dom}(f) \neq K_0\}$$

is countable. **P?** Otherwise, there is an $x \in G_\zeta \backslash K_0$ such that $\{f : f \in R_2, x \in \mathrm{dom}(f)\}$ is uncountable, because G_ζ is countable. Let K be the pure closure of $K_0 + \langle x \rangle$ [34B*c*]. Then $K \subseteq \mathrm{dom}(f)$ whenever $f \in R_2$ and $x \in \mathrm{dom}(f)$, so $\{f : f \in R_2, K \subseteq \mathrm{dom}(f)\}$ is uncountable. But K_0 is a proper subgroup of K, so $r(K) > r(K_0) = m_0$ [34B*e*], contrary to the definition of m_0. **XQ** Consequently there is an $f_\xi \in R_2$ such that $G_\zeta \cap \mathrm{dom}(f_\xi) = K_0$ and $f_\xi \neq f_\eta$ for every $\eta < \xi$.

Take $K_{\xi+1}$ to be the pure closure of $K_\xi + \mathrm{dom}(f_\xi)$ in G. Then $K_{\xi+1} \cap G_\zeta = K_\xi$. **P** If $x \in G_\zeta$ and $nx \in K_\xi + \mathrm{dom}(f_\xi)$, where $n \neq 0$, express nx as $y + z$ where $y \in K_\xi$ and $z \in \mathrm{dom}(f_\xi)$. We have $z = nx - y \in G_\zeta$, so $z \in K_0$ and $nx \in K_\xi$; by the inductive hypothesis K_ξ is pure so (because G is torsion-free) $x \in K_\xi$. **Q** So

$$K_{\xi+1}/K_\xi \cong (K_{\xi+1} + G_\zeta)/G_\zeta.$$

Take $\eta \geq \zeta$ so large that $K_{\xi+1} \subseteq G_\eta$. Then $K_{\xi+1}/K_\xi$ is isomorphic to a subgroup of the free group G_η/G_ζ, so is free. As $K_\xi + \mathrm{dom}(f_\xi)$ is countable, so is $K_{\xi+1}$.

At limit ordinals $\zeta \in \Omega$, set $K_\zeta = \bigcup_{\xi<\zeta} K_\xi$. Because all the quotients $K_{\xi+1}/K_\xi$ are free, we can use 34B*d*(i) to construct an ascending chain $\langle B_\xi \rangle_{\xi<\omega_1}$ such that each B_ξ is a basis of K_ξ. So if $H = \bigcup_{\xi<\omega_1} K_\xi$, $\bigcup_{\xi<\omega_1} B_\xi$ will be a basis of H, and H is free. Also H is pure because every K_ξ is pure. Finally, $\{f : f \in R, \mathrm{dom}(f) \subseteq H\} \supseteq \{f_\xi : \xi < \omega_1\}$ is uncountable.

34I Lemma

P satisfies Knaster's condition upwards.

Proof Let $R \subseteq P$ be an uncountable set. Let $H \subseteq G$ be a pure free subgroup such that $R_3 = \{f : f \in R, \mathrm{dom}(f) \subseteq H\}$ is uncountable [34H]. Let B be a basis for H. For each $f \in R_3$ let B_f be a finite subset of B such that $\mathrm{dom}(f) \subseteq \langle B_f \rangle$. As $\mathrm{dom}(f)$ is pure, $\langle B_f \rangle / \mathrm{dom}(f)$ is free; by 34B*d*(ii), there is an $\bar{f} \in P$ such that $\mathrm{dom}(\bar{f}) = \langle B_f \rangle$ and $\bar{f}$ extends f.

Let $S \subseteq R_3$ be an uncountable set such that $\langle B_f \rangle_{f \in S}$ is a Δ-system with root C say, and $\bar{f} \upharpoonright C$ is constant for $f \in S$ (here using the fact that g has countable kernel, so that $\{f(x) : f \in P, x \in \mathrm{dom}(f)\} \subseteq g^{-1}[\{x\}]$ is countable for every $x \in G$). If e, $f \in S$ then

$$\langle B_e \rangle \cap \langle B_f \rangle = \langle B_e \cap B_f \rangle = \langle C \rangle,$$

and $\tilde{e}$ and $\tilde{f}$ agree on $\langle C \rangle$, so $\tilde{e}$ and $\tilde{f}$ have a common extension h defined on $\langle B_e \rangle + \langle B_f \rangle = \langle B_e \cup B_f \rangle$. Since $h(x) \in g^{-1}[\{x\}]$ for every $x \in B_e \cup B_f$, $gh(x) = x$ for every $x \in \mathrm{dom}(h)$. Since $B_e \cup B_f \subseteq B$, $\mathrm{dom}(h)$ is a direct summand of H; so $\mathrm{dom}(h)$ is pure in H and therefore pure in G. Of course $\mathrm{dom}(h)$ is finitely generated. So $h \in P$ and is a common upper bound of e and f.

Thus S is an uncountable upwards-linked subset of R. As R is arbitrary, P satisfies Knaster's condition upwards.

34J Theorem

$[\mathfrak{m}_K > \omega_1]$ g splits.

Proof For $x \in G$, set $Q_x = \{f : f \in P, x \in \mathrm{dom}(f)\}$. Then Q_x is cofinal with P. **P** Let $f \in P$. By 34F, there is a finitely generated pure subgroup K of G such that $\mathrm{dom}(f) \cup \{x\} \subseteq K$. Now K is free, so $K/\mathrm{dom}(f)$ is free (because $\mathrm{dom}(f)$ is pure), and there is an extension f_1 of f with $\mathrm{dom}(f_1) = K$ and $gf_1(y) = y$ for every $y \in K$ [34B*d*(ii)]. In this case, $f_1 \in Q_x$ and f_1 extends f. **Q**

As $\#(G) = \omega_1$, there is an upwards-directed $R \subseteq P$ meeting every Q_x. There is a unique function $f: G \to M$ extending every member of R, and this is the required splitting of g.

34K Corollary

$[\mathfrak{m}_K > \omega_1]$ There is a pathwise-connected compact Hausdorff abelian topological group which is not a product of copies of the circle group $\mathbf{S}^1 = \mathbf{R}/\mathbf{Z}$.

Proof Take the group G of 34C, with the discrete topology, and let $\hat{G}$ be its character group. Because $(\hat{G})\hat{}$ can be identified with G, which is not a direct sum of copies of $\mathbf{Z}$ [34E], $\hat{G}$ is not a power of $\mathbf{S}^1$. To see that $\hat{G}$ is pathwise-connected, let $\chi \in \hat{G}$. In $\mathbf{R} \times G$ let M be

$$\{(s, x) : s^{\cdot} = \chi(x)\},$$

where $s^{\cdot}$ is the image of s in $\mathbf{S}^1$, and χ is regarded as a homomorphism from G to $\mathbf{S}^1$. Then M is an abelian group and $\pi_2 : M \to G$ is a surjective homomorphism with kernel isomorphic to $\mathbf{Z}$. By 34J there is a homomorphism $f: G \to M$ splitting π_2. For $t \in [0, 1]$, set

$$\chi_t(x) = (t\pi_1 f(x))^{\cdot} \in \mathbf{S}^1.$$

Then $\chi_t \in \hat{G}$ for each $t \in [0, 1]$, and $t \mapsto \chi_t : [0, 1] \to \hat{G}$ is a path from the identity to χ.

Remark Here $\hat{G}$ is just the set of all homomorphisms from G to $\mathbf{S}^1$, with the topology and group structure taken from the power $(\mathbf{S}^1)^G$. It is elementary that $\hat{G}$ is a compact Hausdorff abelian topological group. For the identification of $(\hat{G})\hat{}$ with G (Pontryagin's duality theorem) see, for instance, RUDIN 67 or LOOMIS 53.

34L **Exercise**

Let G be an abelian group, $\#(G) < \mathfrak{m}_K$. Suppose that (α) all countable subgroups of G are free (β) for every countable subgroup H of G there is a countable subgroup K of G including H such that every countable subgroup of G/K is free. (i) If M is any abelian group and $g:M \to G$ is a surjective homomorphism with countable kernel, then g splits. [Hint: use the arguments of 34F–J.] (ii) Any countable subgroup of G is included in a countable direct summand of G. [Hint: apply (i) to a quotient map $G \to G/K$. See EKLOF 80, Ch. 7; MEKLER 80, Theorem 2.5.]

34M **Further results**

(*a*) Let Z be a countable torsion-free abelian group. A **Shelah Z-group** is an abelian group G such that

(i) for every countable subgroup H of G, $\mathrm{Ext}(H, Z) = \{0\}$;
(ii) for every countable pure subgroup H of G, there is a countable subgroup H' of G such that $H \subseteq H'$ and if K is any countable subgroup of G with $K \cap H' = H$, then $\mathrm{Ext}(K\backslash H, Z) = \{0\}$.

(For the definition of $\mathrm{Ext}(H, Z)$, see FUCHS 70, §49. Note that $\mathrm{Ext}(H, Z) = \{0\}$ iff whenever M is an abelian group and $g:M \to H$ is a surjective homomorphism with kernel isomorphic to Z, then g splits.)

If G is a torsion-free abelian group and $\#(G) < \mathfrak{m}_K$, then

$$\mathrm{Ext}(G, Z) = \{0\} \Leftrightarrow G \text{ is a Shelah } Z\text{-group} \Leftrightarrow r(\mathrm{Ext}(G, Z)) \neq \mathfrak{c}.$$

In particular, $\mathrm{Ext}(G, Z) \neq \mathbf{Q}$. [EKLOF & HUBER 80.]

(*b*) Let $\Pi \subseteq \mathbf{N}$ be the set of prime numbers, and $\langle n(p)\rangle_{p\in\Pi}$ any family in $\mathbf{N}$. Let R be the subgroup of $\mathbf{Q}$ generated by $\{p^{-n(p)}: p \in \Pi\}$. Let Z be a countable torsion-free abelian group such that $\mathrm{Ext}(R, Z) = \{0\}$. Let G be a torsion-free abelian group such that $\#(G) < \mathfrak{m}_K$ and, whenever H is a countable subgroup of G, there is a countable subgroup H' of G such that

$H \subseteq H'$;
H' is isomorphic to a direct sum of copies of R;

whenever K is a countable subgroup of G and $H' \subseteq K$
then H' is a direct summand of K.

Then $\text{Ext}(G, Z) = \{0\}$. [EKLOF 77.]

(*c*) $[\mathfrak{m}_K > \omega_1]$ If G is a Whitehead group of cardinal ω_1, then whenever M is an abelian group and $g: M \to G$ a surjective homomorphism with countable kernel, g splits. [SHELAH 79.]

(*d*) $[\mathfrak{m}_K > \omega_1]$ There is an abelian group G such that (i) $\#(G) = \omega_1$, (ii) every countable subgroup of G is free, (iii) if H is subgroup of G and K is a finitely generated pure subgroup of H then K is a direct summand of H (iv) G is not a Whitehead group. [SHELAH 79.]

Notes and comments

The argument of 34C–34J is taken from EKLOF 76; the essential ideas are due to SHELAH 74. For further discussion, with some of Shelah's other results, see EKLOF 76 and EKLOF 80. It is clear that many easy generalizations are possible; for instance, the construction of 34C can be taken up to any cardinal κ, using $\{\xi : \xi < \kappa, \text{cf}(\xi) = \omega\}$ in place of Ω; while 34L(i) is obtained by writing down the properties of the group G used in 34F–J. A more determined analysis leads to 34M*a*–*b*. Some natural questions are clarified in 34M*c*–*d*.

34K is mentioned in EKLOF 76; it is a kind of Whitehead's problem for analysts.

For the other half of Shelah's solution to Whitehead's problem (namely, if $V = L$ then every Whitehead group is free), see SHELAH 75, EKLOF 76, EKLOF 80.

35 Ideals of $\mathscr{P}\omega_1$

I give the theorems of R. Laver and A.D. Taylor, that if $\mathfrak{m}_K > \omega_1$ then a uniform ideal of $\mathscr{P}\omega_1$ is regular if it is either a maximal ideal or a σ-ideal [35I–J]. Most of the section is taken up with the basic theory of regular ideals as developed by J. Ketonen, A. Kanamori and Taylor.

35A Definitions

Let X be an infinite set. An ideal $\mathscr{I}$ of $\mathscr{P}X$ is **uniform** if it contains every subset of X of cardinal less than $\#(X)$. It is **regular** if whenever $\langle A_\xi \rangle_{\xi < \kappa}$ is a family in $\mathscr{P}X \setminus \mathscr{I}$, where $\kappa = \#(X)$, there is a point-finite family $\langle B_\xi \rangle_{\xi < \kappa}$ such that $B_\xi \in \mathscr{P}A_\xi \setminus \mathscr{I}$ for every $\xi < \kappa$. (More explicitly, $\mathscr{I}$ is called '(ω, κ)-regular'.)

Note that a maximal ideal $\mathscr{I} \lhd \mathscr{P}\kappa$ is regular iff there is some point-finite family $\langle B_\xi \rangle_{\xi<\kappa}$ in $\mathscr{P}\kappa \setminus \mathscr{I}$. For more about regular ideals, see 35L*a–b*.

If $\mathscr{I} \lhd \mathscr{P}\kappa$, then $\mathscr{I}$ is **weakly normal** if it is uniform and proper and whenever $f:\kappa \to \kappa$ is such that $f^{-1}[\xi] \in \mathscr{I}$ for every $\xi < \kappa$, then $\{\eta : f(\eta) < \eta\} \in \mathscr{I}$.

35B Lemma

Let $\mathscr{I}$ be a uniform proper σ-ideal of $\mathscr{P}\omega_1$.

(*a*) $\mathscr{I}$ is not ω_1-saturated.

(*b*) If $\langle A_i \rangle_{i \in I}$ is any countable family in $\mathscr{P}\omega_1 \setminus \mathscr{I}$, there is a disjoint family $\langle B_i \rangle_{i\in I}$ such that $B_i \in \mathscr{P}A_i \setminus \mathscr{I}$ for every $i \in I$.

(*c*) If $\mathscr{I}$ is not regular, there is a $Y \in \mathscr{P}\omega_1 \setminus \mathscr{I}$ such that (α) $\mathscr{I} \cap \mathscr{P}Y$ is ω_2-saturated in $\mathscr{P}Y$, (β) $\mathscr{I} \cap \mathscr{P}Z$ is not regular in $\mathscr{P}Z$ for any $Z \in \mathscr{P}Y \setminus \mathscr{I}$.

Proof (*a*) For each $\xi < \omega_1$, let $\varphi_\xi : \xi \to \mathbf{N}$ be injective. For $\eta < \omega_1$, $n \in \mathbf{N}$ set $A(\eta, n) = \{\xi : \xi > \eta, \varphi_\xi(\eta) = n\}$. As $\mathscr{I}$ is uniform and proper, $\bigcup_{n\in\mathbf{N}} A(\eta, n) \notin \mathscr{I}$; as $\mathscr{I}$ is a σ-ideal, there is an $m_\eta \in \mathbf{N}$ such that $A(\eta, m_\eta) \notin \mathscr{I}$. Now there is an $m \in \mathbf{N}$ such that $C = \{\eta : m_\eta = m\}$ is uncountable. But if η, ζ are distinct members of C then

$$A(\eta, m) \cap A(\zeta, m) = \{\xi : \max(\eta, \zeta) < \xi < \omega_1, \varphi_\xi(\eta) = \varphi_\xi(\zeta) = m\}$$
$$= \varnothing.$$

So $\langle A(\eta, m) \rangle_{\eta \in C}$ is an uncountable disjoint family in $\mathscr{P}\omega_1 \setminus \mathscr{I}$, and $\mathscr{I}$ is not ω_1-saturated. [See Jech 78, 27.7.]

(*b*) I take $I = \mathbf{N}$; the finite case is done in the same way. By (*a*), $\mathscr{I} \cap \mathscr{P}A_i$ is never ω_1-saturated in $\mathscr{P}A_i$, so we can choose for each $i \in \mathbf{N}$ a disjoint family $\langle A_i^\xi \rangle_{\xi<\omega_1}$ in $\mathscr{P}A_i \setminus \mathscr{I}$. Choose $\langle \xi(i) \rangle_{i\in\mathbf{N}}$ inductively so that

(i) $A_i^{\xi(i)} \setminus \bigcup_{j<i} A_j^{\xi(j)} \notin \mathscr{I}$;

(ii) $\{\eta : A_k^\eta \setminus (A_i^{\xi(i)} \cup \bigcup_{j<i} A_j^{\xi(j)}) \notin \mathscr{I}\}$ is uncountable for every $k > i$.

This must be possible because when we come to choose $\xi(i)$, the number of choices allowed to us by (i) is uncountable (by the inductive hypothesis), while the number of choices forbidden to us by (ii) is countable (because the A_i^ξ are disjoint).

Now set $B_i = A_i^{\xi(i)} \setminus \bigcup_{j<i} A_j^{\xi(j)}$ for each $i \in \mathbf{N}$.

(*c*) Let $\langle A_\xi \rangle_{\xi<\omega_1}$ be a family in $\mathscr{P}\omega_1 \setminus \mathscr{I}$ witnessing the irregularity of $\mathscr{I}$; i.e., such that whenever $B_\xi \in \mathscr{P}A_\xi \setminus \mathscr{I}$ for every $\xi < \omega_1$, there is an

$\eta<\omega_1$ such that $\{\xi:\eta\in B_\xi\}$ is infinite. Set

$$C=\{\xi:\xi<\omega_1, \exists Z\in\mathscr{P}A_\xi\setminus\mathscr{I}, \mathscr{I}\cap\mathscr{P}Z \text{ is regular in } \mathscr{P}Z\},$$
$$D=\{\xi:\xi\in\omega_1\setminus C, \mathscr{I}\cap\mathscr{P}A_\xi \text{ is not } \omega_2\text{-saturated in } \mathscr{P}A_\xi\}.$$

(i) For $\xi\in C$ choose $Z_\xi\in\mathscr{P}A_\xi\setminus\mathscr{I}$ such that $\mathscr{I}\cap\mathscr{P}Z_\xi$ is regular in $\mathscr{P}Z_\xi$. For $\xi,\eta\in C$ set

$$A_\xi^\eta=Z_\xi\cap Z_\eta \text{ if } Z_\xi\cap Z_\eta\notin\mathscr{I},$$
$$=Z_\xi \text{ if } Z_\xi\cap Z_\eta\in\mathscr{I}.$$

Because $\mathscr{I}\cap\mathscr{P}Z_\xi$ is regular, we can find $B_\xi^\eta\in\mathscr{P}A_\xi^\eta\setminus\mathscr{I}$ such that $\langle B_\xi^\eta\rangle_{\eta\in C}$ is point-finite. For $\eta\in C$, set

$$h(\eta)=\min\{\xi:\xi\in C, Z_\eta\cap Z_\xi\notin\mathscr{I}\},$$
$$B_\eta=B_{h(\eta)}^\eta\setminus\bigcup_{\zeta\in C\cap h(\eta)}Z_\zeta.$$

Then $B_\eta\in\mathscr{P}A_\eta\setminus\mathscr{I}$. If $\eta,\zeta\in C$ and $h(\eta)<h(\zeta)$, then

$$B_\zeta\cap B_\eta\subseteq(B_{h(\zeta)}^\zeta\setminus Z_{h(\eta)})\cap B_{h(\eta)}^\eta=\varnothing.$$

So if $\theta<\omega_1$ and $E=\{\eta:\eta\in C, \theta\in B_\eta\}$, then h must be constant on E; say $h(\eta)=\xi$ for $\eta\in E$. In this case $E\subseteq\{\eta:\eta\in C, \theta\in B_\xi^\eta\}$ is finite. Thus $\langle B_\eta\rangle_{\eta\in C}$ is point-finite.

(ii) For $\xi\in D$ choose a family $\langle A_\xi^\alpha\rangle_{\alpha<\omega_2}$ in $\mathscr{P}A_\xi\setminus\mathscr{I}$ such that $A_\xi^\alpha\cap A_\xi^\beta\in\mathscr{I}$ for all distinct $\alpha,\beta\in\omega_2$. Construct an increasing family $\langle D_\xi\rangle_{\xi<\omega_1}$ of subsets of D, and B'_η for $\eta\in D$, as follows. $D_0=\varnothing$. Given D_ξ, then if $\xi\notin D$ or $\xi\in D_\xi$ set $D_{\xi+1}=D_\xi$. If $\xi\in D\setminus D_\xi$, set

$$V=\{\eta:\eta\in D\setminus D_\xi, \#\{\alpha:\alpha<\omega_2, A_\eta\cap A_\xi^\alpha\notin\mathscr{I}\}\leq\omega_1\},$$
$$U=(D\setminus D_\xi)\setminus V.$$

Note that $\xi\in U$. Take $\beta<\omega_2$ such that $A_\eta\cap A_\xi^\alpha\in\mathscr{I}$ whenever $\eta\in V$ and $\beta\leq\alpha<\omega_2$. Choose $\langle\alpha(\eta)\rangle_{\eta\in U}$, all distinct, such that $\alpha(\eta)\geq\beta$ and $A_\eta\cap A_\xi^{\alpha(\eta)}\notin\mathscr{I}$ for every $\eta\in U$. Set

$$D_{\xi+1}=D_\xi\cup U, B'_\eta=A_\eta\cap A_\xi^{\alpha(\eta)} \text{ for } \eta\in U.$$

At limit ordinals $\xi<\omega_1$ set $D_\xi=\bigcup_{\eta<\xi}D_\eta$.

The effect of this construction is to ensure that

$D=\bigcup_{\xi<\omega_1}D_\xi$;

B'_η is defined for every $\eta\in D$, and $B'_\eta\in\mathscr{P}A_\eta\setminus\mathscr{I}$;

if $\eta\in D_\xi$, $\zeta\in D\setminus D_\xi$ then $A_\zeta\cap B'_\eta\in\mathscr{I}$;

if η,ζ are distinct members of $D_{\xi+1}\setminus D_\xi$ then $B'_\eta\cap B'_\zeta\in\mathscr{I}$.

Set $B_\eta = B'_\eta \setminus \bigcup_{\zeta \in D \cap \eta} B'_\zeta$ for every $\eta \in D$; then $B_\eta \in \mathscr{P}A_\eta \setminus \mathscr{I}$ for every $\eta \in D$ and $\langle B_\eta \rangle_{\eta \in D}$ is disjoint.

(iii) Putting (i) and (ii) together, we see that $\langle B_\xi \rangle_{\xi \in C \cup D}$ is point-finite, while $B_\xi \in \mathscr{P}A_\xi \setminus \mathscr{I}$ for every $\xi \in C \cup D$. By the choice of $\langle A_\xi \rangle_{\xi < \omega_1}$, $C \cup D \neq \omega_1$. Let $\xi \in \omega_1 \setminus (C \cup D)$ and set $Y = A_\xi$; this will serve.

35C Lemma

Suppose that X is a set, $\mathscr{I} \lhd \mathscr{P}X$ an ideal, $A \in \mathscr{P}X \setminus \mathscr{I}$, and that $f: X \to \omega_1$ is a function such that

(i) $f^{-1}[\xi] \in \mathscr{I}$ for every $\xi < \omega_1$;
(ii) whenever $g: X \to \omega_1$ is a function such that $g(x) \leq f(x)$ for every $x \in X$ and $g^{-1}[\xi] \in \mathscr{I}$ for every $\xi < \omega_1$, then $\{x : x \in A, g(x) < f(x)\} \in \mathscr{I}$.

Then $\mathscr{J} = \{B : B \subseteq \omega_1, f^{-1}[B] \cap A \in \mathscr{I}\}$ is a weakly normal ideal of $\mathscr{P}\omega_1$.

Proof $\mathscr{J}$ is an ideal because $\mathscr{I}$ is; it is proper because $A \notin \mathscr{I}$; it is uniform because $f^{-1}[\xi] \in \mathscr{I}$ for every $\xi < \omega_1$. If $h: \omega_1 \to \omega_1$ is such that $h^{-1}[\xi] \in \mathscr{J}$ for every $\xi < \omega_1$, define $g: X \to \omega_1$ by

$$g(x) = h(f(x)) \text{ if } x \in A \text{ and } h(f(x)) \leq f(x),$$
$$g(x) = f(x) \quad \text{otherwise.}$$

Then $g(x) \leq f(x)$ for every $x \in X$, and

$$g^{-1}[\xi] \subseteq f^{-1}[\xi] \cup (A \cap f^{-1}[h^{-1}[\xi]]) \in \mathscr{I}$$

for every $\xi < \omega_1$. So if $B = \{\eta : h(\eta) < \eta\}$, then

$$A \cap f^{-1}[B] = \{x : x \in A, g(x) < f(x)\} \in \mathscr{I},$$

and $B \in \mathscr{J}$. As h is arbitrary, $\mathscr{J}$ is weakly normal.

35D Lemma

If there is a uniform σ-ideal $\mathscr{I} \lhd \mathscr{P}\omega_1$ which is not regular, there is a non-regular weakly normal σ-ideal of $\mathscr{P}\omega_1$.

Proof (*a*) By 35B*c*, we can suppose that $\mathscr{I}$ is ω_2-saturated and that $\mathscr{I} \cap \mathscr{P}Z$ is not regular in $\mathscr{P}Z$ for any $Z \in \mathscr{P}\omega_1 \setminus \mathscr{I}$.

(*b*) There are an $A \in \mathscr{P}\omega_1 \setminus \mathscr{I}$ and an $f: \omega_1 \to \omega_1$ such that

(i) $f^{-1}[\xi] \in \mathscr{I}$ for every $\xi < \omega_1$;
(ii) if $g: \omega_1 \to \omega_1$ is such that $g(\eta) \leq f(\eta)$ for every $\eta < \omega_1$ and

$g^{-1}[\xi]\in\mathscr{I}$ for every $\xi<\omega_1$, then

$$\{\eta:\eta\in A, g(\eta)<f(\eta)\}$$

belongs to $\mathscr{I}$.

P? Suppose this is false. Let $\langle A_\xi\rangle_{\xi<\omega_1}$ be a family in $\mathscr{P}\omega_1\backslash\mathscr{I}$ witnessing the irregularity of $\mathscr{I}$. Choose f_ξ inductively, for $\xi<\omega_1$, as follows. $f_0(\eta)=\eta$ for every $\eta<\omega_1$. Given f_ξ such that $f_\xi^{-1}[\zeta]\in\mathscr{I}$ for every $\zeta<\omega_1$, choose $f_{\xi+1}$ such that

$$f_{\xi+1}(\eta)\leq f_\xi(\eta)\ \forall\eta<\omega_1,$$
$$f_{\xi+1}^{-1}[\zeta]\in\mathscr{I}\ \forall\zeta<\omega_1,$$
$$B_\xi=\{\eta:\eta\in A_\xi, f_{\xi+1}(\eta)<f_\xi(\eta)\}\notin\mathscr{I}.$$

For countable limit ordinals $\xi>0$, set $f_\xi(\eta)=\min_{\zeta<\xi}f_\zeta(\eta)$; then $f_\xi^{-1}[\eta]=\bigcup_{\zeta<\xi}f_\zeta^{-1}[\eta]\in\mathscr{I}$ for each $\eta<\omega_1$, so the induction continues.

Now however $B_\xi\in\mathscr{P}A_\xi\backslash\mathscr{I}$ for each $\xi<\omega_1$, and

$$\{\xi:\xi<\omega_1,\eta\in B_\xi\}\subseteq\{\xi:f_{\xi+1}(\eta)<f_\xi(\eta)\}$$

is finite for every η, i.e. $\langle B_\xi\rangle_{\xi<\omega_1}$ is point-finite, contrary to the choice of $\langle A_\xi\rangle_{\xi<\omega_1}$. **XQ**

(*c*) There is an $A'\in\mathscr{P}A\backslash\mathscr{I}$ such that $A'\cap f^{-1}[\xi]$ is countable for every $\xi<\omega_1$. **P?** Otherwise, for any $g:\omega_1\to\omega_1$, consider $B(g)=\{\xi:gf(\xi)\geq\xi\}$. Then

$$B(g)\cap f^{-1}[\xi]=\{\eta:f(\eta)<\xi, gf(\eta)\geq\eta\}$$
$$\subseteq\{\eta:\exists\zeta<\xi,\eta\leq g(\zeta)\}$$

is countable, for every $\xi<\omega_1$. So $B(g)\cap A\in\mathscr{I}$ and $A\backslash B(g)\notin\mathscr{I}$.

Now let $\langle g_\alpha\rangle_{\alpha<\omega_2}$ be a family of functions from ω_1 to ω_1 such that

$$\alpha<\beta<\omega_2\Rightarrow\{\xi:g_\beta(\xi)=g_\alpha(\xi)\}\text{ is countable.}$$

(It is easy to construct such a family inductively; given $\langle g_\alpha\rangle_{\alpha<\beta}$, take an injective function $\varphi:\beta\to\omega_1$ and set

$$g_\beta(\xi)=\min(\omega_1\backslash\{g_\alpha(\xi):\alpha<\beta,\varphi(\alpha)\leq\xi\})\ \forall\xi<\omega_1.)$$

Define $h_\alpha:A\backslash B(g_\alpha)\to\mathbf{N}$ by writing

$$h_\alpha(\xi)=\varphi_\xi g_\alpha f(\xi)\ \forall\xi\in A\backslash B(g_\alpha),$$

where $\varphi_\xi:\xi\to\mathbf{N}$ is injective, for $\xi<\omega_1$. Let $n(\alpha)\in\mathbf{N}$ be such that

$$C_\alpha=\{\xi:\xi\in A\backslash B(g_\alpha), h_\alpha(\xi)=n(\alpha)\}\notin\mathscr{I};$$

such exists because $A \setminus B(g_\alpha) \notin \mathscr{I}$ and $\mathscr{I}$ is a σ-ideal. Let $m \in \mathbf{N}$ be such that $D = \{\alpha : \alpha < \omega_2, n(\alpha) = m\}$ has cardinal ω_2. Now, for $\alpha < \beta$ in D,

$$\begin{aligned} C_\alpha \cap C_\beta &= \{\xi : \xi \in A \setminus (B(g_\alpha) \cup B(g_\beta)), h_\alpha(\xi) = h_\beta(\xi) = m\} \\ &= \{\xi : \xi \in A, g_\alpha f(\xi) < \xi, g_\beta f(\xi) < \xi, \\ &\qquad \varphi_\xi g_\alpha f(\xi) = \varphi_\xi g_\beta f(\xi) = m\} \\ &\subseteq \{\xi : \xi < \omega_1, g_\alpha f(\xi) = g_\beta f(\xi)\} \\ &= f^{-1}[\{\xi : g_\alpha(\xi) = g_\beta(\xi)\}] \\ &\in \mathscr{I}, \end{aligned}$$

because by the construction of the g_α there is a $\zeta < \omega_1$ such that $\{\xi : g_\alpha(\xi) = g_\beta(\xi)\} \subseteq \zeta$, and $f^{-1}[\zeta] \in \mathscr{I}$ by (*b*) (i) above. Thus $\langle C_\alpha \rangle_{\alpha < \omega_2}$ is an $\mathscr{I}$-almost-disjoint family in $\mathscr{P}\omega_1 \setminus \mathscr{I}$. But $\mathscr{I}$ is supposed to be ω_2-saturated. **XQ**

(*d*) There is an $A'' \in \mathscr{P}A' \setminus \mathscr{I}$ such that $f \restriction A''$ is injective. **P** As $A' \cap f^{-1}[\{\xi\}]$ is countable for each $\xi < \omega_1$, there is a function $\varphi : A' \to \mathbf{N}$ such that $\varphi \restriction A' \cap f^{-1}[\{\xi\}]$ is injective for each $\xi < \omega_1$. Now there is an $n \in \mathbf{N}$ such that $A'' = \varphi^{-1}[\{n\}] \notin \mathscr{I}$. **Q**

(*e*) Set $\mathscr{J} = \{C : C \subseteq \omega_1, A'' \cap f^{-1}[C] \in \mathscr{I}\}$. Then $\mathscr{J}$ is a weakly normal ideal of $\mathscr{P}\omega_1$ by 35C and (*b*) above, and is a σ-ideal because $\mathscr{I}$ is. Finally, $\mathscr{J} \cap \mathscr{P}(f[A''])$ is not regular in $\mathscr{P}(f[A''])$ because $\mathscr{I} \cap \mathscr{P}A''$ is not regular in $\mathscr{P}A''$, by (*a*) above, and $f \restriction A''$ is injective. It follows at once that $\mathscr{J}$ is not regular in $\mathscr{P}\omega_1$.

35E Theorem

Let $\mathscr{I}$ be a uniform maximal ideal of $\mathscr{P}\omega_1$ which is not regular. Let $f : \omega_1 \to \omega_1$ be a function such that $f^{-1}[\xi] \in \mathscr{I}$ for every $\xi < \omega_1$. Then there is a $B \in \mathscr{P}\omega_1 \setminus \mathscr{I}$ such that $B \cap f^{-1}[\{\xi\}]$ is finite for every $\xi < \omega_1$.

Proof For each $\xi < \omega_1$, choose an injection $\varphi_\xi : \xi \to \mathbf{N}$.

(*a*) Let $\langle g_\alpha \rangle_{\alpha < \omega_2}$ be a family of functions from ω_1 to ω_1 such that $\{\xi : g_\alpha(\xi) = g_\beta(\xi)\}$ is countable whenever $\alpha < \beta < \omega_2$ (as in part (*c*) of the proof of 35D). Set

$$H_\alpha = \{\xi : \xi < \omega_1, g_\alpha f(\xi) < \xi\}.$$

Then there is an $\alpha < \omega_2$ such that $H_\alpha \in \mathscr{I}$. **P** Define $k_\alpha : \omega_1 \to \omega + 1$ by

$$\begin{aligned} k_\alpha(\xi) &= \varphi_\xi g_\alpha f(\xi) \; \forall \xi \in H_\alpha, \\ &= \omega \; \forall \xi \in \omega_1 \setminus H_\alpha. \end{aligned}$$

If $\alpha<\beta<\omega_2$ there is a $\zeta_{\alpha\beta}<\omega_1$ such that

$$\{\xi:g_\alpha(\xi)=g_\beta(\xi)\}\subseteq\zeta_{\alpha\beta},$$

and now

$$K_{\alpha\beta}=\{\xi:\xi<\omega_1,k_\alpha(\xi)\neq k_\beta(\xi)\}\supseteq H_\alpha\cap H_\beta\cap\{\xi:g_\alpha f(\xi)\neq g_\beta f(\xi)\}$$
$$\supseteq H_\alpha\cap H_\beta\setminus f^{-1}[\zeta_{\alpha\beta}].$$

Because $\mathscr{I}$ is maximal, the space $X/\sim$ of equivalence classes in

$$X=(\omega+1)^{\omega_1}$$

under the relation

$$g\sim h \text{ if } \{\xi:g(\xi)\neq h(\xi)\}\in\mathscr{I}$$

is totally ordered by saying that

$$g^{\cdot}\leq h^{\cdot} \text{ iff } \{\xi:h(\xi)<g(\xi)\}\in\mathscr{I}.$$

? Suppose, if possible, that no H_α belongs to $\mathscr{I}$. Then $\omega_1\setminus H_\alpha$ belongs to $\mathscr{I}$ for every $\alpha<\omega_2$; since $f^{-1}[\zeta_{\alpha\beta}]\in\mathscr{I}$ whenever $\alpha<\beta<\omega_2$, $K_{\alpha\beta}\notin\mathscr{I}$ for every $\alpha<\beta<\omega_2$, and $k^{\cdot}_\alpha\neq k^{\cdot}_\beta$ in $X/\sim$ for all distinct $\alpha,\beta<\omega_2$. As $X/\sim$ is totally ordered, there must be a $\gamma<\omega_2$ such that

$$D=\{\alpha:\alpha<\omega_2,k^{\cdot}_\alpha<k^{\cdot}_\gamma\}$$

is uncountable. Set $E=\{\alpha:\alpha\in D,\alpha\cap D \text{ is countable}\}$; then $\#(E)=\omega_1$.

For $\alpha\in E$, set

$$A_\alpha=\{\xi:\xi\in H_\gamma,k_\alpha(\xi)<k_\gamma(\xi),f(\xi)\geq\sup_{\beta\in D\cap\alpha}\zeta_{\beta\alpha}\}\notin\mathscr{I}$$

(because $\omega_1\setminus A_\alpha$ can be expressed as the union of three sets in $\mathscr{I}$). If $\beta<\alpha$ in E and $\xi\in A_\alpha\cap A_\beta$, then $f(\xi)\geq\zeta_{\beta\alpha}$ so $g_\alpha f(\xi)\neq g_\beta f(\xi)$; also

$$\max(k_\alpha(\xi),k_\beta(\xi))<k_\gamma(\xi)<\omega,$$

so that $\xi\in H_\alpha\cap H_\beta$ and $k_\alpha(\xi)\neq k_\beta(\xi)$. Thus, for any $\xi\in H_\gamma$,

$$\#(\{\alpha:\alpha\in E,\xi\in A_\alpha\})\leq k_\gamma(\xi)<\omega,$$

and $\langle A_\alpha\rangle_{\alpha\in E}$ is point-finite. But as E is uncountable, this makes $\mathscr{I}$ regular, contrary to the original hypothesis. **X** So there is an $\alpha<\omega_2$ such that $H_\alpha\in\mathscr{I}$. **Q**

(*b*) Set $A=\omega_1\setminus H_\alpha$. Then $A\notin\mathscr{I}$, and $A\cap f^{-1}[\zeta]$ is countable for every $\zeta<\omega_1$.

P $A\cap f^{-1}[\zeta]=\{\xi:\xi<\omega_1,f(\xi)<\zeta,g_\alpha f(\xi)\geq\xi\}$
$\subseteq\{\xi:\exists\eta<\zeta,\xi\leq g_\alpha(\eta)\}$. **Q**

(*c*) Let $\varphi: A \to \mathbf{N}$ be such that $\varphi \restriction A \cap f^{-1}[\{\xi\}]$ is injective for every $\xi < \omega_1$. Set

$$h_\xi(\eta) = \varphi_\eta(\xi) \text{ if } \xi < \eta, 0 \text{ otherwise.}$$

Set

$$L_\xi = \{\eta : \eta \in A, h_\xi f(\eta) < \varphi(\eta)\}$$

for each $\xi < \omega_1$.

? Suppose, if possible, that no L_ξ belongs to $\mathscr{I}$. Set $A_\xi = L_\xi \setminus f^{-1}[\xi + 1] \notin \mathscr{I}$ for each $\xi < \omega_1$. If $\eta \in A_\xi \cap A_\zeta$ where $\xi < \zeta$, then

$$f(\eta) > \xi, \quad f(\eta) > \zeta, \quad h_\xi f(\eta) = \varphi_{f(\eta)}(\xi), \quad \text{and } h_\zeta f(\eta) = \varphi_{f(\eta)}(\zeta);$$

so $h_\xi f(\eta) \neq h_\zeta f(\eta)$. So for any $\eta \in A$,

$$\#(\{\xi : \eta \in A_\xi\}) \leq \varphi(\eta) < \omega.$$

As in (*a*), we find that $\langle A_\xi \rangle_{\xi < \omega_1}$ is a point-finite family in $\mathscr{P}\omega_1 \setminus \mathscr{I}$, which is impossible. **X**

(*d*) So there is a $\zeta < \omega_1$ such that $L_\zeta \in \mathscr{I}$. Set $B = A \setminus L_\zeta \notin \mathscr{I}$. Then, for any $\xi < \omega_1$,

$$\begin{aligned} B \cap f^{-1}[\{\xi\}] &= \{\eta : \eta \in A, h_\zeta f(\eta) \geq \varphi(\eta), f(\eta) = \xi\} \\ &= \{\eta : f(\eta) = \xi, \varphi(\eta) \leq h_\zeta(\xi)\} \end{aligned}$$

is finite because $\varphi \restriction A \cap f^{-1}[\{\xi\}]$ is injective.

35F Theorem

If there is a maximal uniform ideal $\mathscr{I} \lhd \mathscr{P}\omega_1$ which is not regular, there is a weakly normal maximal ideal of $\mathscr{P}\omega_1$.

Proof (*a*) Let Φ be the set of functions $f: \omega_1 \to \omega_1$ such that $f^{-1}[\xi]$ is countable for each $\xi < \omega_1$. Then there is an $f \in \Phi$ such that

$$\forall g \in \Phi, \{\xi : f(\xi) > g(\xi)\} \in \mathscr{I}.$$

P? Otherwise, define $\langle f_\xi \rangle_{\xi < \omega_1}$ in Φ inductively as follows. $f_0(\eta) = \eta$ for every $\eta < \omega_1$. Given $f_\xi \in \Phi$, choose $f_{\xi+1} \in \Phi$ such that

$$\{\eta : f_{\xi+1}(\eta) < f_\xi(\eta)\} \notin \mathscr{I}, \; f_{\xi+1}(\eta) \leq f_\xi(\eta) \, \forall \eta < \omega_1.$$

Given $\langle f_\xi \rangle_{\xi < \zeta}$, where ζ is a non-zero countable limit ordinal, set $f_\zeta(\eta) = \min_{\xi < \zeta} f_\xi(\eta)$ for each η, and check that $f_\zeta \in \Phi$. Now set

$$A_\xi = \{\eta : f_{\xi+1}(\eta) < f_\xi(\eta)\} \in \mathscr{P}\omega_1 \setminus \mathscr{I} \; \forall \xi < \omega_1,$$

and see that $\langle A_\xi \rangle_{\xi < \omega_1}$ is point-finite, so that $\mathscr{I}$ is regular, as in part (*b*) of the proof of 35D. **XQ** (*b*) In this case, if $g:\omega_1 \to \omega_1$ is such that $g(\eta) \le f(\eta)$ for every $\eta < \omega_1$ and $g^{-1}[\xi] \in \mathscr{I}$ for every $\xi < \omega_1$, then $\{\eta : g(\eta) < f(\eta)\} \in \mathscr{I}$. **P** By 35E, there is a $B \in \mathscr{P}\omega_1 \setminus \mathscr{I}$ such that $B \cap g^{-1}[\{\xi\}]$ is finite for every $\xi < \omega_1$. Set

$$h(\eta) = g(\eta) \text{ if } \eta \in B, \ f(\eta) \text{ if } \eta \in \omega_1 \setminus B.$$

Then $h \in \Phi$ and $h \le f$. So

$$\{\eta : g(\eta) < f(\eta)\} \subseteq (\omega_1 \setminus B) \cup \{\eta : h(\eta) < f(\eta)\} \in \mathscr{I}. \quad \mathbf{Q}$$

(*c*) So by 35C the ideal

$$\{C : C \subseteq \omega_1, f^{-1}[C] \in \mathscr{I}\}$$

is a weakly normal ideal of $\mathscr{P}\omega_1$, which is maximal because $\mathscr{I}$ is.

35G Theorem

Let $\mathscr{I}$ be a weakly normal ideal of $\mathscr{P}\omega_1$, and $\langle A_\xi \rangle_{\xi < \omega_1}$ a family in $\mathscr{P}\omega_1 \setminus \mathscr{I}$. Then there is a family $\langle B_\xi \rangle_{\xi < \omega_1}$ such that $B_\xi \in \mathscr{P}A_\xi \setminus \mathscr{I}$ for every $\xi < \omega_1$ and $\{\eta : \eta \le \xi, \zeta \in B_\eta\}$ is finite whenever $\xi < \zeta < \omega_1$.

Proof (*a*) We need to know that if $F \subseteq \omega_1$ is a closed unbounded set, then $\omega_1 \setminus F \in \mathscr{I}$. **P** Define $f:\omega_1 \to \omega_1$ by

$$f(\eta) = \sup(F \cap \eta) \in F \cup \{0\} \ \forall \eta < \omega_1.$$

Then $f^{-1}[\xi] \subseteq \xi + 1$ whenever $\xi \in F$; as F is uncountable, $f^{-1}[\xi]$ is countable and belongs to $\mathscr{I}$ for every $\xi < \omega_1$. As $\mathscr{I}$ is weakly normal, $\mathscr{I}$ contains

$$\{\eta : f(\eta) < \eta\} \supseteq \omega_1 \setminus (F \cup \{0\}),$$

and $\omega_1 \setminus F \in \mathscr{I}$. **Q** In particular, $\omega_1 \setminus \Omega \in \mathscr{I}$ [A2D].

(*b*) Let $\langle \theta(\zeta, n) \rangle_{n \in \mathbf{N}, \zeta \in \Omega}$ be a ladder system on ω_1 [definition: A2G]. Define inductively, for $\xi < \omega_1$, sets $A'_\xi \in \mathscr{P}A_\xi \setminus \mathscr{I}$, countable ordinals α_ξ and α'_ξ, and functions $g_\xi : \omega_1 \to \omega_1$ as follows.

Given $\langle \alpha_\eta \rangle_{\eta < \xi}$, set

$$\alpha'_\xi = \sup_{\eta < \xi} \alpha_\eta < \omega_1,$$
$$A'_\xi = A_\xi \cap \Omega \setminus (\alpha'_\xi + 1) \in \mathscr{P}A_\xi \setminus \mathscr{I},$$
$$g_\xi(\zeta) = \min\{\theta(\zeta, k) : k \in \mathbf{N}, \theta(\zeta, k) \ge \alpha'_\xi\} \text{ if } \zeta \in A'_\xi,$$
$$g_\xi(\zeta) = \zeta \qquad \text{otherwise.}$$

Then $\{\zeta : g_\xi(\zeta) < \zeta\} = A'_\xi \notin \mathscr{I}$; as $\mathscr{I}$ is weakly normal, there is an $\alpha_\xi \ge \alpha'_\xi$

such that $g_\xi^{-1}[\alpha_\xi]\notin\mathcal{I}$. Continue.

(*c*) Now let F be

$$\{\zeta:\zeta<\omega_1,\xi<\zeta\Rightarrow\alpha_\xi<\zeta\}.$$

F is a closed unbounded set [A2D]. If we write

$$B_\xi=\{\eta:\eta\in F\cap A'_\xi,\exists k\in\mathbf{N},\alpha'_\xi\le\theta(\eta,k)<\alpha_\xi\},$$

we see that $B_\xi\subseteq A'_\xi\subseteq A_\xi$ and that

$$g_\xi^{-1}[\alpha_\xi]\subseteq B_\xi\cup\alpha_\xi\cup(\omega_1\backslash F);$$

as α_ξ and $\omega_1\backslash F$ both belong to $\mathcal{I}$, $B_\xi\notin\mathcal{I}$. Suppose now that $\xi<\zeta<\omega_1$. Then $\{\eta:\eta\le\xi,\zeta\in B_\eta\}$ is empty if $\zeta\notin F$, and otherwise is included in

$$I=\{\eta:\eta\le\xi,\exists k\in\mathbf{N},\alpha'_\eta\le\theta(\zeta,k)<\alpha_\eta\}.$$

But since $\alpha_\xi<\zeta$, $\{k:k\in\mathbf{N},\theta(\zeta,k)<\alpha_\xi\}$ is finite, so I is also finite, and $\{\eta:\eta\le\xi,\zeta\in B_\eta\}$ is finite.

35H **Lemma**

$[\mathfrak{m}_K>\omega_1]$ Suppose that $\mathcal{I}$ is a weakly normal ideal of $\mathcal{P}\omega_1$ and that $\langle A_\xi\rangle_{\xi<\omega_1}$ is a family in $\mathcal{P}\omega_1\backslash\mathcal{I}$ such that $\{\eta:\eta\le\xi,\zeta\in A_\eta\}$ is finite whenever $\xi<\zeta<\omega_1$. Then (i) $\mathcal{I}$ cannot be a maximal ideal, (ii) if $\mathcal{I}$ is a σ-ideal there is a point-finite family $\langle B_\xi\rangle_{\xi<\omega_1}$ such that $B_\xi\in\mathcal{P}A_\xi\backslash\mathcal{I}$ for every $\xi<\omega_1$.

Proof (*a*) Set

$$S=\{(\xi,\eta):\eta<\xi<\omega_1,\xi\in A_\eta\}\subseteq\omega_1^2.$$

Then $\mathrm{otp}(S[\{\xi\}])\le\omega$ for every $\xi<\omega_1$. For $\xi<\omega_1$ write

$$U_\xi=\{\zeta:\xi\le\zeta<\omega_1,A_\xi\cap A_\zeta\notin\mathcal{I}\},$$

and set

$$G=\{\xi:\xi<\omega_1,U_\xi\text{ is uncountable}\}.$$

Now from 31F we know that $[\omega_1]^{<\omega}\times[\omega_1]^{<\omega}$, with the S-respecting partial order [31C], satisfies Knaster's condition upwards. Apply 31E*a* with $X=Y=\omega_1$ and $\mathcal{B}=\{U_\xi:\xi\in G\}$ to see that there is an $H\subseteq\omega_1$ such that $H\cap S[\{\xi\}]$ is finite for every $\xi<\omega_1$ and $H\cap U_\xi\neq\varnothing$ for every $\xi\in G$.

(*b*) Note first that $\langle A_\xi\backslash\xi\rangle_{\xi\in H}$ is point-finite, for if $\zeta<\omega_1$ then

$$\{\xi:\xi\in H,\zeta\in A_\xi\backslash\xi\}\subseteq H\cap(S[\{\zeta\}]\cup\{\zeta\})$$

is finite.

(*c*) **?** Suppose, if possible, that $\mathscr{I}$ is a maximal ideal. Then $A_\eta \cap A_\xi \notin \mathscr{I}$ for every $\eta, \xi < \omega_1$ so $G = \omega_1$ and H is uncountable. Define $h:\omega_1 \to \omega_1$ by

$$h(\xi) = \sup(H \cap S[\{\xi\}]) = \sup\{\eta : \eta \in H \cap \xi, \xi \in A_\eta\}.$$

If $\zeta \in H$ then

$$h^{-1}[\zeta] \subseteq \{\eta : \eta \leq \zeta \text{ or } \eta \notin A_\zeta\} \in \mathscr{I}.$$

As H is uncountable, $h^{-1}[\zeta] \in \mathscr{I}$ for every $\zeta < \omega_1$. As $\mathscr{I}$ is weakly normal, $\{\xi : h(\xi) < \xi\} \in \mathscr{I}$. But $h(\xi) < \xi$ for every $\xi > 0$, because $H \cap S[\{\xi\}]$ is always finite. **X** This proves (i).

(*d*) Now suppose that $\mathscr{I}$ is a σ-ideal. We must partition ω_1 into countable sets of two types, as follows. For $\xi \in G$ set

$$\alpha(\xi) = \min(H \cap U_\xi) \in H;$$

this is well-defined because $H \cap U_\xi \neq \emptyset$. Write

$$F = \{\beta : \beta < \omega_1, \text{ if } \xi \in \beta \setminus G \text{ then } U_\xi \subseteq \beta\}.$$

Then F is a closed unbounded set [A2D]. For $\xi \in \omega_1 \setminus G$ set

$$\beta(\xi) = \min(F \setminus (\xi + 1)) \in F.$$

For $\alpha \in H$ set

$$V(\alpha) = \{\xi : \xi \in G, \alpha(\xi) = \alpha\}.$$

Each $V(\alpha)$ is countable because $\xi \leq \alpha(\xi)$ for every $\xi \in G$. So $\langle A_\xi \cap A_\alpha \rangle_{\xi \in V(\alpha)}$ is a countable family in $\mathscr{P}\omega_1 \setminus \mathscr{I}$. By 35B*b* we can find a disjoint family $\langle A'_\xi \rangle_{\xi \in V(\alpha)}$ such that $A'_\xi \in \mathscr{P}(A_\xi \cap A_\alpha \setminus \alpha) \setminus \mathscr{I}$ for every $\xi \in V(\alpha)$. Similarly, for $\beta \in F$,

$$W(\beta) = \{\xi : \xi \in \omega_1 \setminus G, \beta(\xi) = \beta\}$$

is countable, and we can find a disjoint family $\langle A'_\xi \rangle_{\xi \in W(\beta)}$ such that $A'_\xi \in \mathscr{P}A_\xi \setminus \mathscr{I}$ for each $\xi \in W(\beta)$. This defines A'_ξ for every $\xi < \omega_1$.

Now try

$$B_\xi = A'_\xi \setminus \bigcup \{A'_\eta : \eta < \xi, A'_\eta \cap A'_\xi \in \mathscr{I}\} \ \forall \xi < \omega_1,$$

so that $B_\xi \in \mathscr{P}A'_\xi \setminus \mathscr{I} \subseteq \mathscr{P}A_\xi \setminus \mathscr{I}$ for every $\xi < \omega_1$, and $B_\xi \cap B_\eta = \emptyset$ whenever $A_\xi \cap A_\eta \in \mathscr{I}$. Then for any $\zeta < \omega_1$,

$$C = \{\xi : \xi < \omega_1, \zeta \in B_\xi\}$$

is finite. **P** (i) For each $\alpha \in H$, $\langle B_\xi \rangle_{\xi \in V(\alpha)}$ is disjoint, so C can meet $V(\alpha)$ in at most one point. (ii) Similarly C can meet each $W(\beta)$ in at most one point. (iii) If $\xi \in W(\beta)$, $\eta \in W(\gamma)$ and $\beta < \gamma$, then $\xi < \beta \in F$ and $\beta < \gamma = \min(F \setminus (\eta + 1))$, so $\eta \geq \beta$ and (by the definition of F) $\eta \notin U_\xi$ i.e. $A_\xi \cap A_\eta \in \mathscr{I}$; in which case $B_\xi \cap B_\eta = \emptyset$. So C can meet at most one $W(\beta)$. (iv) Finally, if $C \cap V(\alpha) \neq \emptyset$, then $\zeta \in A_\alpha \setminus \alpha$ because $B_\xi \subseteq A'_\xi \subseteq A_\alpha \setminus \alpha$ for $\xi \in V(\alpha)$. So

$$\{\alpha : C \cap V(\alpha) \neq \emptyset\} \subseteq \{\alpha : \alpha \in H, \zeta \in A_\alpha \setminus \alpha\}$$

is finite ((*b*) above). As

$$\omega_1 = \bigcup_{\alpha \in H} V(\alpha) \cup \bigcup_{\beta \in F} W(\beta),$$

this is enough to show that C is finite. **Q**

Thus $\langle B_\xi \rangle_{\xi < \omega_1}$ is point-finite, as required by (ii).

35I Theorem

[$\mathfrak{m}_K > \omega_1$] Let $\mathscr{I}$ be a uniform maximal ideal of $\mathscr{P}\omega_1$. Then $\mathscr{I}$ is regular.

Proof Putting 35G and 35H(i) together, we see that there is no weakly normal maximal ideal of $\mathscr{P}\omega_1$. (For if there were we could start with every $A_\xi = \omega_1$ in 35G.) The result now follows from 35F.

35J Theorem

[$\mathfrak{m}_K > \omega_1$] Let $\mathscr{I}$ be a uniform σ-ideal of $\mathscr{P}\omega_1$. Then $\mathscr{I}$ is regular.

Proof Putting 35G and 35H(ii) together, we see that every weakly normal σ-ideal of $\mathscr{P}\omega_1$ is regular. By 35D it follows that every uniform σ-ideal of $\mathscr{P}\omega_1$ is regular.

35K Sources

Solovay 71 for 35B*a*. Benda & Ketonen 74 for 35E. Kanamori 76 for 35F. Taylor 79 for 35D, 35G and 35J. Laver 82 for 35H(i) and 35I.

35L Exercises

(*a*) If $\mathscr{I}$ is a regular maximal ideal of $\mathscr{P}X$, where X is an infinite set, then for every infinite set Y, $\#(Y^X/\mathscr{I}) = \#(Y^X)$. [Hint: take $\langle B_\xi \rangle_{\xi < \kappa}$ from the second paragraph of 35A. Set $C(x) = \{\xi : x \in B_\xi\}$. Define $h \mapsto f_h : Y^\kappa \to \prod_{x \in X} Y^{C(x)}$ by $f_h(x) = h \restriction C(x)$. See Chang & Keisler 73, Proposition 4.3.7.]

(*b*) If $\mathscr{I}$ is a regular κ-additive ideal of $\mathscr{P}\kappa$, where $\kappa > \omega$, it has **Fodor's property**, i.e. whenever $\langle A_\xi \rangle_{\xi < \kappa}$ is a family in $\mathscr{P}\kappa \setminus \mathscr{I}$ there is a disjoint family $\langle F_\xi \rangle_{\xi<\kappa}$ such that $F_\xi \in \mathscr{P}A_\xi \setminus \mathscr{I}$ for every $\xi<\kappa$. [Hint: take $\langle B_\xi \rangle_{\xi<\kappa}$ from 35A. Set $X_n = \{x : \#(\{\xi : x \in B_\xi\}) = n\}$, $m(\xi) = \min\{n : B_\xi \cap X_n \notin \mathscr{I}\}$, $C_\xi = B_\xi \cap X_{m(\xi)}$,

$$D_\xi = C_\xi \setminus \bigcup\{\bigcap_{\eta\in I} C_\eta : I \in [\xi]^{<\omega}, C_\xi \cap \bigcap_{\eta\in I} C_\eta \in \mathscr{I}\},$$

$f(x) = \#(\{\xi : x \in D_\xi\})$. Choose $x_\xi \in D_\xi$ with $f(x_\xi)$ maximal, set $E_\xi = \bigcap\{D_\eta : x_\xi \in D_\eta\}$. Now choose $F_\xi \subseteq E_\xi$. Or see TAYLOR 79, 5.1.]

(*c*) If $\mathscr{I}$ is a regular σ-ideal of $\mathscr{P}\omega_1$ then $\mathfrak{A} = \mathscr{P}\omega_1/\mathscr{I}$ has no base of cardinal ω_1 i.e. if $\langle a_\xi \rangle_{\xi<\omega_1}$ is any family in $\mathfrak{A}\setminus\{0\}$ there is an $a \in \mathfrak{A}\setminus\{0\}$ not including any a_ξ. [Hint: use (*b*). See BALCAR & VOJTAS 77, and TAYLOR 79, 7.3.]

(*d*) [$\mathfrak{m}_K > \omega_1$] If $\mathscr{I}$ is a uniform σ-ideal of $\mathscr{P}\omega_1$, and $\mathscr{A} \subseteq \mathscr{P}\omega_1$ has cardinal less than or equal to ω_1, then the ideal generated by $\mathscr{I} \cup \mathscr{A}$ is not a maximal ideal. [Hint: use (*c*). See BALCAR & VOJTAS 77.]

(*e*) [$\mathfrak{m}_K > \omega_1$] If $\langle A_\xi \rangle_{\xi<\omega_1}$ is a family of stationary subsets of ω_1, there is a disjoint family $\langle B_\xi \rangle_{\xi<\omega_1}$ of stationary subsets of ω_1 such that $B_\xi \subseteq A_\xi$ for every $\xi<\omega_1$. [BAUMGARTNER, HAJNAL & MÁTÉ 75.]

35M Further results

(*a*) If every σ-ideal of $\mathscr{P}\omega_1$ is regular, then there is no family $\langle \mathscr{I}_\xi \rangle_{\xi<\omega_1}$ of weakly normal σ-ideals of $\mathscr{P}\omega_1$ such that

$$\mathscr{P}\omega_1 = \bigcup_{\xi<\omega_1} (\mathscr{I}_\xi \cup \{\omega_1 \setminus A : A \in \mathscr{I}_\xi\}).$$

[TAYLOR 79.]

(*b*) [$\mathfrak{m}_K = \mathfrak{c}$] (i) If $\kappa < \mathfrak{c}$, there is no family $\langle \mathscr{I}_\xi \rangle_{\xi<\kappa}$ of proper σ-ideals of $\mathscr{P}\mathfrak{c}$ such that

$$\mathscr{P}\mathfrak{c} = \bigcup_{\xi<\kappa} (\mathscr{I}_\xi \cup \{\mathfrak{c} \setminus A : A \in \mathscr{I}_\xi\}).$$

(ii) If $\kappa < \mathfrak{c} < \lambda$, there is no family $\langle \mathscr{I}_\xi \rangle_{\xi<\kappa}$ of proper $\mathfrak{c}$-additive ideals of $\mathscr{P}\lambda$ such that

$$\mathscr{P}\lambda = \bigcup_{\xi<\kappa} (\mathscr{I}_\xi \cup \{\lambda \setminus A : A \in \mathscr{I}_\xi\}).$$

[KRAWCZYK & PELC 80.]

(*c*) If $\kappa < \min(\mathfrak{m}_K, \omega_\omega)$ there is a σ-ideal of $\mathscr{P}\omega_1$ which is not κ-saturated. [BAUMGARTNER & TAYLOR 82*a*.]

35N Problems

(*a*) [$\mathfrak{m} > \omega_1$] Is there a family $\langle \mathscr{I}_\xi \rangle_{\xi<\omega_1}$ of proper σ-ideals of

$\mathscr{P}\omega_1$ such that

$$\mathscr{P}\omega_1 = \bigcup_{\xi < \omega_1}(\mathscr{I}_\xi \cup \{\omega_1 \setminus A : A \in \mathscr{I}_\xi\})?$$

(*b*) [$\mathfrak{m} > \omega_1$] Can $\mathscr{P}\omega_1$ have an ω_2-saturated uniform σ-ideal? [LAVER 82, BAUMGARTNER & TAYLOR 82*b*.]

Notes and comments

In both the last two sections, my aim has been to prove specific major theorems (34C–J, 35I, 35J). I have gone to some trouble to eliminate everything not essential for the proofs of these results; but even so it should be plain that they are embedded in deep and extensive theories, and that this is why they are important.

There are several notable features of the theory of regular ideals. (i) The definition is by no means an obvious one; so I have written out 35L*a*–*b* to provide reasons for trying to think about them. (ii) All the natural questions seem to be hard; of 35B–G, only 35C is straightforward. (I don't think it's even obvious that every infinite set carries a regular maximal ideal. For a proof of this see CHANG & KEISLER 73, Proposition 4.3.5.) (iii) There seem to be two important types of regular ideal of $\mathscr{P}\kappa$: the regular uniform maximal ideals, and the regular κ-additive ideals; and these are in many ways quite different. Most of the time they have to be dealt with separately; thus 35B and 35D deal with σ-ideals, while 35E–F deal with maximal ideals. It is only in 35G and the first part of the proof of 35H that we save time by taking them together. Looked at in another way, it is very remarkable that so much of the proof of 35I should be useful for 35J. (iv) For once, the axiom of constructibility is not opposed to Martin's axiom in this area. In fact, $V = L$ implies that every uniform maximal ideal of $\mathscr{P}\omega_1$ is regular [PRIKRY 70]. It is therefore particularly important to note that there is a model of set theory (due to W.H. Woodin) in which $\diamondsuit$ is true and 35I, 35J and 35L*d* are all false [LAVER 82]. (Thus 35B*c* and 35D–F are not vacuously true.)

When I embarked on this book, my chief delight was in searching out the many ingenious arguments which used Martin's axiom directly by setting up appropriate partially ordered sets. As I write it, however, I find almost equal pleasure in eliminating Martin's axiom from individual arguments by quoting earlier theorems, as I have done in 35H. The slight disadvantage of this procedure is that the exact form of MA required (in the classification of B1A) may be obscured.

Much of the work of this section has been stimulated by an old problem, generally attributed to S. Ulam [35N*a*]. It is known that if there is a Kurepa tree then the answer to 35N*a* is 'no' [PRIKRY 76, TAYLOR 80];

and that the existence of a Kurepa tree seems to be independent of $\mathfrak{m} > \omega_1$ [B2F–G]. But it may be that $\mathfrak{m} > \omega_1$ is already enough to answer 35N*a*. In this context, note 35M*a*–*b*. Actually, 35M*b* has nothing to do with the material here, being a consequence of results in §§22 and 32.

Concerning 35N*b*, note that it is consistent to suppose that $\mathscr{P}\omega_1$ has an ω_2-saturated uniform σ-ideal but no non-regular uniform maximal ideal [LAVER 82]; see also B2I.

4

When $\mathfrak{m} > \omega_1$

I come at last to results which need the full strength of the cardinal $\mathfrak{m}$; that is to say, which involve partially ordered sets which are ccc but may not satisfy Knaster's condition. I divide these into two sections on combinatorics (§§41–42) and two on general topology (§§43–44).

41 Combinatorics I

I base this section on the result that if $\mathfrak{m} > \omega_1$ then every ccc partially ordered set satisfies Knaster's condition [41A*b*]. This is already enough to prove that the product of ccc spaces is ccc [41E], so that Souslin's hypothesis is true [41D, 41F–G], and enables us to apply the results of §31 to ccc sets [41B–C]. I give a result in the partition calculus [41H] with some important corollaries. I conclude with a version of 'Devlin's axiom' [41K] and a description of some principles apparently weaker than $\mathfrak{m} > \omega_1$ [41L].

41A Theorem

$[\mathfrak{m} > \omega_1]$ Let P be an upwards-ccc partially ordered set.

(*a*) If $\langle p_\xi \rangle_{\xi < \omega_1}$ is a family in P, there is an uncountable $A \subseteq \omega_1$ such that $\{p_\xi : \xi \in A\}$ is upwards-centered in P.

(*b*) P satisfies Knaster's condition upwards.

Proof (*a*) For $\xi < \omega_1$ set

$$Q_\xi = \{p : p \in P, \exists \eta \geq \xi \text{ such that } p \geq p_\eta\}.$$

Then each Q_ξ is up-open in P, and $Q_\xi \subseteq Q_\eta$ whenever $\eta \leq \xi$. Now there is a $\zeta < \omega_1$ such that Q_ξ is cofinal with Q_ζ for every $\xi \geq \zeta$. **P?** Otherwise, we can find inductively ordinals $\alpha(\xi) < \omega_1$, for $\xi < \omega_1$, such that

$\alpha(\xi) = \sup_{\eta < \xi} \alpha(\eta)$ for limit ordinals $\xi < \omega_1$.

$\alpha(\xi + 1) \geq \alpha(\xi)$ and $Q_{\alpha(\xi+1)}$ is not cofinal with $Q_{\alpha(\xi)}$, $\forall \xi < \omega_1$.

Now for each $\xi < \omega_1$ we can choose $q_\xi \in Q_{\alpha(\xi)}$ such that no member of $Q_{\alpha(\xi+1)}$ is greater than or equal to q_ξ. As P is upwards-ccc there are $\eta < \xi$ such that q_η and q_ξ have a common upper bound q, say. But in this case $q \in Q_{\alpha(\xi)} \subseteq Q_{\alpha(\eta+1)}$ and $q \geq q_\eta$, contrary to the choice of q_η. **XQ**

As $p_\zeta \in Q_\zeta$, Q_ζ is not empty. As Q_ζ is up-open in P, it is upwards-ccc in its induced ordering. So, because $\mathfrak{m} > \omega_1$, there is an upwards-directed $R \subseteq Q_\zeta$ meeting Q_ξ for every $\xi \geq \zeta$. Set

$$A = \{\eta : \eta < \omega_1, \exists r \in R \text{ such that } r \geq p_\eta\}.$$

Then $A \cap [\xi, \omega_1[\neq \varnothing$ for every $\xi \geq \zeta$, so A is uncountable. Also $\{p_\eta : \eta \in A\}$ is upwards-centered because R is upwards-directed.

(*b*) Follows immediately from (*a*).

41B **Corollary**

Let P be an upwards-ccc partially ordered set. Let $\mathcal{Q}$, $\mathcal{R}$ and $\mathcal{S}$ be families of subsets of P such that

(i) $\#(\mathcal{Q}) < \mathfrak{m}$, $\#(\mathcal{R}) < \mathfrak{m}$, $\#(\mathcal{S}) < \mathfrak{m}$;

(ii) every member of $\mathcal{Q}$ is cofinal with P;

(iii) $S \cap \bigcap \mathcal{R}_0 \neq \varnothing$ for every $S \in \mathcal{S} \cup \{P\}$, $\mathcal{R}_0 \in [\mathcal{R}]^{<\omega}$.

Then there is a sequence $\langle P_i \rangle_{i \in \mathbf{N}}$ of subsets of P such that

(α) every P_i is non-empty and upwards-directed;

(β) $P_i \cap Q \neq \varnothing$ for every $i \in \mathbf{N}$, $Q \in \mathcal{Q}$;

(γ) if $R \in \mathcal{R}$, $\{i : P_i \cap R = \varnothing\}$ is finite;

(δ) if $S \in \mathcal{S}$, $\{i : P_i \cap S \neq \varnothing\}$ is infinite.

Proof If $\mathfrak{m} > \omega_1$ then P satisfies Knaster's condition upwards and the result is immediate from 31A. If $\mathfrak{m} = \omega_1$ then $\mathcal{Q}$, $\mathcal{R}$ and $\mathcal{S}$ are all countable, so there is an easy direct proof by induction.

41C **Corollary**

(*a*) If P is a non-empty upwards-ccc partially ordered set, $\mathcal{Q}$ is a family of cofinal subsets of P with $\#(\mathcal{Q}) < \mathfrak{m}$, and $A \in [P]^{<\mathfrak{m}}$, then there is a sequence $\langle P_i \rangle_{i \in \mathbf{N}}$ of upwards-directed subsets of P, all meeting every member of $\mathcal{Q}$, which covers A.

(*b*) If P is an upwards-ccc partially ordered set and $\#(P) < \mathfrak{m}$ then P is σ-centered upwards.

(*c*) If $\mathfrak{m} > \omega_1$ then $\mathfrak{m}_K = \mathfrak{m}$.

(*d*) $\mathrm{cf}(\mathfrak{m}) > \omega$.

Proof (*a*) From 41B with $\mathcal{S} = \{\{p\} : p \in A\}$ (cf. 31B*a*).

(*b*) From (*a*) (cf. 31B*b*).

(*c*) From 41A*b* and the definitions of $\mathfrak{m}$ and $\mathfrak{m}_K$.

(*d*) From (*c*) and 31K.

41D Corollary

[$\mathfrak{m} > \omega_1$] There is no Souslin tree.

Proof Recall that a Souslin tree is an uncountable upwards-ccc tree without uncountable chains. But if X is an uncountable upwards-ccc tree, it has an uncountable upwards-linked subset, by 41A; and in a tree an upwards-linked subset must be a chain.

41E Theorem

[$\mathfrak{m} > \omega_1$] The product of any family of ccc topological spaces is ccc.

Proof (*a*) Consider first the product of two ccc topological spaces X and Y. If $\langle G_\xi \rangle_{\xi < \omega_1}$ is any family of non-empty open sets in $X \times Y$ then we can choose non-empty open sets $E_\xi \subseteq X$ and $H_\xi \subseteq Y$ such that $E_\xi \times H_\xi \subseteq G_\xi$ for every $\xi < \omega_1$. Now

$$P = \{E : E \subseteq X \text{ is open, } E \neq \emptyset\}$$

is downwards-ccc, so by 41A there is an uncountable $A \subseteq \omega_1$ such that $\{E_\xi : \xi \in A\}$ is downwards-linked in P. As Y is ccc there are distinct ξ, $\eta \in A$ such that $H_\xi \cap H_\eta \neq \emptyset$. As $E_\xi \cap E_\eta \neq \emptyset$ also, $G_\xi \cap G_\eta \neq \emptyset$. As $\langle G_\xi \rangle_{\xi < \omega_1}$ is arbitrary, $X \times Y$ is ccc.

(*b*) It follows at once that the product of any finite family of ccc spaces is ccc. By 12I, any product of ccc spaces is ccc.

41F Corollary

[$\mathfrak{m} > \omega_1$] If X is a totally ordered space which is ccc under its order topology, it is separable.

Proof (*a*) Let $\mathscr{I}$ be the family of non-empty sets of the form $]x, y[$ where $x < y$ in X. Let $\mathscr{E}$ be a maximal disjoint family of sets of the form $I \times J$, where $I, J \in \mathscr{I}$ and $I \cap J = \emptyset$. As $\mathscr{E}$ is a disjoint family of open subsets of X^2, it is countable, by 41E. Let $Y \subseteq X$ be a countable set such that whenever $I \times J \in \mathscr{E}$ then both I and J are expressible as $]x, y[$ with $x, y \in Y$. Let Z be the set of isolated points of X; then Z is also countable.

(*b*) **?** If $Y \cup Z$ is not dense in X, let $G = X \setminus \overline{Y \cup Z}$. As G has no isolated points, there are $s, t, u \in X$ such that $]s, t[$ and $]t, u[$ are non-empty and $]s, u[\subseteq G$. Now $]s, t[\times]t, u[$ must meet some member $I \times J$ of $\mathscr{E}$, by the maximality of $\mathscr{E}$; express I as $]x, y[$ and J as $]z, w[$, where x, y, z, $w \in Y$. Then $x < t$ (because $]s, t[$ meets $]x, y[$) and similarly $t < w$; thus $x < w$; as

$I \cap J = \emptyset$, $y \leq z$; also $s < y$ and $z < u$; so y, z both belong to $]s, u[\subseteq G$. But $G \cap Y$ is supposed to be empty. **X**

(*c*) Thus $Y \cup Z$ is a countable dense set in X, and X is separable.

41G **Corollary**

$[\mathfrak{m} > \omega_1]$ If X is a non-empty Dedekind complete totally ordered set, with no gaps and no greatest or least element, which is ccc under its order topology, then X is order-isomorphic to $\mathbf{R}$.

Sketch of proof

By 41F, X has a countable dense subset Y. Y has no gaps and no greatest or least element, so is order-isomorphic to $\mathbf{Q}$, and the isomorphism extends (uniquely) to an order-isomorphism between X and $\mathbf{R}$.

41H **Theorem**

Let X be a set with $\#(X) < \mathfrak{m}$, and $S \subseteq [X]^2$. Then either X is expressible as $\bigcup_{n \in \mathbf{N}} X_n$ where $[X_n]^2 \cap S = \emptyset$ for every $n \in \mathbf{N}$ or

(i) there is an uncountable $Y \subseteq X$ such that

$$\{x : \{x, y\} \in S \; \forall y \in I\}$$

is uncountable for every finite $I \subseteq Y$;

(ii) there are an infinite $W \subseteq X$ and an uncountable $Z \subseteq X$ such that $\{w, z\} \in S$ for every $w \in W$, $z \in Z$.

Proof If X is countable, the first alternative holds, for trivial reasons. So let us take X to be uncountable, in which case $\mathfrak{m} > \omega_1$.

Let P be the set

$$\{I : I \in [X]^{<\omega}, [I]^2 \cap S = \emptyset\}.$$

Case 1 Suppose that P is upwards-ccc. As $\#(P) = \#(X) < \mathfrak{m}$, P is expressible as $\bigcup_{n \in \mathbf{N}} P_n$ where each P_n is upwards-centered [41C*b*]. Set $X_n = \bigcup P_n$ for each $n \in \mathbf{N}$, and see that $X = \bigcup_{n \in \mathbf{N}} X_n$ and that $[X_n]^2 \cap S = \emptyset$ for each $n \in \mathbf{N}$.

Case 2 Suppose that P is not upwards-ccc. For $W \subseteq X$ write $\tilde{S}(W) = \{x : \exists w \in W, \{x, w\} \in S\}$.

(*a*) In this case, P has an uncountable up-antichain R. Let $\langle I_\xi \rangle_{\xi < \omega_1}$ enumerate a constant-size Δ-system in R with root I. Then for any distinct ξ, $\eta < \omega_1$, $I_\xi \cup I_\eta \notin P$, i.e. $[I_\xi \cup I_\eta]^2 \cap S \neq \emptyset$; as $[I_\xi]^2 \cap S = [I_\eta]^2 \cap S = \emptyset$,

$$(I_\xi \setminus I) \cap \tilde{S}(I_\eta \setminus I) \neq \emptyset.$$

Let m be the common value of $\#(I_\xi \setminus I)$, and enumerate $I_\xi \setminus I$ as $\langle x_i^\xi \rangle_{i<m}$ for each $\xi < \omega_1$.

(*b*) Let $\mathscr{F}$ be a uniform ultrafilter on ω_1. Then, for each $\xi < \omega_1$,

$$\omega_1 \setminus \{\xi\} = \bigcup_{i,j<m} \{\eta : \{x_i^\xi, x_j^\eta\} \in S\},$$

so there are $i(\xi), j(\xi) < m$ such that

$$\{\eta : \{x_{i(\xi)}^\xi, x_{j(\xi)}^\eta\} \in S\} \in \mathscr{F}.$$

Let k, $l < m$ be such that

$$A = \{\xi : i(\xi) = k, j(\xi) = l\}$$

is uncountable. Set $Y = \{x_k^\xi : \xi \in A\}$. As $\langle I_\xi \setminus I \rangle_{\xi<\omega_1}$ is disjoint, Y is uncountable. If $I \in [Y]^{<\omega}$, then

$$B = \{\eta : \{x, x_l^\eta\} \in S \ \forall x \in I\} \in \mathscr{F},$$

and

$$\{y : \{x, y\} \in S \ \forall x \in I\} \supseteq \{x_l^\eta : \eta \in B\}$$

is uncountable. This proves (i).

(*c*) In $\omega \times (\omega_1 \setminus \omega)$ set

$$S_{ij} = \{(n, \xi) : n \in \omega, \xi \in \omega_1 \setminus \omega, \{x_i^n, x_j^\xi\} \in S\}$$

for each $i, j < m$. Then $\bigcup_{i,j<m} S_{ij} = \omega \times (\omega_1 \setminus \omega)$. By 21I (since $\mathfrak{p} \geq \mathfrak{m} > \omega_1$) there are an infinite $M \subseteq \omega$, an uncountable $D \subseteq \omega_1$, and $i, j < m$ such that $M \times D \subseteq S_{ij}$. Set $W = \{x_i^n : n \in M\}$ and $Z = \{x_j^\xi : \xi \in D\}$; then W and Z fulfil the requirements of (ii).

41I **Corollary**

Let X be a partially ordered set of cardinal less than $\mathfrak{m}$ with no uncountable upwards-centered subset. Then X is expressible as a countable union of weak antichains.

Proof In 41H set $S = \{\{x, y\} : x < y\}$. Then a set $A \subseteq X$ is a weak antichain iff $[A]^2 \cap S = \emptyset$. ? If X is not expressible as a countable union of weak antichains, then by 41H(i) there is an uncountable $Y \subseteq X$ such that

$$Z_I = \{x : \{x, y\} \in S \ \forall y \in I\}$$

is uncountable for every finite $I \subseteq Y$. But if $y \in Y$ then $\{x : x < y\}$ is upwards-centered in X, so must be countable. Thus if $I \subseteq Y$ is finite,

$$W_I = \{x : \exists y \in I, x < y\}$$

is countable, and there is a $z \in Z_I \setminus W_I$. In this case z is an upper bound for I. This shows that Y is upwards-centered, contrary to hypothesis. **X**

41J **Corollary**

$[\mathfrak{m} > \omega_1]$ Every Aronszajn tree is special.

Proof Recall that an Aronszajn tree is an uncountable tree without uncountable chains in which every level is countable. As it is a tree without uncountable chains, it has no uncountable upwards-centered set. Also its cardinal must be precisely ω_1. So by 41I it is expressible as a countable union of weak antichains; i.e. it is special.

41K **Theorem**

$[\mathfrak{m} > \omega_1]$ Let P be a partially ordered set such that (i) every up-antichain in P has cardinal less than $\mathfrak{m}$, (ii) if $\langle Q_n \rangle_{n \in \mathbf{N}}$ is a sequence of up-open cofinal subsets of P then $\bigcap_{n \in \mathbf{N}} Q_n \neq \emptyset$. Then if $\mathcal{Q}$ is any family of cofinal subsets of P with $\#(\mathcal{Q}) \leq \omega_1$, there is an upwards-directed $R \subseteq P$ meeting every member of $\mathcal{Q}$.

Proof (*a*) For $p \in P$ write $[p, \infty[= \{q : q \geq p\}$. Set $P_0 = \{p : \exists$ sequence $\langle Q_n \rangle_{n \in \mathbf{N}}$ of up-open cofinal sets in P such that $\bigcap_{n \in \mathbf{N}} Q_n \cap [p, \infty[= \emptyset\}$. Then there is a $p_1 \in P$ such that $P_0 \cap [p_1, \infty[= \emptyset$.

P Let $S \subseteq P_0$ be maximal subject to being an up-antichain in P. For each $r \in S$ let $\langle Q_n^r \rangle_{n \in \mathbf{N}}$ be a sequence of up-open cofinal sets such that $\bigcap_{n \in \mathbf{N}} Q_n^r \cap [r, \infty[= \emptyset$. Set

$$Q_n = \bigcup_{r \in S} Q_n^r \cap [r, \infty[.$$

Then each Q_n is up-open (because all the Q_n^r and $[r, \infty[$ are up-open). As $[r, \infty[\cap [r', \infty[= \emptyset$ whenever r, r' are distinct members of S,

$$\bigcap_{n \in \mathbf{N}} Q_n = \bigcup_{r \in S} (\bigcap_{n \in \mathbf{N}} Q_n^r \cap [r, \infty[) = \emptyset.$$

So there must be some $n \in \mathbf{N}$ such that Q_n is not cofinal with P, by the hypothesis (ii). Let p_1 be such that $Q_n \cap [p_1, \infty[= \emptyset$. If $r \in S$, then $Q_n^r \cap [r, \infty[\cap [p_1, \infty[= \emptyset$; but Q_n^r is cofinal with P, so $[r, \infty[\cap [p_1, \infty[$ must be empty. By the maximality of S, $P_0 \cap [p_1, \infty[= \emptyset$, as required. **Q**

(*b*) Since the cofinality of $\mathfrak{m}$ is not fixed, we need a split argument. If $\mathrm{cf}(\mathfrak{m}) \neq \omega_1$, set $p_0 = p_1$. If $\mathrm{cf}(\mathfrak{m}) = \omega_1$, we can choose $p_0 \geq p_1$ such that

$$\kappa = \sup\{\#(R) : R \subseteq [p_0, \infty[\text{ is an up-antichain}\} < \mathfrak{m}.$$

P? Otherwise, there is certainly an up-antichain $R \subseteq [p_1, \infty[$ with $\#(R) = \omega_1$. Enumerate R as $\langle r_\xi \rangle_{\xi < \omega_1}$, and let $\langle \kappa_\xi \rangle_{\xi < \omega_1}$ be a family of cardinals less than $\mathfrak{m}$ such that $\sup_{\xi < \omega_1} \kappa_\xi = \mathfrak{m}$. For each $\xi < \omega_1$ choose an up-antichain $R_\xi \subseteq [r_\xi, \infty[$ with $\#(R_\xi) = \kappa_\xi$. Then $\bigcup_{\xi < \omega_1} R_\xi$ is an up-antichain in P of cardinal $\mathfrak{m}$, contrary to the hypothesis (i) of the theorem. **XQ**

(*c*) If $[p_0, \infty[$ has a maximal element p we can take $R = \{p\}$; if $\mathcal{Q} = \emptyset$ we can take $R = \emptyset$. So let us suppose that $[p_0, \infty[$ has no maximal element and that there is a family $\langle Q_\xi \rangle_{\xi < \omega_1}$ running over $\mathcal{Q}$. Set

$$Q'_\xi = \{p : \exists q \in Q_\xi, p \geq q\}, \tilde{Q}_\xi = \bigcap_{\eta \leq \xi} Q'_\eta.$$

Then each Q'_ξ is up-open and cofinal with P.

(*d*) Choose inductively a family $\langle T_\xi \rangle_{\xi < \omega_1}$ of subsets of P such that, for each $\xi < \omega_1$, T_ξ is maximal subject to the requirements

$T_\xi \subseteq \tilde{Q}_\xi \cap [p_0, \infty[$;

T_ξ is an up-antichain in P;

$\forall p \in T_\xi, \eta < \xi \; \exists q \in T_\eta, q < p$.

Then if $\xi < \omega_1$ and $p \in [p_0, \infty[$ there is an $r \in T_\xi$ such that $[p, \infty[\cap [r, \infty[\neq \emptyset$.

P Induce on ξ. Assume that the result is true for $\eta < \xi$. For $\eta < \xi$, set

$$S_\eta = \bigcup \{[p', \infty[: p' \in T_\eta\} \cup \{p' : p' \in P, [p_0, \infty[\cap [p', \infty[= \emptyset\}.$$

Then S_η is up-open, and by the inductive hypothesis S_η is cofinal with P for each $\eta < \xi$. So

$$A^\xi_p = [p, \infty[\cap \bigcap_{\eta < \xi} S_\eta \cap \tilde{Q}_\xi \neq \emptyset$$

because $p \notin P_0$. Let $q \in A^\xi_p$. As $[p_0, \infty[$ has no maximal element, there is a $q' > q$. Now for any $\eta < \xi$, $q \in S_\eta \cap [p_0, \infty[$, so there is a $p' \in T_\eta$ such that $p' \leq q < q'$. Thus q' is a candidate for membership of T_ξ; by the maximality of T_ξ, there is an $r \in T_\xi$ such that $[q', \infty[\cap [r, \infty[\neq \emptyset$, and now $[p, \infty[\cap [r, \infty[\neq \emptyset$. **Q**

(*e*) Clearly $T = \bigcup_{\xi < \omega_1} T_\xi$ is a tree in which the T_ξ are the levels. Now $\#(T) < \mathfrak{m}$. **P** (α) If $\mathrm{cf}(\mathfrak{m}) = \omega_1$, then by the choice of p_0 in (b) above, $\#(T_\xi) \leq \kappa$ for each $\xi < \omega_1$ and $\#(T) \leq \max(\kappa, \omega_1) < \mathfrak{m}$. ($\beta$) If $\mathrm{cf}(\mathfrak{m}) \neq \omega_1$, then by 41C*d* $\mathrm{cf}(\mathfrak{m}) > \omega_1$; as $\#(T_\xi) < \mathfrak{m}$ for each $\xi < \omega_1$, by the hypothesis (i), $\#(T) < \mathfrak{m}$. **Q**

(*f*) **?** If T has no uncountable chains, then by 41I it is expressible as $\bigcup_{n \in \mathbb{N}} T^{(n)}$ where each $T^{(n)}$ is a weak antichain; observe that each $T^{(n)}$ is actually an up-antichain in P (because each T_ξ is). Set

$$S_n = \bigcup\{[p, \infty[: p \in T^{(n)}\},$$
$$\tilde{S}_n = S_n \cup \{p : p \in P, [p, \infty[\cap S_n = \emptyset\}.$$

Then each $\tilde{S}_n$ is up-open and cofinal with P. So $\bigcap_{n \in \mathbf{N}} \tilde{S}_n$ meets $[p_0, \infty[$; take $r_0 \in \bigcap_{n \in \mathbf{N}} \tilde{S}_n \cap [p_0, \infty[$.

Set $A = \{q : q \in T, q \leq r_0\}$. Then A is a chain in T, so is countable, and $A \subseteq \bigcup_{\xi < \zeta} T_\xi$ for some $\zeta < \omega_1$. Now by (*d*) there is a $q_0 \in T_\zeta$ such that $[q_0, \infty[\cap [r_0, \infty[\neq \emptyset$; let $n \in \mathbf{N}$ be such that $q_0 \in T^{(n)}$. As $q_0 \notin A$, $r_0 \notin [q_0, \infty[$; while if $q \in T^{(n)} \setminus \{q_0\}$, then $[q_0, \infty[\cap [q, \infty[= \emptyset$, so $r_0 \notin [q, \infty[$. Thus $r_0 \notin S_n$. At the same time,

$$S_n \cap [r_0, \infty[\supseteq [q_0, \infty[\cap [r_0, \infty[\neq \emptyset.$$

So $r_0 \notin \tilde{S}_n$. **X**

(*g*) So T has an uncountable chain C, say, and C meets uncountably many of the $\tilde{Q}_\xi$, because $T_\xi \subseteq \tilde{Q}_\xi$ for every $\xi < \omega_1$. Accordingly C meets every Q'_ξ and

$$R = \{p : p \in P, \exists q \in C, p \leq q\}$$

is an upwards-directed subset of P meeting every member of $\mathcal{Q}$, as required.

41L K, H and $\mathfrak{l}$

(*a*) Consider the following statements:

K: If P is an upwards-ccc partially ordered set it satisfies Knaster's condition upwards.

H: If P is an upwards-ccc partially ordered set and $A \subseteq P$ is uncountable, there is an uncountable subset of A which is upwards-centered in P.

L(κ): If P is an upwards-ccc partially ordered set and $\#(P) \leq \kappa$ then P is σ-centered upwards.

(*b*) L($\mathfrak{c}$) is false. **P** Write

$$P = \{G : G \subseteq [0,1] \text{ is open}, \mu G < 1\}$$

where μ is Lebesgue measure on $[0,1]$. If $\langle P_n \rangle_{n \in \mathbf{N}}$ is any sequence of upwards-centered subsets of P, then for each $n \in \mathbf{N}$ there is an $s_n \in [0,1] \setminus \bigcup P_n$; there is a $G \in P$ such that $\{s_n : n \in \mathbf{N}\} \subseteq G$, and now $G \in P \setminus \bigcup_{n \in \mathbf{N}} P_n$. Thus P is not σ-centered upwards. But P is upwards-ccc [32A] and $\#(P) = \mathfrak{c}$. So L($\mathfrak{c}$) is false. **Q**

Accordingly we may define $\mathfrak{l}$ to be the least cardinal such that L($\mathfrak{l}$) is false, and we have $\mathfrak{l} \leq \mathfrak{c}$.

(*c*) By 41C*b*, $\mathfrak{m} \le \mathfrak{l}$; so by 13A(ii) and Bell's theorem [14C], $\mathfrak{m} = \min(\mathfrak{p}, \mathfrak{l})$. Clearly $\mathfrak{l} > \omega_1 \Rightarrow \mathrm{H} \Rightarrow \mathrm{K}$; it is also true that $\mathrm{H} \Rightarrow \mathfrak{l} > \omega_1$ [FREMLIN *b*]. Beyond these facts the relationships between $\mathfrak{m} > \omega_1$, H and K seem to be quite uncertain [41P*a*]. Compare 13D*b*–*c*.

(*d*) Many of the results given in this book as consequences of $\mathfrak{m} > \omega_1$ can in fact be derived from H or K. In the present section we already have 41D–G as consequences of K, while 33E–F above are consequences of H. It is easy to see that 41H(i) and 41I are true whenever $\#(X) < \mathfrak{l}$, so that 41J is a consequence of H. In 14B, 31G, 31K, 32E–H, 32J–N, 33B–D and 41K also we can, be refining the arguments, replace $\mathfrak{p}$, $\mathfrak{m}_K$ or $\mathfrak{m}$ by $\mathfrak{l}$.

41M Sources

JUHÁSZ 70 for 41A (attributed to K. Kunen, F. Rowbottom and R.M. Solovay). BAUMGARTNER, MALITZ & REINHARDT 70 for 41J. SOLOVAY & TENNENBAUM 71 for 41D. HAJNAL & JUHÁSZ 71 for 41E (attributed to F. Rowbottom). LAVER 75 for 41H(ii). DEVLIN 76 for 41K. JECH 78 for the idea of 41H(i). KUNEN & TALL 79 for H of 41L. S.A. Argyros gave me the proof of 41A*a*.

41N Exercises

(*a*) Let $f:\omega_1 \to \mathbf{R}$ be an injection. Let S be

$$\{\{\eta, \xi\} : \eta < \xi < \omega_1, f(\eta) < f(\xi)\}.$$

Then $[A]^2 \cap S \neq \emptyset$ and $[A]^2 \backslash S \neq \emptyset$ for every uncountable $A \subseteq \omega_1$.

(*b*) $[\mathfrak{m} > \omega_1]$ Let X be a tree of cardinal and height both less than or equal to ω_1, and with at most ω_1 uncountable branches. Then X is expressible as $\bigcup_{n\in\mathbf{N}} X_n$ where each X_n has the property that if $x \in X_n$ then $\{y : y \in X_n, y \ge x\}$ is a chain. [Hint: if the uncountable branches of X can be enumerated as $\langle B_\xi \rangle_{\xi < \omega_1}$, choose $x_\xi \in B_\xi \backslash \bigcup_{\eta < \xi} B_\eta$. Apply 41I to $\{x_\xi : \xi < \omega_1\}$ and then to $X \backslash \bigcup_{\xi < \omega_1} \{x : x \in B_\xi, x \ge x_\xi\}$. See BAUMGARTNER *a*, Corollary 8.4, and B2G; also BAUMGARTNER 83, 7.8.]

(*c*) $[\mathfrak{m} > \omega_1]$ Let X be a tree of cardinal ω_1 in which (giving X the fine tree topology) the set of isolated points is F_σ. Then X is a countable union of weak antichains. [See 25M*m*.]

(*d*) $[\mathfrak{m} > \omega_1]$ Let X be a compact Hausdorff space with $c(X) < \mathfrak{m}$, such that if $\langle \mathscr{G}_n \rangle_{n\in\mathbf{N}}$ is a sequence of collections of open subsets of X and $\bigcup \mathscr{G}_n$ is dense for each $n \in \mathbf{N}$, there is a sequence $\langle G_n \rangle_{n\in\mathbf{N}} \in \prod_{n\in\mathbf{N}} \mathscr{G}_n$ such that $\mathrm{int}(\bigcap_{n\in\mathbf{N}} G_n) \neq \emptyset$. Then X is not the union of ω_1 meagre sets.

(*e*) Show that the principle K of 41L is equivalent to

K′: Let $P \subseteq [\omega_1]^{<\omega}$ be an uncountable upwards-ccc family of sets such that $I \subseteq J \in P \Rightarrow I \in P$. Then there is an uncountable $A \subseteq \omega_1$ such that $[A]^2 \subseteq P$.

[Hint: for K′⇒K, given a partially ordered set P and $\langle p_\xi \rangle_{\xi<\omega_1}$ in P, consider $\{I : I \in [\omega_1]^{<\omega}, \{p_\xi : \xi \in I\}$ is bounded above in $P\}$.]

(*f*) Show that the principle H of 41L is equivalent to

H′: Let $P \subseteq [\omega_1]^{<\omega}$ be an uncountable upwards-ccc family of sets such that $I \subseteq J \in P \Rightarrow I \in P$. Then there is an uncountable $A \subseteq \omega_1$ such that $[A]^{<\omega} \subseteq P$.

[See Kunen & Tall 79, Theorem 2, and Nyikos *a*; also 43P*o*.]

(*g*) Show that $\kappa < \mathfrak{l}$ iff whenever $\mathfrak{A}$ is a Boolean algebra and $\#(\mathfrak{A}) \leq \kappa$ then $\mathfrak{A}$ is isomorphic to a subalgebra of $\mathscr{P}\mathbf{N}$. [Hint: $\mathfrak{A}$ can be embedded in $\mathscr{P}\mathbf{N}$ iff $\mathfrak{A} \setminus \{0\}$ is σ-centered downwards; use the techniques of §13. Compare 26G and 43P*p*.]

41O **Further results**

(*a*) $[\mathfrak{m} > \omega_1]$ Let X be an uncountable Hausdorff space without isolated points. Then X has an uncountable nowhere dense subset. [Kunen 77*a*; see Weiss 83, 7.11.]

(*b*) $[\mathfrak{m} > \omega_1]$ Let E and F be Archimedean Riesz spaces with the countable sup property, and $E \bar{\otimes} F$ their Riesz space tensor product, as described in Fremlin 72. Then $E \bar{\otimes} F$ has the countable sup property.

(*c*) Suppose $\kappa < \mathfrak{m}$. Let $S \subseteq [\kappa]^2$. Then either κ is expressible as $\bigcup_{n\in\mathbf{N}} X_n$ where $[X_n]^2 \cap S = \emptyset$ for each $n \in \mathbf{N}$ or for every $\alpha < \omega_1$ there are A, $B \subseteq \kappa$ such that $\mathrm{otp}(A) = \alpha$, B is uncountable and $\{\xi, \eta\} \in S$ for every $\xi \in A$, $\eta \in B$. [Erdös, Galvin & Hajnal 75.]

(*d*) A consistent theory of a language of cardinal less than $\mathfrak{m}$ which has a ccc Lindenbaum algebra has a countable model. [Blass 72. For explanations of these terms see Chang & Keisler 73, 2.1.15–18.]

(*e*) $[\mathfrak{m} > \omega_1]$ Let X be an uncountable partially ordered set. Then either X has an uncountable weak antichain or there is an uncountable $Y \subseteq X$ such that every countable subset of Y is bounded above in X. [Todorčević *b*. Compare B2J*c*(iii).]

(*f*) Other results connected with those above may be found in Todorčević 83*b*.

41P Problems

(*a*) Consider K and H of 41K. (i) Does K imply H? (ii) Does H imply that $2^{\omega_1} = \mathfrak{c}$? (iii) Does K imply that $\mathfrak{m} > \omega_1$? [See KUNEN & TALL 79 and 13D*b*–*c*.]

(*b*) [$\mathfrak{m} > \omega_1$] Does every uncountable Boolean algebra have an uncountable weak antichain? [See B3C.]

Notes and comments

In accordance with the prominence I have given in this book to Knaster's condition, it is natural to regard the principle K [41A*b*, 41L] as one of the chief consequences of Martin's axiom. It means that henceforth we can apply such results as 31A to all ccc partially ordered sets [41B–C], and by itself it has many important consequences; for instance, it makes Souslin's hypothesis true [41D–G], and it is what is needed for important parts of §44. I have not separated these out partly on account of the many obvious questions which remain open [41P*a*].

Souslin's problem was the original stimulus for the invention of Martin's axiom [SOLOVAY & TENNENBAUM 71]. 41D, 41F and 41G are of course equivalent in ZFC [JECH 78, Lemma 22.1; KUNEN 80, 5.13]. With both 41A and 41E available, however, it is quicker to give separate proofs. Collectively these are known as 'Souslin's hypothesis'; M. Souslin's original question [SOUSLIN 20] amounted to asking 'Is 41G true?', but it is the formulation in terms of trees, due to D. Kurepa, which has led to the most important ideas. For a proof that $\diamondsuit$ implies the existence of a Souslin tree, see DRAKE 74, VII.5.6, LEVY 79, IX.2.45Ac, or KUNEN 80, II.7.8. 41O*a* is equivalent to Souslin's hypothesis when $\mathfrak{p} > \omega_1$.

The topological-space form of the principle K reads

every ccc topological space satisfies Knaster's condition.

Since any product of spaces satisfying Knaster's condition also satisfies Knaster's condition [12M*a*], Theorem 41E is an immediate consequence of K. Subject to the continuum hypothesis, however, the situation is quite different. GALVIN 80 gives a direct construction, using CH, of a pair of ccc partially ordered sets whose product is not ccc; these can of course readily be converted into a pair of ccc (compact Hausdorff) spaces with a non-ccc product. This shows incidentally that K and CH cannot be true together. For a discussion of several properties intermediate between 'ccc' and 'Knaster's condition', subject to the continuum hypothesis, see WAGE 79 or COMFORT & NEGREPONTIS 82, §7. 41O*b* is a functional analyst's version of 41E.

31G, 41H–J, 41O*e*, 42D–G, 42L*c* and B2J*c*(ii)–(iii) can all be regarded as variations on a single theme: the existence of 'free' sets for a relation S on a set X, i.e. sets $A \subseteq X$ such that $(x, y) \notin S$ if x and y are distinct members of A. The natural approach to these problems is to consider the set

$$P = \{I : I \in [X]^{<\omega}, I \text{ is free}\},$$

and to investigate when it is upwards-ccc. I do this explicitly only in 41H; the point being that when S has countable vertical sections and $\mathfrak{m} > \omega_1$ then P is ccc iff the S-respecting partial order on $[X]^{<\omega} \times [X]^{<\omega}$ is ccc [31M*c*], and we get more information from the latter. 41H seems to be the only result available when S does not have countable vertical sections. I observe, however, that the results which I have given as corollaries [41I–J] can both be represented in terms of relations which do have countable sections, and accordingly could be handed with equal efficiency in §42, using 42B(i).

41O*c* is an elaboration of 41H(ii), using 21O*b* instead of 21I.

Let X be a partially ordered set of cardinal less than $\mathfrak{m}$. If X has no uncountable up-antichains, it is a countable union of upwards-centered sets [41C*b*]; if it has no uncountable upwards-centered set it is a countable union of weak antichains [41I]. (See also 41O*e* and B2J*c*(iii).) It follows at once that if $\mathfrak{m} > \omega_1$ then all Aronszajn trees are special [41J]. This is especially interesting in view of the effects of $\mathfrak{p} > \omega_1$ on special Aronszajn trees [25G]. 41N*b*, *c* are consequences of 41I.

41K is a version of DA* in Devlin 76. I include it because it is the only context I know in which we can escape the tyranny of the ccc. The requirement (ii) of 41K to which we must submit instead is of course even worse; indeed, it is not clear that there have to be non-trivial applications (cf. B3I*a*). As usual, there are Boolean-algebra and topological-space versions of this result; I set the latter out in 41N*d*.

The principles K, H and L(κ) of 41L can be thought of as attempts to formalize the vital differences between $\mathfrak{m}$, $\mathfrak{m}_K$ and $\mathfrak{p}$. They cannot be regarded as successful so long as 41P*a*(iii) remains open. Interesting alternative versions of all of these exist (41N*e*–*g*, 43P*o*–*p*].

42 S-respecting partial orders

In this section I discuss results which can be approached using the machinery developed in 31C–E. I begin with a restatement of Lemma 31E [42A] and with criteria for S-respecting partial orders to be ccc, based on ideas of Z. Szentmiklóssy [42B]. The rest of the section is made up of applications of these ideas: U. Abraham's theorem on free sets for relations with closed countable vertical sections [42C–D], corollaries on families of subsets of **N** [42F–G], and M. Wage's theorem on almost-disjoint families of countable sets [42H–I].

42A Lemma

Let X and Y be sets and $S \subseteq X \times Y$ a relation. Take $\mathscr{B} \subseteq \mathscr{P}Y$, $Z \subseteq Y$ and let $f: Z \to X$ be a function. Suppose that

(i) the S-respecting partial order on $[X]^{<\omega} \times [Y]^{<\omega}$ is upwards-ccc;

(ii) $B \setminus S[I] \neq \varnothing$ for every $B \in \mathscr{B}$, $I \in [X]^{<\omega}$;

(iii) $\#(X) < \mathfrak{m}$, $\#(\mathscr{B}) < \mathfrak{m}$ and $\#(Z) < \mathfrak{m}$.

Then there are sequences $\langle Y_n \rangle_{n \in \mathbf{N}}$, $\langle Z_n \rangle_{n \in \mathbf{N}}$ of sets such that the following are true.

(*a*) $Y = \bigcup_{n \in \mathbf{N}} Y_n$; $Y_n \cap S[\{x\}]$ is finite for every $x \in X$, $n \in \mathbf{N}$; and $Y_n \cap B \neq \varnothing$ for every $B \in \mathscr{B}$, $n \in \mathbf{N}$.

(*b*) $Z = \bigcup_{n \in \mathbf{N}} Z_n$; $Z_n \cap S[\{x\}]$ is finite for every $x \in X$, $n \in \mathbf{N}$; and $(f(z), w) \notin S$ whenever $n \in \mathbf{N}$ and z, w are distinct members of Z_n.

Proof As 31E, but using 41C*a* instead of 31B.

42B Szentmiklóssy's lemma

Let X and Y be sets and $S \subseteq X \times Y$ a relation with countable vertical sections. Let P be $[X]^{<\omega} \times [Y]^{<\omega}$ with the S-respecting partial order, and suppose that P is not upwards-ccc. Then there are families $\langle K_\xi \rangle_{\xi < \omega_1}$, $\langle L_\xi \rangle_{\xi < \omega_1}$ such that

(i) $\langle K_\xi \rangle_{\xi < \omega_1}$ is a disjoint family of finite subsets of X, all of the same size; $\langle L_\xi \rangle_{\xi < \omega_1}$ is a disjoint family of finite subsets of Y, all of the same size; $L_\eta \cap S[K_\xi] \neq \varnothing$ whenever $\eta < \xi < \omega_1$;

(ii) for every uncountable $C \subseteq \bigcup_{\xi < \omega_1} L_\xi$ there is a countable $D \subseteq C$ and an $\alpha < \omega_1$ such that $D \cap \bigcap_{\zeta \in M} S[K_\zeta]$ is infinite for every finite $M \subseteq [\alpha, \omega_1[$;

(iii) for every uncountable $C \subseteq \bigcup_{\xi < \omega_1} K_\xi$, $\{\xi : L_\xi \cap S[C] = \varnothing\}$ is countable.

Proof (*a*) For $i, j \in \mathbf{N}$ write

$$\mathbf{H}_{ij} = \{\langle (K_\xi, L_\xi) \rangle_{\xi < \omega_1} : \langle K_\xi \rangle_{\xi < \omega_1} \text{ and } \langle L_\xi \rangle_{\xi < \omega_1} \text{ satisfy (i) and } \#(K_\xi) = i,\ \#(L_\xi) = j\ \forall \xi < \omega_1\}.$$

Then there are $i, j \in \mathbf{N}$ such that $\mathbf{H}_{ij} \neq \varnothing$. **P** We know that there is an uncountable up-antichain in P; let $\langle (I_\xi, J_\xi) \rangle_{\xi < \omega_1}$ enumerate such an antichain. By 31D*b* there is an uncountable $A \subseteq \omega_1$ such that $\langle I_\xi \rangle_{\xi \in A}$ and $\langle J_\xi \rangle_{\xi \in A}$ are constant-size Δ-systems and $(I_\eta, J_\eta) < (I_\eta \cup I_\xi, J_\eta \cup J_\xi)$ whenever $\xi \in A$ and $\eta \in A \cap \xi$. (This is where I use the hypothesis that S has countable vertical sections.) So $(I_\xi, J_\xi) \not\leq (I_\eta \cup I_\xi, J_\eta \cup J_\xi)$ i.e.

$$S[I_\xi] \cap (J_\eta \backslash J_\xi) \neq \emptyset \text{ whenever } \xi \in A,\ \eta \in A \cap \xi.$$

On the other hand, writing $\zeta = \min A$, $I = I_\zeta$, $J = J_\zeta$,

$$S[I] \cap (J_\eta \backslash J) = \emptyset\ \forall \eta \in A \backslash \{\zeta\}.$$

So $S[I_\xi \backslash I] \cap (J_\eta \backslash J) \neq \emptyset$ whenever $\xi \in A$, $\eta \in A \cap \xi \backslash \{\zeta\}$. Thus if $\langle \alpha(\xi) \rangle_{\xi < \omega_1}$ is the increasing enumeration of $A \backslash \{\zeta\}$,

$$\langle (I_{\alpha(\xi)} \backslash I, J_{\alpha(\xi)} \backslash J) \rangle_{\xi < \omega_1} \in \mathbf{H}_{ij},$$

where $i = \#(I_\xi \backslash I)$ and $j = \#(J_\xi \backslash J)$ for $\xi \in A$. **Q**

(*b*) Let $n \in \mathbf{N}$ be minimal subject to $\bigcup_{k \in \mathbf{N}} \mathbf{H}_{kn} \neq \emptyset$, and $m \in \mathbf{N}$ minimal subject to $\mathbf{H}_{mn} \neq \emptyset$; take $\langle (K_\xi, I_\xi) \rangle_{\xi < \omega_1}$ to be any member of $\mathbf{H}_{mn}$. By the definition of $\mathbf{H}_{mn}$, we have already achieved (i).

(*c*) Let $C \subseteq \bigcup_{\xi < \omega_1} L_\xi$ be uncountable. Then there is a countable $D \subseteq C$ and an $\alpha < \omega_1$ such that $\{D \cap S[K_\xi] : \alpha \leq \xi < \omega_1\}$ has the finite intersection property. **P?** Suppose, if possible, otherwise. Observe that, because each L_ξ is finite, C meets $\bigcup_{\alpha \leq \xi < \omega_1} L_\xi$ for every $\alpha < \omega_1$. Accordingly we can choose inductively ordinals $\alpha(\xi) < \omega_1$ and finite sets $M(\xi) \subseteq \omega_1$, for $\xi < \omega_1$, such that
[choosing $\alpha(\xi)$] $\alpha(\xi) > \alpha(\eta)\ \forall \eta < \xi$,

$$\alpha(\xi) > \max M(\eta)\ \forall \eta < \xi,\ C \cap L_{\alpha(\xi)} \neq \emptyset;$$

[choosing $M(\xi)$] $M(\xi) \subseteq [\alpha(\xi), \omega_1[$ is finite,

$$C \cap \bigcup_{\eta < \xi} L_{\alpha(\eta)} \cap \bigcap_{\zeta \in M(\xi)} S[K_\zeta] = \emptyset.$$

Now choose $y_\xi \in C \cap L_{\alpha(\xi)}$ for each $\xi < \omega_1$ and set

$$L'_\xi = L_{\alpha(\xi)} \backslash \{y_\xi\},\quad K'_\xi = \bigcup_{\zeta \in M(\xi)} K_\zeta\ \forall \xi < \omega_1.$$

Because $\langle \alpha(\xi) \rangle_{\xi < \omega_1}$ is strictly increasing, $\langle L'_\xi \rangle_{\xi < \omega_1}$ is disjoint; because $\min M(\xi) \geq \alpha(\xi) > \max M(\eta)$ whenever $\eta < \xi$, $\langle M(\xi) \rangle_{\xi < \omega_1}$ and $\langle K'_\xi \rangle_{\xi < \omega_1}$ and disjoint. If $\eta < \xi < \omega_1$ then $y_\eta \in C \cap L_{\alpha(\eta)}$ so there is a $\zeta \in M(\xi)$ such that $y_\eta \notin S[K_\zeta]$; but as $\alpha(\eta) < \alpha(\xi) \leq \min M(\xi) \leq \zeta$, there is a $y \in L_{\alpha(\eta)} \cap S[K_\zeta]$; y cannot be equal to y_η, so $y \in L'_\eta \cap S[K_\zeta] \subseteq L'_\eta \cap S[K'_\xi]$.

Let $k \in \mathbf{N}$ be such that $B = \{\xi : \#(K'_\xi) = k\}$ is uncountable, and enumerate B in ascending order as $\langle \beta(\xi) \rangle_{\xi < \omega_1}$. Then $\langle (K'_{\beta(\xi)}, L'_{\beta(\xi)}) \rangle_{\xi < \omega_1} \in \mathbf{H}_{k,n-1}$; which contradicts the definition of n. **XQ**

(*d*) It follows that (ii) is true. **P** Let $C \subseteq \bigcup_{\xi < \omega_1} L_\xi$ be uncountable. By (*c*), we can choose inductively, for $k \in \mathbf{N}$, countable sets D_k and ordinals $\alpha_k < \omega_1$ such that

$D_k \subseteq C \setminus \bigcup_{l<k} D_l$ is countable, $\{D_k \cap S[K_\xi] : \alpha_k \leq \xi < \omega_1\}$
has the finite intersection property.

If now $D = \bigcup_{k \in \mathbf{N}} D_k$ and $\alpha = \sup_{k \in \mathbf{N}} \alpha_k$, then for any finite $M \subseteq [\alpha, \omega_1[$ we have $D_k \cap \bigcap_{\zeta \in M} S[K_\zeta] \neq \emptyset$ for every $k \in \mathbf{N}$, so that $D \cap \bigcap_{\zeta \in M} S[K_\zeta]$ is infinite. **Q**

(*e*) **?** Suppose, if possible, that (iii) is false. Then there is an uncountable $C \subseteq \bigcup_{\xi < \omega_1} K_\xi$ such that

$$D = \{\xi : L_\xi \cap S[C] = \emptyset\}$$

is uncountable. Choose inductively ordinals $\gamma(\xi)$, $\delta(\xi) < \omega_1$ such that
[choosing $\gamma(\xi)$] $\quad \gamma(\xi) > \delta(\eta)\ \forall \eta < \xi,\ C \cap K_{\gamma(\xi)} \neq \emptyset$
[choosing $\delta(\xi)$] $\quad \delta(\xi) \geq \gamma(\xi),\ \delta(\xi) \in D$
for each $\xi < \omega_1$. Choose $x_\xi \in C \cap K_{\gamma(\xi)}$ for each $\xi < \omega_1$, and set

$$K'_\xi = K_{\gamma(\xi)} \setminus \{x_\xi\}, \quad L'_\xi = L_{\delta(\xi)}\ \forall \xi < \omega_1.$$

Because $\langle \gamma(\xi) \rangle_{\xi < \omega_1}$ and $\langle \delta(\xi) \rangle_{\xi < \omega_1}$ are strictly increasing, $\langle K'_\xi \rangle_{\xi < \omega_1}$ and $\langle L'_\xi \rangle_{\xi < \omega_1}$ are both disjoint families. If $\eta < \xi$ then $\delta(\eta) < \gamma(\xi)$, so there is a $y \in L_{\delta(\eta)} \cap S[K_{\gamma(\xi)}]$; now $y \notin S[C]$ so $y \in S[K_{\gamma(\xi)} \setminus C] \subseteq S[K'_\xi]$; thus $L'_\eta \cap S[K'_\xi] \neq \emptyset$. This means that $\langle (K'_\xi, L'_\xi) \rangle_{\xi < \omega_1} \in \mathbf{H}_{m-1,n}$, contradicting the definition of m. **X**

So (iii) is true.

42C Lemma

Let Y be a second-countable topological space, X any set, and $S \subseteq X \times Y$ a relation with closed countable vertical sections. Then $[X]^{<\omega} \times [Y]^{<\omega}$, with the S-respecting partial order, is upwards-ccc.

Proof **?** Suppose, if possible, otherwise. Then by Szentmiklóssy's Lemma [42B(ii)] there are disjoint families $\langle K_\xi \rangle_{\xi < \omega_1}$, $\langle L_\xi \rangle_{\xi < \omega_1}$ in $[X]^{<\omega}$ and $[Y]^{<\omega}$ respectively such that

for every uncountable $C \subseteq \bigcup_{\xi < \omega_1} L_\xi\ \exists \alpha < \omega_1$
such that $\{C \cap S[K_\xi] : \alpha \leq \xi < \omega_1\}$ has the finite intersection property.

Let $\langle U_n \rangle_{n \in \mathbf{N}}$ enumerate a base for the topology of Y and set $W = \bigcup_{\xi < \omega_1} L_\xi$. W is uncountable and each $S[K_\xi]$ is closed and countable, so for each $\xi < \omega_1$ we can find in $n(\xi) \in \mathbf{N}$ such that

$$U_{n(\xi)} \cap W \text{ is uncountable, } U_{n(\xi)} \cap S[K_\xi] = \emptyset.$$

Let $m \in \mathbf{N}$ be such that

$$B = \{\xi : \xi < \omega_1, n(\xi) = m\}$$

is uncountable. Set $C = U_m \cap W$. Then C is uncountable, so there is an $\alpha < \omega_1$ such that $\{C \cap S[K_\xi] : \alpha \leq \xi < \omega_1\}$ has the finite intersection property. But now there is a $\xi \in B \setminus \alpha$ and $C \cap S[K_\xi] = \emptyset$. **X**

42D Theorem

Let X be a second-countable space and $S \subseteq X^2$ a relation with closed countable vertical sections. Suppose that $\#(X) < \mathfrak{m}$. Then X is expressible as $\bigcup_{n \in \mathbf{N}} Z_n$ where $(x, y) \notin S$ whenever x and y are distinct points of the same Z_n.

Proof Apply 42A*b* with $X = Y = Z$, f the identity function on X, and $\mathscr{B} = \emptyset$; use 42C to show that 42A(i) is satisfied.

42E Corollary

$[\mathfrak{m} > \omega_1]$ Let $S \subseteq \mathbf{R}^2$ be such that $\{s : (t, s) \in S\}$ is nowhere dense for every $t \in \mathbf{R}$. Then there is an uncountable $A \subseteq \mathbf{R}$ such that $(s, t) \notin S$ for all distinct $s, t \in A$.

Proof Write $S(t) = \{s : (t, s) \in S\}$ for each $t \in \mathbf{R}$. Choose $\langle s_\xi \rangle_{\xi < \omega_1}$ inductively in $\mathbf{R}$ such that

$$s_\xi \notin \bigcup_{\eta < \xi} \overline{S(s_\eta)} \cup \{s_\eta : \eta < \xi\}$$

for every $\xi < \omega_1$. Set $X = \{s_\xi : \xi < \omega_1\}$ and

$$S' = \{(t, u) : t, u \in X, u \in \overline{S(t)}\}.$$

Then, for any $\xi < \omega_1$,

$$\{t : (s_\xi, t) \in S'\} = X \cap \overline{S(s_\xi)} \subseteq \{s_\eta : \eta \leq \xi\}$$

is countable and relatively closed in X. Also $\#(X) = \omega_1$. By 42D there is an uncountable $A \subseteq X$ such that $(s, t) \notin S'$ for all distinct $s, t \in A$, which is what we need.

42F Corollary

$[\mathfrak{m} > \omega_1]$ Let $\mathscr{A}$ be an uncountable family of subsets of $\mathbf{N}$. Then *either* $\mathscr{A}$ includes an uncountable weak antichain *or* there are both strictly increasing and strictly decreasing sequences in $\mathscr{A}$.

Proof It is enough to consider the case $\#(\mathscr{A}) = \omega_1$. Suppose that $\mathscr{A}$ does not include any uncountable weak antichain. Give $\mathscr{A}$ the topology induced by the usual topology of $\mathscr{P}\mathbf{N}$. Set

$$S = \{(A, B): A, B \in \mathscr{A}, A \supseteq B\}.$$

For each $A \in \mathscr{A}$, $S(A) = \{B: B \in \mathscr{A}, B \subseteq A\}$ is closed in $\mathscr{A}$. By Theorem 42D, there must be an $A_0 \in \mathscr{A}$ such that $S(A_0)$ is uncountable.

Repeating this argument on $S(A_0) \setminus \{A_0\}$, there is an $A_1 \subset A_0$ such that $S(A_1)$ is uncountable. Continuing in this way, we find that we have a strictly decreasing sequence $\langle A_n \rangle_{n \in \mathbf{N}}$ in $\mathscr{A}$.

Similarly there is a strictly increasing sequence in $\mathscr{A}$.

42G Corollary

[$\mathfrak{m} > \omega_1$] If $\langle A_\xi \rangle_{\xi < \omega_1}$ is any family of subsets of $\mathbf{N}$ then *either* there is a $\xi < \eta$ such that $A_\xi \subseteq A_\eta$ *or* there is an uncountable $C \subseteq \omega_1$ such that $A_\xi \nsubseteq A_\eta$ whenever ξ, η are distinct members of C.

Proof Take $\mathscr{A} = \{A_\xi : \xi < \omega_1\}$ in 42F. If there is a strictly increasing sequence $\langle A_{\xi(n)} \rangle_{n \in \mathbf{N}}$ in $\mathscr{A}$ then there must be an $n \in \mathbf{N}$ such that $\xi(n) < \xi(n+1)$, $A_{\xi(n)} \subseteq A_{\xi(n+1)}$. If $\mathscr{A}$ is countable, there must be $\xi < \eta$ such that $A_\xi = A_\eta$. Otherwise, there is an uncountable weak antichain $\mathscr{B} \subseteq \mathscr{A}$; take $C \subseteq \omega_1$ such that $\xi \mapsto A_\xi : C \to \mathscr{A}$ is a bijection between C and $\mathscr{B}$.

42H Lemma

[$\mathfrak{p} > \omega_1$] Let X and Y be sets and $S \subseteq X \times Y$ a relation. Suppose that vertical sections of S are countable and that $\{x : x \in X, W \subseteq S[\{x\}]\}$ is countable for every infinite $W \subseteq Y$. Then $[X]^{<\omega} \times [Y]^{<\omega}$ is upwards-ccc under the S-respecting partial order.

Proof ? If not, then by 42B(i) there are disjoint families $\langle K_\xi \rangle_{\xi < \omega_1}$, $\langle L_\xi \rangle_{\xi < \omega_1}$ in $[X]^{<\omega}$, $[Y]^{<\omega}$ respectively such that

$$\#(K_\eta) = \#(K_\xi), \quad \#(L_\eta) = \#(L_\xi), \quad S[K_\xi] \cap L_\eta \neq \emptyset$$

whenever $\eta < \xi < \omega_1$. Let m be the common value of $\#(K_\xi)$ and n the common value of $\#(L_\xi)$; enumerate K_ξ and L_ξ as $\langle x_i^\xi \rangle_{i<m}$ and $\langle y_j^\xi \rangle_{j<n}$ respectively, for each $\xi < \omega_1$. Now set

$$S_{ij} = \{(n, \xi): n \in \omega, \xi \in \omega_1 \setminus \omega, (x_i^\xi, y_j^n) \in S\}.$$

We have $\bigcup_{i<m, j<n} S_{ij} = \omega \times (\omega_1 \setminus \omega)$. By 21I there are an infinite $M \subseteq \omega$, an uncountable $C \subseteq \omega_1 \setminus \omega$, an $i < m$ and a $j < n$ such that $M \times C \subseteq S_{ij}$. But now $W = \{y_j^n : n \in M\}$ is infinite and

$$\{x : W \subseteq S[\{x\}]\} \supseteq \{x_i^\xi : \xi \in C\}$$

is uncountable. **X**

42I Theorem

Let Y be a set and $\mathscr{A}$, $\mathscr{B}$ two families of subsets of Y such that

(i) every member of $\mathscr{A}$ is countable;

(ii) if $W \subseteq Y$ is infinite then $\{A : W \subseteq A \in \mathscr{A}\}$ is countable;

(iii) $B \setminus \bigcup \mathscr{I} \neq \varnothing \, \forall B \in \mathscr{B}, \mathscr{I} \in [\mathscr{A}]^{<\omega}$;

(iv) $\#(\mathscr{A}) < \mathfrak{m}$, $\#(\mathscr{B}) < \mathfrak{m}$.

Then Y is expressible as $\bigcup_{n \in \mathbf{N}} Y_n$ where

$Y_n \cap A$ is finite for every $A \in \mathscr{A}$, $n \in \mathbf{N}$;

$Y_n \cap B \neq \varnothing$ whenever $B \in \mathscr{B}$, $n \in \mathbf{N}$.

Proof Let S be $\{(A, y) : y \in A \in \mathscr{A}\} \subseteq \mathscr{A} \times Y$ and let P be $[\mathscr{A}]^{<\omega} \times [Y]^{<\omega}$ with the S-respecting partial order. Then P is upwards-ccc. **P** If $\mathfrak{m} > \omega_1$ this is a consequence of 42H, since then $\mathfrak{p} > \omega_1$. If $\mathfrak{m} = \omega_1$ then $\mathscr{A}$ is countable and $S[\mathscr{A}] = \bigcup \mathscr{A}$ is countable, so that P is actually σ-centered upwards [31C*b*(iii)]. **Q** So we can apply 42A with $X = \mathscr{A}$, $Z = \varnothing$.

42J Sources

Wage 79 for 42I. Szentmiklóssy 80 for the ideas of 42B. Baumgartner 80 for 42F (attributed to K. Kunen). Avraham 81 for 42D (in essence). Douwen & Kunen 82 for 42G. Balogh *a* for (in effect) the idea of expressing 42B in terms of S-respecting partial orders.

42K Exercises

(*a*) Let X be a second-countable space with $\#(X) < \mathfrak{m}$ and $\mathscr{F} \subseteq \mathscr{P}X$ a point-countable family of closed sets. Then $\mathscr{F}$ is σ-point finite. [Consider $S = \{(x, F) : x \in F \in \mathscr{F}\} \subseteq X \times \mathscr{F}$, using the method of 42C.]

(*b*) Let $\mathscr{U}$ be a point-countable family of sets such that $\bigcap \mathscr{V}$ is countable for every infinite $\mathscr{V} \subseteq \mathscr{U}$ and $\#(\bigcup \mathscr{U}) < \mathfrak{m}$. Then $\mathscr{U}$ is σ-point-finite. [Use 42H.]

42L Further results

(*a*) $[\mathfrak{m} = \mathfrak{c} = \omega_2]$ There is an $S \subseteq [\mathfrak{c}]^2$ such that (i) whenever $\xi < \zeta$ in $\mathfrak{c}$ then $\{\eta : \eta < \xi,\ \{\eta, \xi\} \in S,\ \{\eta, \zeta\} \in S\}$ is finite, (ii) if $A \in [\mathfrak{c}]^{\mathfrak{c}}$ then $[A]^2 \cap S \neq \varnothing$. [Erdös, Hajnal & Máté 73; Laver 75.]

(*b*) $[\mathfrak{m} = \mathfrak{c} = \omega_2]$ There is a disjoint family $\langle \mathscr{H}_\alpha \rangle_{\alpha < \mathfrak{c}}$ in $\mathscr{P}\mathfrak{c}$ such that (i) $\operatorname{otp}(H) = \omega_1$ for every $H \in \mathscr{H} = \bigcup_{\alpha < \mathfrak{c}} \mathscr{H}_\alpha$, (ii) $H \cap K$ is finite for all distinct $H, K \in \mathscr{H}$, (iii) if $A \subseteq \mathfrak{c}$, either $A \subseteq \bigcup \mathscr{C}$ for some countable $\mathscr{C} \subseteq \mathscr{H}$ or there is a $B \subseteq A$ such that $\#(B) = \omega_1$ and $\mathscr{P}B \cap \mathscr{H}_\alpha \neq \varnothing$ for every $\alpha < \mathfrak{c}$. [Wage 79.]

(*c*) $[\mathfrak{m} > \omega_1]$ (i) If $S \subseteq [\omega_1]^2$ then either there is an uncountable

$A \subseteq \omega_1$ such that $[A]^2 \cap S = \emptyset$ or there are an infinite $B \subseteq \omega_1$ and an uncountable $C \subseteq \omega_1$ such that $\{\xi, \eta\} \in S$ whenever $\xi \in B \cup C$ and $\eta \in B \setminus \{\xi\}$. (ii) $\omega_1 \rightarrow (\omega_1, \omega^2)^2$, where ω^2 is the ordinal power. [TODORČEVIĆ *c*. Compare B2J*c*(ii).]

Notes and comments

As I remarked in the notes to §31, my motive in extracting the concept of 'S-respecting' partial order is the hope of making the mechanisms of these arguments more accessible. For the same reason I have detached arguments showing that the partial orders are ccc [42B, 42C, 42H] from the basic application of Martin's axiom in this context [42A], so that the original results [42D, 42I] become elementary corollaries. As it happens, they use different halves of 42A; so readers can amuse themselves by writing out versions of these theorems which use the opposite halves.

The combinatorial ideas of 42B(ii)–(iii) are taken from SZENTMIKLÓSSY 80, where they were used to analyse partially ordered sets of the form

$$\{I : I \in [\omega_1]^{<\omega}, [I]^2 \cap S = \emptyset\}$$

for $S \subseteq [\omega_1]^2$. I therefore think of these, rather than of the older 42B(i), as 'Szentmiklóssy's lemma'. The steps from an arbitrary $S \subseteq [\omega_1]^2$ to an $S \subseteq \omega_1^2$ with countable vertical sections, and thence to the S-respecting partial order, are technical details which do not affect the arguments. (See 31M*b*–*c*.) It will be clear that I have not yet used the full strength of 42B; this will be needed when we come closer to Szentmiklóssy's own applications in 44B and 44E.

Both 42D and 42I can be regarded as applying 42A–B to relations of the form $S = \{(A, x) : x \in A \in \mathscr{A}\}$ where $\mathscr{A}$ is a family of countable sets. So another way of doubling the number of theorems proved by these methods is to look at relations $S = \{(x, A) : x \in A \in \mathscr{A}\}$ where $\mathscr{A}$ is a point-countable family of sets. This is what I do in 42K; see also 31M*d*.

Both 42L*a* and 42L*b* can be derived as corollaries of 42I. S. Todorčević has shown me the proof of 42L*c*, which is elegant and ingenious, but has asked me not to reproduce it here.

43 Martin's axiom and separability

The central result of this section is Theorem 43E. This is, among other things, a kind of strong Baire theorem (see 43F*a*); it is therefore natural to try to express it in terms of 'Martin-completeness', a new weak completeness property for topological spaces which fits naturally among those which have been devised in the past as conditions sufficient to make spaces Baire. 43A–D and 43O give the definition, elementary properties and most important examples of Martin-complete spaces.

The second half of the section consists of corollaries of 43E, mostly in the form of conditions under which ccc spaces will be separable [43F*b*, 43I*a*–*c*, 43K–M], but with some discussion of calibers [43F*c*, 43G(ii)], spaces with countable π-bases [43G–I], and other cardinal functions [43G, 43I*d*, 43J].

43A Definition

Let κ be a cardinal. A topological space X is **Martin-κ-complete** if there exist $\preccurlyeq$, $\mathbf{U}$ such that

(i) $\preccurlyeq$ is a partial order on the non-empty open subsets of X such that $G \preccurlyeq H \Rightarrow G \subseteq H$ and $G \subseteq G' \prec H' \subseteq H \Rightarrow G \prec H$;

(ii) If G_1, G_2 are open sets and $\varnothing \neq G_1 \subset G_2$ then there is a non-empty open set H such that $H \subseteq G_1$ and $H \prec G_2$;

(iii) $\mathbf{U}$ is a collection of π-bases for the topology of X;

(iv) if $\mathscr{G}$ is a family of non-empty open subsets of X such that $\mathscr{G}$ is downwards-directed for $\preccurlyeq$ and meets every member of $\mathbf{U}$, then $\bigcap \mathscr{G} \neq \varnothing$;

(v) $\#(\mathbf{U}) \leq \kappa$.

If X is Martin-0-complete (i.e. the conditions above can be satisfied with $\mathbf{U} = \varnothing$) then I shall say that X is **Martin-complete**. I shall say that X is **Martin-$<\kappa$-complete** if it is Martin-λ-complete for some $\lambda < \kappa$.

43B Remarks

(*a*) I regret the complexity of this definition, but I have been unable to find anything simpler which will cover the relevant material. The best aid to digestion which I can offer is a glance at 43D. The definition is of course designed primarily for use in Theorem 43E; but certain features are adopted in order to prove 43C and 43O*a*–*c*.

(*b*) Since the structures set up in 43A are likely to be unfamiliar, it is perhaps worth making some elementary facts explicit. Observe first that if we have a relation $\prec$ on the non-empty open sets of X which has the properties

$$G \prec H \Rightarrow G \subset H, \quad G \subseteq G' \prec H' \subseteq H \Rightarrow G \prec H$$

then we can define $\preccurlyeq$ by writing

$$G \preccurlyeq H \Leftrightarrow G = H \quad \text{or} \quad G \prec H,$$

and $\preccurlyeq$ will be a partial order satisfying 43A(i).

(*c*) If we have a relation $\preccurlyeq$ satisfying 43A(i)–(ii), then whenever G_1 and G_2 are open sets and $G_1 \cap G_2 \neq \varnothing$ there is a non-empty open H such that $H \preccurlyeq G_1$ and $H \preccurlyeq G_2$. **P** (i) If $G_2 \nsubseteq G_1$, then $G_1 \cap G_2 \subset G_2$ so by 43A(ii) there is a non-empty open $H_1 \subseteq G_1 \cap G_2$ such that $H_1 \prec G_2$. If

$H_1 = G_1$ take $H = H_1$. If $H_1 \subset G_1$ use 43A(ii) again to find a non-empty open $H \subseteq H_1$ such that $H \prec G_1$. Then $H \subseteq H_1 \prec G_2$ so $H \prec G_2$ by 43A(i). (ii) If $G_1 \nsubseteq G_2$ the same argument works; while if $G_1 = G_2$ we can take $H = G_1 = G_2$. **Q**

It follows that in the set P of non-empty open subsets of X, the downwards-linked and downwards-centered sets, and the down-antichains, are the same for both $\preccurlyeq$ and $\subseteq$; also the minimal elements of P are the same for both orders. This means that P is downwards-ccc, or σ-centered downwards, or satisfies Knaster's condition downwards, for $\preccurlyeq$ iff it has the same property for $\subseteq$; i.e. iff the topological space X is ccc, or σ-centered, or satisfies Knaster's condition.

We find also that if $\mathscr{U}$ is a π-base for the topology of X (i.e. is coinitial with P for $\subseteq$), then $\mathscr{U}$ is coinitial with P for $\preccurlyeq$. **P** Take $G \in P$. Then there is a $U \in \mathscr{U}$ such that $U \subseteq G$. If $U = G$ then $U \preccurlyeq G$. Otherwise there is an $H \in P$ such that $H \subseteq U$ and $H \prec G$, by 43A(ii). Now there is a $V \in \mathscr{U}$ such that $V \subseteq H$, and $V \preccurlyeq G$ by 43A(i). **Q**

(*d*) Evidently, when we seek to verify the condition 43A(iv), it will be enough to consider non-empty families $\mathscr{G}$ with no smallest element. Now, given that $\preccurlyeq$ satisfies 43A(i), and $\mathscr{G}$ is a non-empty collection of open sets with no smallest element (I mean, smallest for the usual ordering $\subseteq$), we find that

$\mathscr{G}$ is downwards-directed for $\preccurlyeq$

$\Leftrightarrow \mathscr{G}$ is downwards-directed for $\subseteq$ and

$$\forall G \in \mathscr{G}\ \exists H \in \mathscr{G},\ H \prec G$$

$\Leftrightarrow \mathscr{G}^* = \{H : H \subseteq X \text{ is open}, \exists G \in \mathscr{G}, H \supseteq G\}$

is downwards-directed for $\preccurlyeq$.

Since $\bigcap \mathscr{G} \neq \varnothing$ iff $\bigcap \mathscr{G}^* \neq \varnothing$, 43A(iv) is equivalent to

(iv)′ whenever $\mathscr{G}$ is a non-empty collection of non-empty open sets, downwards-directed for $\subseteq$, such that

$\forall G \in \mathscr{G}\ \exists H \in \mathscr{G},\ H \prec G$

and also

$\forall \mathscr{U} \in \mathbf{U}\ \exists G \in \mathscr{G},\ U \in \mathscr{U}$ such that $G \subseteq U$, then $\bigcap \mathscr{G} \neq \varnothing$.

(*e*) Finally, you may be surprised that in 43A(iv) I look at $\bigcap \mathscr{G}$, not $\bigcap \{\overline{G} : G \in \mathscr{G}\}$. The point is that in almost all cases,

$G \prec H \Rightarrow \overline{G} \subseteq H$,

so that it makes no difference.

43C **Lemma**

(*a*) If X is Martin-κ-complete and $\lambda \geq \kappa$ then X is Martin-λ-complete.

(*b*) If X is Martin-ω-complete then X is Martin-complete.

(*c*) If X is Martin-κ-complete and quasi-regular, and $\mathscr{H}$ is a non-empty family of open sets, with $\#(\mathscr{H}) \leq \kappa$, such that $Y = \bigcap \mathscr{H}$ is dense in X, then Y is Martin-κ-complete.

(*d*) If, for each ι in an index set I, we have a Martin-κ_ι complete space X_ι, then $X = \prod_{\iota \in I} X_\iota$ is Martin-κ-complete where κ is the cardinal sum $\sum_{\iota \in I} \kappa_\iota$. In particular, any product of Martin-complete spaces is Martin-complete, and the product of κ or fewer Martin-κ-complete spaces in Martin-κ-complete.

Proof (*a*) This is trivial, since κ enters into 43A only in part (v).

(*b*) Suppose that $\preccurlyeq$ and $\mathbf{U}$ satisfy the conditions of 43A, with $\mathbf{U}$ countable. If $\mathbf{U} = \varnothing$ then of course X is Martin-complete. Otherwise, we can express $\mathbf{U}$ as $\{\mathscr{U}_n : n \in \mathbf{N}\}$. For a non-empty open set $G \subseteq X$ set

$$\theta(G) = \inf \{n : n \in \mathbf{N}, \nexists U \in \mathscr{U}_n, G \subseteq U\}$$

writing ∞ for $\inf \varnothing$. Then $G \subseteq H \Rightarrow \theta(G) \geq \theta(H)$. Now define $\preccurlyeq_1$ by saying that $G \prec_1 H$ if $G \prec H$ and either $\theta(H) = \infty$ or $\theta(G) > \theta(H)$, and that $G \preccurlyeq_1 H$ if $G = H$ or $G \prec_1 H$. Then $\preccurlyeq_1$ satisfies 43A(i) [43B*b*]. To see that it satisfies 43A(ii), take open sets G_1, G_2 with $\varnothing \neq G_1 \subset G_2$. Then there is a non-empty open $H \subseteq G_1$ such that $H \prec G_2$. If $\theta(G_2) = \infty$, then $H \prec_1 G_2$. If $\theta(G_2) = n < \infty$, then there is a $U \in \mathscr{U}_n$ such that $U \subseteq H$, because $\mathscr{U}_n$ is a π-base; and in this case $U \prec G_2$ and $\theta(U) > n$, so that $U \prec_1 G_2$.

Now consider 43A(iv). If $\mathscr{G}$ is a non-empty family of non-empty open sets, downwards-directed for $\preccurlyeq_1$, and with no least element, then $\mathscr{G}$ is downwards-directed for $\preccurlyeq$ and

$$\mathscr{G}^* = \{H : H \subseteq X \text{ is open}, \quad \exists G \in \mathscr{G}, G \subseteq H\}$$

is downwards-directed for $\preccurlyeq$ [43B*d*]. Now there is a sequence $\langle G_n \rangle_{n \in \mathbf{N}}$ in $\mathscr{G}$ which is strictly decreasing for $\preccurlyeq_1$, and it follows by induction on n that $\theta(G_n) \geq n$ for every $n \in \mathbf{N}$. So $\mathscr{G}^*$ meets every $\mathscr{U}_n$. As $\preccurlyeq$, $\{\mathscr{U}_n : n \in \mathbf{N}\}$ satisfy the conditions of 43A, $\bigcap \mathscr{G}^* \neq \varnothing$ and $\bigcap \mathscr{G} \neq \varnothing$. This shows that $\preccurlyeq_1$, $\varnothing$ satisfy the conditions 43A(i)–(iv), so that X is Martin-complete.

(*c*) Suppose that $\preccurlyeq$ and $\mathbf{U}$ satisfy the conditions of 43A, with $\#(\mathbf{U}) \leq \kappa$. Write $\mathfrak{T}_Y$ for the topology of Y. For $A \in \mathfrak{T}_Y$ write

$$W_A = \bigcup \{G : G \subseteq X \text{ open}, \ G \cap Y = A\}.$$

Define $\preccurlyeq_Y$ by writing

$$A \preccurlyeq_Y B \quad \text{iff} \quad W_A \prec W_B \quad \text{or} \quad A = B$$

for non-empty $A, B \in \mathfrak{T}_Y$. Then $\preccurlyeq_Y$ satisfies 43A(i) [see 43B*b*]. If $A, B \in \mathfrak{T}_Y$ and $\varnothing \neq A \subset B$ then $\varnothing \neq W_A \subset W_B$, so there is a non-empty open $G \subseteq W_A$ with $G \prec W_B$. As X is quasi-regular there is a non-empty open $H \subseteq X$ such that $\bar{H} \subseteq G$. Now

$$W_{H \cap Y} \subseteq \overline{H \cap Y} \subseteq G \prec W_B,$$

and $H \cap Y \neq \varnothing$ (because Y is dense), so $W_{H \cap Y} \prec W_B$ and $H \cap Y \prec_Y B$. Also $H \cap Y \subseteq G \cap Y \subseteq A$. So $\preccurlyeq_Y$ satisfies 43A(ii).

For $\mathscr{U} \in \mathbf{U}$ set

$$\mathscr{U}_Y = \{A : A \in \mathfrak{T}_Y, \exists U \in \mathscr{U}, \bar{A} \subseteq U\},$$

and for $H \in \mathscr{H}$ set

$$\mathscr{V}_H = \{A : A \in \mathfrak{T}_Y, \bar{A} \subseteq H\}.$$

Again, because X is quasi-regular and Y is dense, each $\mathscr{U}_Y$ and each $\mathscr{V}_H$ is a π-base for $\mathfrak{T}_Y$. Set

$$\mathbf{V} = \{\mathscr{U}_Y : \mathscr{U} \in \mathbf{U}\} \cup \{\mathscr{V}_H : H \in \mathscr{H}\},$$

so that $\mathbf{V}$ is a collection of π-bases for $\mathfrak{T}_Y$, as required by 43A(iii).

Now let $\mathscr{A} \subseteq \mathfrak{T}_Y$ be a non-empty collection of non-empty sets, downwards-directed for $\preccurlyeq_Y$ and with no least element, meeting every member of $\mathbf{V}$. Then $\{W_A : A \in \mathscr{A}\}$ is downwards-directed for $\preccurlyeq$ and has no least element. If $\mathscr{U} \in \mathbf{U}$, there is an $A \in \mathscr{A} \cap \mathscr{U}_Y$; now there is a $U \in \mathscr{U}$ such that $\bar{A} \subseteq U$, so that $W_A \subseteq U$. By 43B*d*, $Z = \bigcap\{W_A : A \in \mathscr{A}\} \neq \varnothing$. Also, if $H \in \mathscr{H}$, there is an $A \in \mathscr{A} \cap \mathscr{V}_H$, in which case $Z \subseteq W_A \subseteq \bar{A} \subseteq H$. So $Z \subseteq \bigcap \mathscr{H} = Y$ and $\bigcap \mathscr{A} = Y \cap Z \neq \varnothing$. As $\mathscr{A}$ is arbitrary, $\preccurlyeq_Y$ and $\mathbf{V}$ satisfy all the conditions of 43A, and Y is Martin-$\#(\mathbf{V})$-complete.

If κ is infinite, than $\#(\mathbf{V}) \leq \kappa$ so Y is Martin-κ-complete. If κ is finite, so is $\mathbf{V}$, and Y is Martin-ω-complete, therefore Martin-complete [(*b*) above] and (if you like) Martin-κ-complete.

(*d*) For each $\iota \in I$ let $\preccurlyeq_\iota$ and $\mathbf{U}_\iota$ satisfy the conditions of 43A for X_ι, with $\#(\mathbf{U}_\iota) \leq \kappa_\iota$.

Let us say that a *cylinder set* in X is a set of the form $C = \prod_{\iota \in I} C_\iota$ where $C_\iota \subseteq X_\iota$ for each $\iota \in I$ and $\{\iota : C_\iota \neq X_\iota\}$ is finite; for the rest of this proof I will reserve the subscript ι for the factors of cylinder sets. If G, H are non-empty open sets in X, say that

> $G \prec H$ if $G \subset H$ and there are open cylinder sets C, C' such that $G \subseteq C \subseteq C' \subseteq H$ and, for every $\iota \in I$, either $C_\iota \prec_\iota C'_\iota$ or $C_\iota = X_\iota$ or C'_ι is a minimal non-empty open set.

(Recall from 43B*c* that the minimal open sets for $\preccurlyeq_\iota$ are the minimal open sets for $\subseteq$.) Say that $G \preccurlyeq H$ if $G = H$ or $G \prec H$. Then $\preccurlyeq$ satisfies 43A(i) [43B*b* agatin]. For 43A(ii), suppose that G and H are open sets in X and that $\varnothing \neq G \subset H$. Then there is a non-empty open cylinder set $C' \subseteq G$. Let

$$J = \{\iota : \iota \in I,\ C'_\iota \neq X_\iota,\ C'_\iota \text{ is not minimal}\};$$

then J is finite. Take an open cylinder set C such that

$$\varnothing \neq C_\iota \prec_\iota C'_\iota\ \forall \iota \in J,$$
$$C_\iota = C'_\iota\ \forall \iota \in I \setminus J.$$

Then $\varnothing \neq C \subseteq G$ and $C \prec H$. So $\preccurlyeq$ satisfies 43A(ii).

For $\iota \in I$, $\mathscr{U} \in \mathbf{U}_\iota$ let $\mathscr{V}^\iota_{\mathscr{U}}$ be the family of open non-empty cylinder sets $C \subseteq X$ such that $C_\iota \in \mathscr{U}$. Write

$$\mathbf{V} = \{\mathscr{V}^\iota_{\mathscr{U}} : \iota \in I,\ \mathscr{U} \in \mathbf{U}_\iota\};$$

then $\mathbf{V}$ is a collection of π-bases for the topology of X.

Let $\mathscr{G}$ be a non-empty collection of open sets in X, downwards-directed for $\preccurlyeq$ and with no least element, meeting every member of $\mathbf{V}$. For $\iota \in I$ set

$$\mathscr{G}_\iota = \{B : B \subseteq X_\iota \text{ open},\ \exists G \in \mathscr{G},\ \pi_\iota[G] \subseteq B\},$$

where $\pi_\iota : X \to X_\iota$ is the coordinate map. Then $\mathscr{G}_\iota$ is downwards-directed for $\subseteq$, because $\mathscr{G}$ is. And $\bigcap \mathscr{G}_\iota \neq \varnothing$. **P** If $\mathscr{G}_\iota$ has a least element B, then $\bigcap \mathscr{G}_\iota = B \neq \varnothing$. So suppose that $\mathscr{G}_\iota$ has no least element. Take any $B \in \mathscr{G}_\iota$ other than X_ι. There is an $H \in \mathscr{G}$ such that $\pi_\iota[H] \subseteq B$. Now there is a $G \in \mathscr{G}$ such that $G \prec H$. Let C, C' be open cylinder sets as in the definition of $\prec$ above. Then $C'_\iota = \pi_\iota[C'] \subseteq \pi_\iota[H] \subseteq B$, so $C'_\iota \neq X_\iota$.

As $\pi_\iota[G] \subseteq C'_\iota$, $C'_\iota \in \mathscr{G}_\iota$. Now we are supposing that $\mathscr{G}_\iota$ has no least element, so C'_ι cannot be a minimal non-empty open set, and $C_\iota \prec_\iota C'_\iota$. As $G \subseteq C$, $C_\iota \in \mathscr{G}_\iota$ and $C_\iota \prec_\iota B$.

This proves that $\mathscr{G}_\iota$ is downwards-directed for $\preccurlyeq_\iota$ [43B*d*]. If $\mathscr{U} \in \mathbf{U}_\iota$ then $\mathscr{G}$ meets $\mathscr{V}^\iota_{\mathscr{U}}$; let $C \in \mathscr{G} \cap \mathscr{V}^\iota_{\mathscr{U}}$; then $C_\iota \in \mathscr{G}_\iota \cap \mathscr{U}$. Accordingly $\mathscr{G}_\iota$ meets every member of $\mathbf{U}_\iota$ and $\bigcap \mathscr{G}_\iota \neq \varnothing$ in this case also. **Q**

This is true for every $\iota \in I$. Now (again because $\mathscr{G}$ has no least element) we see that whenever $H \in \mathscr{G}$ there is a $G \in \mathscr{G}$ and an open cylinder set C such that $G \subseteq C \subseteq H$. In this case $C_\iota \in \mathscr{G}_\iota$ for every $\iota \in I$. So

$$\varnothing \neq \textstyle\prod_{\iota \in I} (\bigcap \mathscr{G}_\iota) \subseteq \bigcap \mathscr{G}.$$

As $\mathscr{G}$ is arbitrary, 43A(iv) is satisfied.

So X is Martin-$\#(\mathbf{V})$-complete. But $\#(\mathbf{V}) \leq \sum_{\iota \in I} \#(\mathbf{U}_\iota) \leq \kappa$, so X is Martin-κ-complete.

For the final remark (on the product of κ or fewer Martin-κ-complete spaces) we need part (*b*) again to deal with the case in which κ is finite.

43D **Examples**

(*a*) Quasi-regular locally compact spaces are Martin-complete; in particular, compact Hausdorff spaces are Martin-complete.

(*b*) Absolute G_κ spaces are Martin-κ-complete; Cech-complete spaces are Martin-complete; complete metric spaces are Martin-complete.

(*c*) 'Base-compact' spaces and 'almost subcompact' spaces [AARTS & LUTZER 74, 2.1.2] are Martin-complete.

(*d*) Copolish spaces are Martin-complete.

Proof (*a*) If X is locally compact and quasi-regular, define $\preccurlyeq$ by saying that for non-empty open sets G, H

$$G \preccurlyeq H \text{ if } G = H \text{ or } \bar{G} \text{ is compact and } \bar{G} \subseteq H.$$

It is easy to check the conditions of 43A; 43A(ii) is almost the definition of 'quasi-regular'.

(*b*) This follows at once from (*a*) and 43C*b*–*c*.

(*c*) Recall that a regular space X is **base-compact** if there is a base $\mathcal{U}$ for the topology of X such that if $\mathcal{G} \subseteq \mathcal{U}$ has the finite intersection property then $\bigcap\{\bar{G} : G \in \mathcal{U}\} \neq \emptyset$. A quasi-regular space X is **almost subcompact** if there is a π-base $\mathcal{U}$ for the topology of X such that

$$\text{whenever } \mathcal{G} \subseteq \mathcal{U} \text{ is such that } \forall G_1, G_2 \in \mathcal{G}\ \exists G \in \mathcal{G}, \bar{G} \subseteq G_1 \cap G_2, \text{ then } \bigcap \mathcal{G} \neq \emptyset. \qquad (*)$$

Plainly base-compact spaces are almost subcompact. If X is almost subcompact, with π-base $\mathcal{U}$, define $\preccurlyeq$ by saying

$$G \preccurlyeq H \quad \text{if} \quad G = H \quad \text{or} \quad \exists U \in \mathcal{U}, G \subseteq U \subseteq \bar{U} \subseteq H.$$

Because $\mathcal{U}$ is a π-base and X is quasi-regular, 43A(i)–(ii) are satisfied. If $\mathcal{G}$ is non-empty, downwards-directed for $\preccurlyeq$ and has no least element, then $\bigcap \mathcal{G} = \bigcap \mathcal{G}'$ where

$$\mathcal{G}' = \{U : U \in \mathcal{U}, \exists G \in \mathcal{G}, G \subseteq U\},$$

and $\mathcal{G}'$ satisfies (*) above, so $\bigcap \mathcal{G}' \neq \emptyset$.

(*d*) Let $(X, \mathfrak{T})$ be a copolish space, and let $\mathfrak{S}$ be the associated Polish topology on X. For non-empty $G, H \in \mathfrak{T}$ say that $G \preccurlyeq H$ if $G = H$ or $\bar{G}^{\mathfrak{S}} \subseteq H$. Then $\preccurlyeq$ satisfies 43A(i)–(ii). For $n \in \mathbb{N}$ set

$$\mathcal{U}_n = \{G : G \in \mathfrak{T} \setminus \{\emptyset\}, \operatorname{diam}(G) \leq 2^{-n}\}$$

where $\operatorname{diam}(G)$ is the diameter of G for a complete metric defining $\mathfrak{S}$. Then each $\mathscr{U}_n$ is a π-base for $\mathfrak{T}$. Also $\preccurlyeq$ and $\{\mathscr{U}_n : n \in \mathbf{N}\}$ satisfy 43A(iv). This shows that X is Martin-ω-complete under $\mathfrak{T}$; by 43C*b*, it is Martin-complete.

43E Theorem

Let X be a ccc Martin-$<\mathfrak{m}$-complete space. Let $\mathscr{G}$, $\mathscr{H}$, $\mathscr{V}$ be three families of open subsets of X such that

$$\#(\mathscr{G}) < \mathfrak{m}, \quad \#(\mathscr{H}) < \mathfrak{m}, \quad \#(\mathscr{V}) < \mathfrak{m};$$

every member of $\mathscr{V}$ is dense in X;

$$H \cap \bigcap \mathscr{G}_0 \neq \varnothing \, \forall H \in \mathscr{H} \cup \{X\}, \quad \text{finite} \quad \mathscr{G}_0 \subseteq \mathscr{G}.$$

Then there is a sequence $\langle x_n \rangle_{n \in \mathbf{N}}$ in $X \cap \bigcap \mathscr{V}$ such that

$\forall\, G \in \mathscr{G}$, $\{n : x_n \notin G\}$ is finite;
$\forall\, H \in \mathscr{H}$, $\{n : x_n \in H\}$ is infinite.

Proof (*a*) Take $\preccurlyeq$, $\mathbf{U}$ satisfying the conditions of 43A, with $\#(\mathbf{U}) < \mathfrak{m}$. Let P be the set of non-empty open sets in X, ordered by $\preccurlyeq$. Then P is downwards-ccc [43B*c*]. For $G \in P$ write $Q_G = \{E : E \in P, E \subseteq G\}$.

(*b*) Set $\mathscr{Q} = \mathbf{U} \cup \{Q_V : V \in \mathscr{V}\}$. Then every member of $\mathscr{Q}$ is a π-base, so is coinitial with P [43B*c*]. Set $\mathscr{R} = \{Q_G : G \in \mathscr{G}\}$, $\mathscr{S} = \{Q_H : H \in \mathscr{H}\}$. If $\mathscr{G}_0 \subseteq \mathscr{G}$ is finite, $H \in \mathscr{H} \cup \{X\}$ then

$$H \cap \bigcap \mathscr{G}_0 \in Q_H \cap \bigcap \{Q_G : G \in \mathscr{G}_0\},$$

so $S \cap \bigcap \mathscr{R}_0 \neq \varnothing$ for every $S \in \mathscr{S} \cup \{P\}$ and finite $\mathscr{R}_0 \subseteq \mathscr{R}$.

(*c*) By 41B, there is a sequence $\langle P_n \rangle_{n \in \mathbf{N}}$ of subsets of P such that
(α) every P_n is non-empty, downwards-directed for $\preccurlyeq$;
(β) $P_n \cap Q_V \neq \varnothing$, $P_n \cap \mathscr{U} \neq \varnothing \, \forall n \in \mathbf{N}$, $V \in \mathscr{V}$, $\mathscr{U} \in \mathbf{U}$;
(γ) $\{n : P_n \cap Q_G = \varnothing\}$ is finite, $\forall G \in \mathscr{G}$;
(δ) $\{n : P_n \cap Q_H \neq \varnothing\}$ is infinite, $\forall H \in \mathscr{H}$.

Then $\bigcap P_n \neq \varnothing$ for each $n \in \mathbf{N}$, because $\preccurlyeq$ and $\mathbf{U}$ satisfy 43A(iv). Choose $x_n \in \bigcap P_n$ for each $n \in \mathbf{N}$. Then

$x_n \in \bigcap \mathscr{V}$ because $P_n \cap Q_V \neq \varnothing \, \forall V \in \mathscr{V}$, $n \in \mathbf{N}$;
$\{n : x_n \notin G\} \subseteq \{n : P_n \cap Q_G = \varnothing\}$ is finite, $\forall G \in \mathscr{G}$;
$\{n : x_n \in H\} \supseteq \{n : P_n \cap Q_H \neq \varnothing\}$ is infinite, $\forall H \in \mathscr{H}$.

So $\langle x_n \rangle_{n \in \mathbf{N}}$ has the properties required.

43F Corollary

Let X be a ccc Martin-$<\mathfrak{m}$-complete space.

(*a*) The intersection of fewer than $\mathfrak{m}$ dense open sets in X is dense.

(*b*) If $\pi(X) < \mathfrak{m}$ then X is separable.
(*c*) If $\omega < \mathrm{cf}(\kappa) \le \kappa < \mathfrak{m}$ then κ is a caliber of X.

Proof (*a*) Apply 43E with $\mathscr{V}$ a family of fewer than $\mathfrak{m}$ dense open sets, $\mathscr{G} = \emptyset$, $\mathscr{H} = \{H\}$ where H is an arbitrary non-empty open set, to see that $H \cap \bigcap \mathscr{V} \neq \emptyset$.

(*b*) Apply 43E with $\mathscr{H}$ a π-base for the topology of X of cardinal less than $\mathfrak{m}$ and with $\mathscr{G} = \mathscr{V} = \emptyset$, and see that $\{x_n : n \in \mathbf{N}\}$ will be dense in X.

(*c*) Let $\langle H_\xi \rangle_{\xi < \kappa}$ be any family of non-empty open sets. Apply 43E with $\mathscr{H} = \{H_\xi : \xi < \kappa\}$ to see that $\{\xi : x_n \in H_\xi\}$ will have cardinal κ for some $n \in \mathbf{N}$.

43G **Proposition**
Let X be a compact Hausdorff space. Then
(i) $\pi(X) \le \max(t^+(X), hc(X))$;
(ii) if κ is a caliber of X and $t^+(X) \le \mathrm{cf}(\kappa)$, then $\pi(X) < \kappa$;
(iii) $\pi(X) < \max(t^+(X), d(X)^+)$.

Proof (*a*) The arguments for parts (i) and (ii) may be taken together, as follows. Write $\lambda = \max(t^+(X), hc(X))^+$ for part (i), $\lambda = \mathrm{cf}(\kappa)$ for part (ii). **?** If $\pi(X) \ge \lambda$, choose inductively continuous functions $g_\xi : X \to [0,1]$, for $\xi < \lambda$, as follows. Given $\langle g_\eta \rangle_{\eta < \xi}$, set $f_\xi(x) = \langle g_\eta(x) \rangle_{\eta < \xi} \in [0,1]^\xi$. We have $\pi(X) \ge \lambda \ge t^+(X) \ge \omega$, so X is infinite and $t^+(X) \ge \omega_1$ and $\lambda > \omega$; accordingly $\pi(X) > \max(\omega, \#(\xi)) \ge \pi([0,1]^\xi)$ and f cannot be irreducible. Let $G_\xi \subseteq X$ be a non-empty open set such that $f_\xi[X \setminus G_\xi] = f_\xi[X]$. Choose g_ξ to be non-constant and zero on $X \setminus G_\xi$, and set $H_\xi = \{x : g_\xi(x) > 0\}$. Continue.

(*b*) Let $x \in X$. Then $\#(\{\xi : x \in H_\xi\}) < t^+(X)$. **P** It is enough to show that if $\theta \le \lambda$ and $\mathrm{cf}(\theta) = t^+(X)$, there is a $\zeta < \theta$ such that $g_\eta(x) = 0$ for $\zeta < \eta < \theta$. For each $\xi \le \zeta \le \theta$, $\xi < \theta$ choose $x_{\xi\zeta}$ as follows. $x_{\xi\xi} = x$. Given $x_{\xi\zeta}$, choose $x_{\xi,\zeta+1} \in X \setminus G_\zeta$ such that $f_\zeta(x_{\xi,\zeta+1}) = f_\zeta(x_{\xi\zeta})$. For limit ordinals $\zeta \in]\xi, \theta]$ choose $x_{\xi\zeta}$ to be any cluster point of $\langle x_{\xi\eta} \rangle_{\eta \uparrow \zeta}$. Then we see by induction on ζ that $g_\eta(x_{\xi\zeta}) = g_\eta(x)$ if $\eta < \xi \le \zeta \le \theta$, while $g_\eta(x_{\xi\zeta}) = 0$ if $\xi \le \eta < \zeta \le \theta$. Now let y be any cluster point of $\langle x_{\xi\theta} \rangle_{\xi \uparrow \theta}$. We shall have $g_\xi(y) = g_\xi(x)$ for every $\xi < \theta$, and $y \in \overline{\{x_{\xi\theta} : \xi < \theta\}}$. As $t^+(X) = \mathrm{cf}(\theta)$, there is a $\zeta < \theta$ such that $y \in \overline{\{x_{\xi\theta} : \xi \le \zeta\}}$. Now if $\eta \in]\zeta, \theta[$, $g_\eta(x_{\xi\theta}) = 0$ for every $\xi \le \zeta$, so that $g_\eta(x) = g_\eta(y) = 0$, as required. **Q**

(*c*) This finishes with part (ii), because $\lambda = \mathrm{cf}(\kappa)$ is supposed to be a caliber of X, so that there ought to be a point belonging to λ of the non-empty open sets H_ξ. For part (i) we have a little more work to do. Set $\alpha = \max(t^+(X), hc(X))$.

(*d*) Choose inductively sets $Y_\xi \subseteq X$ and $A_\xi \subseteq \lambda$, for $\xi \leq \alpha$, as follows. Given $\langle Y_\eta \rangle_{\eta < \xi}$, where $\xi \leq \alpha$, set

$$A_\xi = \{\zeta : \zeta < \lambda, H_\zeta \cap \bigcup_{\eta < \xi} Y_\eta = \varnothing\}.$$

Given A_ξ, choose a subset Y_ξ of $\bigcup \{H_\zeta : \zeta \in A_\xi\}$ which is maximal subject to

$$\#(Y_\xi \cap H_\zeta) \leq 1 \; \forall \zeta \in A_\xi.$$

Continue. Each Y_ξ is discrete so $\#(Y_\xi) \leq hc(X) \leq \alpha$ for every $\xi \leq \alpha$. As $\#(\{\zeta : \zeta < \lambda, x \in H_\zeta\}) < t^+(X) \leq \alpha$ for each $x \in X$,

$$\#(\{\zeta : \zeta < \lambda, H_\zeta \cap \bigcup_{\xi < \alpha} Y_\xi \neq \varnothing\}) \leq \alpha,$$

and $A_\alpha \neq \varnothing$. Now take any $\zeta \in A_\alpha$ and any $x \in H_\zeta$. If $\xi < \alpha$, then $\zeta \in A_\xi$, but $x \notin Y_\xi$, because $H_\zeta \cap Y_\xi = \varnothing$; so there is a $\theta(\xi) \in A_\xi$ and a $y_\xi \in Y_\xi$ such that both x and y_ξ belong to $H_{\theta(\xi)}$, by the maximality of Y_ξ. If $\eta < \xi < \alpha$ then $\theta(\xi) \in A_\xi$, so $H_{\theta(\xi)}$ cannot meet Y_η; as $y_\eta \in H_{\theta(\eta)} \cap Y_\eta$, $\theta(\eta) \neq \theta(\xi)$. Thus

$$\#(\{\xi : x \in H_\xi\}) \geq \#(\{\theta(\xi) : \xi < \alpha\}) = \alpha \geq t^+(X),$$

which again contradicts (*b*) above. Thus in part (i) also we reach a contradiction. **X**

(*e*) Finally, to prove (iii), apply (ii) with $\kappa = \max(t^+(X), d(X)^+)$.

43H **Corollary**

If X is a separable countably tight compact Hausdorff space, then it has a countable π-base.

Proof For $\max(t^+(X), d(X)^+) \leq \omega_1$.

43I **Corollary**

(*a*) Let X be a ccc compact Hausdorff space such that $t^+(X) < \mathfrak{m}$. Then $\pi(X) < t^+(X)$ and X is separable.

(*b*) $[\mathfrak{m} > \omega_1]$ If X is a countably tight ccc compact Hausdorff space, it has a countable π-base.

(*c*) $[\mathfrak{m} > \omega_1]$ If X is a first-countable ccc compact Hausdorff space it is separable.

(*d*) $[\mathfrak{m} > \omega_1]$ If X is a countably tight compact Hausdorff space then $hd(X) = hc(X)$.

Proof (*a*) By 43F*c*, $t^+(X)$ is a caliber of X. (This is still true if $t^+(X) = \omega$, for in this case X is finite.) So by 43G(ii) $\pi(X) < t^+(X) < \mathfrak{m}$, and now X is separable by 43F*b*.

(*b*) We have $t^+(X) \leq \omega_1$; use (*a*).

(*c*) Immediate from (*b*).

(*d*) If X is finite, this is trivial; suppose that X is infinite. Of course $hc(X) \leq hd(X)$ [A4B*a*]. For the reverse inequality, if $Y \subseteq X$, then $\bar{Y}$ is also a countably tight compact Hausdorff space. If $\bar{Y}$ is ccc then

$$d(Y) \leq \pi(Y) = \pi(\bar{Y}) \leq \omega \leq hc(X)$$

by (*b*) above. If $\bar{Y}$ is not ccc then $t^+(\bar{Y}) \leq \omega_1$ so

$$d(Y) \leq \pi(Y) = \pi(\bar{Y}) \leq \max(\omega_1, hc(\bar{Y})) = hc(\bar{Y}) \leq hc(X)$$

by 43G(i). So in all cases $d(Y) \leq hc(X)$ and $hd(X) \leq hc(X)$.

43J **Proposition**

Let X be a ccc absolute G_κ space and Y a continuous image of X. If $\max(\omega_1, t^+(Y)) \leq \mathrm{cf}(\kappa) \leq \kappa < \mathfrak{m}$ then $d(Y) < \kappa$.

Proof Express X as $\bigcap_{\xi<\kappa} H_\xi$ where each H_ξ is an open set in a compact Hausdorff space Z; we can suppose that $Z = \bar{X}$, so that Z is also ccc. Let $\mathfrak{T}$ be the topology of Z. Let $\varphi: X \to Y$ be a continuous surjection.

? Suppose, if possible, that $d(Y) \geq \kappa$. Construct inductively, for $\xi < \kappa$, countable sets $A_\xi \subseteq X$ and $\mathcal{U}_\xi \subseteq \mathfrak{T}$ such that $\#(\mathcal{U}_\xi) \leq \max(\omega, \#(\xi))$, as follows. Given $\langle A_\eta \rangle_{\eta<\xi}$, where $\xi < \kappa$, set $B_\xi = \bigcup_{\eta<\xi} A_\eta$. Then $\#(B_\xi) < \kappa$, so $\varphi[B_\xi]$ is not dense in Y, and $E_\xi = \overline{\varphi^{-1}[\varphi[B_\xi]]} \neq X$. Write $F_\xi = \bar{E}_\xi$, the closure of E_ξ in Z; then $F_\xi \cap X = E_\xi$ (because φ is continuous), so $F_\xi \neq Z$. Let $\mathcal{U}_\xi \subseteq \mathfrak{T}$ be such that

$Z \backslash F_\xi \in \mathcal{U}_\xi$, $H_\xi \in \mathcal{U}_\xi$;
if $U, V \in \bigcup_{\eta<\xi} \mathcal{U}_\eta$ then $U \cap V \in \mathcal{U}_\xi$;
if $U \in \bigcup_{\eta<\xi} \mathcal{U}_\eta$ and $x \in U \cap B_\xi$ there is a V such that $x \in V \subseteq \bar{V} \subseteq U$ and V, $Z \backslash \bar{V}$ both belong to $\mathcal{U}_\xi$;
$\#(\mathcal{U}_\xi) \leq \max(\omega, \#(\xi))$.

(This is possible because B_ξ and $\bigcup_{\eta<\xi} \mathcal{U}_\eta$ both have cardinals less than or equal to $\max(\omega, \#(\xi))$.) Now by 43E there is a countable $A_\xi \subseteq X$ such that $A_\xi \cap U \neq \emptyset$ for every non-empty $U \in \mathcal{U}_\xi$. Continue.

Set $\mathcal{U} = \bigcup_{\xi<\kappa} \mathcal{U}_\xi$; then $\mathcal{U}$ is a base for a topology $\mathfrak{S}$ on Z, coarser than $\mathfrak{T}$. As $\mathfrak{T}$ is ccc, so is $\mathfrak{S}$. Let $B = \bigcup_{\xi<\kappa} A_\xi$, let F be the $\mathfrak{T}$-closure of B, and let $\mathfrak{S}_F$ be the topology induced on F by $\mathfrak{S}$. As F meets every non-empty member of

$\mathscr{U}$, F is $\mathfrak{S}$-dense in Z, and is ccc under $\mathfrak{S}_F$. As F is $\mathfrak{T}$-compact, it is $\mathfrak{S}_F$-compact.

Now $\mathfrak{S}_F$ is quasi-regular. **P** If $S \in \mathfrak{S}_F$ and $S \neq \emptyset$, there is a $U \in \mathscr{U}$ such that $\emptyset \neq F \cap U \subseteq S$. Let $\xi < \kappa$ be such that $U \in \mathscr{U}_\xi$. Then $A_\xi \cap U \neq \emptyset$ and there is a $V \in \mathscr{U}_{\xi+1}$ such that $A_\xi \cap V \neq \emptyset$, $\bar{V}^{\mathfrak{T}} \subseteq U$, and $Z \backslash \bar{V}^{\mathfrak{T}} \in \mathscr{U}_{\xi+1}$; so that $\bar{V}^{\mathfrak{S}} = \bar{V}^{\mathfrak{T}} \subseteq U$. Now $\emptyset \neq F \cap V \subseteq F \cap \overline{F \cap V}^{\mathfrak{S}} \subseteq F \cap U \subseteq S$. **Q**

So F is Martin-complete for $\mathfrak{S}_F$ [43D*a*]. Applying 43E to F, $\mathfrak{S}_F$ with $\mathscr{V} = \{F \cap H_\xi : \xi < \kappa\}$ and

$$\mathscr{H} = \{F \cap U : U \in \mathscr{U} \backslash \{\emptyset\}\},$$

we see that there is a countable $C \subseteq \bigcap \mathscr{V} = F \cap X$ such that C meets every non-empty member of $\mathscr{U}$, i.e. $\bar{C}^{\mathfrak{S}} = Z$. Now consider

$$\varphi[C] \subseteq \varphi[F \cap X] = \varphi[X \cap \bar{B}^{\mathfrak{T}}] \subseteq \overline{\varphi[B]}$$
$$= \overline{\bigcup_{\xi<\kappa} \varphi[B_\xi]} = \bigcup_{\xi<\kappa} \overline{\varphi[B_\xi]}$$

because $t^+(Y) \leq \mathrm{cf}(\kappa)$. Because also $\omega < \mathrm{cf}(\kappa)$, there is a $\zeta < \kappa$ such that $\varphi[C] \subseteq \overline{\varphi[B_\zeta]}$ and $C \subseteq E_\zeta$. But in this case $\bar{C}^{\mathfrak{S}} \subseteq F_\zeta \neq Z$ because $Z \backslash F_\zeta \in \mathscr{U}_\zeta \subseteq \mathscr{U} \subseteq \mathfrak{S}$. This is the contradiction we seek. **X**

43K Corollary

[$\mathfrak{m} > \omega_1$] If X is a countably tight continuous image of a ccc Cech-complete space it is separable.

43L Proposition

Let X be a ccc Hausdorff space such that every closed subspace of X is Martin-$<\mathfrak{m}$-complete. Suppose there is a regular cardinal κ such that $t^+(X) < \kappa \leq \mathfrak{m}$ and $Y = \{x : x \in X, \pi\chi(x, X) < \kappa\}$ is dense. Then X is separable.

Proof Write $\lambda = t^+(X)$. If $\lambda = \omega$, X is finite and the result is trivial; so suppose that $\lambda \geq \omega_1$. For each non-empty open $G \subseteq X$ choose $y_G \in Y \cap G$, and let $y_\emptyset$ be any member of Y. For each $y \in Y$ choose a family $\mathscr{U}_y$ of non-empty open sets such that $\#(\mathscr{U}_y) < \kappa$ and every neighbourhood of y includes some member of $\mathscr{U}_y$. For $A \subseteq Y$ set

$$A' = A \cup \{y_{X \backslash A}\} \cup \{y_{U \cap V} : U, V \in \bigcup_{z \in A} \mathscr{U}_z\} \subseteq Y.$$

Finally, define $\langle F_\xi \rangle_{\xi \leq \lambda}$ by writing

$$F_0 = \emptyset,\ F_{\xi+1} = F'_\xi\ \forall \xi < \lambda,$$
$$F_\xi = \bigcup_{\eta<\xi} F_\eta \text{ for non-zero limit ordinals } \xi \leq \lambda.$$

Because κ is regular we see that $\#(A') < \kappa$ whenever $\#(A) < \kappa$, so that $\#(F_\xi) < \kappa$ whenever $\xi \le \lambda$.

We find now that (α) F_λ is ccc, (β) $\pi(\bar{F}_\lambda) < \kappa$, ($\gamma$) $\bar{F}_\lambda$ is separable, (δ) $\bar{F}_\lambda = X$.
P (α) Let $\mathscr{H}$ be an uncountable family of non-empty relatively open subsets of F_λ. For each $H \in \mathscr{H}$ choose an open $G_H \subseteq X$ such that $H = F_\lambda \cap G_H$, and a point $x(H) \in H$. Then there is a $U(H) \in \mathscr{U}_{x(H)}$ such that $U(H) \subseteq G_H$. Because X is ccc, there are distinct $H, H' \in \mathscr{H}$ such that $U(H) \cap U(H') \neq \emptyset$. Because λ is a limit ordinal, there is a $\xi < \lambda$ such that $x(H)$ and $x(H')$ both belong to F_ξ. In this case $U(H)$ and $U(H')$ both belong to $\bigcup\{\mathscr{U}_y : y \in F_\xi\}$, so $y_{U(H) \cap U(H')} \in F'_\xi = F_{\xi+1} \subseteq F_\lambda$; as $U(H) \cap U(H') \neq \emptyset$, $y_{U(H) \cap U(H')} \in U(H) \cap U(H') \cap F_\lambda \subseteq H \cap H'$. Thus $\mathscr{H}$ is not disjoint. As $\mathscr{H}$ is arbitrary, F_λ is ccc. (β) If $y \in F_\lambda$ and $U \in \mathscr{U}_y$, there is a $\xi < \lambda$ such that $y \in F_\xi$, and $y_U \in U \cap F_{\xi+1}$; so U meets F_λ. Set $\mathscr{V} = \{\bar{F}_\lambda \cap U : \exists y \in F_\lambda, U \in \mathscr{U}_y\}$. Then (again because κ is regular) $\#(\mathscr{V}) < \kappa$. Since every open set which meets F_λ includes a member of $\mathscr{V}$, and $\emptyset \notin \mathscr{V}$, $\mathscr{V}$ is a π-base for the topology of $\bar{F}_\lambda$, and $\pi(\bar{F}_\lambda) \le \#(\mathscr{V}) < \kappa$. ($\gamma$) Now $c(\bar{F}_\lambda) = c(F_\lambda) \le \omega$ and $\pi(\bar{F}_\lambda) < \kappa \le \mathfrak{m}$, so by 43F*b* $\bar{F}_\lambda$ is separable. (δ) As $F_\lambda = \bigcup_{\xi<\lambda} F_\xi$ and $\lambda = t^+(X)$, $\bar{F}_\lambda = \bigcup_{\xi<\lambda} \bar{F}_\xi$. We know that $\bar{F}_\lambda$ is separable; let $C \subseteq \bar{F}_\lambda$ be a countable dense set. As $\lambda = \mathrm{cf}(\lambda) > \omega$, there is a $\xi < \lambda$ such that $C \subseteq \bar{F}_\xi$. Now we have $F'_\xi = F_{\xi+1} \subseteq F_\lambda \subseteq \bar{F}_\xi$. Since $y_{X \setminus \bar{A}} \in A'$ for every $A \subseteq Y$, we can have $A' \subseteq \bar{A}$ only when $\bar{A} = X$. So $\bar{F}_\xi = X$ and $\bar{F}_\lambda = X$. **Q**

Thus $X = \bar{F}_\lambda$ is separable.

43M Corollary

[$\mathfrak{m} > \omega_1$] Let X be a first-countable ccc Hausdorff space in which every closed subspace is Martin-ω_1-complete. Then X is separable.

Proof Take $\kappa = \omega_2$ above.

43N Sources

Most of the results above were proved first for compact Hausdorff spaces or for one of the other classes of Baire spaces (Cech-complete, cocompact, base-compact, or absolute G_κ spaces). Among the principal results, 43E is based on ideas in MALYHIN & SAPIROVSKII 73 and ARHANGEL'SKII 76; 43G on SAPIROVSKII 80; 43J on SAPIROVSKII 72*a*; and 43L on HAJNAL & JUHÁSZ 71, TALL 74*a* and JUHÁSZ 77. Of the others, I find versions of 43F*b*, 43I*c* and 43M in HAJNAL & JUHÁSZ 71; of 43F*c* and 43K in SAPIROVSKII 72*a*; of 43I*b* in ARHANGEL'SKII 72; of 43F*a* in MALYHIN & SAPIROVSKII 73; and of 43H, 43I*a*–*b* and 43I*d* in SAPIROVSKII 80.

43O Exercises on Martin-completeness

(*a*) (i) Let X be a quasi-regular space and $Z \subseteq X$ a dense Martin-κ-

complete subspace. Then X is Martin-κ-complete. (ii) Let X be a Martin-κ-complete space, $G \subseteq X$ an open set. Then G and $\bar{G}$ are Martin-κ-complete. (iii) Any disjoint union of Martin-κ-complete spaces is Martin-κ-complete. (iv) If X is a topology space such that every point of X has a Martin-κ-complete neighbourhood then X is Martin-κ-complete. (v) If X is a quasi-regular topological space and is expressible as a finite union of Martin-κ-complete subspaces then X is Martin-κ-complete.

(*b*) If X is a Martin-κ-complete space, Y is a topological space, and $f:X \to Y$ is a continuous surjection such that $\operatorname{int} f[G] \neq \emptyset$ for every non-empty open $G \subseteq X$, then Y is Martin-κ-complete.

(*c*) Any box product of Martin-κ-complete spaces is Martin-κ-complete.

(*d*) For any topological space X, consider the following infinite game. White chooses a non-empty open set $G_0 \subseteq X$; Black chooses a non-empty open set $G_1 \subseteq G_0$; White chooses a non-empty open set $G_2 \subseteq G_1$; Black chooses a non-empty open set $G_3 \subseteq G_2$; and so on. White wins if $\bigcap_{n \in \mathbf{N}} G_n = \emptyset$; otherwise Black wins. X is **α-favourable** if Black has a stationary winning strategy [A3L]. Now a Martin-complete space is α-favourable, therefore Baire.

(*e*) JUHÁSZ 77 defines a regular Hausdorff space X to be **π-complete** if there is a family U of π-bases for the topology of X such that (i) $\#(\mathrm{U}) < \mathfrak{c}$ (ii) whenever $\mathscr{G}$ is a family of non-empty open sets in X, meeting every member of U, such that $\#(\mathscr{G}) < \mathfrak{c}$ and for all $G_1, G_2 \in \mathscr{G}$ there is a $G \in \mathscr{G}$ with $\bar{G} \subseteq G_1 \cap G_2$, then $\bigcap \mathscr{G} \neq \emptyset$. A π-complete space is Martin-$<\mathfrak{c}$-complete.

(*f*) Let X be a quasi-regular topological space with a countable π-base. Then X is Martin-complete iff it is **pseudo-complete** [OXTOBY 61, AARTS & LUTZER 74, 4.1.2]; i.e. there is a sequence $\langle \mathscr{U}_n \rangle_{n \in \mathbf{N}}$ of π-bases for the topology of X such that if $\langle U_n \rangle_{n \in \mathbf{N}} \in \prod_{n \in \mathbf{N}} \mathscr{U}_n$ and $\bar{U}_{n+1} \subseteq U_n$ for every $n \in \mathbf{N}$ then $\bigcap_{n \in \mathbf{N}} U_n \neq \emptyset$.

(*g*) Let $(X, \mathfrak{S})$ be a Polish space. Let $\mathfrak{T}$ be the topology on X generated by $\mathfrak{S} \cup \{E : X = \overline{\operatorname{int} E}\}$. Then $\mathfrak{T}$ is Martin-complete. [Hint: argument of 43D*d*.]

(*h*) Let $\mathfrak{S}$ be the usual topology on $\mathbf{R}$, $\mathfrak{T}$ the density topology [25J]. (i) Every $\mathfrak{T}$-closed subset of X is Martin-complete under the topology induced by $\mathfrak{T}$. [Hint: 25J*b*, 43D*d*, *a*(ii) and *a*(v) above.] (ii) [$\mathfrak{p} = \mathfrak{c}$] There is a non-separable first-countable topology coarser than $\mathfrak{T}$ and finer than $\mathfrak{S}$; so that 43J is false for copolish X.

(*i*) Let S be a topological space. (i) $\mathscr{P}S$ is Martin-complete under its Pixley-Roy topology. [Say $V \preccurlyeq W$ if $V = W$ or, in the notation of 25H, $V \subseteq U_{\mathscr{P}S}(A, C) \subseteq W$ for some A, C.] (ii) If S is regular, the set of closed subsets of S is Martin-complete under its Pixley-Roy topology. [Require C to be closed in the last formula.] (iii) If S is a metric space, the set of totally bounded subsets of S is Martin-complete under its Pixley-Roy topology.

(*j*) Let X be

$$\{x : x \in \{0,1\}^{\omega_1}, \exists \xi < \omega_1, x(\eta) = 0 \; \forall \eta \geq \xi\}.$$

(i) X is pseudo-complete, therefore α-favourable and Baire [(*d*), (*f*) above]. (ii) [CH] X is almost subcompact. (iii) [$\mathfrak{p} > \omega_1$] X is not π-complete, nor Martin-ω_1-complete.

43P Exercises on 43E–43M

(*a*) If $\kappa < \mathfrak{m}$, Theorem 43E applies to any space X which is a product of ccc Martin-κ-complete spaces. [Hint: if $\kappa \geq \omega_1$, reduce to the case in which every member of $\mathscr{G} \cup \mathscr{H} \cup \mathscr{V}$ factors through a countable set of coordinates, and use 43C*d*.]

(*b*) Let $X \subseteq \mathbf{R}$ be a set such that both X and $\mathbf{R} \setminus X$ meet every uncountable Borel set in $\mathbf{R}$. (i) The conclusions of 43E, with $\mathfrak{m}$ replaced by $\mathfrak{p}$, are valid for X. [Hint: by 22B, $X \cap \bigcap \mathscr{V}$ is dense in X; now use 21A on a countable dense subset of $X \cap \bigcap \mathscr{V}$.] (ii) X is not Martin-$<\mathfrak{p}$-complete. [Hint: suppose $\preccurlyeq$, $\mathbf{U}$ satisfied 43A, with $\#(\mathbf{U}) < \mathfrak{p}$. Let $\mathscr{V}$ be a countable base for the topology of $\mathbf{R}$. For $V \in \mathscr{V}$, $\mathscr{U} \in \mathbf{U}$ set $A_V = \bigcup \{W : W \in \mathscr{V}, W \cap X \prec V \cap X\}$,

$$B_{\mathscr{U}}^V = \bigcup \{W : W \in \mathscr{V}, \exists U \in \mathscr{U}, W \cap X \subseteq V \cap U\}.$$

Take

$$s \in \mathbf{R} \setminus (X \cup \bigcup_{V \in \mathscr{V}} (V \setminus (A_V \cap \bigcap_{\mathscr{U} \in \mathbf{U}} B_{\mathscr{U}}^V)))$$

Try $\mathscr{G} = \{G \cap X : G \subseteq \mathbf{R} \text{ open}, s \in G\}$.]

(*c*) Let X be a ccc Martin-$<\mathfrak{m}$-complete T_1 space. (i) If $x \in X$ and $1 < \chi(x, X) < \mathfrak{m}$ then there is a sequence in $X \setminus \{x\}$ converging to x. [Use 43E.] (ii) If X is Hausdorff and extremally disconnected and $x \in X$ is not isolated, $\chi(x, X) \geq \mathfrak{m}$. [Malyhin & Sapirovskii 73, Theorem 1.2 and Corollary 1.2.]

(*d*) If X is a ccc Martin-$<\mathfrak{m}$-complete space and $\omega < \mathrm{cf}(d(X)) < \mathfrak{m}$ then $\mathrm{cf}(d(X)) \leq t(X)$. [Hint: let $\langle x_\xi \rangle_{\xi < d(X)}$ enumerate a dense subset of X. Use 43E to find $\langle y_n \rangle_{n \in \mathbf{N}}$ such that

$$\forall \xi < d(X) \, \exists n \in \mathbf{N}, \, y_n \notin \overline{\{x_\eta : \eta \leq \xi\}}.$$

See ARHANGEL'SKII 71, Proposition 6; TALL 74*a*, Theorem 4.16.]

(*e*) Let X be a countably compact ccc Martin-$<\mathfrak{m}$-complete space. (i) If $\mathscr{G}$ is a family of open sets in X with the finite intersection property and $\#(\mathscr{G})<\mathfrak{m}$, then $\bigcap_{G\in\mathscr{G}}\bar{G}\neq\emptyset$. [Use 43E.] (ii) If $\mathscr{G}$ is an open cover of X and $\#(\mathscr{G})<\mathfrak{m}$ then there is a finite $\mathscr{G}_0\subseteq\mathscr{G}$ such that $X=\overline{\bigcup\mathscr{G}_0}$. [Cf. 24H, 24N*e*.] (iii) If X is regular and $\hat{L}(X)\leq\mathfrak{m}$ then X is compact. [Cf. 24I.] (iv) If X is normal and $F\subseteq X$ is closed and $\psi(F,X)<\mathfrak{m}$ then $\chi(F,X)=\psi(F,X)$. [Use 24N*g* and 43O*a*(ii). See MALYHIN & SAPIROVSKII 73, Theorem 1.5.]

(*f*) [$\mathfrak{m}=\mathfrak{c}$] Let X be a ccc Martin-$<\mathfrak{c}$-complete space in which singleton sets are nowhere dense, with $\pi(X)\leq\mathfrak{c}$. (i) There is a $\mathfrak{c}$-Lusin set $Y\subseteq X$. (ii) If X is cometrizable, Y is perfectly normal. [Use 43F*a*, A3F*b*, 25E.]

(*g*) [$\mathfrak{m}>\omega_1$] Let X be a ccc compact Hausdorff space in which all separable subspaces are metrizable. Then X is metrizable. [Hint: by 43F*b*, any continuous image of X of weight less than or equal to ω_1 is separable, therefore metrizable. Compare 22N*j*(ii).]

(*h*) Suppose that $\omega<\mathrm{cf}(\kappa)\leq\kappa<\mathfrak{m}$. Let X be a ccc Martin-$<\mathfrak{m}$-complete space and $\mathscr{U}$ a π-base for the topology of X such that $\#(\{U:x\in U\in\mathscr{U}\})<\kappa$ for every $x\in X$. Then X is separable. [MALYHIN & SAPIROVSKII 73, Corollary 3.2(3).]

(*i*) A compact Hausdorff space is **Corson-compact** if it can be embedded into

$$\{x:x\in\mathbf{R}^I,\ \{\iota:x(\iota)\neq 0\}\text{ is countable}\}$$

for some set I. [$\mathfrak{m}>\omega_1$] A ccc Corson-compact space is metrizable. [See COMFORT & NEGREPONTIS 82, pp. 204–7, and (*g*) above.]

(*j*) If $\kappa<\mathfrak{m}$, 43F*b*–*c* apply to continuous images of products of ccc Martin-κ-complete spaces. [Use (*a*) above.]

(*k*) [$\mathfrak{m}=\mathfrak{c}$] Let X be a ccc Cech-complete space with $\#(X)<2^{\mathfrak{c}}$. (i) If $\max(d(X),\ t(X))<\mathfrak{c}$ then X is separable. [By A4J*b*, $\{x:\chi(x,X)<\mathfrak{c}\}$ is dense in X, and $\mathfrak{c}$ is regular, so $\pi(X)<\mathfrak{c}$.] (ii) If $hd(X)<\mathfrak{c}$ then X is separable. [For $t(X)\leq hd(X)$. See TALL 74*a*, 4.14.]

(*l*) Suppose $\omega<\mathrm{cf}(\kappa)\leq\kappa<\mathfrak{m}$. If X is a ccc absolute G_κ space in which sequentially closed sets are closed, then $\pi(X)<\mathfrak{c}$. [By 43J, $d(X)<\kappa$, so $\#(X)\leq\mathfrak{c}$. Now $A=\{x:\chi(x,X)<\mathfrak{c}\}$ is dense in X by the remark following 2.22 of JUHÁSZ 71 (compare A4J*b*), and $d(A)\leq\max(d(X),\ t(X))<\mathfrak{m}$, while $\mathrm{cf}(\mathfrak{c})\geq\mathfrak{m}$. See SAPIROVSKII 72*a*, Theorem 1.4′.]

(*m*) Let X be a ccc compact Hausdorff space such that $w(X) < \mathfrak{m}$. Then X is sequentially separable. If X is countably tight then it has a dense Polish subspace. [Use 43F*b*, 24A, 43I*b*, 22N*d*.]

(*n*) Use 43O*i*(ii) to prove 22E.

(*o*) The principle H of 41L is equivalent to each of the following:
(i) ω_1 is a caliber of every ccc compact Hausdorff space;
(ii) If X is a ccc topological space and $\mathscr{G}$ is an uncountable collection of open subsets of X there is an uncountable subset of $\mathscr{G}$ with the finite intersection property ('ω_1 is a **precaliber** of X');
(iii) If X is a ccc compact Hausdorff space and $\langle G_\xi \rangle_{\xi < \omega_1}$ is a decreasing family of dense open sets in X, then $\bigcap_{\xi < \omega_1} G_\xi \neq \emptyset$.
[See Tall 74*a*, §5.]

(*p*) For any cardinal κ the following are equivalent:
(i) $\kappa < \mathfrak{t}$ [41L];
(ii) every ccc compact Hausdorff space of weight less than or equal to κ is separable;
(iii) If X is a regular ccc base-compact space and $\mathscr{G}$ is a family of non-empty open sets in X with $\#(\mathscr{G}) \leq \kappa$, there is a countable $Y \subseteq X$ meeting every member of $\mathscr{G}$;
(iv) every ccc topological space of π-weight less than or equal to κ is σ-centered;
(v) If X is a ccc topological space and $\mathscr{G}$ is a family of non-empty open subsets of X with $\#(\mathscr{G}) \leq \kappa$, then $\mathscr{G}$ is expressible as $\bigcup_{n \in \mathbf{N}} \mathscr{G}_n$ where each $\mathscr{G}_n$ has the finite intersection property.

[Hint: (i) $\Rightarrow$ (iii) $\Rightarrow$ (ii) and (i) $\Rightarrow$ (v) $\Rightarrow$ (iv) $\Rightarrow$ (i) are easy. For (ii) $\Rightarrow$ (iv) consider the Stone space of a subalgebra of the regular open algebra of X.]

(*q*) In 43L–M the hypothesis 'Hausdorff' may be omitted. [Show that if $t^+(X) = \omega$ and every closed subspace of X is Baire then the topology of X has a smallest π-base.]

43Q Further results

(*a*) [$\mathfrak{m} > \omega_1$] A locally ccc, locally compact, regular metalindelöf space is paracompact. [Tall 74*a*.]

(*b*) [$\mathfrak{m} > \omega_1$] Suppose that there is no proper ω_1-saturated σ-ideal of $\mathscr{P}\mathfrak{c}$ containing all singletons. [See B2D.] Let X be a ccc compact Hausdorff space and Y a complete metric space. Let Σ be the algebra of subsets of X with the Baire property. Let $R \subseteq X \times Y$ be a relation with closed separable vertical sections such that $R^{-1}[G] \in \Sigma$ for every open $G \subseteq Y$. Then R has a selector f such that $f^{-1}[G] \in \Sigma$ for every open $G \subseteq Y$. [FREMLIN 82. Compare 32Q*k*.]

(*c*) Let X be a non-empty ccc absolute G_κ space without isolated points, where $\kappa < \mathfrak{m}$. Then X has a non-empty compact perfect subset.

43R **Problems**

(*a*) Are all Martin-complete spaces pseudo-complete? [43O*f*.]

(*b*) If X is a countably tight separable Cech-complete space does it have a countable π-base?

(*c*) [$\mathfrak{m} > \omega_1$] If X is a ccc compact Hausdorff space and $\pi\chi(x, X) \leq \omega$ for every $x \in X$, is X separable? [JUHÁSZ 77.]

(*d*) Let X be a non-empty almost subcompact ccc regular Hausdorff space such that the set of non-empty open subsets of X is downwards-stable in the sense of B1E. Assume that $\mathfrak{m}_{\text{stable}}$, as defined in B1A/B1E, is greater than ω_ω. Does it follow that X is not the union of ω_ω nowhere-dense sets?

Notes and comments

Definition 43A is an attempt to describe a large class of spaces for which Theorem 43E will work. Since any version of 43E must be proved by examining some partially ordered set, it is not wholly unreasonable to allow a partial order into the definition, even though existential definitions of this kind are often difficult to handle. When seeking to prove that a space is Martin-κ-complete, we have to guess at suitable $\preccurlyeq$ and $\mathbf{U}$; in 43C–D I describe a handful of simple techniques for this. Similarly straightforward constructions suffice for 43O*a*(i)–(iii), 43O*b*–*c*, 43O*f*–*g* and 43O*i*. To prove that a space is *not* Martin-complete, the simplest method (when applicable) is to prove that it is not Baire [43O*d*] or fails some other known property [43O*j*(iii)]; in 43P*b* I suggest a direct approach.

Martin-completeness behaves very much like some of the other completeness properties discussed by AARTS & LUTZER 74. In their classification, it is an 'almost completeness' property. Consequently it is inherited by open subspaces [43O*a*(ii)], products [43C*d*], and closures [43O*a*(i)], but not as a rule by closed subsets or continuous images. (Note, however, 43O*a*(ii) and 43O*b*.) It is weaker than 'almost subcompactness' [43D*c*] but conceivably stronger than pseudo-completeness [43O*f*, 43O*j*, 43R*a*]. Pseudo-completeness is of course a 'countable' property in Aarts' and Lutzer's classification. If we replace the downwards-directed set $\mathscr{G}$ of 43A(iv) by a decreasing sequence, we obtain a 'countable' version of Martin-completeness, which in fact coincides with G. Choquet's concept of α-favourable space [43O*d*]; but 43O*j* shows that this is not what is needed for 43E.

Martin-κ-completeness, for uncountable κ, does not belong in the same scheme, as it leads straight to undecidable questions. **Q**, for instance, is Martin-$\mathfrak{c}$-complete, and can be Martin-ω_1-complete even when $\omega_1 < \mathfrak{c}$. Martin-κ-completeness is important to us because it is close to Martin-completeness when $\kappa < \mathfrak{m}$, and is what we need in order to be able to discuss the absolute G_κ spaces [43D*b*, 43J, 43P*k*–*l*].

43D amounts to a list of most of the spaces for which versions of 43E–F have been proved in the past. If you have not seen any of these versions before, I suggest writing out 43E in one of its special cases (Cech-complete spaces, or locally compact Hausdorff spaces) using 43C–D for hints on what to take for $\preccurlyeq$ and **U**.

That 43A is not fully successful in its aim of describing the spaces to which 43E can be applied is shown by 43P*a*–*b*. Another gap is shown by the π-complete spaces of Juhász 77 [43O*e*]. The extra idea is that (in effect) the downwards-directed set $\mathscr{G}$ of 43A(iv) is required to have cardinal less than or equal to κ. This does not interfere significantly with the proof of 43E, but it spoils 43C*a*, and leads to extra difficulties of the type in 43O*j*(ii)–(iii); since I have no examples in which it really seems to matter, I prefer the simpler definition.

The proof of 43E is short because the work has already been done in 31A and we have only to check that the natural identifications are valid. I have already remarked in 43B*c* that the effect of 43A(i)–(ii) is to ensure that $\preccurlyeq$, on the non-empty open sets, has many of the same properties as $\subseteq$. It will therefore sometimes be possible to show that $\preccurlyeq$ satisfies Knaster's condition, or is σ-centered downwards; in which case 43E will proceed with $\mathfrak{m}_K$ or $\mathfrak{p}$ in place of $\mathfrak{m}$. In the applications I give in this section I use the hypothesis 'ccc' throughout, and consequently must use $\mathfrak{m}$; but if we were to replace this with 'Knaster's condition' anywhere in 43F–43M, we would obtain a result for which $\mathfrak{m}_K$ will be sufficient; and in concrete situations it is quite likely that $\mathfrak{m}_K$ or $\mathfrak{p}$ will be enough [25I *f*, 32P*i*, 43O*j*(iii), 43P*b*, 43P*n*].

Among the applications of Martin's axiom, Theorem 43E is quite exceptional in that it demands *directed* sets rather than *centered* sets. From 13A and 14C we see that in the forms of Martin's axiom treated in this book there is no distinction to trouble us; and the same arguments apply if we repeat this work with one of the cardinals of B1B–D in place of $\mathfrak{m}$. Moreover, for the great majority of the applications of 43E, it will be enough if we can find downwards-centered sets $P_n \subseteq P$. It seems, however, that a difficulty arises in the rather far-fetched circumstances of 43R*d*. Other points where the distinction between 'centered' and 'directed' sets is important are in 13D*b*, 41L and 43P*p*(iii).

43F sets out the most frequently used constituents of 43E; historically, all parts of 43F have versions antedating 31A/43E. Observe that, for compact Hausdorff spaces X, we can replace $\mathfrak{m}$ in 43F*b–c* by the cardinal $\mathfrak{l}$ of 41L [43P*o–p*]. On the other hand, it is plain that 43F*a* characterizes $\mathfrak{m}$ [13A(iii)], so that 43E also gives an alternative definition of $\mathfrak{m}$. There are applications of 43E–F in 43P*c–i* and 43Q*a–b*, and an extension of 43F in 43P*j*.

43G, 43J and 43L can be regarded as three different proofs of 43I*c*. I have set out all three in the hope that the proofs may turn out to have further applications. The original approach was that of HAJNAL & JUHÁSZ 71 in 43L–M, but I have given first place to B.E. Sapirovskii's method in 43G, because a greater part of the argument can be carried through before appealing to Martin's axiom. To compare the three approaches, let us put 43I*b* side by side with the corresponding corollaries of 43J and 43L, all subject to $\mathfrak{m} > \omega_1$. The hypotheses on X are that it is ccc and countably tight and either

(α) it is compact and Hausdorff

or (β) it is a continuous image of a ccc absolute G_{ω_1} space

or (γ) all its closed subspaces are Martin-ω_1-complete and it has a dense set of points with π-character less than or equal to ω_1;

and in each case we conclude that X is separable. (In 43I*b* the conclusion reads '$\pi(X) \leq \omega$', but this can be regarded as a consequence of 43H, and nothing to do with Martin's axiom.) Before concluding that the argument for (β) must be more powerful than that for (α), we should note that (α) can be proved from H alone, in the form 43P*o*(*i*); while for (β) we seem to need the apparently stronger principle $\mathfrak{m} > \omega_1$. There are plenty of further questions begged by this comparison; e.g. 43R*b–c*.

43J is remarkable in giving information about an image Y of a space X, subject to conditions on both. For instance, Y could be any analytic space, and need not itself be Martin-$<\mathfrak{m}$-complete; while X need not be countably tight nor separable. As in 43G, it seems that some appeal to compactness is essential [43O*h*]. In 43L, by way of contrast, we work more directly from 43F*b*. The form I have given, with a postulated cardinal κ, seems strained; but since $\mathfrak{m} = \omega_{\omega+1}$ and $\mathfrak{m} = \omega_{\omega_1}$ are both possible, it is not clear how to remove it without discarding some of the strength of the argument. The condition 'every closed subspace is Martin-$<\mathfrak{m}$-complete' is inelegant but natural in the context; the important examples are of course the absolute G_κ spaces (but see 43O*h* again). The next condition, the existence of lots of points of small π-character, can sometimes be shown to hold using the Cech–Pospíšil theorem, as in 43P*k*–*l*. There is a variant of 43L in JUHÁSZ 77.

In view of 43O*i*, 22E/43P*n*, 23M*h*(β) and 25I*f* it seems that it might be worth while to look for other examples of ccc Pixley-Roy topologies; see 31M*i*–*j* and HAJNÁL & JUHASZ 82.

Finally, M.G. Bell's example, mentioned in 24O*m*, shows that in 43I*b*–*c*, 43K and 43M we cannot replace 'countably tight' or 'first-countable' by '$\chi(X) \leq \omega_1$'; and in 43G(ii) and 43L we need $t^+(X)$ or $t(X)^+$, not $t(X)$.

44 Hereditarily ccc spaces

This section is based on partial answers to three old problems. (i) Does a hereditarily ccc regular Hausdorff space have to be separable? Many examples are now known to show that the answer can be 'no' [44J*a*, 44J*c*, B3E*a*], but all depend on special axioms, and it remains open whether one can be constructed with MA alone. In 44A–B I give theorems showing that if $\mathfrak{m} > \omega_1$ then there is no such example in any of a variety of familiar classes of topological spaces. (ii) Does a hereditarily ccc regular Hausdorff space have to be Lindelöf? This time it is known that MA & not-CH do not settle the question [B3E*b*, B3F*b*]; but, as before, there are important special cases in which we can answer 'yes' [44E, 44P*c*–*d*]. In the approach to Theorem 44E I use a lemma on locally countable spaces [44C] which has other interesting consequences [44D]. Combining 44B and 44E with some further results on hereditarily ccc spaces [44F–G] we arrive at a long list of properties of hereditarily ccc compact Hausdorff spaces [44H]. (iii) The third problem is: does a perfectly normal manifold have to be metrizable? Yet again, if $\mathfrak{c} = \omega_1$, the answer is no [RUDIN & ZENOR 76]; but now $\mathfrak{m} > \omega_1$ does give a definite answer [44M]. I approach this by the method of Z. Balogh [44K–L], again relying on 44C.

44A Lemma

[$\mathfrak{m} > \omega_1$] Suppose that Z is a regular topological space and that $Y \subseteq Z$ is a subspace which is hereditarily ccc but not separable. Then there is a non-empty $D \subseteq Y$ such that $\pi\chi(z, \bar{D}) > \omega$ for every $z \in \bar{D}$, the closure being taken in Z.

Proof (*a*) As Y is not separable, we can choose inductively a family $\langle x_\xi \rangle_{\xi < \omega_1}$ in Y such that $x_\xi \notin \overline{\{x_\eta : \eta < \xi\}}$ for each $\xi < \omega_1$. As Z is regular, we can choose for each $\xi < \omega_1$ an open set $G_\xi \subseteq Z$ such that $x_\xi \in G_\xi$ and $\bar{G}_\xi \cap \overline{\{x_\eta : \eta < \xi\}} = \emptyset$.

(*b*) Set $W=\{x_\xi:\xi<\omega_1\}$. Then W is uncountable and $W\cap\bar{C}$ is countable for every countable $C\subseteq W$. Now if $E\subseteq W$ is uncountable there is an uncountable $D\subseteq E$ such that $D\cap H$ is uncountable whenever $H\subseteq Z$ is open and $D\cap H\neq\emptyset$. **P?** Otherwise, we can choose inductively for $\xi<\omega_1$ open set $H_\xi\subseteq Z$ such that $(E\setminus\overline{\bigcup_{\eta<\xi}(E\cap H_\eta)})\cap H_\xi$ is countable and not empty for every ξ; the point being that $E\cap\overline{\bigcup_{\eta<\xi}(E\cap H_\eta)}$ is always countable, so that $E\setminus\overline{\bigcup_{\eta<\xi}(E\cap H_\eta)}$ is always uncountable. Now if we choose $w_\xi\in H_\xi\cap E\setminus\overline{\bigcup_{\eta<\xi}(E\cap H_\eta)}$ for each $\xi<\omega_1$.

$$\overline{\{w_\eta:\eta\neq\xi\}}\subseteq\overline{\bigcup_{\eta<\xi}(E\cap H_\eta)}\cup(X\setminus H_\xi)$$

does not contain w_ξ for any $\xi<\omega_1$, so $\{w_\xi:\xi<\omega_1\}$ is discrete and Y is not hereditarily ccc. **XQ**

(*c*) Set $S=\{(w,\xi):\xi<\omega_1, w\in G_\xi\cap W\}\subseteq W\times\omega_1$. Let P be $[W]^{<\omega}\times[\omega_1]^{<\omega}$ with the S-respecting partial order [31C]. Then P is not upwards-ccc **P?** Otherwise, by 41A*b*, there is an uncountable $C\subseteq\omega_1$ such that $R=\{(\{x_\xi\},\{\xi\}):\xi\in C\}$ is upwards-linked in P. Now by 31C*b*(ii) $(x_\xi,\eta)\notin S$ whenever η and ξ are distinct members of C i.e. $x_\xi\notin G_\eta$ whenever η and ξ are distinct members of C. But now $\{x_\xi:\xi\in C\}$ is an uncountable discrete subset of Y. **XQ**

(*d*) Since the vertical sections of S are of the form

$$\{\xi:x_\zeta\in W\cap G_\xi\}\subseteq\zeta+1$$

for $\zeta<\omega_1$, they are countable. So we can apply 42B(iii) to see that there are disjoint families $\langle K_\xi\rangle_{\xi<\omega_1}$ in $[W]^{<\omega}$ and $\langle L_\xi\rangle_{\xi<\omega_1}$ in $[\omega_1]^{<\omega}$ such that

$$\forall \text{ uncountable } C\subseteq\bigcup_{\xi<\omega_1}K_\xi,\ \{\xi:L_\xi\cap S[C]=\emptyset\} \text{ is countable.}$$

By (*b*), there is an uncountable $D\subseteq\bigcup_{\xi<\omega_1}K_\xi$ such that $D\cap H$ is uncountable whenever H is an open set in Z meeting D. Set $V_\xi=\bigcup\{G_\alpha:\alpha\in L_\xi\}$ for $\xi<\omega_1$. Then we see that, for $C\subseteq W$,

$$C\cap V_\xi=\emptyset\Leftrightarrow C\cap G_\alpha=\emptyset\,\forall\alpha\in L_\xi\Leftrightarrow L_\xi\cap S[C]=\emptyset.$$

It follows that, if $H\subseteq Z$ is open and $D\cap H\neq\emptyset$,

$$\{\xi:D\cap H\cap V_\xi=\emptyset\}=\{\xi:L_\xi\cap S[D\cap H]=\emptyset\}$$

is countable.

(*e*) Now take any $z\in\bar{D}$. **?** Suppose, if possible, that $\pi\chi(z,\bar{D})\leqq\omega$. Then there is a sequence $\langle H_n\rangle_{n\in\mathbf{N}}$ of open sets in Z such that $\bar{D}\cap H_n\neq\emptyset$ for each $n\in\mathbf{N}$ and every neighbourhood of z includes $\bar{D}\cap H_n$ for some n. It follows at once that $z\in\bar{C}$ for some countable $C\subseteq D$. Now $\{\xi:\bar{G}_\xi\cap\bar{C}\neq\emptyset\}$ is countable, because $C \subseteq \{x_\eta : \eta < \zeta\}$ for some countable ζ, so $\{\xi : \bar{V}_\xi \cap \bar{C} \neq \varnothing\}$ is countable, because $\langle L_\xi \rangle_{\xi < \omega_1}$ is disjoint. At the same time, $\{\xi : D \cap V_\xi \cap H_n = \varnothing\}$ is countable for each $n \in \mathbb{N}$. So there is a $\xi < \omega_1$ such that

$$\bar{V}_\xi \cap \bar{C} = \varnothing, \quad \bar{D} \cap \bar{V}_\xi \cap H_n \neq \varnothing \forall n \in \mathbb{N},$$

which is impossible, because $Z \backslash \bar{V}_\xi$ is a neighbourhood of z. **X**

Thus this set D has the property sought.

44B Theorem

[$\mathfrak{m} > \omega_1$] Suppose that X is a hereditarily ccc regular space. If

either (i) X can be embedded in a countably tight Cech-complete space

or (ii) X is itself Cech-complete

or (iii) X is locally countably compact

or (iv) X is first-countable

or (v) X can be embedded in a hereditarily normal compact Hausdorff space,

then X is hereditarily separable.

Proof (*a*) I argue in reverse. Suppose that X is not hereditarily separable; I have to show that (i)–(v) are all false. Let $Y \subseteq X$ be a non-separable subspace. Choose $\langle x_\xi \rangle_{\xi < \omega_1}$ inductively so that $x_\xi \in Y \backslash F_\xi$ for every $\xi < \omega_1$, where $F_\xi = \overline{\{x_\eta : \eta < \xi\}}$. Set $F = \bigcup_{\xi < \omega_1} F_\xi$. As X is hereditarily ccc, F is ccc.

(*b*) Suppose that X is embedded in a Cech-complete space Z. Write $\bar{F}_\xi^Z, \bar{F}^Z$ for the closures of F_ξ, F in Z. As F is ccc, $\bar{F}^Z$ is ccc; moreover, $\bar{F}^Z$ is Cech-complete. By 43F*c*, ω_1 is a caliber of $\bar{F}^Z$. As $x_\xi \in \bar{F}^Z \backslash \bar{F}_\xi^Z$ for every $\xi < \omega_1$, the sets $\bar{F}^Z \backslash \bar{F}_\xi^Z$, which are relatively open in $\bar{F}^Z$, are all non-empty. So there is a $z \in \bar{F}^Z$ such that $\{\xi : z \notin \bar{F}_\xi^Z\}$ is uncountable. As $F_\eta \subseteq F_\xi$ for $\eta < \xi$, $z \notin \bigcup_{\xi < \omega_1} \bar{F}_\xi^Z$. Consequently z is not in the closure (in Z) of any countable subset of F, and Z is not countably tight.

This shows that (i) is false.

(*c*) It follows at once that X cannot itself be a countably tight Cech-complete space; so that in fact X cannot be Cech-complete, because a hereditarily ccc Cech-complete space is countably tight [A4K]. So (ii) is also false.

(*d*) **?** Suppose, if possible, that X is locally countably compact. Then we can choose open sets $G_\xi \subseteq X$ such that $x_\xi \in G_\xi$, $\bar{G}_\xi$ is countably compact, and $\bar{G}_\xi \cap F_\xi = \varnothing$ for each $\xi < \omega_1$. As F is ccc, we can apply 41A*a* to the family of non-empty relatively open subsets of F to see that there is an uncountable $A \subseteq \omega_1$ such that $\{G_\xi \cap F : \xi \in A\}$ has the finite intersection property. We see that $\bar{G}_\xi \cap F$ is always countably compact, since any sequence in $\bar{G}_\xi \cap F$ will lie within $\bar{G}_\xi \cap F_\zeta$ for some $\zeta < \omega_1$, and will have a cluster point in $\bar{G}_\xi \cap F_\zeta \subseteq \bar{G}_\xi \cap F$. Accordingly $F \cap \bigcap_{\xi \in B} \bar{G}_\xi$ will be non-empty for any countable $B \subseteq A$. Choose inductively, for $\xi < \omega_1$, points $y_\xi \in X$ and ordinals $\alpha(\xi) \in A$ as follows. Given $\langle \alpha(\eta) \rangle_{\eta < \xi}$, choose $y_\xi \in F \cap \bigcap_{\eta < \xi} \bar{G}_{\alpha(\eta)}$; now choose $\alpha(\xi) \in A$ such that $y_\xi \in F_{\alpha(\xi)}$ and $\alpha(\xi) > \alpha(\eta)$ for every $\eta < \xi$.

In this case, $y_{\xi+1} \in \bar{G}_{\alpha(\xi)} \cap F_{\alpha(\xi+1)}$ for every $\xi < \omega_1$. So, for $\xi < \omega_1$,

$$\overline{\{y_{\eta+1} : \eta \neq \xi\}} \subseteq F_{\alpha(\xi)} \cup \bar{G}_{\alpha(\xi+1)}$$

does not contain $y_{\xi+1}$. Thus $\{y_{\xi+1} : \xi < \omega_1\}$ is an uncountable discrete set, and X is not hereditarily ccc. **X** So (iii) is false.

(*e*) Applying 44A with $Z = X$, we have a non-empty $D \subseteq Y$ such that $\omega < \pi\chi(x, \bar{D}) \leq \chi(x, X)$ for every $x \in \bar{D}$. So X is not first-countable.

(*f*) Suppose that Z is a compact Hausdorff space in which X is embedded. By 44A, we have a non-empty set $D \subseteq Y$ such that $\pi\chi(z, \bar{D}^Z) > \omega$ for every $z \in \bar{D}^Z$, the closure of D in Z. So $\bar{D}^Z$ is not hereditarily normal [A4L*g*], and Z is not hereditarily normal.

44C **Lemma**

Let X be a locally countable topological space with $\#(X) < \mathfrak{m}$. Suppose that X can be embedded in a countably tight compact Hausdorff space. Then X can be expressed as a countable union of discrete subsets which are closed in X.

Proof (*a*) Take X to be a subset of a countably tight compact Hausdorff space Z. Set

$$\mathscr{W} = \{W : W \subseteq Z \text{ is open}, W \cap X \text{ is countable}\},$$
$$\mathscr{V} = \{V : V \subseteq Z \text{ is open}, \exists W \in \mathscr{W}, \bar{V} \subseteq W\}.$$

Then $\mathscr{W}$ and $\mathscr{V}$ are families of open subsets of Z closed under finite unions, and $X \subseteq \bigcup \mathscr{V}$. Choose $\mathscr{U} \subseteq \mathscr{V}$ such that $X \subseteq \bigcup \mathscr{U}$ and $\#(\mathscr{U}) < \mathfrak{m}$. Set

$$S = \{(U, x) : U \in \mathscr{U}, x \in X \cap U\} \subseteq \mathscr{U} \times X.$$

Then the vertical sections of S are countable. Let P be $[\mathscr{U}]^{<\omega} \times [X]^{<\omega}$ with the S-respecting partial order [31C].

(*b*) P is upwards-ccc. **P?** If not, then by 42B(ii) there are disjoint families $\langle \mathscr{K}_\xi \rangle_{\xi < \omega_1}$ and $\langle L_\xi \rangle_{\xi < \omega_1}$ in $[\mathscr{U}]^{<\omega}$ and $[X]^{<\omega}$ respectively such that

> for every uncountable $C \subseteq \bigcup_{\xi<\omega_1} L_\xi$, $\exists \alpha < \omega_1$ such that $\{C \cap S[\mathscr{K}_\xi] : \alpha \leq \xi < \omega_1\}$ has the finite intersection property.

Set $V_\xi = \bigcup \mathscr{K}_\xi \in \mathscr{V}$ for $\xi < \omega_1$, so that $S[\mathscr{K}_\xi] = V_\xi \cap X$, and

$$F_\alpha = \bigcap_{\xi>\alpha} \bar{V}_\xi, \ F = \bigcup_{\alpha<\omega_1} F_\alpha.$$

As $\langle F_\alpha \rangle_{\alpha<\omega_1}$ is an increasing family of closed subsets in Z and Z is countably tight, F is closed in Z, therefore compact. As $F_\alpha \subseteq \bar{V}_\alpha \subseteq \bigcup \mathscr{W}$ for each $\alpha < \omega_1$, $F \subseteq \bigcup \mathscr{W}$ and there is a $W \in \mathscr{W}$ such that $F \subseteq W$. Now $W \cap X$ is countable, so $C = \bigcup_{\xi<\omega_1} L_\xi \backslash W$ is uncountable, and there is an $\alpha < \omega_1$ such that $\{C \cap V_\xi : \alpha \leq \xi < \omega_1\}$ has the finite intersection property. It follows that there is a $z \in \bigcap_{\xi \geq \alpha} \overline{C \cap V_\xi}$. But now $z \in F_\alpha \backslash W$ and $F \nsubseteq W$. **XQ**

(*c*) By 42A*a*, X is expressible as $\bigcup_{n\in\mathbb{N}} Y_n$ where $Y_n \cap U = Y_n \cap S[\{U\}]$ is finite for every $U \in \mathscr{U}$ and $n \in \mathbb{N}$. As $\mathscr{U}$ is an open cover of X, this is enough to show that each Y_n is discrete and relatively closed in X, as required.

44D Corollary

Let X be a locally compact Hausdorff space with $\#(X) < \mathfrak{m}$. Then the following are equivalent:

(i) X is locally countable and the one-point compactification of X is countably tight.
(ii) Open sets in X are F_σ.
(iii) X is expressible as a countable union of closed discrete sets.
(iv) X is a Moore space.

Proof (*a*) (i)$\Rightarrow$(iii) by 44C.

(*b*) (iii)$\Rightarrow$(ii) If $X = \bigcup_{n\in\mathbb{N}} X_n$ where each X_n is closed and discrete, and $A \subseteq X$, then $A = \bigcup_{n\in\mathbb{N}} (A \cap X_n)$, which is F_σ.

(*c*) (ii)$\Rightarrow$(i) If X is countable, so is its one-point compactification, which will then be countably tight. If X is uncountable then $\omega_1 \leq \#(X) < \mathfrak{m} \leq \mathfrak{p}$; so by 24L the one-point compactification of X is countably tight. Also $\chi(x, X) = \psi(x, X) \leq \omega$ for every $x \in X$, so X is first-countable. By A4H*c*, since $\#(X) < \mathfrak{m} \leq \mathfrak{c}$, compact sets in X are countable. So X is locally countable.

(*d*) (iii)$\Rightarrow$(iv) Suppose X is expressible as a countable union of closed discrete sets. By (*b*)–(*c*) above, X is first-countable. So by A4N*f* it is a Moore space.

(*e*) (**iv**)⇒(**ii**) by A4N*b*.

44E **Theorem**

$[\mathfrak{m} > \omega_1]$ Let X be a countably tight compact Hausdorff space and $Y \subseteq X$ a hereditarily ccc subspace. Then Y is hereditarily Lindelöf and hereditarily separable.

Proof By 44B(i), Y is hereditarily separable. ? Suppose, if possible, that Y is not hereditarily Lindelöf. Let Z be a non-Lindelöf subspace of Y. Let $\mathscr{G}$ be a cover of Z by open sets in X which has no countable subcover. Then we can choose $\langle z_\xi \rangle_{\xi < \omega_1}$ and $\langle G_\xi \rangle_{\xi < \omega_1}$ inductively so that

$$z_\xi \in Z \setminus \bigcup_{\eta < \xi} G_\eta, \quad z_\xi \in G_\xi \in \mathscr{G} \; \forall \xi < \omega_1.$$

Consider $W = \{z_\xi : \xi < \omega_1\}$. For each $\xi < \omega_1$, $z_\xi \in G_\xi$ and $G_\xi \cap W$ is countable. By 44C, W is expressible as $\bigcup_{n \in \mathbf{N}} W_n$ where each W_n is discrete. But now some W_n must be uncountable, so Y is not hereditarily ccc. **X**

44F **Proposition**

Let X be a hereditarily ccc K-analytic space and $\mathscr{E}$ a cover of X by closed sets with $\#(\mathscr{E}) < \mathfrak{m}$. Then there is a countable $\mathscr{E}_0 \subseteq \mathscr{E}$ covering X.

Proof If $\mathfrak{m} = \omega_1$ this is trivial; so I suppose that $\omega_1 < \mathfrak{m} \le \mathfrak{p}$.

(*a*) X is hereditarily Lindelöf. **P?** Otherwise, as in 44E, there is a family $\langle x_\xi \rangle_{\xi < \omega_1}$ of points in X and a family $\langle G_\xi \rangle_{\xi < \omega_1}$ of open sets in X such that $x_\xi \in G_\xi \setminus \bigcup_{\eta < \xi} G_\eta$ for every $\xi < \omega_1$. By 23H*b*, there is a compact $K \subseteq X$ such that $\{\xi : x_\xi \in K\}$ is uncountable. As K is hereditarily ccc it is countably tight [A4K]. Also

$$K \cap \bigcup_{\xi < \omega_1} G_\xi \neq K \cap \bigcup_{\xi < \zeta} G_\xi$$

for every $\zeta < \omega_1$, so K is not hereditarily Lindelöf. But this contradicts 44E. **XQ**

(*b*) Now let $R \subseteq \mathbf{N}^{\mathbf{N}} \times X$ be an usco-compact relation such that $\pi_2[R] = X$. By A5C*c*, R is Cech-complete. Set $\mathscr{F} = \{\pi_2^{-1}[E] : E \in \mathscr{E}\}$, so that $\mathscr{F}$ is a cover of R by closed sets and $\#(\mathscr{F}) < \mathfrak{m}$. Let $\mathscr{G}$ be

$$\{G : G \subseteq \mathbf{N}^{\mathbf{N}} \times X \text{ is open, } \exists \text{ countable } \mathscr{F}_0 \subseteq \mathscr{F} \text{ such that } R \cap G \subseteq \bigcup \mathscr{F}_0\}.$$

As X is hereditarily Lindelöf, so is $\mathbf{N}^{\mathbf{N}} \times X$ [A4D*e*], and there is a countable $\mathscr{G}_0 \subseteq \mathscr{G}$ such that $G_0 = \bigcup \mathscr{G}_0 = \bigcup \mathscr{G}$; now $G_0 \in \mathscr{G}$.

? If $G_0 \not\supseteq R$, then $R \setminus G_0$ is a ccc Cech-complete space, and $\{F \cap R \setminus G_0 : F \in \mathscr{F}\}$

is a cover of $R\backslash G_0$ by fewer than $\mathfrak{m}$ closed sets. By 43F*a*, $R\backslash G_0$ is not the union of fewer than $\mathfrak{m}$ nowhere-dense sets, so there is an $F\in\mathscr{F}$ such that $F\cap R\backslash G_0$ has non-empty relative interior in $R\backslash G_0$; i.e. there is an open set $H\subseteq \mathbf{N}^{\mathbf{N}}\times X$ such that $\varnothing\neq H\cap R\backslash G_0\subseteq F\backslash G_0$. But now $R\cap H\subseteq G_0\cup F$ so $H\in\mathscr{G}$ and $H\subseteq G_0$. **X**

Thus R is covered by a countable subset of $\mathscr{F}$ and X is covered by the corresponding countable subset of $\mathscr{E}$, as required.

44G Theorem

Let X be a hereditarily ccc compact Hausdorff space. If $\mathscr{E}$ is a partition of X into G_δ sets and $\#(\mathscr{E})<\mathfrak{m}$, then $\mathscr{E}$ is countable.

Proof If $\mathfrak{m}=\omega_1$ this is trivial; so I take $\mathfrak{m}>\omega_1$. In this case we see that X is countably tight [A4K], hereditarily Lindelöf [44E] and perfectly normal [A4L*d*]. Write $\kappa=\max(\omega, \#(\mathscr{E}))$.

(*a*) Express each $E\in\mathscr{E}$ as $\bigcap_{k\in\mathbf{N}} G_E^k$ where each G_E^k is an open set. Let $\mathscr{I}$ be the σ-ideal of subsets of X generated by $\mathscr{E}$; I am trying to prove that $X\in\mathscr{I}$. Let $\mathfrak{A}$ be the smallest collection of subsets of X such that

$G_E^k\in\mathfrak{A}\ \forall k\in\mathbf{N},\ E\in\mathscr{E}$;
$B\cap C\in\mathfrak{A}$, $X\backslash B\in\mathfrak{A}$ whenever $B, C\in\mathfrak{A}$;
if $B\in\mathfrak{A}$, then $W_B=\bigcup\{G: G \text{ open}, B\cap G\in\mathscr{I}\}\in\mathfrak{A}$.

Then $\mathfrak{A}$ is an algebra of sets and $\#(\mathfrak{A})\leq\kappa$. Moreover, every set in $\mathfrak{A}$ is both F_σ and G_δ (because X is perfectly normal, so open sets in X are F_σ). Set $\mathfrak{A}_0=\mathfrak{A}\cap\mathscr{I}$ and $Y=X\backslash\bigcup\mathfrak{A}_0$. As $\#(\mathfrak{A}_0)\leq\kappa$, Y is expressible as the intersection of κ or fewer open sets in X, and is in itself an absolute G_κ space.

(*b*) We need a couple of easy observations. (i) If $B\in\mathfrak{A}$, then $B\cap Y$ is an absolute G_κ space, and is ccc, so is a Baire space in its induced topology [43F*a*]. (ii) If $A\subseteq X$ is K-analytic and $A\cap Y\in\mathscr{I}$ then $A\in\mathscr{I}$. **P** Let $\mathscr{E}_0\subseteq\mathscr{E}$ be a countable set such that $A\cap Y\subseteq\bigcup\mathscr{E}_0$. Then $B=A\backslash\bigcup\mathscr{E}_0$ is the intersection of A with a $K_{\sigma\delta}$ set, so is K-analytic [A5D*b*], and $B\subseteq\bigcup\mathfrak{A}_0$. As each member of $\mathfrak{A}_0$ is F_σ, 44F above shows that there is a countable $\mathfrak{A}_0'\subseteq\mathfrak{A}_0$ covering B. Now $\bigcup\mathfrak{A}_0'\in\mathscr{I}$ so $B\in\mathscr{I}$ and $A\in\mathscr{I}$. **Q** (iii) If $B\in\mathfrak{A}$ then $\overline{B\cap Y}\in\mathfrak{A}$. **P** We have $W_B\in\mathfrak{A}$. Because X is hereditarily Lindelöf, $B\cap W_B\in\mathscr{I}$, so $B\cap W_B\in\mathfrak{A}_0$ and $B\cap Y\cap W_B=\varnothing$ and $\overline{B\cap Y}\cap W_B=\varnothing$. At the same time, $W=X\backslash\overline{B\cap Y}$ is F_σ, therefore K-analytic, and $W\cap B\cap Y=\varnothing$, so $W\cap B\in\mathscr{I}$ by (ii), and $W\subseteq W_B$. This shows that $W=W_B$ and $\overline{B\cap Y}=X\backslash W_B\in\mathfrak{A}$. **Q**

(*c*) Now construct subalgebras $\mathfrak{B}_n$ of $\mathfrak{A}$ inductively, for $n\in\mathbf{N}$, as follows. $\mathfrak{B}_0 = \{\emptyset, X\}$. Given $\mathfrak{B}_n$, let $\mathfrak{B}_{n+1}$ be the subalgebra of $\mathfrak{A}$ generated by

$$\mathfrak{B}_n \cup \{G_E^k : k \in \mathbf{N},\ E \in \mathscr{E}, \exists B \in \mathfrak{B}_n \text{ and open } H \subseteq X \text{ such that } \emptyset \neq H \cap B \cap Y \subseteq \overline{E \cap B \cap Y}\}.$$

Then each $\mathfrak{B}_n$ is countable. **P** Induce on n. The point is that if $B \in \mathfrak{B}_n$ then $B \cap Y$ is Baire, as observed in (*b*)(i) above. So the disjoint relative G_δ sets $E \cap B \cap Y$, as E runs through $\mathscr{E}$, must be dense in disjoint relatively open sets of $B \cap Y$. But also $B \cap Y$ is ccc, so that

$$\{E : E \in \mathscr{E},\ E \cap B \cap Y \text{ is somewhere dense in } B \cap Y\}$$

must be countable. Thus if $\mathfrak{B}_n$ is countable, $\mathfrak{B}_{n+1}$ will be the algebra generated by a countable set and will also be countable. **Q**

Set $\mathfrak{B} = \bigcup_{n \in \mathbf{N}} \mathfrak{B}_n$; then $\mathfrak{B}$ is a countable subalgebra of $\mathfrak{A}$ with the property that

if $B \in \mathfrak{B}$, $E \in \mathscr{E}$ and $E \cap B \cap Y$ is somewhere dense in $B \cap Y$, then $G_E^k \in \mathfrak{B}$ for every $k \in \mathbf{N}$.

(*d*) Let P be

$$\{\overline{B \cap Y \cap G} : B \in \mathfrak{B},\ G \in \mathfrak{A} \text{ is open},\ B \cap Y \cap G \neq \emptyset\}.$$

Then P is downwards-ccc. **P** Let $R \subseteq P$ be uncountable. As $\mathfrak{B}$ is countable, there is a $B \in \mathfrak{B}$ such that

$$R_B = \{\overline{B \cap Y \cap G} : G \in \mathfrak{A} \text{ is open},\ \overline{B \cap Y \cap G} \in R\}$$

is uncountable. Now $B \cap Y$ is ccc, so there must be open $G_1, G_2 \in \mathfrak{A}$ such that

$\overline{B \cap Y \cap G_1}$ and $\overline{B \cap Y \cap G_2}$ are distinct members of R_B,
$B \cap Y \cap G_1 \cap G_2 \neq \emptyset$.

In this case $\overline{B \cap Y \cap G_1 \cap G_2}$ is a common lower bound for $\overline{B \cap Y \cap G_1}$ and $\overline{B \cap Y \cap G_2}$ in P. **Q**

(*e*) For $E \in \mathscr{E}$ set

$$Q_E = \{F : F \in P,\ F \cap E = \emptyset\}.$$

Then Q_E is coinitial with P. **P** Let $F \in P$. Express F as $\overline{B \cap Y \cap G}$ where $B \in \mathfrak{B}$ and $G \in \mathfrak{A}$ is open. Consider two cases:

(i) $E \cap B \cap Y$ is somewhere dense in $B \cap Y$. In this case $G_E^k \in \mathfrak{B}$ for every $k \in \mathbf{N}$. Now $B \cap Y \cap G \neq \emptyset$, so $B \cap G \notin \mathfrak{A}_0$ and $B \cap G \notin \mathscr{I}$ and $B \cap Y \cap G \notin \mathscr{I}$, by (*b*)(ii) above. So $B \cap Y \cap G \nsubseteq E$ and there is a $k \in \mathbf{N}$ such that $B \cap$

$Y\cap G\setminus G_E^k \neq \emptyset$. We have $B\setminus G_E^k \in \mathfrak{B}$ so that

$$F_1 = \overline{(B\setminus G_E^k)\cap Y\cap G} \in P;$$

now $F \supseteq F_1 \in Q_E$.

(ii) $E\cap B\cap Y$ is nowhere dense in $B\cap Y$; in particular, $E\cap B\cap Y\cap G$ is not dense in $B\cap Y\cap G$. In this case, because $B\cap Y\cap G$ is Baire, $G_E^k\cap B\cap Y\cap G$ cannot always be dense in $B\cap Y\cap G$; let $k\in\mathbf{N}$ be such that $B\cap Y\cap G \not\subseteq \overline{G_E^k\cap B\cap Y\cap G}$. $G_E^k\cap B\cap G\in\mathfrak{A}$, so by (b)(iii) above $\overline{G_E^k\cap B\cap Y\cap G}\in\mathfrak{A}$, so $H = G\setminus\overline{G_E^k\cap B\cap Y\cap G}\in\mathfrak{A}$, and $B\cap Y\cap H\neq\emptyset$. Now $F_1=\overline{B\cap Y\cap H}\in P$; $B\cap Y\cap H\cap G_E^k=\emptyset$, so $F_1\in Q_E$; and $F_1\subseteq F$. **Q**

(*f*) **?** If $X\notin\mathcal{I}$, then by (*b*)(ii) above $X\cap Y\notin\mathcal{I}$, so $Y\neq\emptyset$, $\bar{Y}\in P$ and $P\neq\emptyset$. So, because $\kappa<\mathfrak{m}$, there is a downwards-directed $R\subseteq P$ meeting Q_E for every $E\in\mathcal{E}$. But now R is a downwards-directed family of closed sets in the compact space X, so there is an $x\in\bigcap R$. As R meets every Q_E, $x\notin\bigcup\mathcal{E}$, and $\mathcal{E}$ does not cover X. **X**

Thus $X\in\mathcal{I}$, as required.

44H Theorem

[$\mathfrak{m}>\omega_1$] Let X be a compact Hausdorff space. Then the following are equivalent:

(i) X is hereditarily ccc;

(ii) X is hereditarily Lindelöf;

(iii) X is hereditarily separable;

(iv) no Borel set in X has cardinal ω_1;

(v) every uncountable Borel set in X has cardinal $\mathfrak{c}$;

(vi) every uncountable Borel set in X has a non-empty compact perfect subset;

(vii) every uncountable Borel set in X has a non-Borel subset;

(viii) no closed set in X can be partitioned into ω_1 non-empty closed subsets;

(ix) no Borel set in X can be partitioned into ω_1 non-empty relative G_δ sets;

(x) if $\mathcal{E}$ is a cover of X by closed sets with $\#(\mathcal{E})\leq\omega_1$, then $\mathcal{E}$ has a countable subcover;

(xi) every closed set in X is the support of a Radon measure;

(xii) every closed set in X is expressible as the union of a perfect set and a countable set;

(xiii) if $f:X\to\mathbf{R}$ is a Borel measurable function, then $f[X]$ is an analytic set.

Proof Without special axioms, the following are fairly easy to establish:

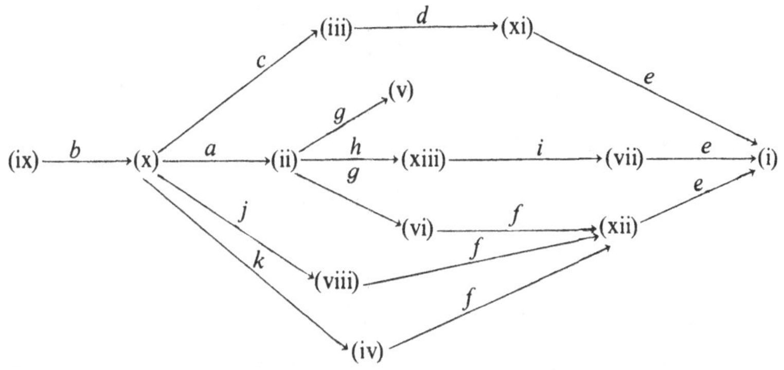

Using $\mathfrak{m} > \omega_1$, I shall show that (i) $\overset{l}{\rightarrow}$ (ii) $\overset{m}{\rightarrow}$ (ix) and that (v) $\overset{n}{\rightarrow}$ (iv). The letters above the arrows give the sections below in which each implication is proved.

(*a*) **not (ii)⇒not(x) & not (ix)** Suppose that (ii) fails; that X is not hereditarily Lindelöf. As in the proof of 44E, we can choose $x_\xi \in X$ and open sets $G_\xi \subseteq X$, for $\xi < \omega_1$, in such a way that $x_\xi \in G_\xi \setminus \bigcup_{\eta<\xi} G_\eta$ for every $\xi < \omega_1$. Now choose continuous functions $f_\xi : X \to [0,1]$ such that

$$f_\xi(x_\xi) = 1,\ f_\xi(x) = 0\ \forall x \in X \setminus G_\xi$$

for each $\xi < \omega_1$.

(α) Setting $F_\xi^n = \{x : f_\xi(x) \geq 2^{-n}\}$, $F = \{x : f_\xi(x) = 0\ \forall \xi < \omega_1\}$, we see that $\mathscr{E} = \{F\} \cup \{F_\xi^n : \xi < \omega_1,\ n \in \mathbf{N}\}$ is a cover of X by closed sets with no countable subcover; so (x) fails.

(β) Moreover, the open set $X \setminus F$ can be partitioned into the G_δ sets

$$F_\xi^n \setminus (\bigcup_{i\in\mathbf{N},\eta<\xi} F_\eta^i \cup \bigcup_{i<n} F_\xi^i)$$

of which uncountably many are non-empty; so that (ix) fails.

So if any of (ii), (ix) or (x) is true then X is perfectly normal and every closed subset of X is G_δ [A4L*d*].

(*b*) **(ix)⇒(x)** Assume (ix). We have just seen that every closed set in X is G_δ. So if $\langle F_\xi \rangle_{\xi<\omega_1}$ enumerates a cover of X by closed sets, $\langle F_\xi \setminus \bigcup_{\eta<\xi} F_\eta \rangle_{\xi<\omega_1}$ is a partition of X into G_δ sets; by (ix), only countably many of these can be non-empty; so that $X = \bigcup_{\eta<\xi} F_\eta$ for some $\xi < \omega_1$. This proves (x).

(*c*) **(x)⇒(iii)** Assume (x). We know by (*a*) above that X is hereditarily Lindelöf. **?** Suppose, if possible, that X is not hereditarily separable. As in 44B prf *a*, there is a family $\langle x_\xi \rangle_{\xi<\omega_1}$ in X such that $x_\xi \notin \overline{\{x_\eta : \eta < \xi\}} = F_\xi$ for each $\xi < \omega_1$. Because X is hereditarily Lindelöf, it is first-countable, and $F = \bigcup_{\xi<\omega_1} F_\xi$ is closed; also F is G_δ. Consequently $X \setminus F$ can be expressed as the union of a sequence $\langle H_n \rangle_{n\in\mathbf{N}}$ of closed sets and

$$\{H_n : n\in\mathbf{N}\} \cup \{F_\xi : \xi < \omega_1\}$$

is a cover of X by closed sets which has no countable subcover. **X** So (iii) is true.

(*d*) **(iii)⇒(xi)** Trivial; any separable closed set will be the support of a Radon measure which is the sum of a sequence of point masses.

(*e*) **not(i)⇒not(xi) & not(xii) & not(iv) & not(vii) & not(viii)** Suppose that X is not hereditarily ccc; then it has a discrete subset A of cardinal ω_1 [A4E*b*]. Set $F = \bar{A}$. Then F is not ccc, so is not the support of a Radon measure, and (xi) fails. At the same time, no perfect subset of F meets A, so F cannot be the union of a perfect set and a countable set, and (xii) fails. A is relatively open in F so is a Borel set in X, so (iv) fails; but every subset of A is also Borel, by the same argument, so (vii) fails. Finally, F can be partitioned into $\{F\setminus A\} \cup \{\{x\} : x\in A\}$, so (viii) fails.

(*f*) **not(xii)⇒not(vi) & not(viii) & not(iv)** Suppose that $F \subseteq X$ is a closed set not expressible as the union of a perfect set and a countable set. Consider its derived sets, given by

$$F_0 = F,\ F_{\xi+1} = F_\xi \setminus \bigcup \{G : G \text{ open, } G \cap F_\xi \text{ a singleton}\},$$
$$F_\xi = \bigcap_{\eta<\xi} F_\eta \text{ for non-zero limit ordinals } \xi.$$

? If $F\setminus F_{\omega_1}$ is countable, then F_{ω_1} must be equal to F_ξ for some $\xi < \omega_1$; but in this case $F_{\xi+1} = F_\xi$, so F_ξ is perfect, and $F = F_\xi \cup (F\setminus F_\xi)$ is the union of a perfect set and a countable set. **X** Thus $F\setminus F_{\omega_1}$ is uncountable.

(α) If $K \subseteq F$ is perfect, then $K \subseteq F_\xi$ for every ordinal ξ; so the Borel set $F\setminus F_{\omega_1}$ has no non-empty perfect subset, and (vi) is false.

(β) Suppose that $F_\xi \setminus F_{\xi+1}$ is uncountable for some $\xi < \omega_1$. Then $F_\xi \setminus F_{\xi+1}$ is an uncountable discrete set, so (i) fails; by (e) above, (viii) and (iv) both fail.

(γ) Suppose that $F_\xi \setminus F_{\xi+1}$ is countable for every $\xi < \omega_1$. Then $F\setminus F_{\omega_1}$ has cardinal precisely ω_1, so (iv) fails. Also F has a partition

$$\{F_{\omega_1}\} \cup \{\{x\} : x\in F\setminus F_{\omega_1}\}$$

into precisely ω_1 closed sets, so (viii) fails too.

(*g*) **(ii)⇒(vi) & (v)** Assume (ii). Let $E \subseteq X$ be an uncountable Borel set. Then E has an uncountable compact subset K. **P** As X is perfectly normal, every open set in X is a cozero set [A4L*c*]. Now the collection

$$\{g^{-1}[H]: H \subseteq [0,1]^{\mathbb{N}} \text{ is Borel}, \quad g: X \to [0,1]^{\mathbb{N}} \text{ is continuous}\}$$

is a σ-algebra of subsets of X containing every open set, so contains E; let $g: X \to [0,1]^{\mathbb{N}}$ be a continuous function and $H \subseteq [0,1]^{\mathbb{N}}$ a Borel set such that $E = g^{-1}[H]$. In this case $g[E] = H \cap g[X]$ is Borel. If $g[E]$ is countable, there is a $z \in g[E]$ such that $K = g^{-1}[\{z\}] \subseteq E$ is uncountable. If $g[E]$ is uncountable, there is an uncountable compact $F \subseteq g[E]$ [A5F*f*], so that $K = g^{-1}[F]$ is an uncountable compact subset of E. **Q**

As X is first-countable, K must have a non-empty perfect subset L say [A4H*c*], which is now a non-empty compact perfect subset of E. As E is arbitrary, (vi) is true. At the same time, we see from A4H that X and L, being non-scattered first-countable compact Hausdorff spaces, both have cardinal $\mathfrak{c}$; so that $\#(E) = \mathfrak{c}$, and (v) is true.

(*h*) **(ii)$\Rightarrow$(xiii)** Assume (ii), and let $f: X \to \mathbf{R}$ be a Borel measurable function. Then the graph Γ of f is a Borel set in $X \times \bar{\mathbf{R}}$, where $\bar{\mathbf{R}}$ is the two-point compactification of $\mathbf{R}$ [A5A*f*]. As $X \times \bar{\mathbf{R}}$ is hereditarily Lindelöf [A4D*e*], Γ is K-analytic [A5D*e*]; so $f[X] = \pi_2[\Gamma]$ is K-analytic [A5D*b*], therefore analytic [A5F*a*].

(*i*) **(xiii)$\Rightarrow$(vii)** Assume (xiii), and let $E \subseteq X$ be an uncountable Borel set. Let $A \subseteq [0,1]$ be of cardinal ω_1 and not analytic [A5F*h*]. Let $f: X \to \mathbf{R}$ be any function such that $f[E] = A$ and f is constant on $X \setminus E$. Then $f[X]$ is not analytic so f is not Borel measurable and there is a non-Borel subset of E.

(*j*) **(x)$\Rightarrow$(viii)** The arguments of (*a*) and (*c*) above make it plain that if (x) is true of X, it is true of every closed subset of X. So of course no closed subset of X can be partitioned into ω_1 non-empty closed subsets.

(*k*) **(x)$\Rightarrow$(iv)** Assume (x). We already know that (x)$\Rightarrow$(ii)$\Rightarrow$(v) & (vi). **?** If $E \subseteq X$ is a Borel set of cardinal ω_1, then by (v) we have $\omega_1 = \mathfrak{c}$, and by (vi) X has a non-empty perfect subset. So there is a continuous surjection $f: X \to [0,1]$ [A4H*a*]. But now $\{f^{-1}[\{s\}]: s \in [0,1]\}$ is a cover of X by ω_1 closed sets which has no countable subcover. **X** This proves (iv).

(*l*) $[\mathfrak{m} > \omega_1]$ **(i)$\Rightarrow$(ii)** If X is hereditarily ccc, it is countably tight [A4K]. So by 44E it is hereditarily Lindelöf.

(*m*) $[\mathfrak{m} > \omega_1]$ **(ii)$\Rightarrow$(ix)** Assume (ii). Let $H \subseteq X$ be a Borel set and $\mathscr{E}$ a partition of H into non-empty relative G_δ sets with $\#(\mathscr{E}) \leq \omega_1$. Then H is K-analytic [A5D*e*]. **?** If $\#(\mathscr{E}) = \omega_1$, then for each $E \in \mathscr{E}$ choose $x_E \in E$. By

23H*b*, there is a compact $K \subseteq H$ such that $\{E : x_E \in K\}$ is uncountable. Now K is a hereditarily Lindelöf compact Hausdorff space, and $\{E \cap K : E \in \mathscr{E}\}$ is partition of K into just ω_1 relative G_δ sets; which by 44G is impossible. **X** So $\mathscr{E}$ is countable. As H and $\mathscr{E}$ are arbitrary, (ix) is true.

(*n*) $[\mathfrak{m} > \omega_1]$ (**v**)$\Rightarrow$(**iv**) Trivial, since $\mathfrak{c} > \omega_1$.

44I **Corollary**

$[\mathfrak{m} > \omega_1]$ Let X be a compact Hausdorff space and suppose that the σ-algebra $\mathscr{B}(X)$ of Borel subsets of X is isomorphic, as Boolean algebra, to $\mathscr{B}(Y)$ for some compact metric space Y. Then X is metrizable.

Proof Because $\omega_1 < \mathfrak{c}$, Y can have no Borel set of cardinal ω_1. So X has no Borel set of cardinal ω_1; by 44H, it follows that X is hereditarily Lindelöf. Because the points of Y can be separated by a sequence of Borel sets, so can the points of X. So there is a sequence $\langle G_n \rangle_{n \in \mathbf{N}}$ of open sets in X such that whenever x, y are distinct points of X there is a G_n containing one but not the other. As X is hereditarily Lindelöf, each G_n is a cozero set, so there is a sequence of continuous real-valued functions on X separating the points of X. These induce a metrizable topology on X which (because X is compact) must be the original topology of X.

44J **Examples**

(*a*) If Souslin's hypothesis is false, there is a **Souslin line**, i.e. a non-empty Dedekind complete totally ordered set X with no gaps and no greatest or least element, which is ccc under its order topology, but in which no non-trivial interval is separable. [See Kunen 80, Theorem II.5.13 and Exercise II.30.] Now any non-trivial closed interval of X satisfies (i)(ii)(v)(vi)(vii)(xii)(xiii) of 44H and not (iii)(ix)(x)(xi). It will satisfy (iv) iff the continuum hypothesis is false. If the continuum hypothesis is true, it will not satisfy (viii).

(*b*) If the continuum hypothesis is true, there is a locally compact, locally countable, hereditarily separable topology on [0, 1], finer than the usual topology, with the same Borel sets; this can be embedded as an open set in a first-countable hereditarily separable compact Hausdorff space X which has the Borel structure of $[0, 1] \times \{0, 1\}$. [Juhász, Kunen & Rudin 76, §1.] Now X satisfies (i)(iii)(v)(vii)(xi)(xiii) of 44H and not (ii)(iv)(vi)(viii)(ix)(x)(xii). X is also copolish, perfectly normal, hereditarily collectionwise Hausdorff but not hereditarily metalindelöf; see 44K.

(*c*) If the continuum hypothesis is true, there is a hereditarily

Lindelöf compact Hausdorff space X carrying a Radon probability measure μ such that, for $Y \subseteq X$, Y is negligible iff Y is nowhere dense iff Y is separable iff Y is metrizable [KUNEN 81]. [See 22N*j*, 43P*g*.] X satisfies (i)(ii)(v)(vi)(vii)(xi)(xii)(xiii) of 44H and not (iii)(iv)(viii)(ix)(x). The construction can be adapted so that $L^1(X)$ is either separable or non-separable [see 32N].

44K **Theorem**

[$\mathfrak{m} > \omega_1$] Let X be a locally compact collectionwise Hausdorff space in which open sets are F_σ. Then X is expressible as a disjoint union of open K_σ sets; in particular, it is paracompact.

Proof (*a*) Note first that compact sets in X are perfectly normal, therefore hereditarily Lindelöf and (by 44B) separable; and that by 24L the one-point compactification of X is countably tight.

(*b*) To begin with, let us consider the case in which X can be covered by a family $\langle G_\xi \rangle_{\xi < \omega_1}$ of relatively compact open sets. Set

$$C = \{\xi : \xi < \omega_1, \ \bigcup_{\eta < \xi} G_\eta \text{ is not closed}\}.$$

Then C is not stationary in ω_1. **P** For each $\xi \in C$ choose

$$x_\xi \in \overline{\bigcup_{\eta < \xi} G_\eta} \setminus \bigcup_{\eta < \xi} G_\eta,$$

and let $\alpha(\xi) < \omega_1$ be such that $x_\xi \in G_{\alpha(\xi)}$. Write

$$F = \{\xi : \xi < \omega_1, \ \alpha(\eta) < \xi \ \forall \eta \in C \cap \xi\}.$$

Then F is a closed unbounded set in ω_1 [A2D]. **?** If C is stationary, then $C \cap F$ is stationary. But $\{x_\xi : \xi \in C \cap F\}$ is locally countable (because if $\eta < \xi$ in $C \cap F$ then $x_\eta \in G_{\alpha(\eta)}$ but $x_\xi \notin G_{\alpha(\eta)}$). So, by 44C, $C \cap F$ is expressible as $\bigcup_{n \in \mathbf{N}} C_n$ where $\{x_\xi : \xi \in C_n\}$ is discrete for each $n \in \mathbf{N}$. Because open sets in X are F_σ, each C_n is expressible as $\bigcup_{m \in \mathbf{N}} C_{nm}$ where $\{x_\xi : \xi \in C_{nm}\}$ is discrete and closed for each $m, n \in \mathbf{N}$ [A4E*c*]. There must now be some m, $n \in \mathbf{N}$ such that $D = C_{nm}$ is stationary. As X is collectionwise Hausdorff, there is a disjoint family $\langle H_\xi \rangle_{\xi \in D}$ of open sets in X such that $x_\xi \in H_\xi$ for every $\xi \in D$. For each $\xi \in D$, $x_\xi \in \overline{\bigcup_{\eta < \xi} G_\eta}$, so there is an $f(\xi) < \xi$ such that $H_\xi \cap G_{f(\xi)} \neq \emptyset$. By the pressing-down lemma [A2F] there is a $\zeta < \omega_1$ such that $D' = \{\xi : \xi \in D, f(\xi) = \zeta\}$ is uncountable. But now $\{H_\xi \cap G_\zeta : \xi \in D'\}$ is an uncountable disjoint collection of open sets in G_ζ and G_ζ is not ccc, contradicting (*a*) above. **XQ**

(*c*) It follows that in this case the theorem is true. **P** Let

$E \subseteq \omega_1 \setminus C$ be a closed unbounded set; we can suppose that $0 \in E$. Enumerate E is ascending order as $\langle \beta(\xi) \rangle_{\xi < \omega_1}$. Set

$$Y_\xi = \bigcup_{\eta < \beta(\xi+1)} G_\eta \setminus \bigcup_{\eta < \beta(\xi)} G_\eta.$$

Then $\langle Y_\xi \rangle_{\xi < \omega_1}$ is a partition of X into open-and-closed subsets; as each G_η is open and relatively compact, it is K_σ, and each Y_ξ is also K_σ. **Q**

(*d*) For the general case, therefore, it will be enough to show that X is expressible as a disjoint union of open subspaces each of which can be covered by ω_1 or fewer relatively compact open sets. To see that this is the case, define inductively a family $\langle \mathscr{G}_\xi^n \rangle_{n \in \mathbf{N}, \xi < \omega_1}$ of disjoint collections of open sets in X, as follows.

Given $\langle \mathscr{G}_\eta^n \rangle_{n \in \mathbf{N}, \eta < \xi}$, set

$$F_\xi = X \setminus \bigcup_{n \in \mathbf{N}, \eta < \xi} \bigcup \mathscr{G}_\eta^n.$$

Let $\mathscr{H}_\xi$ be a maximal disjoint family of non-empty relatively open relatively compact subsets of F_ξ. Each $H \in \mathscr{H}_\xi$ is separable; let $\langle x_i^H \rangle_{i \in \mathbf{N}}$ run over a dense subset of H. Set $A_\xi^i = \{x_i^H : H \in \mathscr{H}_\xi\}$. Then A_ξ^i is discrete, so may be expressed as $\bigcup_{j \in \mathbf{N}} A_\xi^{ij}$ where each A_ξ^{ij} is discrete and closed [A4E*c* again]. Let $\langle B_\xi^n \rangle_{n \in \mathbf{N}}$ be a re-indexing of $\langle A_\xi^{ij} \rangle_{i,j \in \mathbf{N}}$, so that each B_ξ^n is closed and discrete and

$$\bigcup_{n \in \mathbf{N}} B_\xi^n = \bigcup_{i,j \in \mathbf{N}} A_\xi^{ij} = \{x_i^H : i \in \mathbf{N}, H \in \mathscr{H}_\xi\}$$

is dense in $\bigcup \mathscr{H}_\xi$; by the maximality of $\mathscr{H}_\xi$, $\bigcup_{n \in \mathbf{N}} B_\xi^n$ is dense in F_ξ. As X is locally compact and collectionwise Hausdorff, we can find for each $n \in \mathbf{N}$ a disjoint family $\mathscr{G}_\xi^n$ of relatively compact open sets such that $B_\xi^n \subseteq \bigcup \mathscr{G}_\xi^n$. Continue.

Now $\mathscr{G} = \bigcup_{n \in \mathbf{N}, \xi < \omega_1} \mathscr{G}_\xi^n$ is a cover of X. **P?** Otherwise, there is an $x \in \bigcap_{\xi < \omega_1} F_\xi$. Let V be a compact neighbourhood of x. Then V is hereditarily Lindelöf, and $\langle F_\xi \rangle_{\xi < \omega_1}$ is a decreasing family of closed sets, so there is a $\zeta < \omega_1$ such that $V \cap F_\zeta = V \cap F_{\zeta+1}$. In this case $V \cap F_\zeta \cap \bigcup_{n \in \mathbf{N}} \bigcup \mathscr{G}_\zeta^n = \varnothing$ and $V \cap \bigcup_{n \in \mathbf{N}} B_\zeta^n = \varnothing$. But $\bigcup_{n \in \mathbf{N}} B_\zeta^n$ is supposed to be dense in F_ζ. **XQ**

If $H \in \mathscr{G}$, it is relatively compact, therefore ccc, so that

$$\{G : G \in \mathscr{G}_\xi^n, H \cap G \neq \varnothing\}$$

is countable for each $n \in \mathbf{N}$, $\xi < \omega_1$, and

$$\#(\{G : G \in \mathscr{G}, H \cap G \neq \varnothing\}) \leq \omega_1.$$

It follows at once that if $\sim$ is the smallest equivalence relation on $\mathscr{G}$ such that

$$G \cap H \neq \varnothing \Rightarrow G \sim H,$$

then the equivalence classes under $\sim$ have ω_1 or fewer members each. For each equivalence class $\mathscr{E} \subseteq \mathscr{G}$, set $X_{\mathscr{E}} = \bigcup \mathscr{E}$. The $X_{\mathscr{E}}$ form a partition of X into open sets, each covered by ω_1 or fewer relatively compact open sets. By (*b*)–(*c*), each $X_{\mathscr{E}}$ has a partition into open K_σ sets; putting these together, X has a partition into open K_σ sets.

(*e*) Finally, X is paracompact because it is a topological sum of K_σ subspaces, each of which is paracompact [A4M*a*].

44L **Corollary**

$[\mathfrak{m} > \omega_1]$ Let X be a perfectly normal locally connected locally compact Hausdorff space. Then all the components of X are open K_σ sets.

Proof The point is that X is collectionwise Hausdorff. **P** Let $A \subseteq X$ be discrete and closed. As X is perfectly normal, A is a zero set, and there is a sequence $\langle H_n \rangle_{n \in \mathbb{N}}$ of open sets in X such that $\bar{H}_{n+1} \subseteq H_n$ for each $n \in \mathbb{N}$ and $A = \bigcap_{n \in \mathbb{N}} H_n$. For each $x \in A$ there is a relatively compact open neighbourhood U_x of x such that $\bar{U}_x \cap A = \{x\}$. Then ∂U_x is a compact set not meeting $A = \bigcap_{n \in \mathbb{N}} \bar{H}_n$, so there is an $n(x) \in \mathbb{N}$ such that

$$\partial U_x \cap \bar{H}_{n(x)} = \partial U_x \cap \bigcap_{i \leq n(x)} \bar{H}_i = \varnothing.$$

Let G_x be the component of $U_x \cap H_{n(x)}$ containing x; because X is locally connected, G_x is open. If x and y are distinct points of A with $n(x) \leq n(y)$, then $G_y \subseteq H_{n(y)} \subseteq H_{n(x)}$, so $G_y \cap \partial U_x = \varnothing$; as G_y is connected, and $G_y \not\subseteq U_x$ because $y \notin U_x$, $G_y \cap U_x$ must be empty and $G_y \cap G_x = \varnothing$. This shows that $\langle G_x \rangle_{x \in A}$ is disjoint. As A is arbitrary, X is collectionwise Hausdorff. **Q**

Accordingly, by 44K, X can be partitioned into open K_σ sets, and every point $x \in X$ belongs to an open-and-closed K_σ set H, say. The component C of x in X is now a closed subset of H, so is also a K_σ set. Again because X is locally connected, C is also open.

44M **Corollary**

$[\mathfrak{m} > \omega_1]$ Let X be a perfectly normal manifold. Then X is metrizable.

Proof I leave the definition of 'manifold' to specialists; all we need to know here is that a manifold is locally compact, locally connected, locally metrizable and Hausdorff. As X is locally connected, it is the topological sum of its components, and it will be enough to show that its components are metrizable. If C is any component of X, then by 44L C is K_σ, therefore Lindelöf. As X is locally metrizable and locally compact, C is expressible as

$\bigcup_{n\in\mathbb{N}} G_n$ where each G_n is open and each $\bar{G}_n$ is metrizable and compact. As each G_n is second-countable, so is C; as X is also normal and Hausdorff, and C is closed (therefore normal), C is metrizable.

44N Sources

JUHÁSZ 70 for 44H(ii)$\Rightarrow$(iii). TALL 74*a* for 44B(ii). ALSTER & ZENOR 77 for the argument given in the proof of 44L. WEISS 78 for 44B(iii). FREMLIN & SHELAH 79 for the argument of 44G. RUDIN 79 for 44M. SZENTMIKLÓSSY 80 for 44B(i) (in essence), 44B(iv)–(v), and the ideas of 44A. LANE 80 for 44L. GRUENHAGE 80 for 44D(i)$\Rightarrow$(iv). BALOGH *a* for 44C and the rest of 44D, and 44K.

44O Exercises

(*a*) [$\mathfrak{m} > \omega_1$] Suppose that X is a countably tight hereditarily ccc space, and all its closed subsets are Martin-$<\mathfrak{m}$-complete. Then X is hereditarily separable. [TALL 74*a*, 4.1.]

(*b*) [$\mathfrak{m} > \omega_1$] Let X be a countably tight compact Hausdorff space, $Y \subseteq X$ an uncountable scattered subset. Then Y has an uncountable discrete subset. [Hint: Y has an uncountable, locally countable subset.]

(*c*) Let X be a locally compact Hausdorff space. Then the following are equivalent: (i) the one-point compactification of X is countably tight; (ii) if $Y \subseteq X$ and $f: Y \to \omega_1$ is a continuous surjection there is a $\xi < \omega_1$ such that $f^{-1}[\xi + 1]$ is not compact; (iii) X can be embedded in a countably tight compact Hausdorff space. [Hint for (ii)$\Rightarrow$(i): A4K prf. See BALOGH *a*, 2.1.] So if X is metalindelöf and countably tight, its one-point compactification is countably tight.

(*d*) [$\mathfrak{m} > \omega_1$] If X is a compact Hausdorff space and X^2 is hereditarily ccc then X is metrizable. [BELL & GINSBERG 82, 2.4.]

(*e*) [$\mathfrak{m} > \omega_1$] Let X be a Hausdorff space. Let $\mathscr{C}$ be the space of closed subsets of X with the Vietoris topology [A4T]. If $\mathscr{C}$ is hereditarily ccc then X is compact and metrizable. [Hint: show (i) X is countably compact (ii) X is Lindelöf (iii) $\mathscr{C}$ is compact, Hausdorff and perfectly normal (iv) X has a G_δ diagonal. See BELL & GINSBERG 82, 2.1.]

(*f*) [$\mathfrak{m} > \omega_1$] If X is a K-analytic space, then the conditions (i)–(xiii) of Theorem 44H are equivalent. [Hint: in most cases it is easy to show that X has property (*j*) iff every compact subset of X has property (*j*); see 44F prf *a*.]

(*g*) [$\mathfrak{m} > \omega_1$] If X is a K-analytic space and the algebra $\mathscr{B}(X)$ of

Borel subsets of X is isomorphic, as Boolean algebra, to $\mathscr{B}(Y)$ for some analytic space Y, then X is analytic. [Hint: 23N*d*.]

(*h*) [$\mathfrak{m} > \omega_1$] If X is a locally compact Hausdorff space then the conditions (i)–(xiii) of Theorem 44H are equivalent. [In most cases it is easy to show that X has property (*j*) iff its one-point compactification has.]

(*i*) [$\mathfrak{m} > \omega_1$] If X is a locally compact Hausdorff space and the algebra $\mathscr{B}(X)$ of Borel sets of X is isomorphic, as Boolean algebra, to $\mathscr{B}(Y)$ for some analytic space Y, then X is separable and metrizable.

(*j*) [$\mathfrak{m} > \omega_1$] Let X be a locally compact collectionwise Hausdorff space in which open sets are F_σ. If X is scattered it is metrizable. [Use 44H to show that X is locally countable.]

(*k*) [$\mathfrak{m} > \omega_1$] Let X be a hereditarily collectionwise Hausdorff, locally ccc, locally compact space. Then X is metalindelöf iff it is paracompact iff it has a partition into open K_σ sets iff its one-point compactification is countably tight. [Hint: show that X is locally hereditarily Lindelöf; then use 44O*c*, and the arguments of 44K. See Balogh *a*, 3.3. Compare 43Q*a*.]

(*l*) (i) If X is a locally countable Moore space it is expressible as the union of a sequence of closed discrete subsets. [Use A4O.] (ii) If X is a collectionwise Hausdorff locally compact Moore space it is metrizable. [Use the ideas of 44K.] (iii) If X is a locally connected locally compact normal Moore space it is metrizable. [Use the ideas of 44L, or see Rudin 75, VIII(3), or Reed & Zenor 76.]

44P **Further results**

(*a*) Let X be a locally compact Hausdorff space with a G_δ diagonal. Then the one-point compactification of X is countably tight. [Balogh *a*.]

(*b*) [$\mathfrak{m} > \omega_1$] If X is a regular Hausdorff space, and X^n is hereditarily ccc for every $n \in \mathbb{N}$, then X^n is hereditarily separable and hereditarily Lindelöf for every $n \in \mathbb{N}$. [Kunen 77*b*; see Roitman 83, 6.3.]

(*c*) [$\mathfrak{m} > \omega_1$] If X is a hereditarily normal compact Hausdorff space, then every hereditarily ccc subset of X is hereditarily Lindelöf. [Szentmiklóssy 80.]

(*d*) [$\mathfrak{m} > \omega_1$] Let X be a hereditarily ccc regular Hausdorff space such that whenever $E, F \subseteq X$ are countable and $E \cap \bar{F} = \bar{E} \cap F = \emptyset$ then $\bar{E} \cap \bar{F} = \emptyset$. Then X is hereditarily Lindelöf. In particular, an extremally disconnected hereditarily separable regular Hausdorff space is hereditarily Lindelöf. [Szymański 80*a*.]

(*e*) Other work connected with these topics may be found in BURKE & HODEL 76 and ISMAIL & NYIKOS 82.

44Q Problems

(*a*) Is there is an example in ZFC of a hereditarily Lindelöf regular Hausdorff space with is not hereditarily separable?

(*b*) Let Z be a countably tight compact Hausdorff space and $X \subseteq Z$ a scattered subspace with $\#(X) < \mathfrak{m}$. Is X expressible as a countable union of discrete sets?

(*c*) $[\mathfrak{m} > \omega_1]$ Let X be a hereditarily Lindelöf compact Hausdorff space. Is there a compact metric space Z and a continuous $f: X \to Z$ such that $f^{-1}[\{z\}]$ is finite for every $z \in Z$?

Notes and comments

The first version of 44B to be found was that of JUHÁSZ 70: a hereditarily Lindelöf compact Hausdorff space is separable. Parts (*a*) and (*c*) of the proof of 44B come from this source, and both 44B(ii) and 44B(iii) can be regarded as generalizations of Juhász' theorem. As far as I know, both 44B(i) and 44B(iv) were first published by SZENTMIKLÓSSY 80; but of course it was the latter that was the real advance. There is an alternative version of 44B(i) in 44O*a*. A short proof of 44B(iv) may be found in ABRAHAM & TODORČEVIĆ 83.

44C is derived from Szentmiklóssy's proof of 44E; the version here owes a good deal to BALOGH *a*. I do not know whether the condition 'locally countable' can be usefully relaxed [44O*b*, 44Q*b*]. 44B and 44D–E give reasons to search for conditions under which a topological space X will be embeddable in a countably tight compact Hausdorff space; 24L, 44O*c* and 44P*a* are attempts on the locally compact case.

44E describes a context in which 'hereditarily separable', 'hereditarily Lindelöf' and 'hereditarily ccc' become equivalent. Others are in 44P*b*–*c*. Another result of this kind is 44P*d* (which depends on 42I).

I have presented 44F as if it were a corollary of 44E, but this is used only in order to relax the hypothesis from 'hereditarily Lindelöf' to 'hereditarily ccc'; and it is also a little misleading to quote 43F*a*, because we can easily argue directly from 13A(iii). I state 44F in terms of K-analytic spaces because I need it in this form in the proof of 44G, but the essential idea appears, rather more clearly, if X is taken to be compact. 44G is an elaborated version of 23N*e*. The idea needed to link them is that if in 44G the space X is metrizable, then the partially ordered set of part (*d*) of the proof can be replaced by a countable partially ordered set, so that $\#(\mathscr{E}) < \mathfrak{p}$ becomes sufficient.

On collecting the results of 44E and 44G together, and using A4K, we see that for a compact Hausdorff space X the following are equivalent: (α) X is hereditarily separable; (β) X is hereditarily Lindelöf; (γ) X cannot be partitioned into ω_1 closed sets. Relatively elementary arguments now show that these three can be extended to the thirteen properties listed in 44H. I think it likely that there are more to come; 44Q*c* is an optimistic suggestion. I do not have a complete analysis of the relations between these properties which are valid in ZFC: for instance, I have not seen an example to separate (viii) from (ix). However, the three examples mentioned in 44J, with various modifications, settle the most important questions. 44I is a simple consequence of 44H(iv)$\Rightarrow$(ii) (which of course is immediate from 44E). There are corresponding results for K-analytic spaces [44O*f*–*g*] and for locally compact spaces [44O*h*–*i*]. 44O*d*–*e* are easy consequences of 44H(i)$\Rightarrow$(ii).

In 44K–M I turn to a different group of ideas. Observe that $\mathfrak{m} > \omega_1$ is used repeatedly in parts (*a*) and (*b*) of the proof of 44K, but not thereafter. The inspiration behind this work was of course M.E. Rudin's proof of 44M. 44O*j*–*l* are obtained by reworking the ideas of 44K–L.

It is perhaps worth noting which of the main results of this section can be derived from the principles of 41L. K is enough for 44A, 44B(iv)–(v), 44E and 44I. H is sufficient for the other parts of 44B as well.

APPENDIX A

Useful facts

In this appendix I write out the definitions and theorems which I use in the pages above, and for which no natural place presented itself in the main line of the exposition. In general I give proofs only when I have been unable to find satisfactory references in hard covers. I hope that the index will prove adequate and that there will be no need for you to read systematically through this appendix; but perhaps a preliminary glance at §A1 will be useful. Some of the material which you might look for here is in §12.

A1 Notation

Here I list some of the special symbols I use, and indicate the ways in which I think of some of the fundamental concepts of set theory. I have tried to express these in terms which are readily translatable into the formulae of any conventional description of Zermelo–Fraenkel set theory, though it will be clear that this particular framework is not the only possible one. Note that I use the axiom of choice without scruple and without comment.

A1A Reserved symbols

(*a*) **N**, **Z**, **Q**, **R** represent respectively the sets of non-negative integers, integers, rational numbers and real numbers.

(*b*) ω is the first infinite ordinal. ω_1 is the first uncountable ordinal. $\mathfrak{c} = 2^{\omega} = \#(\mathbf{R})$, the cardinal of the continuum. κ and λ always stand for cardinals. $\mathfrak{m}$, $\mathfrak{m}_K$ and $\mathfrak{p}$ stand for the special cardinals defined in §11.

(*c*) If (x, y) is an ordered pair then (unless otherwise indicated) $\pi_1(x, y) = x$ and $\pi_2(x, y) = y$.

(*d*) On is the class of all ordinals. Ω is the set of non-zero countable limit ordinals.

(*e*) $\mathbf{Z}_2$ is the field $\mathbf{Z}/2\mathbf{Z}$, identified with $\{0, 1\}$.

(*f*) The symbols **P**...**Q** are used to mark off a section of text comprising a proof of the statement immediately preceding the **P**. The symbols **?**...**X** enclose a refutation, by reductio ad absurdum, of the statement immediately following the **?**.

A1B **Remarks**

(*a*) 'Ordinals' are von Neumann ordinals. Thus, for ordinals ξ, η,

$$\xi < \eta \Leftrightarrow \xi \in \eta; \quad \xi \leq \eta \Leftrightarrow \xi \subseteq \eta; \quad \xi + 1 = \xi \cup \{\xi\};$$

and for a set A of ordinals

$$\sup A = \bigcup A, \ \min A = \bigcap A \quad \text{if} \quad A \neq \emptyset.$$

(*b*) As I assume the axiom of choice throughout this book, I can and do identify cardinals with initial ordinals, so that

$$\#(X) = \min\{\xi : \xi \in \mathrm{On}, \exists \text{ bijection } f: X \to \xi\}.$$

Thus ω and ω_1 are the first and second infinite cardinals; $\mathfrak{c}$ is an ordinal; and the continuum hypothesis becomes just '$\omega_1 = \mathfrak{c}$'.

(*c*) It is quite often convenient to identify **N** with ω, so that the natural numbers become the finite ordinals. Thus if A is a set and $n \in \mathbf{N}$, $A \cap n = \{i : i < n, \ i \in A\}$. When it suits me, I am willing simultaneously to regard **N** as a subset of **R** (consider the formula in A3M*d*).

A1C **Functions and relations**

(*a*) A **relation** is a class R of ordered pairs. In this case

$$R[A] = \{y : \exists x \in A, (x, y) \in R\},$$
$$R^{-1}[B] = \{x : \exists y \in B, (x, y) \in R\}$$

for any classes A, B. The **vertical sections** of R are the sets of the form $R[\{x\}] = \{y : (x, y) \in R\}$.

(*b*) If f is a function, I write $\mathrm{dom}(f)$ for its domain. Now, for any classes A and B,

$$f[A] = \{f(x) : x \in A \cap \mathrm{dom}(f)\},$$
$$f^{-1}[B] = \{x : x \in \mathrm{dom}(f), f(x) \in B\}.$$

f is **injective** or an **injection** if it is one-to-one (i.e. its inverse is a function). $f: X \to Y$ is **surjective** or a **surjection** if $f[X] = Y$; f is **bijective** or a **bijection** if it is injective and surjective. When convenient, I identify a function with the relation which is its graph; thus '$f \subseteq g$' means 'g extends f'.

(*c*) If f is any function and A is any class, then $f \restriction A$ is that function such that

$$\mathrm{dom}(f \restriction A) = A \cap \mathrm{dom}(f), \ (f \restriction A)(x) = f(x) \quad \text{if} \quad x \in A \cap \mathrm{dom}(f).$$

(*d*) A **sequence** is a function with domain a subset of **N**; unless otherwise indicated, the domain will be $\mathbf{N}$ itself. A **finite sequence** is a function with domain $\{i : i < n\}$ for some $n \in \mathbf{N}$. If f is a sequence and $n \in \mathbf{N}$, then $f \restriction n$ is the finite sequence $\langle f(i) \rangle_{i<n}$. If $\langle x_n \rangle_{n \in \mathbf{N}}$ is a sequence, then I shall use the word **subsequence** either for $\langle x_n \rangle_{n \in I} = \langle x_n \rangle_{n \in \mathbf{N}} \restriction I$, where $I \subseteq \mathbf{N}$ is infinite, or for $\langle x_{n(k)} \rangle_{k \in \mathbf{N}}$ where $\langle n(k) \rangle_{k \in \mathbf{N}}$ is a strictly increasing sequence in $\mathbf{N}$.

(*e*) If R is a relation then a **selector** for R is a function f with domain $\pi_1[R] = \{x : \exists y, (x, y) \in R\}$ such that $(x, f(x)) \in R$ for every $x \in \pi_1[R]$–if you like, such that $f \subseteq R$.

A1D Powers

(*a*) If κ is a cardinal, then 2^κ is the cardinal $\#(\mathscr{P}\kappa)$ of the power set of κ.

(*b*) In all other cases, unless otherwise indicated [see A2A*b*], X^Y is the set of functions from Y to X. (Thus $\{0, 1\}^\kappa$ is the set of functions from κ to $\{0, 1\}$, while 2^κ is the cardinal of this set.) In particular, $X^\varnothing = \{\varnothing\}$; while X^n is the set of finite sequences $\langle x_i \rangle_{i<n}$ such that $x_i \in X$ for each $i < n$.

A1E Miscellaneous notation

(*a*) If X is a set and $A \subseteq X$, then A is **cofinite** in X if $X \setminus A$ is finite.

(*b*) If X is a set and $A \subseteq X$, then χA is the function with domain X such that

$$(\chi A)(x) = 1 \quad \text{if} \quad x \in A, 0 \quad \text{if} \quad x \in X \setminus A.$$

(*c*) If X is a set and $\sim$ is an equivalence relation on X, I often write $x^{\cdot}$ for the equivalence class of an $x \in X$.

(*d*) If P is a partially ordered set, then

$$[p, q] = \{r : p \leq r \leq q\},$$
$$[p, q[= \{r : p \leq r < q\},$$
$$]p, q[= \{r : p < r < q\}$$

for all $p, q \in P$. (For most of my terminology concerning partially ordered sets, see 11A.)

A2 Cardinals and ordinals

I briefly state the essential facts about cardinals and ordinals which I use repeatedly. In many cases it will be plain that the theorems I give are special cases of more general results; for a more systematic analysis of these topics see the books referred to below.

A2A **Ordinals** [See A1B*a*]

(*a*) If X is any well-ordered set, there is a unique ordinal which is order-isomorphic to X [KUNEN 80, I.7.6; LEVY 79, II.3.23; HALMOS 60, §20]. I denote this ordinal $\operatorname{otp}(X)$.

(*b*) Suppose that ξ and η are ordinals. Let F be the set of functions $f:\eta\to\xi$ such that $\{\zeta:f(\zeta)\neq 0\}$ is finite. Say that $f<g$ if there is a $\zeta_0<\eta$ such that $f(\zeta_0)<g(\zeta_0)$ and $f(\zeta)=g(\zeta)$ for $\zeta_0<\zeta<\eta$. Then F is well-ordered. The **ordinal power** ξ^η is $\operatorname{otp}(F)$. [See also KUNEN 80, I.9.5 or LEVY 79, IV.2.4 or HALMOS 60, §21 for an equivalent alternative definition.]

A2B **Cardinals**

(*a*) If κ is any cardinal, then κ^+ is the smallest cardinal greater than κ, and 2^κ is $\#(\mathscr{P}\kappa)$.

(*b*) If ξ is a non-zero limit ordinal, then its **cofinality** $\operatorname{cf}(\xi)$ is $\min\{\operatorname{otp}(A):A\subseteq\xi,\ \xi=\sup A\}$.

(*c*) A **regular** cardinal is an infinite cardinal κ such that $\operatorname{cf}(\kappa)=\kappa$. For any infinite κ, κ^+ is regular. Note that $\operatorname{cf}(\xi)$ is a regular cardinal for every non-zero limit ordinal ξ [LEVY 79, IV.3.8; JECH 78, Lemma 3.5; JUHÁSZ 71, AO.1.5.]

(*d*) $\langle\omega_\xi\rangle_{\xi\in On}$ is the increasing enumeration of the well-ordered class of infinite cardinals; thus $\omega_0=\omega$, $\omega_{\xi+1}=\omega_\xi^+$ for every $\xi\in On$, and $\omega_\xi=\sup_{\eta<\xi}\omega_\eta$ for limit ordinals $\xi>0$.

(*e*) If $\langle\kappa_\iota\rangle_{\iota\in I}$ is an indexed family of cardinals, its **cardinal sum** $\sum_{\iota\in I}\kappa_\iota$ is $\#(\{(\iota,\xi):\iota\in I,\ \xi<\kappa_\iota\})$.

A2C **König's theorem**

For any infinite cardinal κ, $\operatorname{cf}(2^\kappa)>\kappa$.

Proof KUNEN 80, I.10.41; LEVY 79, V.5.2; JUHÁSZ 71, AO.1.10a.

A2D **Closed unbounded sets**

Let ξ be a non-zero limit ordinal. A subset F of ξ is closed for the order topology of ξ iff $\sup A\in F$ whenever $\emptyset\neq A\subseteq F$ and $\sup A<\xi$. A **closed unbounded** set in ξ is a closed set $F\subseteq\xi$ such that $\sup F=\xi$.

If ξ has uncountable cofinality (i.e. $\operatorname{cf}(\xi)>\omega$) then the intersection of countably many closed unbounded sets is a closed unbounded set [KUNEN 80, II.6.8; LEVY 79, IV.4.15; JECH 78, Lemma 7.4.]

If $A \subseteq \omega_1$ and $f:A \to \omega_1$ is any function then

$$\{\zeta : \zeta < \omega_1, f[A \cap \zeta] \subseteq \zeta\}$$

is a closed unbounded set in ω_1 [KUNEN 80, II.6.13].

Ω is a closed unbounded set in ω_1 [definition: A1A*d*].

A2E Stationary sets

Let ξ be a non-zero limit ordinal. A subset of ξ is **stationary** in ξ if it meets every closed unbounded subset of ξ.

If $\mathrm{cf}(\xi) > \omega$, then every closed unbounded set in ξ is stationary, and the family of non-stationary subsets of ξ is a σ-ideal of $\mathscr{P}\xi$; turning this round, if $\mathscr{A}$ is countable and $\bigcup \mathscr{A}$ is a stationary subset of ξ, there is a stationary $A \in \mathscr{A}$ [KUNEN 80, II.6.9].

A2F Pressing-down lemma

If $D \subseteq \omega_1$ is a stationary set and $f:D \to \omega_1$ is such that $f(\xi) < \xi$ for every $\xi \in D$, then there is a $\zeta < \omega_1$ such that $\{\xi : \xi \in D, f(\xi) = \zeta\}$ is stationary.

Proof KUNEN 80, II.6.15; LEVY 79, IV.4.41; JECH 78, Theorem 22; JUHÁSZ 71, A1.5.

A2G Ladder systems

Let Ω be the set of non-zero countable limit ordinals. A **ladder system** on ω_1 is an indexed family $\langle \theta(\zeta, n) \rangle_{n \in \mathbf{N}, \zeta \in \Omega}$ such that, for each $\zeta \in \Omega$, $\langle \theta(\zeta, n) \rangle_{n \in \mathbf{N}}$ is a strictly increasing sequence of ordinals with supremum ζ.

A2H Square brackets notation

Let X be a set, λ a cardinal. Then

$$[X]^{\lambda} = \{A : A \subseteq X, \ \#(A) = \lambda\},$$
$$[X]^{<\lambda} = \{A : A \subseteq X, \ \#(A) < \lambda\},$$
$$[X]^{\leq\lambda} = \{A : A \subseteq X, \ \#(A) \leq \lambda\},$$
$$[X]^{\geq\lambda} = \{A : A \subseteq X, \ \#(A) \geq \lambda\},$$

so that

$$[X]^{<\omega} = \{A : A \subseteq X, A \text{ is finite}\},$$
$$[X]^2 = \{\{x, y\} : x, y \in X, x \neq y\}.$$

A2I Lemma

For each $n \in \mathbf{N}$ there is a cofinal $\mathscr{C}_n \subseteq [\omega_n]^{\omega}$ such that $\#(\mathscr{C}_n) \leq \omega_n$.

Proof Induce on n. Take $\mathscr{C}_0 = \{\omega\}$. Given $\mathscr{C}_n$, choose for each $\xi < \omega_{n+1}$ an injection $\theta_\xi : \xi \to \omega_n$, and set

$$\mathscr{C}_{n+1} = \{\theta_\xi^{-1}[C] : \xi < \omega_{n+1}, C \in \mathscr{C}_n\}.$$

A2J **Ramsey's theorem**

If X is an infinite set, $r \in \mathbb{N}$ and $\mathscr{S}$ is a finite cover of $[X]^r$, then there is an infinite $Y \subseteq X$ and an $S \in \mathscr{S}$ such that $[Y]^r \subseteq S$.

Proof Levy 79, IX.3.7; Jech 78, Lemma 29.1; Comfort & Negrepontis 74, Corollary 8.8; Juhász 71, A.4.6.

A2K **Lemma**

Let $S \subseteq [\omega_1]^2$. Then for every stationary $C \subseteq \omega_1$ *either* there is a strictly increasing sequence $\langle \zeta_n \rangle_{n \in \mathbb{N}}$ in C, with supremum $\zeta \in C$, such that $[\{\zeta_n : n \in \mathbb{N}\} \cup \{\zeta\}]^2 \subseteq S$ *or* there is a stationary $A \subseteq C$ such that $[A]^2 \cap S = \varnothing$.

Proof Let F be $\{\zeta : 0 < \zeta < \omega_1, \zeta = \sup(C \cap \zeta)\}$. Then F is a closed unbounded set in ω_1, so $F \cap C$ is stationary. Suppose, for each $\zeta \in F \cap C$, we make a simple-minded effort to find a sequence $\langle \zeta_n \rangle_{n \in \mathbb{N}}$ satisfying the first alternative. The obvious method is to take some sequence $\langle \theta(\zeta, n) \rangle_{n \in \mathbb{N}}$ increasing strictly to ζ and to try to choose $\langle \zeta_n \rangle_{n \in \mathbb{N}}$ inductively so that

$$\zeta_n \in C \cap \zeta \setminus \theta(\zeta, n), \quad \zeta_n > \zeta_i \; \forall i < n,$$
$$\{\zeta_i, \zeta_n\} \in S \; \forall i < n, \; \{\zeta_n, \zeta\} \in S.$$

If this works, we have the first alternative. If it always fails, we must have, for each $\zeta \in F \cap C$, a $\theta_\zeta < \zeta$ and a finite set $I_\zeta \subseteq \theta_\zeta$ such that

$$[I_\zeta \cup \{\zeta\}]^2 \subseteq S, \quad [I_\zeta \cup \{\xi, \zeta\}]^2 \nsubseteq S \quad \text{if} \quad \xi \in C \cap \zeta \setminus \theta_\zeta.$$

By the pressing-down lemma [A2F], there is a $\theta < \omega_1$ such that $A_0 = \{\zeta : \zeta \in F \cap C, \theta_\zeta = \theta\}$ is stationary. Now $[\theta]^{<\omega}$ is countable, so there is a finite $I \subseteq \theta$ such that $A_1 = \{\zeta : \zeta \in A_0, I_\zeta = I\}$ is stationary. Try $A = A_1 \setminus \theta$. If $\xi < \zeta$ in A, then

$$[I \cup \{\xi\}]^2 \subseteq S, \quad [I \cup \{\zeta\}]^2 \subseteq S, \quad [I \cap \{\xi, \zeta\}]^2 \nsubseteq S$$

because $\theta_\zeta \leq \xi < \zeta$. But this must be because $\{\xi, \zeta\} \notin S$. As ξ and ζ are arbitrary, $[A]^2 \cap S = \varnothing$ and we have the second alternative.

Remark Versions of this result have been given by many authors. The form here is taken from Laver 75.

A2L Lemma

Let $S \subseteq [\omega_1]^2$, and let $C \subseteq \omega_1$ be uncountable. If there is an $\alpha < \omega_1$ such that $\text{otp}\{\eta : \eta \in C \cap \zeta, \{\eta, \zeta\} \in S\} \leq \alpha$ for every $\zeta \in C$, then there is an uncountable $A \subseteq C$ such that $[A]^2 \cap S = \emptyset$.

Proof Comfort & Negrepontis 82, 1.11.

A2M Arrow notation

Let ξ, η, ζ be ordinals. Then

$$\xi \rightarrow (\eta, \zeta)^2$$

means: whenever $S \subseteq [\xi]^2$, either there is an $A \subseteq \xi$ such that $\text{otp}(A) = \eta$ and $[A]^2 \subseteq S$, or there is a $B \subseteq \xi$ such that $\text{otp}(B) = \zeta$ and $[B]^2 \cap S = \emptyset$.

The important cases to us are $\omega \rightarrow (\omega, \omega)^2$ (which is a special case of A2J with $r = 2$), $\omega_1 \rightarrow (\omega_1, \omega + 1)^2$ [A2K] and $\mathfrak{c} \not\rightarrow (\omega_1, \omega_1)^2$ [41N*a*]. For these, note that if $A \subseteq \omega$ then $\text{otp}(A) = \omega$ iff A is infinite, and if $A \subseteq \omega_1$ then $\text{otp}(A) = \omega_1$ iff A is uncountable.

A2N Large cardinals

(*a*) A cardinal κ is called **strongly inaccessible** (or, by many authors, just **inaccessible**) if κ is regular and uncountable and $2^\lambda < \kappa$ for any $\lambda < \kappa$.

(*b*) A cardinal κ is called **strongly Mahlo** if it is strongly inaccessible and $\{\lambda : \lambda < \kappa, \lambda \text{ is regular}\}$ is stationary in κ [Kunen 80, Ex. 2.48; Jech 78, §7.]

(*c*) A cardinal κ is called **weakly compact** (or **Ramsey**) if it is uncountable and $\kappa \rightarrow (\kappa, \kappa)^2$ [Jech 78, §32; Comfort & Negrepontis 74, §8; Juhász 71, §A6].

(*d*) A cardinal κ is called **two-valued-measurable** (or, by many authors, just **measurable**) if it is uncountable and there is a non-principal ultrafilter $\mathcal{F}$ on κ such that $\bigcap \mathcal{A} \in \mathcal{F}$ whenever $\mathcal{A} \subseteq \mathcal{F}$ and $0 < \#(\mathcal{A}) < \kappa$ [Levy 79, §IX.4; Jech 78, §27; Comfort & Negrepontis 74, §8; Juhász 71, §A6; Drake 74, Ch. 6. See B2D.]

A3 Set theory: other topics

I collect miscellaneous items which do not seem to belong anywhere else in this appendix.

A3A Collections of sets

(*a*) Let $\mathcal{A}$ be a set of sets. $\mathcal{A}$ is **disjoint** if $A \cap B = \emptyset$ whenever A

and B are distinct members of $\mathscr{A}$. $\mathscr{A}$ is **almost disjoint** if $A \cap B$ is finite whenever A, B are distinct members of $\mathscr{A}$. $\mathscr{A}$ is a **cover** of a set X if $X \subseteq \bigcup \mathscr{A}$, $\mathscr{A}$ is a **partition** of X if it is a disjoint cover of X and does not contain $\varnothing$. $\mathscr{A}$ has the **finite intersection property** if $\bigcap \mathscr{A}_0 \neq \varnothing$ for every finite $\mathscr{A}_0 \subseteq \mathscr{A}$. $\mathscr{A}$ is **point-finite** if $\{A : x \in A \in \mathscr{A}\}$ is finite for every x. $\mathscr{A}$ is **σ-point-finite** if it is expressible as the union of a sequence of point-finite families. $\mathscr{A}$ is **point-countable** if $\{A : x \in A \in \mathscr{A}\}$ is countable for every x.

(*b*) An indexed family $\langle A_\xi \rangle_{\xi \in I}$ is **disjoint** (resp. **almost disjoint**) if $A_\xi \cap A_\eta$ is empty (resp. finite) for all distinct $\xi, \eta \in I$. It is **point-finite** if $\{\xi : \xi \in I, x \in A_\xi\}$ is finite for every x.

(*c*) If $\mathscr{A}$ and $\mathscr{B}$ are two families of sets, I say that $\mathscr{B}$ **refines** $\mathscr{A}$, or is a **refinement** of $\mathscr{A}$, if for every $B \in \mathscr{B}$ there is an $A \in \mathscr{A}$ with $B \subseteq A$.

(*d*) There is an almost-disjoint family $\mathscr{A}$ of infinite subsets of $\mathbf{N}$ with $\#(\mathscr{A}) = \mathfrak{c}$. **P** For each infinite $C \subseteq \mathbf{N}$ set

$$A_C = \{\Sigma_{i \in C, i \leq n} 2^i : n \in \mathbf{N}\}.$$

Take $\mathscr{A} = \{A_C : C \in [\mathbf{N}]^\omega\}$. **Q**

(*e*) If $\mathscr{A}$ is any family of sets, write $\mathrm{St}(x, \mathscr{A}) = \bigcup \{A : x \in A \in \mathscr{A}\}$ for every x.

A3B Filters

(*a*) If $\mathscr{F}$ is a filter on a set X and $f : X \to Y$ is a function, then $f[[\mathscr{F}]]$ is the filter $\{B : B \subseteq Y, f^{-1}[B] \in \mathscr{F}\}$ on Y.

A filter $\mathscr{F}$ on a set X is **uniform** if $\#(A) = \#(X)$ for every $A \in \mathscr{F}$ [cf. 35A].

(*b*) An ultrafilter $\mathscr{F}$ on a set X is **principal** if it contains $\{x\}$ for some $x \in X$; otherwise it is **non-principal**. (Observe that an ultrafilter on $\mathbf{N}$ is non-principal iff it is uniform.)

A3C Types of ultrafilter on N

(*a*) If κ is an infinite cardinal, a **$p(\kappa)$-point ultrafilter** on $\mathbf{N}$ is a non-principal ultrafilter which is a $p(\kappa)$-point in $\beta\mathbf{N} \setminus \mathbf{N}$ i.e. whenever $\mathscr{A} \subseteq \mathscr{F}$, $\#(\mathscr{A}) < \kappa$ there is an $I \in \mathscr{F}$ such that $I \setminus A$ is finite for every $A \in \mathscr{A}$.

A **p-point** (or 'δ-stable') ultrafilter is a $p(\omega_1)$-point ultrafilter.

(*b*) A **Ramsey** or **selective** or **absolute** ultrafilter on $\mathbf{N}$ is a non-principal ultrafilter $\mathscr{F}$ such that, whenever $r \in \mathbf{N}$ and $\mathscr{S}$ is a finite cover of $[\mathbf{N}]^r$, there is an $S \in \mathscr{S}$ and an $I \in \mathscr{F}$ such that $[I]^r \subseteq S$.

For the many necessary and sufficient conditions known for a filter to be Ramsey, see BOOTH 70 and COMFORT & NEGREPONTIS 74.

(*c*) If κ is an infinite cardinal, I shall say that a non-principal ultrafilter $\mathscr{F}$ on $\mathbf{N}$ is **(ω, κ)-saturating** if whenever $\mathscr{B} \subseteq \mathscr{P}(\mathbf{N} \times \mathbf{N})$, $\#(\mathscr{B}) < \kappa$, and $\pi_1[\bigcap \mathscr{B}_0] \in \mathscr{F}$ for every non-empty finite $\mathscr{B}_0 \subseteq \mathscr{B}$, then there is an $f:\mathbf{N} \to \mathbf{N}$ such that

$$\{i:(i, f(i)) \in B\} \in \mathscr{F} \; \forall B \in \mathscr{B}.$$

I do not have an example of a $p(\kappa)$-point ultrafilter which is not (ω, κ)-saturating [26M*a*]. (Every ultrafilter is (ω, ω_1)-saturating, so such an example would have to contradict the continuum hypothesis.)

I have chosen the name '(ω, κ)-saturating' on account of the following characterization. For the terminology of the next lemma, and for the basic theory of ultraproducts, see Chang & Keisler 73.

A3D Lemma

Let $\mathscr{F}$ be a non-principal ultrafilter on $\mathbf{N}$. Then the following are equivalent:

(i) $\mathscr{F}$ is (ω, κ)-saturating;

(ii) whenever $\mathscr{L}$ is a language of cardinal less than κ, and $\langle \mathfrak{A}_i \rangle_{i \in \mathbf{N}}$ is a sequence of countable models of $\mathscr{L}$, then the ultraproduct $\mathfrak{A} = \prod_{\mathscr{F}} \mathfrak{A}_i$ is κ-saturated.

Proof (*a*) (i)$\Rightarrow$(ii) Assume (i), and let $\mathscr{L}$, $\langle \mathfrak{A}_i \rangle_{i \in \mathbf{N}}$ and $\mathfrak{A}$ be as given in (ii). It will be enough to show that

for every set $\Sigma(x)$ of formulae in $\mathscr{L}$, if each finite subset of $\Sigma(x)$ is satisfiable in $\mathfrak{A}$, then $\Sigma(x)$ is satisfiable in $\mathfrak{A}$.

[Cf. Chang & Keisler 73, 6.1.1.] For each $i \in \mathbf{N}$, let A_i be the underlying set of $\mathfrak{A}_i$, and $h_i : A_i \to \mathbf{N}$ an injection. For each $\sigma(x) \in \Sigma(x)$ set

$$B(\sigma) = \{(i, h_i(a)) : i \in \mathbf{N}, \mathfrak{A}_i \models \sigma[a]\}.$$

If $\sigma_0(x), \ldots, \sigma_k(x) \in \Sigma(x)$ then

$$\pi_1[\bigcap_{j \leq k} B(\sigma_j)] = \{i : \exists a \in A_i, \mathfrak{A}_i \models \sigma_j[a] \; \forall j \leq k\}$$
$$= \{i : \mathfrak{A}_i \models (\exists x)(\sigma_0(x) \wedge \ldots \wedge \sigma_k(x))\} \in \mathscr{F}$$

because $\{\sigma_0(x), \ldots, \sigma_k(x)\}$ is satisfiable in $\mathfrak{A}$. So by (i) there is an $f:\mathbf{N} \to \mathbf{N}$ such that

$$\{i:(i, f(i)) \in B(\sigma)\} \in \mathscr{F} \; \forall \sigma(x) \in \Sigma(x).$$

Choose $g \in \prod_{i \in \mathbf{N}} A_i$ such that

$$g(i) = h_i^{-1}(f(i)) \quad \text{if} \quad f(i) \in h_i[A_i].$$

Then

$$\{i:\mathfrak{A}_i \models \sigma[g(i)]\} \supseteq \{i:(i,f(i))\in B(\sigma)\}\in\mathscr{F}$$

for each $\sigma(x)\in\Sigma(x)$. So $g_{\mathscr{F}}$ satisfies $\Sigma(x)$ in $\mathfrak{A}$.

(*b*) **(ii)⇒(i)** Assume (ii), and let $\mathscr{B}\subseteq\mathscr{P}(\mathbf{N}\times\mathbf{N})$ be a set of cardinal less than κ such that $\pi_1[\bigcap\mathscr{B}_0]\in\mathscr{F}$ for every non-empty finite set $\mathscr{B}_0\subseteq\mathscr{B}$. Let $\mathscr{L}$ be the language with one one-place relation symbol P_B for each $B\in\mathscr{B}$ (and no functions or constants). For each $i\in\mathbf{N}$, let $\mathfrak{A}_i$ be the model of $\mathscr{L}$ with underlying set $\mathbf{N}$ in which each relation symbol P_B is assigned to the set, or one-place relation, $\{n:(i,n)\in B\}$. Form the ultraproduct $\mathfrak{A}=\prod_{\mathscr{F}}\mathfrak{A}_i$ and consider the set $\Sigma(x)=\{P_B(x):B\in\mathscr{B}\}$ of formulae in $\mathscr{L}$. If $B(0),\ldots,B(k)\in\mathscr{B}$ then $\{P_{B(j)}(x):j\le k\}$ is satisfiable in $\mathfrak{A}$. **P** Choose $f\in\mathbf{N}^{\mathbf{N}}$ so that $(i,f(i))\in\bigcap_{j\le k}B(j)$ whenever $i\in\pi_1[\bigcap_{j\le k}B(j)]$. Then

$$\{i:\mathfrak{A}_i\models(P_{B(0)}\wedge\ldots\wedge P_{B(k)})[f(i)]\}=\pi_1[\bigcap_{j\le k}B(j)]\in\mathscr{F}$$

so $\mathfrak{A}\models(P_{B(0)}\wedge\ldots\wedge P_{B(k)})[f_{\mathscr{F}}]$. **Q**

By (ii), there is an $f\in\mathbf{N}^{\mathbf{N}}$ such that

$$\mathfrak{A}\models P_B[f_{\mathscr{F}}]\,\forall B\in\mathscr{B}.$$

But this is saying just that, for each $B\in\mathscr{B}$, $\mathscr{F}$ contains

$$\{i:\mathfrak{A}_i\models P_B[f(i)]\}=\{i:(i,f(i))\in B\},$$

as required by (i).

A3E Lusin sets

(*a*) Let X be a set, Σ a subalgebra of $\mathscr{P}X$, $\mathscr{I}$ an ideal of Σ, κ an infinite cardinal. Then a set $A\subseteq X$ is a $(\Sigma,\mathscr{I},\kappa)$-**Lusin** set if, for $E\in\Sigma$,

$$E\in\mathscr{I}\Leftrightarrow \#(E\cap A)<\kappa.$$

(*b*) Take X, Σ, $\mathscr{I}$, κ as in (*a*). Write

$$\hat{\mathscr{I}}=\{B:\exists E\in\mathscr{I},\,B\subseteq E\},$$
$$\hat{\Sigma}=\{E\triangle B:E\in\Sigma,\,B\in\hat{\mathscr{I}}\}.$$

Then $\hat{\Sigma}$ is a subalgebra of $\mathscr{P}X$ and $\hat{\mathscr{I}}$ is an ideal of $\hat{\Sigma}$. Now a set $A\subseteq X$ is a $(\hat{\Sigma}$, $\hat{\mathscr{I}}$, $\kappa)$-Lusin set iff it is a $(\Sigma$, $\mathscr{I}$, $\kappa)$-Lusin set.

(*c*) Let X be a topological space, κ an infinite cardinal. A set $A\subseteq X$ is a **κ-Lusin** set if it is a $(\mathscr{B},\mathscr{M},\kappa)$-Lusin set, where $\mathscr{B}$ is the algebra of Borel subsets of X and $\mathscr{M}$ is the ideal of meagre Borel sets. Observe that A is a κ-Lusin set iff it is a $(\hat{\mathscr{B}},\hat{\mathscr{M}},\kappa)$-Lusin set, where $\hat{\mathscr{B}}$ is the algebra of sets with the Baire property [A5B*b*] and $\hat{\mathscr{M}}$ is the ideal of all meagre sets.

(*d*) Let (X, Σ, μ) be a measure space, κ an infinite cardinal. A set $A \subseteq X$ is a κ-**Sierpiński** set if it is a $(\Sigma, \mathcal{N}, \kappa)$-Lusin set, where $\mathcal{N}$ is the ideal of sets of measure zero. Observe that (X, Σ, μ) and its completion [A6D] have the same κ-Sierpiński sets.

Remark
The term 'Lusin set' is often used for what I call an 'ω_1-Lusin' set.

A3F **Lemma**

(*a*) Let X be a set, Σ an algebra of subsets of X, $\mathcal{I}$ an ideal of Σ, κ an infinite cardinal. Suppose that (i) $\bigcup \mathcal{I} = X$, (ii) $\kappa \geq \#(\Sigma)$, (iii) no member of $\Sigma \backslash \mathcal{I}$ can be covered by fewer than κ members of $\mathcal{I}$. Then there is a $(\Sigma, \mathcal{I}, \kappa)$-Lusin set $Y \subseteq X$.

(*b*) Let X be a ccc topological space with $\pi(X) \leq \mathfrak{c}$, and suppose that no non-meagre open set in X can be covered by fewer than $\mathfrak{c}$ meagre sets, and that singleton sets are nowhere dense. Then X has a $\mathfrak{c}$-Lusin set Y.

Proof (*a*) If $X \in \mathcal{I}$ then take $Y = \emptyset$. Otherwise, let $\langle E_\xi \rangle_{\xi < \kappa}$ be a listing of Σ such that each member of Σ recurs κ times in the listing. Set $\mathcal{I}_\xi = \{E_\eta : \eta \leq \xi, E_\eta \in \mathcal{I}\}$ for each $\xi < \kappa$. Choose $\langle x_\xi \rangle_{\xi < \kappa}$ inductively such that

$$x_\xi \in X \backslash (\bigcup \mathcal{I}_\xi \cup \{x_\eta : \eta < \xi\}), \quad x_\xi \in E_\xi \quad \text{if} \quad E_\xi \notin \mathcal{I}.$$

This is possible because $X = \bigcup \mathcal{I}$, so that $\{x_\eta : \eta < \xi\}$ can be covered by fewer than κ members of $\mathcal{I}$, and no member of $\Sigma \backslash \mathcal{I}$ can be included in $\bigcup \mathcal{I}_\xi \cup \{x_\eta : \eta < \xi\}$, if $\xi < \kappa$.

Set $Y = \{x_\xi : \xi < \kappa\}$. If $E \in \Sigma \backslash \mathcal{I}$, then

$$\kappa = \#(\{\xi : E = E_\xi\}) \leq \#(E \cap Y) \leq \kappa.$$

If $E \in \mathcal{I}$, then there is a $\xi < \kappa$ such that $E = E_\xi$; now $E \in \mathcal{I}_\zeta$ for every $\zeta \geq \xi$, so $E \cap Y \subseteq \{x_\eta : \eta < \xi\}$ and $\#(E \cap Y) < \kappa$. So Y is a $(\Sigma, \mathcal{I}, \kappa)$-Lusin set.

(*b*) Let $\mathcal{U}$ be a π-base for the topology of X with $\#(\mathcal{U}) \leq \mathfrak{c}$, and let Σ be the σ-algebra of subsets of X generated by $\mathcal{U}$. Then $\#(\Sigma) \leq \mathfrak{c}$. Let $\mathcal{I}$ be the set of meagre members of Σ; then $\mathcal{I}$ is a σ-ideal of Σ. If $E \in \Sigma \backslash \mathcal{I}$, then E is Borel, and there is an open set G such that $E \triangle G$ is meagre [A5B]; as E is not meagre, nor is G; so G cannot be covered by fewer than $\mathfrak{c}$ meagre sets; so E cannot be covered by fewer than $\mathfrak{c}$ meagre sets; so E cannot be covered by fewer than $\mathfrak{c}$ members of $\mathcal{I}$. Thus (ii) and (iii) of (*a*) above are true, with $\kappa = \mathfrak{c}$.

Now observe that if $G \subseteq X$ is open, there is an open $H \subseteq G$, dense in G, such that $H \in \Sigma$. **P** Let $\mathcal{U}_0$ be a maximal disjoint collection of members of

$\mathscr{U}$ which are subsets of G. As X is ccc, $\mathscr{U}_0$ is countable, and $H = \bigcup \mathscr{U}_0 \in \Sigma$. As $\mathscr{U}$ is a π-base for the topology of X, H is dense in G. **Q** It follows that if $F \subseteq X$ is nowhere dense, there is an $E \in \mathscr{I}$ such that $F \subseteq E$. **P** There is an open $H \in \Sigma$ which is a dense subset of $X \backslash \bar{F}$; take $E = X \backslash H$. **Q**

In particular, every singleton set is included in a member of $\mathscr{I}$, and $\bigcup \mathscr{I} = X$. So by part (*a*) there is a $(\Sigma, \mathscr{I}, \mathfrak{c})$-Lusin set $Y \subseteq X$. By A3E*b*, Y is $(\hat{\Sigma}, \hat{\mathscr{I}}, \mathfrak{c})$-Lusin, where

$$\hat{\mathscr{I}} = \{F : \exists E \in \mathscr{I}, F \subseteq E\},$$
$$\hat{\Sigma} = \{E \triangle F : E \in \Sigma, F \in \hat{\mathscr{I}}\}.$$

By the last paragraph, every nowhere-dense set belongs to $\hat{\mathscr{I}}$; as $\mathscr{I}$ is a σ-ideal of $\Sigma, \hat{\mathscr{I}}$ is a σ-ideal of $\mathscr{P}X$, and must be precisely the ideal of all meagre sets. If $G \subseteq X$ is open, then there is an open $H \in \Sigma$ such that $H \subseteq G \subseteq \bar{H}$, so that $G \triangle H$ is nowhere dense, and $G \in \hat{\Sigma}$; it follows that $\hat{\Sigma}$ must be precisely the algebra of sets with the Baire property. As remarked in A3E*c*, the $(\hat{\Sigma}, \hat{\mathscr{I}}, \mathfrak{c})$-Lusin set Y is therefore $\mathfrak{c}$-Lusin.

A3G Souslin's operation

(*a*) Let X be a set, $\mathscr{A} \subseteq \mathscr{P}X$. Write $\mathscr{S}(\mathscr{A})$ for the collection of subsets of X expressible in the form

$$\bigcup\{\bigcap_{n \geq 1} A_{f \restriction n} : f \in \mathbf{N}^{\mathbf{N}}\},$$

where $A_g \in \mathscr{A}$ for every $g \in \bigcup_{n \geq 1} \mathbf{N}^n$. Then $\mathscr{S}(\mathscr{S}(\mathscr{A})) = \mathscr{S}(\mathscr{A})$ [Jayne & Rogers 80*b*, 2.3.1]. I say that members of $\mathscr{S}(\mathscr{A})$ are **obtainable by Souslin's operation** from members of $\mathscr{A}$; if $\mathscr{S}(\mathscr{A}) = \mathscr{A}$, then $\mathscr{A}$ is **closed under Souslin's operation.** Note that $\mathscr{S}(\mathscr{A})$ is always closed under countable unions and intersections [Jayne & Rogers 80*b*, 2.3.3].

(*b*) Let X be a set and Σ a σ-algebra of subsets of X. Suppose that for every $A \subseteq X$ there is an $E \in \Sigma$ such that (α) $A \subseteq E$ (β) whenever $F \in \Sigma$ and $A \subseteq F$, then every subset of $E \backslash F$ belongs to Σ. Then Σ is closed under Souslin's operation. [Jayne & Rogers 80*b*, 2.9.2; Kuratowski 66, §11.VII.] In particular, (i) if there is an ω_1-saturated σ-ideal $\mathscr{I}$ of Σ such that $E \subseteq F \in \mathscr{I} \Rightarrow E \in \mathscr{I}$, then Σ is closed under Souslin's operation; (ii) if (X, Σ, μ) is a complete locally determined measure space, then Σ is closed under Souslin's operation; (iii) if Σ is the algebra of sets with the Baire property for some topology on X, then Σ is closed under Souslin's operation.

A3H Trees

(*a*) A **tree** is a partially ordered set T such that, for any $x \in T$, the set $\{y : y \in T, y \leq x\}$ is well ordered.

(*b*) Any tree T is **well founded** i.e. every non-empty subset of T has a minimal element. Consequently there is a function $r:T\to \mathrm{On}$ defined by writing

$$r(x)=\sup\{r(y)+1:y<x\}\ \forall x\in T.$$

[KUNEN 80, III.5.6; JECH 78, Theorem 5; LEVY 79, II.5.2.] We say that $r(x)$ is the **rank** of x. Observe that $r(x)\le\xi$ iff $r(y)<\xi$ for every $y<x$.

(*c*) The **height** of T is $\sup_{x\in T}(r(x)+1)$. The **levels** of T are the sets $r^{-1}[\{\xi\}]$ as ξ runs through the ordinals.

(*d*) A **branch** of a tree T is a maximal chain.

(*e*) Observe that in a tree T (i) a subset of T is upwards-linked iff it is a chain (ii) a subset of T is an up-antichain iff it is a weak antichain [definitions: 11A].

A3I Varieties of tree

(*a*) A **Souslin tree** is an uncountable upwards-ccc tree in which every branch is countable. (Also called an 'ω_1-Souslin' tree.)

(*b*) An **Aronszajn** tree is an uncountable tree in which every branch and every level is countable. (Also called an 'ω_1-Aronszajn' tree.)

(*c*) A **Kurepa** tree is a tree of height ω_1 in which every level is countable and which has more than ω_1 uncountable branches. (Also called an 'ω_1-Kurepa' tree.)

(*d*) A **special** tree is one expressible as a countable union of weak antichains.

(*e*) A **Hausdorff** tree is a tree T such that if $x, y\in T$ and $r(x)=r(y)$ is a non-zero limit ordinal, and $\{z:z<x\}=\{z:z<y\}$, then $x=y$.

A3J Lemma

There is a special Hausdorff Aronszajn tree.

Proof JECH 78, Lemma 22.3; TODORČEVIĆ 83*c*, 5.2.

A3K Tree topologies

If T is a tree, the **fine tree topology** on T is that generated by sets of the form

$$\{x:x\le y\},\quad \{x:x>y\}$$

as y runs through T.

Observe that (i) the fine tree topology is Hausdorff iff T is a Hausdorff tree in the sense of A3I*e* (ii) on any branch of T, the fine tree topology induces the order topology.

A3L **Infinite games**

These appear in 22O*i*, 43O*d*. They can both be represented as follows. We are given a set X and a set $W \subseteq X^{\mathbf{N}}$. Two players choose points in X alternately; White chooses $x_0, x_2, x_4, \ldots$ while Black chooses $x_1, x_3, \ldots$. White wins if the resulting sequence $\langle x_n \rangle_{n \in \mathbf{N}}$ belongs to W; otherwise Black wins. (Usually there are 'rules' limiting the points that players may select. In the cases under discussion these can be accommodated in the framework above by adjustments to the set W.) A **strategy** for White is a function $f: \bigcup_{n \in \mathbf{N}} X^{2n} \to X$; it is a **winning** strategy if $\langle x_n \rangle_{n \in \mathbf{N}} \in W$ whenever $x_{2n} = f(\langle x_i \rangle_{i < 2n})$ for every $n \in \mathbf{N}$. Similarly, a winning strategy for Black is a function $f: \bigcup_{n \in \mathbf{N}} X^{2n+1} \to X$ such that $\langle x_n \rangle_{n \in \mathbf{N}} \notin W$ whenever $x_{2n+1} = f(\langle x_i \rangle_{i \leq 2n})$ for every $n \in \mathbf{N}$. The game is **determined** if one of the players has a winning strategy. A strategy f for Black is **stationary** if there is a $g: X \to X$ such that $f(\langle x_i \rangle_{i \leq 2n}) = g(x_{2n})$ whenever $n \in \mathbf{N}$, $\langle x_i \rangle_{i \leq 2n} \in X^{2n+1}$.

A3M **Miscellaneous definitions**

(*a*) In 21O*h* I mention **constructible** sets and $\omega_1^{L(A)}$. These refer to basic concepts of model theory; see JECH 78, §15.

(*b*) If X is a totally ordered set and ζ is an ordinal, the **lexicographic ordering** on X^ζ is given by saying that $f < g$ iff there is a $\xi < \zeta$ such that $f \upharpoonright \xi = g \upharpoonright \xi$ and $f(\xi) < g(\xi)$.

(*c*) A **graph** is a set X of **vertices** together with a set $S \subseteq [X]^2$ of **edges**. A **colouring** of X is a partition $\mathscr{A}$ of X such that $S \cap [A]^2 = \emptyset$ for every $A \in \mathscr{A}$. The **chromatic number** of X is the least cardinal of any colouring.

(*d*) A set $A \subseteq \mathbf{N}$ has **asymptotic density 0**, if $\lim_{n \to \infty} \#(A \cap n)/n = 0$.

(*e*) A **lattice** is a partially ordered set P such that $x \vee y = \sup\{x, y\}$ and $x \wedge y = \inf\{x, y\}$ exist for all $x, y \in P$. P is **Dedekind complete** if $\sup A$ exists in P whenever $A \subseteq P$ is non-empty and has an upper bound in P.

(*f*) The following words are used by many authors. (i) 'Filter', to mean a downwards-directed, or up-open downwards-directed, or (rarely) upwards-directed subset of a partially ordered set. (ii) 'Dense', to describe a coinitial or (occasionally) cofinal subset of a partially ordered set.

(iii) 'Generic', to describe a set which meets each of a given family of coinitial or cofinal sets.

A4 General topology

My principal sources are KELLEY 55, ENGELKING 77 and JUHÁSZ 80*b*; I try to be explicit about any point on which my notation conflicts with any of these. I follow Kelley, but not Engelking, in using the terms 'regular', 'completely regular', 'normal' and 'compact' in their wider senses, that is, admitting non-Hausdorff spaces. I have split certain topological concepts off into §A5.

A4A Cardinal functions

[see JUHÁSZ 80*b*] Let $(X, \mathfrak{T})$ be a topological space.

(*a*) The **weight** of X, $w(X)$, is the smallest cardinal of any base for $\mathfrak{T}$.

(*b*) A **π-base** for $\mathfrak{T}$ is a set $\mathscr{U} \subseteq \mathfrak{T} \setminus \{\emptyset\}$ such that every member of $\mathfrak{T} \setminus \{\emptyset\}$ includes a member of $\mathscr{U}$. The **π-weight** of X, $\pi(X)$, is the smallest cardinal of any π-base for $\mathfrak{T}$.

(*c*) The **density** of X, $d(X)$, is the smallest cardinal of any dense subset of X.

(*d*) The **hereditary density** of X, $hd(X)$, is

$$\sup\{d(Y): Y \subseteq X\}.$$

(JUHÁSZ 80*b* calls this the 'width' of X and denotes it $z(X)$.)

(*e*) The **cellularity** of X, $c(X)$, is

$$\sup\{\#(\mathscr{G}): \mathscr{G} \text{ is a disjoint family of non-empty open sets}\}.$$

(ENGELKING 77 calls this the 'Souslin number' of X.)

(*f*) The **hereditary cellularity** of X, $hc(X)$, is

$$\sup\{c(Y): Y \subseteq X\}.$$

(JUHÁSZ 80*b* calls this the 'spread' of X and denotes it $s(X)$.)

(*g*) The **Lindelöf degree** of X, $L(X)$, is the smallest cardinal κ such that every open cover of X has a subcover of cardinal less than or equal to κ.

(*h*) The **hereditary Lindelöf degree** of X, $hL(X)$, is

$$\sup\{L(Y): Y \subseteq X\}.$$

(JUHÁSZ 80*b* calls this the 'height' of X and denotes it $h(X)$.)

(*i*) The **compactness degree** of X, $\hat{L}(X)$, is the smallest cardinal κ such that every open cover of X has a subcover of cardinal less than κ.

(*j*) If $F \subseteq X$, then $\chi(F, X)$ is the smallest cardinal of any collection $\mathscr{G}$ of neighbourhoods of F such that every open set including F includes a member of $\mathscr{G}$. If $x \in X$, the $\chi(x, X) = \chi(\{x\}, X)$ is the smallest cardinal of any base of neighbourhoods of x. The **character** of X, $\chi(X)$, is $\sup_{x \in X} \chi(x, X)$.

(*k*) If $F \subseteq X$ is expressible as the intersection of a family of open sets, then $\psi(F, X)$ is the smallest cardinal of any such family. If $x \in X$, then $\psi(x, X)$ is $\psi(\{x\}, X)$ if this is defined.

(*l*) If $x \in X$ then $\pi\chi(x, X)$, the **π-character** of x in X, is the smallest cardinal of any family $\mathscr{G}$ of non-empty open sets such that every neighbourhood of x includes some member of $\mathscr{G}$.

(*m*) The **tightness** of X, $t(X)$, is the smallest cardinal such that whenever $A \subseteq X$ and $x \in \bar{A}$ there is a $B \subseteq A$ such that $x \in \bar{B}$ and $\#(B) \leq \kappa$. (ENGELKING 77 denotes this $\tau(X)$.)

(*n*) The **augmented tightness** of X, $t^+(X)$, is the smallest regular infinite cardinal κ such that whenever $A \subseteq X$ and $x \in \bar{A}$ there is a $B \subseteq A$ such that $x \in \bar{B}$ and $\#(B) < \kappa$.

(*o*) A **caliber** of X (following JUHÁSZ 80*b*, but not ENGELKING 77) is an infinite cardinal κ such that whenever $\langle G_\xi \rangle_{\xi < \kappa}$ is a family of non-empty open subsets of X there is an $x \in X$ such that

$$\#(\{\xi : x \in G_\xi\}) = \kappa.$$

Remark

Note that these definitions differ from the customary ones in that I permit finite values, except for t^+ and for calibers. For Hausdorff spaces, of course, this makes a difference only in trivial cases.

A4B Elementary relationships

[see JUHÁSZ 80*b*, 2.1]

(*a*) For any topological space X,

$$c(X) \leq d(X) \leq \pi(X) \leq w(X);$$
$$c(X) \leq hc(X) \leq hd(X) \leq w(X);$$
$$L(X) \leq hL(X) \leq w(X);\ hc(X) \leq hL(X) \leq \#(X);$$
$$t(X) \leq \chi(X) \leq w(X);\ t(X) \leq hd(X);\ d(X) \leq hd(X) \leq \#(X).$$

Diagrammatically

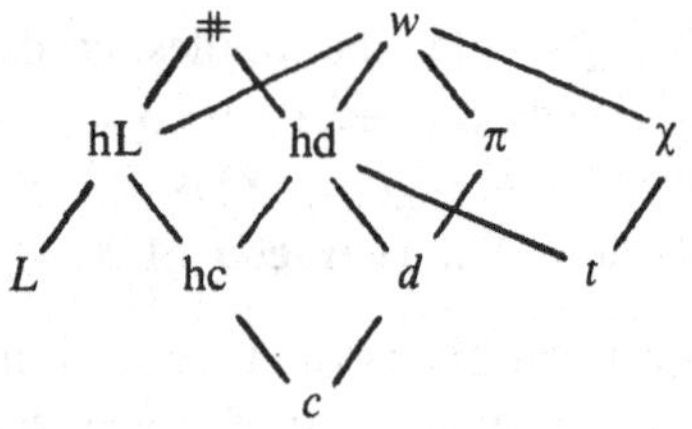

[See also A4G*d*.]

(*b*) For any topological space X, $\max(\omega, t(X)) \leq t^+(X) \leq \max(\omega, t(X)^+)$; if $t(X)$ is ω or an infinite successor cardinal, $t^+(X) = t(X)^+$. Similarly, $L(X) \leq \hat{L}(X) \leq L(X)^+$, and $\hat{L}(X) = L(X)^+$ if $L(X)$ is a successor cardinal. [See Juhász 80*b*, 1.22.]

If κ is a caliber of X, so is $\mathrm{cf}(\kappa)$. Any infinite cardinal κ with $\mathrm{cf}(\kappa) > d(X)$ is a caliber of X; in particular, $\max(\omega, d(X)^+)$ is always a caliber of X. If κ is a caliber of X, $c(X) < \mathrm{cf}(\kappa)$. Note that a regular cardinal κ is a caliber of X iff whenever $\mathscr{G}$ is a family of open sets in X with $\#(\mathscr{G}) = \kappa$, there is an $x \in X$ such that $\#(\{G : x \in G \in \mathscr{G}\}) = \kappa$. In particular, ω_1 is a caliber of X iff every point-countable family of open sets in X is countable.

(*c*) If X is T_1, then $\psi(F, X)$ is defined and $\psi(F, X) \leq \chi(F, X)$ for every $F \subseteq X$; so that $\psi(x, X) \leq \chi(x, X)$ for every $x \in X$.

(*d*) If X is compact and Hausdorff, $\psi(F, X) = \chi(F, X)$ for every closed $F \subseteq X$ [Juhász 80*b*, 1.15]. If X is locally compact and Hausdorff, $\psi(x, X) = \chi(x, X)$ for every $x \in X$.

(*e*) If X is metrizable then $t(X) \leq \omega$, $\chi(X) \leq \omega$,

$$w(X) = \pi(X) = d(X) = hd(X) = c(X) = hc(X) = L(X) = hL(X),$$

and an infinite cardinal κ is a caliber of X iff $d(X) < \mathrm{cf}(\kappa)$.

(*f*) If X is regular, then $\psi(F, X)$ is defined and $\psi(F, X) \leq hL(X)$ for every closed $F \subseteq X$ [Juhász 80*b*, 2.10]. So if X is locally compact and Hausdorff, $\chi(X) \leq hL(X)$.

(*g*) For any topological space X, if $\langle F_\xi \rangle_{\xi < \kappa}$ is an increasing family of closed sets in X and $\mathrm{cf}(\kappa) \geq t^+(X)$, then $\bigcup_{\xi < \kappa} F_\xi$ is closed. (This is the chief use of t^+.)

(*h*) $hL(X)$ is the smallest cardinal κ such that whenever $\mathscr{G}$ is a family of open sets in X there is an $\mathscr{H} \subseteq \mathscr{G}$ such that $\bigcup \mathscr{H} = \bigcup \mathscr{G}$ and $\#(\mathscr{H}) \leq \kappa$.

(*i*) $\#(C(X)) \leq 2^{\max(\omega,d(X))}$. **P** Let $D \subseteq X$ be a dense set of cardinal $d(X)$. Then $f \to f \restriction D : C(X) \to \mathbf{R}^D$ is injective, so

$$\#(C(X)) \leq \#(\mathbf{R}^D) = \#(\{0,1\}^{\mathbf{N}\times D}) \leq 2^{\max(\omega,d(X))}. \quad \mathbf{Q}$$

(*j*) If $\mathscr{U}$ is any base for the topology of X, there is a base $\mathscr{V} \subseteq \mathscr{U}$ with $\#(\mathscr{V}) = w(X)$. **P** Let $\mathscr{W}$ be a base with $\#(\mathscr{W}) = w(X)$. (α) If $w(X) < \omega$ then $\mathscr{W} \subseteq \mathscr{U}$. ($\beta$) Otherwise, set $\mathscr{S} = \{(V, W): V, W \in \mathscr{W}, \exists U \in \mathscr{U}$ such that $V \subseteq U \subseteq W\}$; for $(V, W) \in \mathscr{S}$, choose $U_{VW} \in \mathscr{U}$ such that $V \subseteq U_{VW} \subseteq W$; set $\mathscr{V} = \{U_{VW}:(V, W) \in \mathscr{S}\}$. **Q**

A4C The countable cases

(*a*) X is called **second-countable** if $w(X) \leq \omega$. X is called **separable** if $d(X) \leq \omega$. X is called **hereditarily separable** if $hd(X) \leq \omega$. X is called **ccc** if $c(X) \leq \omega$. X is called **hereditarily ccc** if $hc(X) \leq \omega$. X is called **Lindelöf** if $L(X) \leq \omega$. X is called **hereditarily Lindelöf** ('strongly Lindelöf' in SCHWARTZ 73) if $hL(X) \leq \omega$. X is called **compact** ('quasi-compact' in ENGELKING 77, BOURBAKI 66) if $\hat{L}(X) \leq \omega$. X is called **first-countable** if $\chi(X) \leq \omega$. X is called **countably tight** if $t(X) \leq \omega$, i.e. $t^+(X) \leq \omega_1$. X is called **discrete** if it is T_1 and $t^+(X) = \omega$; equivalently, if it is T_1 and $t(X) \leq 1$; equivalently, if it is T_1 and $\chi(X) < \omega$. Observe that a hereditarily Lindelöf locally compact Hausdorff space is first-countable [A4B*f*].

(*b*) If X is Hausdorff, then X is finite iff one of $w(X)$, $\pi(X)$, $d(X)$, $hd(X)$, $c(X)$, $hc(X)$, $L(X)$, $hL(X)$, $\hat{L}(X)$ is finite, iff ω is a caliber of X. In this case

$$\#(X) = w(X) = \ldots = c(X) = \ldots = hL(X)$$

as in A4B*e*.

A4D Subspaces, products, continuous images

(*a*) If Y is a subspace of X, then $w(Y) \leq w(X)$, $hd(Y) \leq hd(X)$, $hc(Y) \leq hc(X)$, $hL(Y) \leq hL(X)$, $\chi(Y) \leq \chi(X)$, $t(Y) \leq t(X)$ and $t^+(Y) \leq t^+(X)$.

(*b*) Suppose that Y is a dense subspace of a topological space X. Then $c(Y) = c(X)$ and $\pi(Y) \leq \pi(X)$; if X is regular, $\pi(Y) = \pi(X)$. [JUHASZ 80*b*, 2.6–2.7.]

(*c*) If Y is a continuous image of X then $d(Y) \leq d(X)$, $hd(Y) \leq hd(X)$, $c(Y) \leq c(X)$, $hc(Y) \leq hc(X)$, $L(Y) \leq L(X)$, $hL(Y) \leq hL(X)$, $\hat{L}(Y) \leq \hat{L}(X)$ and any caliber of Y is a caliber of X.

(*d*) If X and Y are compact Hausdorff spaces and Y is a continuous image of X, then $t(Y) \le t(X)$, $t^+(Y) \le t^+(X)$ and $w(Y) \le w(X)$. **P** For the first two inequalities, observe that if $f:X \to Y$ is any continuous surjection then $\overline{f[A]} = f[\bar{A}]$ for every $A \subseteq X$; see JUHÁSZ 80*b*, 1.17. For the third, use ENGELKING 77, 3.1.21 or JUHÁSZ 80*b*, 3.32. **Q**

(*e*) If X is hereditarily Lindelöf and Y is second-countable then $X \times Y$ is hereditarily Lindelöf. **P** Let $\mathscr{H}$ be a family of open sets in $X \times Y$. Let $\mathscr{U}$ be a countable base for the topology of Y. For $U \in \mathscr{U}$ set

$$\mathscr{G}_U = \{G : G \subseteq X \text{ open}, \quad \exists H \in \mathscr{H}, G \times U \subseteq H\}.$$

For each $U \in \mathscr{U}$ we can choose a countable $\mathscr{G}'_U \subseteq \mathscr{G}_U$ such that $\bigcup \mathscr{G}'_U = \bigcup \mathscr{G}_U$ (because X is hereditarily Lindelöf). For each $U \in \mathscr{U}$, $G \in \mathscr{G}'_U$ choose an $H^G_U \in \mathscr{H}$ such that $G \times U \subseteq H^G_U$. Then $\mathscr{H}' = \{H^G_U : U \in \mathscr{U}, G \in \mathscr{G}'_U\}$ is countable and $\bigcup \mathscr{H}' = \bigcup \mathscr{H}$. **Q**

(*f*) If $\langle X_\iota \rangle_{\iota \in I}$ is any family of topological spaces, then $w(\prod_{\iota \in I} X_\iota) \le \max(\omega, \sum_{\iota \in I} w(X_\iota))$. [JUHÁSZ 80*b*, 5.3(*a*).]

A4E **Discrete sets**

Let X be a topological space.

(*a*) A set $A \subseteq X$ is **discrete** if it is discrete in its induced topology i.e. $x \notin \overline{A \setminus \{x\}}$ for every $x \in A$.

(*b*) $hc(X) = \sup\{\#(A) : A \subseteq X \text{ is discrete}\}$. **P** If $A \subseteq X$ is discrete, $\#(A) = c(A) \le hc(X)$. If Y is any subset of X and $\mathscr{G}$ is a disjoint family of non-empty relatively open sets in Y, let $A \subseteq Y$ be a set consisting of just one point chosen from each member of $\mathscr{G}$; then A is discrete and $\#(A) = \#(\mathscr{G})$; as Y and $\mathscr{G}$ are arbitrary, $hc(X) \le \sup\{\#(A) : A \subseteq X \text{ discrete}\}$. **Q**

(*c*) If every open set in X is an F_σ set, and $A \subseteq X$ is discrete, then A is expressible as $\bigcup_{n \in \mathbb{N}} A_n$ where each A_n is closed and discrete. **P** A is relatively open in $\bar{A}$; say $A = G \cap \bar{A}$ where G is open in X. Express G as $\bigcup_{n \in \mathbb{N}} F_n$ where each F_n is closed in X. Then $A = \bigcup_{n \in \mathbb{N}} (A \cap F_n)$, and $A \cap F_n = \bar{A} \cap F_n$ is closed and discrete for each $n \in \mathbb{N}$. **Q**

(*d*) A family $\mathscr{A}$ of subsets of X is **discrete** if every point of X has a neighbourhood meeting at most one element of $\mathscr{A}$. In this case

$$\overline{\bigcup \mathscr{B}} = \bigcup \{\bar{B} : B \in \mathscr{B}\} \; \forall \mathscr{B} \subseteq \mathscr{A}.$$

A4F **Limits and cluster points**

Let X be a topological space.

(*a*) If $\mathscr{F}$ is a filter on X, I say that $\mathscr{F} \to x$, '$\mathscr{F}$ converges to x', 'x is a limit of $\mathscr{F}$' if $\mathscr{F}$ contains every neighbourhood of x. If X is Hausdorff, I write $x = \lim \mathscr{F}$.

(*b*) If $\langle x_n \rangle_{n \in \mathbf{N}}$ is a sequence in X, I say that $\langle x_n \rangle_{n \in \mathbf{N}} \to x$ if the filter on X generated by $\{\{x_i : i \geq n\} : n \in \mathbf{N}\}$ converges to x.

(*c*) If X is Hausdorff, $\langle x_n \rangle_{n \in \mathbf{N}}$ is a sequence in X, and $\mathscr{F}$ is a filter on $\mathbf{N}$, I say that $x = \lim_{n \to \mathscr{F}} x_n$ if x is the limit of the filter on X generated by $\{\{x_n : n \in A\} : A \in \mathscr{F}\}$.

(*d*) If $\mathscr{F}$ is a filter on X, a **cluster point** of $\mathscr{F}$ is a point in $\bigcap_{F \in \mathscr{F}} \bar{F}$; equivalently, a limit of some filter on X which includes $\mathscr{F}$.

(*e*) If $\langle x_n \rangle_{n \in \mathbf{N}}$ is a sequence in X, a **cluster point** of $\langle x_n \rangle_{n \in \mathbf{N}}$ is a point in $\bigcap_{n \in \mathbf{N}} \overline{\{x_i : i \geq n\}}$.

(*f*) If $A \subseteq X$, an **accumulation point** of A is any $x \in X$ such that $x \in \overline{A \setminus \{x\}}$.

A4G **Compactness and related concepts**
Let X be a topological space.

(*a*) X is **sequentially compact** if every sequence in X has a convergent subsequence. X is **countably compact** if every countable open cover of X has a finite subcover; equivalently, if every sequence in X has a cluster point [Engelking 77, 3.10.3]. X is **locally countably compact** if every point of X has a countably compact neighbourhood. A subset A of X is **relatively countably compact** if every sequence in A has a cluster point in X. (Do not confuse with the related concepts '$\bar{A}$ is countably compact', 'A is included in some countably compact set'.) X is **initially κ-compact** if every cover of X by κ or fewer open sets has a finite subcover.

(*b*) X is **angelic** if (i) X is regular and Hausdorff, (ii) every relatively countably compact subset of X is relatively compact, (iii) whenever $A \subseteq X$ is relatively compact and $x \in \bar{A}$ then there is a sequence in A coverging to x. [See Pryce 71.]

(*c*) X is compact iff it is Lindelöf and countably compact [immediate from the definitions] iff every ultrafilter on X has a limit in X [Engelking 77, 3.1.24 and 1.6.8].

(*d*) For compact Hausdorff spaces, the diagram of A4B*a* becomes

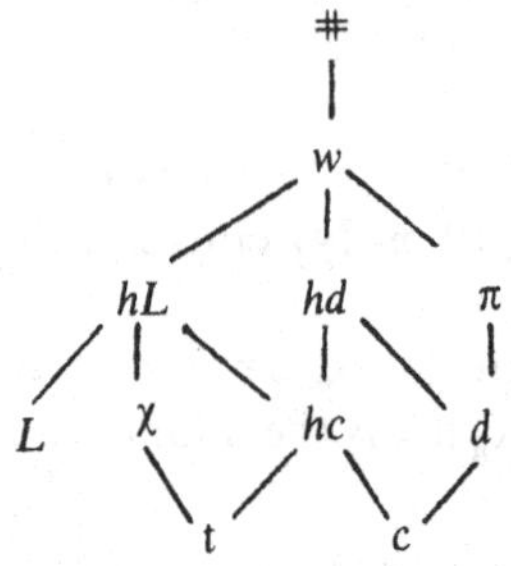

[Use the argument of A4K for $t \leq hc$, and A4B*f* for $\chi \leq hL$.]

A4H **Proposition**

Let X be a compact Hausdorff space.

(*a*) If X has a non-empty perfect subset, there is a continuous surjection from X onto $[0,1]$.

(*b*) X is metrizable iff $\Delta = \{(x,x): x \in X\}$ is a G_δ set in X^2.

(*c*) If X is uncountable and first-countable then it has a non-empty perfect subset and $\#(X) = \mathfrak{c}$.

Proof (*a*) JUHÁSZ 80*b*, 3.18(iii)$\Rightarrow$(i); cf. A4J*a*.

(*b*) JUHÁSZ 80*b*, 3.32.

(*c*) JUHÁSZ 80*b*, 3.17 Corollary.

A4I **Cech-complete spaces**

(*a*) A **Cech-complete** space is a topological space X which is homeomorphic to a G_δ subset of some compact Hausdorff space. Any complete metric space is Cech-complete [ENGELKING 77, 4.3.26]. Any locally compact Hausdorff space is Cech-complete. Any closed subspace of a Cech-complete space is Cech-complete.

(*b*) An **absolute G_κ space** is a topological space which is homeomorphic to a subspace of a compact Hausdorff space which is expressible as the intersection of κ or fewer open sets. (Thus a Cech-complete space is an 'absolute G_ω space'.) Any closed subspace of an absolute G_κ space is an absolute G_κ space.

A4J **Proposition**

Let X be a non-empty Cech-complete space.

(*a*) If X has no isolated points, there is a compact $K \subseteq X$ and a continuous surjection $f: K \to \{0,1\}^{\mathbf{N}}$.

(*b*) If $\chi(x,X) \geq \kappa \geq \omega$ for every $x \in X$, then $\#(X) \geq 2^{\kappa}$.

Proof (*a*) We may suppose that X is actually a G_δ set in a compact Hausdorff space Z; suppose that $X = \bigcap_{n\in\mathbf{N}} G_n$ where each $G_n \subseteq Z$ is open. For each open $H \subseteq Z$ and $n \in \mathbf{N}$, chosen open sets $\psi_0^n(H)$, $\psi_1^n(H) \subseteq Z$, with disjoint closures, such that $\overline{\psi_i^n(H)} \subseteq H \cap G_n$ for each i, and if $H \cap X \neq \emptyset$ then $\psi_i^n(H) \cap X \neq \emptyset$ for both i; such exist because if $H \cap X = \emptyset$ then we can take both $\psi_i^n(H)$ to be empty, while if $H \cap X \neq \emptyset$ it contains more than one point, and we can take the $\psi_i^n(H)$ to be neighbourhoods of two of them. For $w \in \{0,1\}^{\mathbf{N}}$ define $\langle H_n^w \rangle_{n\in\mathbf{N}}$ by writing

$$H_0^w = Z, \quad H_{n+1}^w = \psi_{w(n)}^n(H_n^w) \ \forall n \in \mathbf{N}.$$

Then (by induction on n) $\emptyset \neq \bar{H}_{n+1}^w \subseteq G_n \cap H_n^w$ for each $n \in N$; $\bar{H}_n^w = \bar{H}_n^v$ if $w \restriction n = v \restriction n$; and $\bar{H}_n^w \cap \bar{H}_n^v = \emptyset$ if $w \restriction n \neq v \restriction n$. Set

$$F_w = \bigcap_{n\in\mathbf{N}} \bar{H}_n^w \ \forall w \in \{0,1\}^{\mathbf{N}},$$
$$K = \bigcup \{F_w : w \in \{0,1\}^{\mathbf{N}}\} = \bigcap_{n\in\mathbf{N}} \bigcup \{\bar{H}_n^w : w \in \{0,1\}^{\mathbf{N}}\}.$$

Because Z is compact, no F_w is empty, and K is compact; also $K \subseteq \bigcap_{n\in\mathbf{N}} G_n = X$. Define $f: K \to \{0,1\}^{\mathbf{N}}$ by writing $f(x) = w$ if $x \in F_w$. Then f is surjective. Also, f is continuous, because

$$\{x : f(x) \restriction n = w \restriction n\} = K \cap H_n^w$$

is relatively open in K for every $w \in \{0,1\}^{\mathbf{N}}$, $n \in \mathbf{N}$.

(*b*) (i) If X is compact, this is given in Juhász 80*b*, 3.16, and Engelking 77, 3.12.11. (ii) If $\kappa = \omega$, it is a consequence of (*a*). (iii) For the general case, if $\kappa > \omega$, again express X as $\bigcap_{n\in\mathbf{N}} G_n$ where each G_n is an open set in a compact Hausdorff space Z. Start from any $x_0 \in X$. Because Z is regular, there is for each $n \in \mathbf{N}$ a closed G_δ set $F_n \subseteq Z$ such that $x_0 \in F_n \subseteq G_n$. Set $K = \bigcap_{n\in\mathbf{N}} F_n$; then K is a non-empty compact subset of X and $\psi(K,Z) \leq \omega$. Now if $x \in K$

$$\begin{aligned}\omega < \kappa \leq \chi(x,X) \leq \chi(x,Z) = \psi(x,Z) \\ \leq \max(\omega, \psi(x,K), \psi(K,Z)) \leq \max(\omega, \chi(x,K)).\end{aligned}$$

So $\chi(x,K) \geq \kappa$ for every $x \in K$, and by case (i) we have

$$\#(X) \geq \#(K) \geq 2^{\kappa}.$$

Remark (*b*) above is given in Cech & Pospíšil 38; I will call it the 'Cech-Pospíšil theorem'.

A4K **Lemma** [Sapirovskii 72*b*]

If X is a hereditarily ccc Cech-complete space, it is countably tight.

Proof Let us suppose that X is itself a G_δ subset of a compact Hausdorff space Z; say $X = \bigcap_{n\in\mathbf{N}} G_n$ where each G_n is open in Z. Let A be any subset of X. Write

$$A^* = \bigcup\{\bar{B}\cap X : B\subseteq A \text{ is countable}\}.$$

(Here, as throughout the proof, closures are taken in Z.) Observe that if $B\subseteq A^*$ is countable, then $\bar{B}\cap X\subseteq A^*$. Suppose that $x\in\bar{A}\cap X$, the closure of A in X.

If $\langle H_n\rangle_{n\in\mathbf{N}}$ is any sequence of neighbourhoods of x in Z, then $A^*\cap\bigcap_{n\in\mathbf{N}} H_n \neq \emptyset$. **P** Choose $\langle U_n\rangle_{n\in\mathbf{N}}$ inductively such that U_n is always a neighbourhood of x and $\bar{U}_{n+1}\subseteq U_n\cap H_n\cap G_n$ for each $n\in\mathbf{N}$. Choose $y_n\in A\cap U_n$ for each $n\in\mathbf{N}$. Let y be any cluster point of $\langle y_n\rangle_{n\in\mathbf{N}}$ in Z. Then $y\in\bigcap_{n\in\mathbf{N}}\bar{U}_n\subseteq\bigcap_{n\in\mathbf{N}} H_n\cap\bigcap_{n\in\mathbf{N}} G_n = X\cap\bigcap_{n\in\mathbf{N}} H_n$. As also $y\in\overline{\{y_n : n\in\mathbf{N}\}}$, $y\in A^*$. **Q**

? If $x\notin A^*$, then choose points x_ξ in A^* and neighbourhoods V_ξ of x_ξ in Z inductively, for $\xi<\omega_1$, as follows. Given $\{x_\eta : \eta<\xi\}$, our hypothesis is that $x\notin\overline{\{x_\eta : \eta<\xi\}}$, so we can choose V_ξ such that $\bar{V}_\xi\cap\overline{\{x_\eta : \eta<\xi\}} = \emptyset$. Now, given $\langle V_\eta\rangle_{\eta\leq\xi}$, we can (by the preceding paragraph) choose $x_\xi\in A^*\cap\bigcap_{\eta\leq\xi} V_\eta$.

In this case we see that, for any $\xi<\omega_1$, $x_\xi\notin\overline{\{x_\eta : \eta<\xi\}}$ because $x_\xi\in V_\xi$; while $x_\xi\notin\overline{\{x_\eta : \eta>\xi\}}$ because $x_\xi\notin\bar{V}_{\xi+1}$. So $\{x_\xi : \xi<\omega_1\}$ is an uncountable discrete set in X and X is not hereditarily ccc. **X**

Thus $x\in A^*$. As x and A are arbitrary, X is countably tight.

A4L **Normality and related concepts**

Let X be a topological space.

(*a*) I say that two sets A, $B\subseteq X$ can be **separated by open sets** if there are disjoint open sets G, $H\subseteq X$ such that $A\subseteq G$ and $B\subseteq H$.

(*b*) X is **hereditarily normal** (sometimes 'completely normal') if every subspace of X is normal; equivalently, if whenever A, $B\subseteq X$ and $\bar{A}\cap B = A\cap\bar{B} = \emptyset$ then A and B can be separated by open sets [Engelking 77, 2.1.7].

(*c*) X is **perfectly normal** if it is normal and every open set is an F_σ set; equivalently, if every closed set is a zero set [ENGELKING 77, 1.5.19]. If X is perfectly normal it is hereditarily normal [ENGELKING 77, 2.1.6]. A metrizable space is perfectly normal.

(*d*) If X is compact and Hausdorff, it is perfectly normal iff it is hereditarily Lindelöf [see A4B*d*, A4B*f*].

(*e*) If X^n is perfectly normal for every $n \in \mathbf{N}$ then $X^{\mathbf{N}}$ is perfectly normal. **P** Let $F \subseteq X^{\mathbf{N}}$ be closed. For each $n \in \mathbf{N}$ let $\pi_n : X^{\mathbf{N}} \to X^n$ be the canonical map. Then $F = \bigcap_{n \in \mathbf{N}} \pi_n^{-1}[\overline{\pi_n[F]}]$. As each $\overline{\pi_n[F]}$ is a zero set, so is F. **Q**

(*f*) X is **pseudonormal** if E and F can be separated by open sets whenever E and F are disjoint closed sets in X and E is countable.

(*g*) If X is a non-empty hereditarily normal compact Hausdorff space there is an $x \in X$ such that $\pi\chi(x, X) \leq \omega$. **P?** If $\pi\chi(x, X) \geq \omega_1$ for every $x \in X$ there is a continuous surjection from X onto $[0, 1]^{\omega_1}$ [JUHÁSZ 80*b*, 3.20]. So $[0, 1]^{\omega_1}$ must be hereditarily normal [ENGELKING 77, remark following 2.1.6]. But the Tychonov plank [ENGELKING 77, 3.12.19] is not normal and can be embedded in $[0, 1]^{\omega_1}$. **XQ**

A4M **Strong separation axioms**

Let $(X, \mathfrak{T})$ be a topological space.

(*a*) X is **paracompact** if for every open cover $\mathscr{G}$ of X there is a locally finite open cover of X refining $\mathscr{G}$. X is **countably paracompact** if for every countable open cover $\mathscr{G}$ of X there is a locally finite open cover refining $\mathscr{G}$. A **Dowker space** is a normal Hausdorff space which is not countably paracompact. If X is regular and Lindelöf it is paracompact [ENGELKING 77, 5.1.2]. If X is pseudometrizable it is paracompact [BOURBAKI 66, IX.4.5; KELLEY 55, 5.35].

(*b*) X is **collectionwise Hausdorff** if it is Hausdorff and whenever $A \subseteq X$ is a closed discrete set there is a disjoint family $\langle G_x \rangle_{x \in A}$ of open sets such that $x \in G_x$ for every $x \in A$. A paracompact Hausdorff space is collectionwise Hausdorff [ENGELKING 77, 5.1.18]. X is **hereditarily collectionwise Hausdorff** if every subspace of X is collectionwise Hausdorff; equivalently, if X is Hausdorff and whenever $A \subseteq X$ is discrete there is a disjoint family $\langle G_x \rangle_{x \in A}$ of open sets such that $x \in G_x$ for every $x \in A$. If X is regular, Hausdorff and hereditarily ccc it is hereditarily collectionwise Hausdorff.

(*c*) X is **metacompact** if for every open cover $\mathscr{G}$ of X there is a point-finite open cover of X refining $\mathscr{G}$. X is **hereditarily metacompact** if every subspace of X is metacompact i.e. whenever $\mathscr{G}$ is a collection of open sets in X there is a point-finite collection $\mathscr{H}$ of open sets refining $\mathscr{G}$ with $\bigcup\mathscr{H} = \bigcup\mathscr{G}$. X is **countably metacompact** if for every countable open cover $\mathscr{G}$ of X there is a point-finite open cover of X refining $\mathscr{G}$.

(*d*) X is **metalindelöf** if for every open cover $\mathscr{G}$ of X there is a point-countable open cover of X refining $\mathscr{G}$. X is **hereditarily metalindelöf** if every subspace of X is metalindelöf.

A4N Moore spaces

(*a*) Let X be a topological space. A **development** for the topology of X is a sequence $\langle\mathscr{U}_n\rangle_{n\in\mathbf{N}}$ of open covers of X such that

$$\{\mathrm{St}(x,\mathscr{U}_n):n\in\mathbf{N}\}$$

is a base of neighbourhoods of x for every $x\in X$. Observe that if $\langle\mathscr{U}_n\rangle_{n\in\mathbf{N}}$ is a development, and $\mathscr{U}'_n\subseteq\mathscr{U}_n$ and $\bigcup\mathscr{U}'_n = X$ for each $n\in\mathbf{N}$, then $\bigcup_{n\in\mathbf{N}}\mathscr{U}'_n$ is a base for the topology of X.

A **Moore** space is a regular Hausdorff space whose topology has a development. Clearly Moore spaces are first-countable.

(*b*) If X is a Moore space then every open set in X is F_σ. **P** If $\langle\mathscr{U}_n\rangle_{n\in\mathbf{N}}$ is a development for the topology of X and $G\subseteq X$ is open, set

$$F_n = \{x:\mathrm{St}(x,\mathscr{U}_n)\subseteq G\} = X\setminus\bigcup_{x\in X\setminus G}\mathrm{St}(x,\mathscr{U}_n);$$

then each F_n is closed and $G = \bigcup_{n\in\mathbf{N}}F_n$. **Q**

(*c*) If X is a Moore space then $w(X) = L(X)$. **P** (i) For any topological space, $L(X)\le w(X)$. (ii) Let $\langle\mathscr{U}_n\rangle_{n\in\mathbf{N}}$ be a development for the topology of X. Then for each $n\in\mathbf{N}$ there is a $\mathscr{U}'_n\subseteq\mathscr{U}_n$ such that $\bigcup\mathscr{U}'_n = X$ and $\#(\mathscr{U}'_n)\le L(X)$. Now $\bigcup_{n\in\mathbf{N}}\mathscr{U}'_n$ is a base for the topology of X so $w(X)\le \max(\omega, L(X))$. (iii) If $L(X)<\omega$ use A4C*b*. **Q**

(*d*) Any subspace of a Moore space is a Moore space; the product of countably many Moore spaces is a Moore space.

(*e*) A topological space X is metrizable iff it is a paracompact Moore space. [ENGELKING 77, 5.4.1 and 5.1.18.] In particular, compact Moore spaces are metrizable; it follows at once that compact subsets of Moore spaces are metrizable.

(*f*) Let X be a first-countable regular Hausdorff space expressible as $\bigcup_{n\in\mathbf{N}}X_n$ where each X_n is closed and discrete. Then X is a Moore space. **P** For each $x\in X$ choose a decreasing sequence $\langle U_n(x)\rangle_{n\in\mathbf{N}}$ of open neighbourhoods of x such that (i) $\{U_n(x): n\in\mathbf{N}\}$ is a base of neighbourhoods of x, (ii) $U_n(x)\cap X_n \subseteq \{x\}$ for each $n\in\mathbf{N}$. Set $\mathcal{U}_n = \{U_n(x): x\in X\}$. Then each $\mathcal{U}_n$ is an open cover of X. If $x\in X$, there is an $m\in\mathbf{N}$ such that $x\in X_m$. Now if $n\geq m$ and $x\in U_n(y)$, we have $x\in U_m(y)\cap X_m \subseteq \{y\}$, so $x=y$; accordingly $\mathrm{St}(x,\mathcal{U}_n)=U_n(x)$ for every $n\geq m$, and $\{\mathrm{St}(x,\mathcal{U}_n): n\in\mathbf{N}\}\supseteq\{U_n(x): n\geq m\}$ is a base of neighbourhoods of x. This shows that $\{\mathcal{U}_n: n\in\mathbf{N}\}$ is a development for the topology of X, so that X is a Moore space. **Q**

A4O Lemma

Let X be a Moore space. Then there is a sequence $\langle\mathcal{E}_n\rangle_{n\in\mathbf{N}}$ of discrete families of closed subsets of X such that if $x\in X$ and U is a neighbourhood of x there is an $E\in\bigcup_{n\in\mathbf{N}}\mathcal{E}_n$ such that $x\in E\subseteq U$.

Proof [see ENGELKING 77, 5.4.1] Let $\langle\mathcal{U}_n\rangle_{n\in\mathbf{N}}$ be a development for the topology of X. For each $n\in\mathbf{N}$ set $\kappa(n)=\#(\mathcal{U}_n)$ and enumerate $\mathcal{U}_n$ as $\langle U_{n\xi}\rangle_{\xi<\kappa(n)}$. For k, $n\in\mathbf{N}$ and $\xi<\kappa(n)$ set

$$E^k_{n\xi} = \{x: x\in X, \mathrm{St}(x,\mathcal{U}_k)\subseteq U_{n\xi}\}\setminus\bigcup_{\eta<\xi}U_{n\eta},$$

so that each $E^k_{n\xi}$ is closed. Set

$$\mathcal{E}_{kn} = \{E^k_{n\xi}: \xi<\kappa(n)\}.$$

Then $\mathcal{E}_{kn}$ is discrete. **P** Let $z\in X$. Let ζ be the least ordinal such that $z\in U_{n\zeta}$, and let $V\in\mathcal{U}_k$ be such that $z\in V$. If $\xi>\zeta$ then $U_{n\zeta}\cap E^k_{n\xi}=\emptyset$. If $\xi<\zeta$ and $x\in E^k_{n\xi}$, then $\mathrm{St}(x,\mathcal{U}_k)\subseteq U_{n\xi}$ so $z\notin\mathrm{St}(x,\mathcal{U}_k)$ and $x\notin V$; thus $V\cap E^k_{n\xi}=\emptyset$. So there is only one set that can possibly belong to $\mathcal{E}_{kn}$ and meet $U_{n\zeta}\cap V$, namely $E^k_{n\zeta}$. **Q**

If $x\in X$ and V is a neighbourhood of X, let $n\in\mathbf{N}$ be such that $\mathrm{St}(x,\mathcal{U}_n)\subseteq V$. Let ξ be the least ordinal such that $x\in U_{n\xi}$, and let k be such that $\mathrm{St}(x,\mathcal{U}_k)\subseteq U_{n\xi}$. Then $x\in E^k_{n\xi}\subseteq V$.

So $\langle\mathcal{E}_{kn}\rangle_{k,n\in\mathbf{N}}$, suitably re-indexed, is an appropriate sequence.

A4P Proposition

Let X be a Moore space.

(*a*) If $d(X)\leq\mathfrak{c}$ then X is sub-second-countable.

(*b*) If X is normal then $2^{d(X)}=2^{w(X)}$.

(*c*) If $d(X)\leq\mathfrak{c}$ and X is normal there is a separable metrizable topology on X coarser than the given topology.

Proof Take $\langle\mathcal{E}_n\rangle_{n\in\mathbf{N}}$ from A4O; we may suppose that no $\mathcal{E}_n$ contains $\emptyset$. Set $\mathcal{E}=\bigcup_{n\in\mathbf{N}}\mathcal{E}_n$.

(*a*) As X is Hausdorff and first-countable, $\#(\mathscr{E}_n) \leq \#(X) \leq \#(d(X)^{\mathbf{N}}) \leq \mathfrak{c}$ for each $n \in \mathbf{N}$. So $\#(\mathscr{E}) \leq \mathfrak{c}$. Let $\mathscr{A} \subseteq \mathscr{P}\mathbf{N}$ be an almost-disjoint family of infinite sets with cardinal $\mathfrak{c}$ [A3A*d*]. Let $\varphi:\mathscr{E} \to \mathscr{A}$ be any injection. For n, $r \in \mathbf{N}$ set

$$H_{nr} = \bigcup\{E : E \in \mathscr{E}_n, r \in \varphi(E)\}.$$

As $\mathscr{E}_n$ is a discrete family of closed sets, H_{nr} is closed. Let $\mathfrak{S}_0$ be the topology on X generated by $\{X \setminus H_{nr} : n, r \in \mathbf{N}\}$. Then $\mathfrak{S}_0$ is a second-countable topology coarser than the given topology of X. Also $\mathfrak{S}_0$ is T_1. **P** Let x, y be distinct points of X. Let U be a neighbourhood of x not containing y. Then there is an $E \in \mathscr{E}$ such that $x \in E \subseteq U$; suppose $E \in \mathscr{E}_n$. If $y \notin \bigcup \mathscr{E}_n$ take any $r \in \varphi(E)$; if $y \in F \in \mathscr{E}_n$ take $r \in \varphi(E) \setminus \varphi(F)$. In either case we see that $x \in H_{nr}$ but $y \notin H_{nr}$. As y is arbitrary, $\{x\}$ is $\mathfrak{S}_0$-closed; as x is arbitrary, $\mathfrak{S}_0$ is T_1. **Q** So the original topology on X is sub-second-countable.

(*b*) For each $n \in \mathbf{N}$ and $\mathscr{D} \subseteq \mathscr{E}_n$, we see that $\bigcup \mathscr{D}$ and $\bigcup(\mathscr{E}_n \setminus \mathscr{D})$ are both closed. As X is normal, there is a continuous function $f_{\mathscr{D}} : X \to [0, 1]$ which is 1 on $\bigcup \mathscr{D}$ and 0 on $\bigcup(\mathscr{E}_n \setminus \mathscr{D})$. As $\varnothing \notin \mathscr{E}_n$, $\mathscr{D} \mapsto f_{\mathscr{D}} : \mathscr{P}\mathscr{E}_n \to C(X)$ is injective. So

$$\#(\mathscr{P}\mathscr{E}_n) \leq \#(C(X)) \leq 2^{\max(\omega, d(X))}$$

[A4B*i*]. Accordingly

$$\#(\mathscr{P}\mathscr{E}) \leq \#(\textstyle\prod_{n \in \mathbf{N}}(\mathscr{P}\mathscr{E}_n)) \leq 2^{\max(\omega, d(X))}.$$

Clearly $L(X) \leq \#(\mathscr{E})$; also $w(X) = L(X)$, by A4N*c*. So

$$2^{w(X)} \leq \#(\mathscr{P}\mathscr{E}) \leq 2^{\max(\omega, d(X))}.$$

Of course, if $d(X)$ is finite, then $w(X) = \#(X) = d(X)$; so in fact $2^{w(x)} \leq 2^{d(X)}$. On the other hand, $d(X) \leq w(X)$ so certainly $2^{d(X)} \leq 2^{w(X)}$ and $2^{d(X)} = 2^{w(X)}$.

(*c*) Finally, if X is normal and $d(X) \leq \mathfrak{c}$, return to $\mathfrak{S}_0$ and H_{nr} of (*a*). Each H_{nr} is a zero set [A4L*c*]; express H_{nr} as $f_{nr}^{-1}[\{0\}]$ where $f_{nr} \in C(X)$. Let $\mathfrak{S}_1$ be the coarsest topology on X for which every f_{nr} is continuous. Then $\mathfrak{S}_1$ is separable and pseudometrizable and coarser than the given topology on X. But also $\mathfrak{S}_1 \supseteq \mathfrak{S}_0$ so $\mathfrak{S}_1$ is T_1 and is metrizable.

A4Q ***σ*-centered spaces**

(*a*) A topological space $(X, \mathfrak{T})$ is ***σ*-centered** if there is a sequence $\langle \mathscr{G}_n \rangle_{n \in \mathbf{N}}$ of families of open sets, each with the finite intersection property, such that every non-empty open set belongs to some $\mathscr{G}_n$; i.e. if $\mathfrak{T} \setminus \{\varnothing\}$ is σ-centered downwards [14A].

(*b*) Let $(X, \mathfrak{T})$ be a topological space and $\mathscr{U}$ a π-base for $\mathfrak{T}$. Then $\mathscr{U}$ is σ-centered downwards iff X is σ-centered. **P** (i) If X is σ-centered, express $\mathfrak{T}\setminus\{\emptyset\}$ as $\bigcup_{n\in\mathbf{N}}\mathscr{G}_n$ where each $\mathscr{G}_n$ has the finite intersection property. Set $\mathscr{U}_n = \mathscr{U}\cap\mathscr{G}_n$; then $\mathscr{U} = \bigcup_{n\in\mathbf{N}}\mathscr{U}_n$ and each $\mathscr{U}_n$ has the finite intersection property, so is downwards-centered in $\mathscr{U}$, and $\mathscr{U}$ is σ-centered downwards. (ii) If $\mathscr{U}$ is σ-centered downwards, express $\mathscr{U}$ as $\bigcup_{n\in\mathbf{N}}\mathscr{U}_n$ where each $\mathscr{U}_n$ is downwards-centered. Set $\mathscr{G}_n = \{G : G\in\mathfrak{T}, \exists U\in\mathscr{U}_n, G\supseteq U\}$; then $\mathfrak{T}\setminus\{\emptyset\} = \bigcup_{n\in\mathbf{N}}\mathscr{G}_n$ and each $\mathscr{G}_n$ has the finite intersection property, so X is σ-centered. **Q**

(*c*) Let $(X, \mathfrak{T})$ be a σ-centered compact Hausdorff space. Then X is separable. **P** Express $\mathfrak{T}\setminus\{\emptyset\}$ as $\bigcup_{n\in\mathbf{N}}\mathscr{G}_n$ where each $\mathscr{G}_n$ has the finite intersection property. For each $n\in\mathbf{N}$ choose $x_n\in X\cap\bigcap\{\bar{G} : G\in\mathscr{G}_n\}$. (I pass over the trivial case $X=\emptyset$.) If $H\subseteq X$ is a non-empty open set, there is a non-empty open set G such that $\bar{G}\subseteq H$; now there is an $n\in\mathbf{N}$ such that $G\in\mathscr{G}_n$ and $x_n\in H$. So $\{x_n : n\in\mathbf{N}\}$ is dense in X and X is separable. **Q** [Of course the converse is also true.]

A4R **Proposition**

Let $(X, \mathfrak{T})$ be a topological space, $Y\subseteq X$ a dense subset; write $\mathscr{G}$ for the regular open algebra of X and Z for the Stone space of $\mathscr{G}$. Then the following are equivalent:

(i) X is σ-centered;

(ii) Y is σ-centered;

(iii) there is a π-base for $\mathfrak{T}$ which is σ-centered downwards;

(iv) every π-base for $\mathfrak{T}$ is σ-centered downwards;

(v) X is expressible as a dense subset of a separable topological space;

(vi) $\mathscr{G}\setminus\{\emptyset\}$ is σ-centered downwards;

(vii) Z is separable.

Proof (*a*) For (i)$\Leftrightarrow$(ii), observe just that a set $\mathscr{H}\subseteq\mathfrak{T}\setminus\{\emptyset\}$ is downwards-centered in $\mathfrak{T}\setminus\{\emptyset\}$ iff $\{H\cap Y : H\in\mathscr{H}\}$ is downwards-centered in the set of non-empty relatively open subsets of Y. Next, (i)$\Rightarrow$(iv)$\Rightarrow$(iii)$\Rightarrow$(i) by A4Q*b*.

(*b*)**(i)$\Rightarrow$(v)** We need consider only non-empty X, and can suppose that $X\cap\mathbf{N}=\emptyset$. Express $\mathfrak{T}\setminus\{\emptyset\}$ as $\bigcup_{n\in\mathbf{N}}\mathscr{H}_n$ where each $\mathscr{H}_n$ has the finite intersection property. Let $\mathscr{F}_n$ be the filter on X generated by $\mathscr{H}_n$. Set $W = X\cup\mathbf{N}$ and

$$\mathscr{W} = \{G : G\subseteq W, G\cap\mathbf{N} = \{n : G\cap X\in\mathscr{F}_n\}, G\cap X\in\mathfrak{T}\}.$$

Then $\mathscr{W}$ is a topology base on W; give W the topology it generates. It is easy to see that both X and $\mathbf{N}$ are dense in W.

(*c*)(**v**)$\Rightarrow$(**i**) If X is dense in a separable space W, let $D \subseteq W$ be a countable dense set. Clearly D is σ-centered. Because (i) and (ii) are equivalent, it follows that W and X are σ-centered.

(*d*)(**i**)$\Rightarrow$(**vi**) Express $\mathfrak{T}\setminus\{\emptyset\}$ as $\bigcup_{n\in\mathbf{N}}\mathscr{H}_n$ where each $\mathscr{H}_n$ has the finite intersection property. Let $\mathscr{G}_n = \mathscr{G} \cap \mathscr{H}_n$ for each $n \in \mathbf{N}$. If $\mathscr{H} \subseteq \mathscr{G}_n$ is a non-empty finite set, then $\bigcap \mathscr{H} \neq \emptyset$; but also $\bigcap \mathscr{H} \in \mathscr{G}$, so $\mathscr{G}_n$ is downwards-centered in $\mathscr{G}\setminus\{\emptyset\}$. As $\mathscr{G}\setminus\{\emptyset\} = \bigcup_{n\in\mathbf{N}}\mathscr{G}_n$, $\mathscr{G}\setminus\{\emptyset\}$ is σ-centered downwards.

(*e*)(**vi**)$\Rightarrow$(**i**) Express $\mathscr{G}\setminus\{\emptyset\}$ as $\bigcup_{n\in\mathbf{N}}\mathscr{G}_n$ where each $\mathscr{G}_n$ is downwards-centered in $\mathscr{G}\setminus\{\emptyset\}$ i.e. has the finite intersection property. Set

$$\mathscr{H}_n = \{H : H \in \mathfrak{T}, \operatorname{int} \bar{H} \in \mathscr{G}_n\}.$$

Then $\bigcup_{n\in\mathbf{N}}\mathscr{H}_n = \mathfrak{T}\setminus\{\emptyset\}$. If $\mathscr{H} \subseteq \mathscr{H}_n$ is finite and not empty, then

$$G = \bigcap_{H\in\mathscr{H}} \operatorname{int} \bar{H} \neq \emptyset.$$

But now $H \cap G$ is dense in G for each $H \in \mathscr{H}$, so $G \cap \bigcap \mathscr{H} \neq \emptyset$. Thus each $\mathscr{H}_n$ has the finite intersection property and X is σ-centered.

(*f*)(**vi**)$\Leftrightarrow$(**vii**) As the algebra of open-and-closed subsets of Z is a base for the topology of Z, and is isomorphic (as partially ordered set) to $\mathscr{G}$, we see that $\mathscr{G}\setminus\{\emptyset\}$ is σ-centered downwards iff Z is σ-centered, by A4Q*b*. If Z is separable, it is σ-centered, because (v)$\Rightarrow$(i); while if it is σ-centered it is separable, by A4Q*c*.

A4S Connectedness [Engelking 77]

(*a*) A topological space X is **connected** if its only open-and-closed subsets are $\emptyset$ and X.

(*b*) A **component** of X is a maximal non-empty connected set. Each point of X belongs to just one component. All the components of X are closed.

(*c*) If X is a compact Hausdorff space and $x \in X$, then the component of X containing x is precisely the intersection of all the open-and-closed sets containing x.

(*d*) X is **locally connected** if every point has a neighbourhood base consisting of connected sets; equivalently if every open subspace of X has open components [Csaszar 78, 10.2.1; Bourbaki 66, I.11.6].

(*e*) X is **zero-dimensional** if its topology has a base consisting of open-and-closed sets. (I allow $X = \emptyset$.) A compact Hausdorff space is zero-dimensional iff all its components are singletons.

(*f*) X is **extremally disconnected** if the closure of any open set in X is open.

(*g*) X is **pathwise-connected** if for any $x, y \in X$ there is a continuous $f:[0,1] \to X$ such that $f(0) = x$ and $f(1) = y$.

A4T **The Vietoris topology**

(*a*) Let X be a Hausdorff space, $\mathscr{C}$ the set of closed subsets of X, $\mathscr{K}$ the set of compact subsets of X. The **Vietoris topology** on $\mathscr{C}$ is that generated by sets of the form

$$\{F : F \in \mathscr{C}, F \subseteq G\}, \quad \{F : F \in \mathscr{C}, F \cap G \neq \emptyset\}$$

where G runs through the open subsets of X. The induced subspace topology is the Vietoris topology on $\mathscr{K}$.

(*b*) Let X be a metric space. Then the Vietoris topology on $\mathscr{K}$ is metrizable. **P** Let ρ be a metric on X defining its topology. For $x \in X$, $K \in \mathscr{K}$ write $\rho(x, K) = \inf_{y \in K} \rho(x, y)$ (taking $\inf \emptyset = \infty$). For $K, L \in \mathscr{K}$ write

$$d_0(K, L) = \sup_{x \in K} \rho(x, L) \text{ (taking } \sup \emptyset = 0),$$
$$d(K, L) = \min(1, d_0(K, L) + d_0(L, K)).$$

It is straightforward to confirm that (i) $\rho(x, L) \leq \rho(x, K) + d_0(K, L)$, (ii) $d_0(K, M) \leq d_0(K, L) + d_0(L, M)$, (iii) $\rho(x, K) = 0$ iff $x \in K$, (iv) $d_0(K, L) = 0$ iff $K \subseteq L$, (v) d is a metric on $\mathscr{K}$, (vi) because every $K \in \mathscr{K}$ is compact, d defines the Vietoris topology on $\mathscr{K}$. **Q**

(*c*) If X is separable, so are $\mathscr{K}$ and $\mathscr{C}$. **P** If $D \subseteq X$ is dense then $[D]^{<\omega}$ is dense in both. **Q**

(*d*) If X is complete and metrizable so is $\mathscr{K}$. **P** In (*b*), we can suppose that X is complete under ρ. If $\langle K_n \rangle_{n \in \mathbf{N}}$ is a sequence in $\mathscr{K}$ such that $\sum_{n \in \mathbf{N}} d(K_{n+1}, K_n) < \infty$, then

$$K = \bigcap_{n \in \mathbf{N}} \overline{\bigcup_{m \geq n} K_m}$$

is totally bounded, so belongs to $\mathscr{K}$. Also, if $\langle x_n \rangle_{n \geq r} \in \prod_{n \geq r} K_n$ and $\sum_{n \geq r} \rho(x_{n+1}, x_n) < \infty$, $\lim_{n \to \infty} x_n$ exists in X and belongs to K. From this it is easy to show that $d(K_r, K) \leq \sum_{n \geq r} d(K_{n+1}, K_n)$ for every $r \in \mathbf{N}$. So $\mathscr{K}$ is complete under d. **Q** Accordingly $\mathscr{K}$ is Polish if X is.

A4U **Miscellaneous definitions**

(*a*) If X is a topological space and $A \subseteq X$, then ∂A, the **boundary** of A, is $\bar{A} \setminus \mathrm{int}\, A$.

(*b*) A subset of X is **perfect** if it is closed and has no isolated points.

(*c*) X is **scattered** if every non-empty subset of X has an isolated point; equivalently, if X has no non-empty perfect subset.

(*d*) Let X and Y be topological spaces. A function $f:X \to Y$ is **sequentially continuous** if $\langle f(x_n)\rangle_{n\in\mathbf{N}} \to f(x)$ in Y whenever $\langle x_n\rangle_{n\in\mathbf{N}} \to x$ in X.

(*e*) If (X, ρ) is a metric space, a set $A \subseteq C(X)$ is **uniformly equicontinuous** if for every $\varepsilon > 0$ there is a $\delta > 0$ such that $|f(x) - f(y)| \le \varepsilon$ whenever $f \in A$ and $\rho(x, y) \le \delta$.

(*f*) Let X be a topological space. A set $A \subseteq X$ is **sequentially closed** if $x \in A$ whenever there is a sequence in A converging to x. X is **sequentially separable** if there is a countable $A \subseteq X$ such that every point of X is a limit of a sequence in A.

(*g*) A topological space X is **quasi-regular** if whenever G is a non-empty open set in X there is a non-empty open set H such that $\bar{H} \subseteq G$.

(*h*) X is **locally countable** if every point in X has a countable neighbourhood.

(*i*) X is **homogeneous** if for all $x, y \in X$ there is a homeomorphism $f:X \to X$ such that $f(x) = y$.

(*j*) A set $A \subseteq X$ is a **$p(\kappa)$-set** if whenever $\mathscr{G}$ is a non-empty collection of open sets, all including A, and $\#(\mathscr{G}) < \kappa$, then $A \subseteq \mathrm{int}(\bigcap \mathscr{G})$. A point $x \in X$ is a **$p(\kappa)$-point** if $\{x\}$ is a $p(\kappa)$-set, i.e. the intersection of fewer than κ neighbourhoods of x is a neighbourhood of x.

(*k*) Let X and Y be topological spaces. A continuous function $f:X \to Y$ is **irreducible** if $f[F] \ne f[X]$ for any proper closed subset F of X. If X and Y are compact Hausdorff spaces and $f:X \to Y$ is continuous, there is a closed $F_0 \subseteq X$ such that $f[F_0] = f[X]$ and $f \restriction F_0$ is irreducible [apply Zorn's lemma to $\{F : F \subseteq X \text{ closed}, f[F] = f[X]\}$].

(*l*) Let $\langle X_\iota\rangle_{\iota\in I}$ be a family of topological spaces. Their **box product** is $X = \prod_{\iota\in I} X_\iota$ with the topology generated by

$$\{\textstyle\prod_{\iota\in I} G_\iota : G_\iota \subseteq X_\iota \text{ is open } \forall \iota \in I\}.$$

(*m*) If X is a topological space, a set $A \subseteq X$ is a **retract** of X if there is a continuous function $f:X \to A$ such that $f(x) = x$ for every $x \in A$.

(*n*) A **Q-space** is a topological space in which every subset is F_σ (and therefore G_δ).

(*o*) An **L-space** is a hereditarily Lindelöf regular Hausdorff space which is not hereditarily separable. An **S-space** is a hereditarily *s*eparable regular Hausdorff space which is not hereditarily Lindelöf.

A5 Descriptive set theory

Under this heading I have collected the material I need on Borel and analytic sets. My principal reference is Kuratowski 66; I also use Rogers 80 and Moschovakis 80.

A5A Borel sets and functions

Let X be a topological space.

(*a*) Let $E \subseteq X$. E is a G_δ set if it is expressible as the intersection of a sequence of open sets. E is an F_σ set if it is expressible as the union of a sequence of closed sets, i.e. its complement is a G_δ set. E is a K_σ set if it is expressible as the union of a sequence of compact sets. E is a $G_{\delta\sigma}$ set if it is expressible as the union of a sequence of G_δ sets. E is an $F_{\sigma\delta}$ set if it is expressible as the intersection of a sequence of F_σ sets, i.e. its complement is a $G_{\delta\sigma}$ set.

(*b*) For ordinals $\xi \geq 1$, define $\boldsymbol{\Sigma}^0_\xi(X) \subseteq \mathscr{P}X$ inductively by writing

$\boldsymbol{\Sigma}^0_1(X) = \{G : G \subseteq X \text{ is open}\}$;

if $\xi > 1$, $\boldsymbol{\Sigma}^0_\xi(X)$ is $\{\bigcup_{n\in\mathbf{N}}(X \setminus E_n) : E_n \in \bigcup_{1\leq\eta<\xi}\boldsymbol{\Sigma}^0_\eta(X)\ \forall n \in \mathbf{N}\}$.

Thus $\boldsymbol{\Sigma}^0_2(X)$ is the set of F_σ sets and, if every open set in X is an F_σ set, $\boldsymbol{\Sigma}^0_3(X)$ is the set of $G_{\delta\sigma}$ sets. (It is customary to reserve $\boldsymbol{\Sigma}^0_0$ for a base for the topology of X.)

(*c*) The **Borel** sets of X are the members of the smallest σ-algebra of subsets of X containing all the open sets, viz. $\boldsymbol{\Sigma}^0_{\omega_1}(X) = \bigcup_{\xi<\omega_1}\boldsymbol{\Sigma}^0_\xi(X)$.

(*d*) If Y is a subspace of X, then a subset of Y is a Borel set in Y iff it is of the form $Y \cap E$ where E is a Borel subset of X [Kuratowski 66, §5.VI].

(*e*) If Y is another topological space, a function $f:X \to Y$ is **Borel measurable** if $f^{-1}[E]$ is a Borel set in X whenever E is a Borel set in Y; equivalently, if $f^{-1}[G]$ is a Borel set in X whenever G is an open set in Y. f is a **Borel isomorphism** if it is a bijection and f, f^{-1} are both Borel measurable.

(*f*) If $f:X\to\mathbf{R}$ is Borel measurable, then its graph $\Gamma=\{(x,f(x)):x\in X\}$ is a Borel set in $X\times\mathbf{R}$. **P** Continuous functions are Borel measurable; in particular, $\pi_1:X\times\mathbf{R}\to X$ and $\pi_2:X\times\mathbf{R}\to\mathbf{R}$ are Borel measurable. So if $E\subseteq X$ and $U\subseteq\mathbf{R}$ are Borel, $E\times U=\pi_1^{-1}[E]\cap\pi_2^{-1}[U]$ is Borel. Now let $\langle U_n\rangle_{n\in\mathbf{N}}$ enumerate a base for the topology of $\mathbf{R}$. Then

$$\Gamma=\bigcap_{n\in\mathbf{N}}(f^{-1}[\mathbf{R}\setminus U_n]\times\mathbf{R})\cup(X\times U_n)$$

is Borel. **Q**

(*g*) A set $E\subseteq X$ is a **zero set** if it is expressible as $f^{-1}[\{0\}]$ where $f:X\to\mathbf{R}$ is continuous. E is a **cozero set** if its complement is a zero set. A countable union of cozero sets is a cozero set. A zero set is always a closed G_δ set. In a normal space, every closed G_δ set is a zero set [Kuratowski 66, §14.VI].

A5B **The Baire property**

Let X be a topological space.

(*a*) A set $A\subseteq X$ is **nowhere dense** if $\operatorname{int}\bar{A}=\varnothing$. A is **meagre** if it is expressible as the union of a sequence of nowhere dense sets. A is **comeagre** if its complement is meagre.

(*b*) A set $E\subseteq X$ has the **Baire property** in X if there is an open set $G\subseteq X$ such that $G\triangle E$ is meagre. The family of sets with the Baire property is a σ-subalgebra of $\mathcal{P}X$ containing all the Borel sets [Kuratowski 66, §11.III; Oxtoby 71, 4.3].

(*c*) If Y is a second-countable space and $f:X\to Y$ is a function, then the following are equivalent: (i) $f^{-1}[G]$ has the Baire property in X for each open $G\subseteq Y$, (ii) there is a comeagre $E\subseteq X$ such that $f\restriction E$ is continuous [Kuratowski 66, §32.II; Oxtoby 71, 8.1]. In this case I shall say that f has the **Baire property.**

(*d*) A set $E\subseteq X$ has the **strong Baire property** (or, 'the Baire property in the restricted sense') if $E\cap F$ has the Baire property in F for every closed $F\subseteq X$ [Kuratowski 66, §11.VI].

(*e*) If X is a ccc completely regular space and $A\subseteq X$ is nowhere dense, there is a nowhere-dense zero set $F\supseteq A$. **P** Let $\mathcal{G}$ be a maximal disjoint family of cozero sets not meeting A. Then $\mathcal{G}$ is countable so $F=X\setminus\bigcup\mathcal{G}$ is a zero set. As the cozero sets form a base for the topology of X, $\operatorname{int}(F\setminus A)=\varnothing$ and F is nowhere dense. **Q**

(*f*) X is a **Baire space** if the intersection of any sequence of dense open sets is dense; i.e. no non-empty open set is meagre, and every comeagre set is dense.

A5C **Usco-compact relations**

Let X and Y be Hausdorff spaces, $R \subseteq X \times Y$ a relation.

(*a*) R is **usco-compact** if (i) the vertical sections of R are compact in Y, (ii) $R^{-1}[F]$ is closed in X whenever F is closed in Y.

(*b*) If R is usco-compact and $K \subseteq X$ is compact then $R[K]$ is compact in Y [Jayne & Rogers 80*b*, 2.7.2].

(*c*) If $X = \mathbf{N}^{\mathbf{N}}$ and R is usco-compact then R is Cech-complete [Jayne & Rogers 80*b*, 2.8.1–2].

A5D ***K*-analytic spaces**

(*a*) Let X be a Hausdorff space. Then X is ***K*-analytic** if there is an usco-compact $R \subseteq \mathbf{N}^{\mathbf{N}} \times X$ such that $\pi_2[R] = X$; equivalently, if there is a compact Hausdorff space Z, an $F_{\sigma\delta}$ set $E \subseteq Z$, and a continuous surjection $f: E \to X$ [Jayne & Rogers 80*b*, 2.8.1].

(*b*) A Hausdorff continuous image of a K-analytic space is K-analytic. A closed subspace of a K-analytic space is K-analytic. The family of K-analytic subsets of a given Hausdorff space is closed under Souslin's operation [Jayne & Rogers 80*b*, 2.5.4]. In particular, an $F_{\sigma\delta}$ set in a K-analytic space is K-analytic.

(*c*) K-analytic spaces are Lindelöf [Jayne & Rogers 80*b*, 2.7.1].

(*d*) If X is a Hausdorff space and A, B are disjoint K-analytic subsets of X, there is a Borel set $E \subseteq X$ such that $A \subseteq E$ and $B \subseteq X \setminus E$ [Jayne & Rogers 80*b*, 3.3.1].

(*e*) If X is a K-analytic space in which all open sets are K-analytic, then the σ-algebra

$$\{E : E \subseteq X,\ E \text{ and } X \setminus E \text{ both } K\text{-analytic}\}$$

includes the Borel sets (so, by A5D*d*, is precisely the algebra of Borel sets). In particular, if X is a perfectly normal K-analytic space (e.g. a hereditarily Lindelöf compact Hausdorff space), then every Borel set in X is K-analytic.

A5E **Polish spaces**

(*a*) A **Polish** space is a separable topological space whose topology can be defined by a complete metric.

(*b*) Let X and Y be Polish spaces, $E \subseteq X$ a Borel set, $f: E \to Y$ an injective function. (i) If f is continuous, then $f[E]$ is Borel. (ii) f is Borel measurable iff its graph is a Borel set in $X \times Y$. (iii) If f is Borel measurable it is a Borel isomorphism between E and $f[E]$. [Kuratowski 66, §39.IV–V.]

(*c*) If X is Polish and $E \subseteq X$ is an uncountable Borel set, then $\#(E) = \mathfrak{c}$ [KURATOWSKI 66, §37.I]. If X and Y are Polish and $E \subseteq X$, $F \subseteq Y$ are uncountable Borel sets, there is a Borel isomorphism $f: E \to F$ [KURATOWSKI 66, §37.II].

A5F **Analytic spaces**

(*a*) An **analytic** space ('Suslin space' in SCHWARTZ 73) is a Hausdorff continuous image of a Polish space; equivalently, it is either $\emptyset$ or a Hausdorff continuous image of $\mathbf{N}^{\mathbf{N}}$ [JAYNE & ROGERS 80*b*, 2.4.3]. Every analytic space is K-analytic; a metrizable K-analytic space is analytic; any K-analytic subspace of an analytic space is analytic [JAYNE & ROGERS 80*b*, 5.5.1].

(*b*) If X is Polish, a subset of X is analytic iff it is obtainable by Souslin's operation from the closed subsets of X [JAYNE & ROGERS 80*b*, 2.5.2–3].

(*c*) The product of two analytic spaces is analytic; any open subset of an analytic space is analytic (because an open subset of a Polish space is Polish [KURATOWSKI 66, §33.VI]). So every Borel set in an analytic space is analytic [A5D*e*]. An analytic space is hereditarily Lindelöf (being a continuous image of a hereditarily Lindelöf space).

(*d*) A compact analytic space is metrizable [SCHWARTZ 73, p. 106, Corollary 2]. So any compact subset of an analytic space is metrizable.

(*e*) If $(X, \mathfrak{T})$ is an analytic space and $\mathfrak{S}$ is a Hausdorff topology on X coarser than $\mathfrak{T}$, then the $\mathfrak{S}$-Borel sets and the $\mathfrak{T}$-Borel sets are the same. [SCHWARTZ 73, p. 101, Corollary 2.]

(*f*) If X is an uncountable analytic space then it has a subset homeomorphic to $\{0, 1\}^{\mathbf{N}}$ [JAYNE & ROGERS 80*b*, 3.5.2].

(*g*) If X is a Polish space, there is an analytic set $R \subseteq \mathbf{N}^{\mathbf{N}} \times X$ such that the analytic subsets of X are precisely the vertical sections of R [JAYNE & ROGERS 80*b*, 4.2.2; KURATOWSKI 66, §38.V].

(*h*) There is a non-analytic subset of $[0, 1]$ which has cardinal ω_1. **P** If the continuum hypothesis is true, take any non-analytic set [KURATOWSKI 66, §38.VI]; if it is false, take any set of cardinal ω_1. **Q**

A5G **Projective sets**

(*a*) For Polish spaces X, the classes $\Sigma_n^1(X)$, $\Pi_n^1(X)$ of subsets of X are defined by induction on n, simultaneously for all Polish spaces, by saying that

$$\Sigma_0^1(X) = \{G : G \subseteq X \text{ is open}\}$$
$$\Pi_n^1(X) = \{X \setminus E : E \in \Sigma_n^1(X)\},$$
$$\Sigma_{n+1}^1(X) = \{E : E \subseteq X, \exists \text{ Polish } Z, F \in \Pi_n^1(Z), \text{ and a continuous surjection } f : F \to E\}.$$

Thus $\Pi_0^1(X)$ is the family of closed subsets of X; $\Sigma_1^1(X)$ the family of analytic sets. In this book I shall normally use the notation of KURATOWSKI 66, so that $\Pi_1^1(X)$ is the family of **coanalytic** sets, $\Sigma_2^1(X)$ is the family of **PCA** sets, $\Pi_2^1(X)$ is the family of **CPCA** sets, and $\Sigma_3^1(X)$ is the family of **PCPCA** sets. A subset of X is **projective** if it belongs to $\bigcup_{n \in \mathbf{N}} \Sigma_n^1(X)$. If X is infinite, it has just $\mathfrak{c}$ projective sets [KURATOWSKI 66, §38.VI].

(*b*) Let X and Y be Polish spaces. (i) The intersection and union of any sequence of coanalytic sets in X is coanalytic [KURATOWSKI 66, 38.III.3]. (ii) If $E \subseteq X$ and $F \subseteq Y$ are coanalytic then $E \times F$ is coanalytic in $X \times Y$ [KURATOWSKI 66, 38.III.1]. (iii) If $E \subseteq X$ is coanalytic and $f : E \to Y$ is continuous then the graph of f is coanalytic in $X \times Y$. [Use the formula of A5A*f*.] (iv) If $f : X \to Y$ is Borel measurable and $F \subseteq Y$ is coanalytic, then $f^{-1}[F]$ is coanalytic in X [KURATOWSKI 66, 38.III.5].

(*c*) Any PCA set (in particular, any analytic or coanalytic set) in a Polish space is expressible as the union of ω_1 Borel sets [KURATOWSKI 66, §39.III]. The families of PCA and CPCA sets in any Polish space are closed under Souslin's operation [KURATOWSKI 66, §38.IX]. In particular, the intersection of two PCA sets is a PCA set.

A5H Kondô's theorem

Let X and Y be Polish spaces, $R \subseteq X \times Y$ a coanalytic set. Then there is a selector f for R such that the graph of f is coanalytic in $X \times Y$.

Proof MOSCHOVAKIS 80, 4E.4; DELLACHERIE 80, III.39.

A6 Measure theory

I follow FREMLIN 74. When I find rival terminology in common use I try to give translations.

A6A Measure theory

Throughout this book, a **measure space** is a triple (X, Σ, μ) where X is a set, Σ is a σ-subalgebra of $\mathscr{P}X$, and $\mu : \Sigma \to [0, \infty]$ is a countably additive functional.

If (X, Σ, μ) is a measure space I shall write

$$\mu^* A = \inf\{\mu E : A \subseteq E \in \Sigma\}$$

for every $A \subseteq X$. A subset A of X is **negligible** if $\mu^* A = 0$. I shall say that a proposition $\varphi(.)$ referring to points of X is true **almost everywhere** (**a.e.**) if $\{x : \varphi(x) \text{ is false}\}$ is negligible. In particular, two functions f, g defined on X are 'equal a.e.' if $\{x : f(x) \neq g(x)\}$ is negligible.

A6B **Taxonomy of measure spaces**

[FREMLIN 74, §62 & §64] A measure space (X, Σ, μ) is **complete** if all negligible sets are measurable. It is **semi-finite** if whenever $E \in \Sigma$ and $\mu E = \infty$ there is an $F \in \Sigma$ such that $F \subseteq E$ and $0 < \mu F < \infty$. It is **locally determined** if it is semi-finite and, for any set $E \subseteq X$,

$$\text{if } E \cap F \in \Sigma \text{ whenever } F \in \Sigma \text{ and } \mu F < \infty \text{ then } E \in \Sigma.$$

(In the language of SCHWARTZ 73, (X, Σ, μ) is complete and locally determined iff Σ is the set of 'μ-measurable' sets and μ is the 'essential measure' associated with itself.) (X, Σ, μ) is **decomposable** if there is a partition $\langle X_\iota \rangle_{\iota \in I}$ of X into measurable sets of finite measure such that, for $E \subseteq X$,

$$\begin{aligned} E \in \Sigma &\Leftrightarrow E \cap X_\iota \in \Sigma \ \forall \iota \in I \\ &\Rightarrow \mu E = \textstyle\sum_{\iota \in I} \mu(E \cap X_\iota). \end{aligned}$$

(The 'strictly localizable' spaces of IONESCU TULCEA 69 are in effect the complete decomposable spaces.)

(X, Σ, μ) is a **probability space** if $\mu X = 1$; it is **totally finite** (or 'of finite magnitude') if $\mu X < \infty$; it is **σ-finite** (or 'totally σ-finite', or 'of countable magnitude') if X can be covered by a sequence of sets of finite measure.

A σ-finite measure space is decomposable; a decomposable measure space is locally determined [FREMLIN 74, 64H].

A6C **Function spaces**

If (X, Σ, μ) is a measure space, I write $\mathscr{L}^1$ for the space of integrable real-valued functions on X, $\mathscr{L}^\infty$ for the space of bounded measurable real-valued functions on X, and $\mathscr{L}^2$ for the space of square-integrable real-valued functions on X; and L^1, L^∞ and L^2 for their Banach space quotients under the equivalence relation

$$f \sim g \quad \text{iff} \quad f = g \text{ a.e.}$$

(For definiteness, let me say that I require integrable functions to be measurable.)

A6D **Completions**

If (X, Σ, μ) is any measure space, set

$$\hat{\Sigma} = \{E \triangle A : E \in \Sigma,\ A \text{ negligible}\},$$
$$\hat{\mu} = \mu^* \restriction \hat{\Sigma}.$$

Then $(X, \hat{\Sigma}, \hat{\mu})$ is a complete measure space, the **completion** of (X, Σ, μ) [Halmos 50, §13; Dunford & Schwartz 5, III.5.17]. We have $\hat{\mu} \restriction \Sigma = \mu$, so that $\mathscr{L}^1(X, \Sigma, \mu) \subseteq \mathscr{L}^1(X, \hat{\Sigma}, \hat{\mu})$; this inclusion induces an isomorphism between $L^1(X, \Sigma, \mu)$ and $L^1(X, \hat{\Sigma}, \hat{\mu})$.

A6E **Measure algebras**

(*a*) Let (X, Σ, μ) be a measure space. Its **measure algebra** is the Boolean algebra quotient $\mathfrak{A} = \Sigma / \{E : \mu E = 0\}$ together with the countably additive functional $\mathfrak{A} \to [0, \infty]$ induced by μ; I shall normally use the same symbol for this new functional, so that $\mu E^{\cdot} = \mu E$ for every $E \in \Sigma$.

(*b*) If X is totally finite, we have a natural metric ρ on $\mathfrak{A}$ defined by

$$\rho(E^{\cdot}, F^{\cdot}) = \mu(E \triangle F) = \int |\chi E - \chi F| \, d\mu \, \forall E, F \in \Sigma$$

[Halmos 50, §40]. In this case, $\max(\omega, d(\mathfrak{A})) = \max(\omega, d(L^1(X)))$; in particular, $\mathfrak{A}$ is separable iff L^1 is separable.

(*c*) If X is totally finite, the **Maharam type** of X is $d(\mathfrak{A})$. (This definition disagrees with Lacey 74 iff $\mathfrak{A}$ is finite.) X is **homogeneous** if $d(\mathfrak{B}) = d(\mathfrak{A})$ for every non-zero principal ideal $\mathfrak{B}$ of $\mathfrak{A}$.

A6F **Maharam's theorem**

(*a*) If (X, μ) is a homogeneous probability space, and $\mathfrak{A}$ its measure algebra, then either $\mathfrak{A} = \{0, 1\}$ or $\lambda = d(\mathfrak{A}) \geq \omega$. In the latter case $\mathfrak{A}$ is isomorphic, as measure algebra, to the measure algebra of $\{0, 1\}^\lambda$ where $\{0, 1\}^\lambda$ is given its usual measure [A6K*c*]. In this case, $L^1(X) \cong L^1(\{0, 1\}^\lambda)$ and $L^\infty(X) \cong L^\infty(\{0, 1\}^\lambda)$, both isomorphisms being order-preserving linear isometries.

(*b*) If (X, μ) is any totally finite measure space, there is a countable partition $\langle X_i \rangle_{i \in I}$ of X into measurable sets such that each (X_i, μ_i) is homogeneous, where $\mu_i = \mu \restriction \mathscr{P} X_i$ for each $i \in I$.

Proof Lacey 74, §14.

A6G **σ-finite measure spaces**

(*a*) If (X, Σ, μ) is σ-finite and $\mathscr{E} \subseteq \Sigma$, there is a countable $\mathscr{E}_0 \subseteq \mathscr{E}$

such that $\mu(E\setminus\bigcup\mathscr{E}_0)=0$ for every $E\in\mathscr{E}$. **P** Assume $\mathscr{E}\neq\emptyset$. Let $\langle X_n\rangle_{n\in\mathbf{N}}$ be a sequence of sets of finite measure covering X. For each $n\in\mathbf{N}$ set

$$s_n=\sup\{\mu(X_n\cap\bigcup\mathscr{D}):\mathscr{D}\in[\mathscr{E}]^{<\omega}\}.$$

Choose finite sets $\mathscr{D}_{nm}\subseteq\mathscr{E}$ such that

$$\mu(X_n\cap\bigcup\mathscr{D}_{nm})\geq s_n-2^{-m}\ \forall m,n\in\mathbf{N}.$$

Set $\mathscr{E}_0=\bigcup_{m,n\in\mathbf{N}}\mathscr{D}_{nm}$. **Q**

(*b*) If (X,μ) is a semi-finite measure space and $L^1(X)$ is separable, then X is σ-finite. **P** Let $\langle X_\xi\rangle_{\xi\in I}$ be a maximal disjoint family of measurable sets of non-zero finite measure. Then $\|(\chi X_\xi)^\cdot-(\chi X_\eta)^\cdot\|_1\geq\mu X_\xi$ for $\xi\neq\eta$, so that I must be countable, and $\bigcup_{\xi\in I}X_\xi$ is measurable. As X is semi-finite, $\mu(X\setminus\bigcup_{\xi\in I}X_\xi)$ must be 0, and $\{X\setminus\bigcup_{\xi\in I}X_\xi\}\cup\{X_\xi:\xi\in I\}$ is a countable cover of X by sets of finite measure. **Q**

(*c*) If (X,Σ,μ) is σ-finite and Σ is countably generated as σ-algebra (i.e. there is a countable $\mathscr{A}\subseteq\Sigma$ such that Σ is the σ-subalgebra of $\mathscr{P}X$ generated by $\mathscr{A}$), then $L^1(X)$ is separable. [Cf. HALMOS 50, Theorem 40B.]

A6H Measurable functions

(*a*) Let (X,Σ,μ) be a measure space, Y a topological space. I say that $f:X\to Y$ is **measurable** ('Borel measurable' in SCHWARTZ 73) if $f^{-1}[G]\in\Sigma$ for every open $G\subseteq Y$; equivalently, if $f^{-1}[E]\in\Sigma$ for every Borel $E\subseteq Y$.

(*b*) Let (X,Σ,μ) and (Y,T,ν) be measure spaces. I say that a function $f:X\to Y$ is **inverse-measure-preserving** if $f^{-1}[F]\in\Sigma$ and $\mu f^{-1}[F]=\nu F$ for every $F\in T$.

A6I Atomless measure spaces

(*a*) A measure space (X,Σ,μ) is **atomless** (or 'non-atomic') if its measure algebra is atomless; i.e. whenever $E\in\Sigma$ and $\mu E>0$, there is an $F\in\Sigma$ such that $F\subseteq E$ and neither F nor $E\setminus F$ is negligible.

(*b*) Let (X,Σ,μ) be an atomless σ-finite measure space.

Then there is a function

$$f:X\to\{s:s\in\mathbf{R},\ 0\leq s\leq\mu X\}=I$$

such that $\mu f^{-1}[E]$ is the Lebesgue measure of E for every Borel set $E\subseteq I$. If (X,Σ,μ) is complete, then f will be inverse-measure-preserving for μ and Lebesgue measure on I. **P** (sketch) (i) Consider first the case in which $\mu X=1$. (α) If $\mu E>0$, there is an $F\subseteq E$ such that $0<\mu F\leq\frac{1}{2}\mu E$, because (X,μ) is atomless. (β) If $\mu E>0$ and $s>0$, there is an $F\subseteq E$ such that

$0 < \mu F \leq s$. (γ) If $0 \leq s \leq \mu E$ there is an $F \subseteq E$ such that $\mu F = s$. [Use A6G*a* with $\mathscr{E}$ a maximal set such that $\mu(\bigcup \mathscr{D}) \leq s$ for every finite $\mathscr{D} \subseteq \mathscr{E}$.] ($\delta$) Enumerate $\mathbf{Q} \cap [0,1]$ as $\langle q_n \rangle_{n \in \mathbf{N}}$ and use (γ) to choose inductively $E_n \in \Sigma$ such that $\mu E_n = q_n$ and $E_n \subseteq E_m$ if $q_n \leq q_m$. (ε) Find $f: X \to [0,1]$ such that $\{x: f(x) < q_n\} \subseteq E_n \subseteq \{x: f(x) \leq q_n\}$. Show that $\mu f^{-1}[F]$ is the Lebesgue measure of F for rational intervals F, therefore for Borel sets $F \subseteq [0,1]$. (ii) In general, express X as $\bigcup_{n \in \mathbf{N}} X_n$ where each X_n is of finite measure and the X_n are disjoint, and use the method of (i) to define $f: X_n \to I_n$ for each n, where $I_n \subseteq \mathbf{R}$ are suitable intervals. (iii) Finally, consider the case in which (X, Σ, μ) is complete. **Q**

A6J **Subspaces**

Let (X, Σ, μ) be any measure space, $Y \subseteq X$ any subset. Set

$$\Sigma_Y = \{Y \cap E : E \in \Sigma\}, \quad \mu_Y = \mu^* \restriction \Sigma_Y.$$

Then (Y, Σ_Y, μ_Y) is a measure space. [Compare HALMOS 50, Theorem 17A.] If $L^1(X)$ is separable, so is $L^1(Y)$. [The surjection $f \mapsto f \restriction Y : \mathscr{L}^1(X) \to \mathscr{L}^1(Y)$ induces a surjection from $L^1(X)$ onto $L^1(Y)$.]

A6K **Product measures**

(*a*) Let (X, Σ, μ) and (Y, T, ν) be measure spaces. Let us define outer measures θ_0^*, θ^* on $X \times Y$ by writing

$$\theta_0^*(A) = \inf\{\textstyle\sum_{n \in \mathbf{N}} \mu E_n \cdot \nu F_n : E_n \in \Sigma \quad \text{and} \quad F_n \in T \ \forall n \in \mathbf{N}, \quad \text{and} \quad A \subseteq \bigcup_{n \in \mathbf{N}} E_n \times F_n\}$$

(conventionally taking $0.\infty = 0$),

$$\theta^*(A) = \sup\{\theta_0^*(A \cap (E \times F)) : E \in \Sigma, F \in T, \mu E < \infty, \nu F < \infty\}.$$

What I call the **c.l.d. product measure** on $X \times Y$ is the measure θ defined by Carathéodory's method from θ^*. (This θ is an extension of the measure defined by BERBERIAN 62, §39. If X and Y are σ-finite, it is the measure defined by MUNROE 53, §28, and the completion of the measure defined by HALMOS 50, §35.)

(*b*) Let $\langle (X_\iota, \Sigma_\iota, \mu_\iota) \rangle_{\iota \in I}$ be any family of probability spaces. Let $\mathscr{C}$ be the family of subsets of $X = \prod_{\iota \in I} X_\iota$ expressible as $E = \prod_{\iota \in I} E_\iota$ where $E_\iota \in \Sigma_\iota$ for each $\iota \in I$ and $\{\iota : E_\iota \neq X_\iota\}$ is finite. Define an outer measure θ^* on X by

$$\theta^*(A) = \inf\{\textstyle\sum_{n \in \mathbf{N}} \prod_{\iota \in I} \mu_\iota E_\iota^{(n)} : E^{(n)} \in \mathscr{C} \ \forall n \in \mathbf{N}, \quad A \subseteq \bigcup_{n \in \mathbf{N}} E^{(n)}\}.$$

Then the **c.l.d. product measure** on X is the measure θ defined by Carathéodory's method from θ^*. (This θ is the completion of the measure defined by DUNFORD & SCHWARTZ 57, III.11.20; if I is countable, it is the completion of the measure defined by HALMOS 50, §38.)

(*c*) If, in (*b*), every X_ι is $\{0,1\}$, and $\mu_\iota\{0\} = \mu_\iota\{1\} = \frac{1}{2}$ for every ι, then $X = \{0,1\}^I$ and θ is the 'usual' measure on X.

(*d*) In (*a*), if $L^1(X)$ and $L^1(Y)$ are separable, so is $L^1(X \times Y)$. In (*b*), if $L^1(X_\iota)$ is separable for each $\iota \in I$ and I is countable, then $L^1(X)$ is separable. **P** In the notation of (*b*), observe that if $E \subseteq X$ is measurable and $\varepsilon > 0$, there is a finite set $\{E^{(0)}, \ldots, E^{(n)}\}$ in $\mathscr{C}$ such that $\theta(E \Delta \bigcup_{i \leq n} E^{(i)}) \leq \varepsilon$. So if the measure algebra of each X_ι is separable, and I is countable, the measure algebra of X is separable. The same argument works for the product of two factors, if we take $\mathscr{C}$ to be $\{E \times F : \mu E < \infty, \nu F < \infty\}$ and look at sets of finite measure in $X \times Y$. **Q**

(*e*) If X and Y are sets, $\Sigma \subseteq \mathscr{P}X$ and $T \subseteq \mathscr{P}Y$ are algebras, I write $\Sigma \hat{\otimes}_\sigma T$ for the σ-algebra of subsets of $X \times Y$ generated by $\{E \times F : E \in \Sigma, F \in T\}$. (Most authors prefer '$\Sigma \times T$'.)

A6L Properly based measure spaces

(*a*) Let (X, Σ, μ) be a measure space and κ a cardinal. I shall say that (X, Σ, μ) is **properly κ-based** if there is a family $\mathscr{K} \subseteq \mathscr{P}X$ such that (i) no member of $\mathscr{K}$ can be covered by fewer than κ negligible sets, (ii) every non-negligible measurable set includes a member of $\mathscr{K}$, (iii) $\#(\mathscr{K}) \leq \kappa$. I shall say that (X, Σ, μ) is **properly based** if it is properly κ-based for some cardinal κ.

(*b*) (X, Σ, μ) is properly κ-based iff its completion is. **P** It is easy to see that $\mathscr{K} \subseteq \mathscr{P}X$ satisfies each of (i)–(iii) above with respect to the completion iff it satisfies the same condition with respect to Σ, μ. **Q**

(*c*) Let (X, Σ, μ) be a properly based probability space. Let $E \subseteq X^2$ be a set of measure 1 for the c.l.d. product measure on X^2. Then there is a set $A \subseteq X$ such that $\mu^* A = 1$ and $(x, y) \in E$ whenever x and y are distinct members of A. [Use the argument of FREMLIN & TALAGRAND 79, Lemma 3D, or TALAGRAND *a*.]

(*d*) Let (X, Σ, μ) be a properly based probability space. Let $\mathscr{E}$ be a point-countable cover of X by negligible sets. Then there is an $\mathscr{E}' \subseteq \mathscr{E}$ such that $\bigcup \mathscr{E}' \notin \Sigma$. [FREMLIN 82.]

(*e*) Let (X, Σ, μ) be a complete properly based probability space, $K \subseteq \mathbf{R}^X$ a countable set of functions such that $\sup_{f \in K} |f(x)| < \infty$ for every

$x \in X$. Suppose that $\bar{K}$, taken in $\mathbf{R}^X$, consists entirely of measurable functions. Then the closed convex hull of K in $\mathbf{R}^X$ also consists entirely of measurable functions. [TALAGRAND 82, TALAGRAND *a*.]

A6M **The lifting theorem**

(*a*) Let (X, Σ, μ) be a measure space. **A multiplicative lifting** of (X, Σ, μ) is a Boolean homomorphism $\theta: \Sigma \to \Sigma$ such that $\theta E = \varnothing$ whenever $\mu E = 0$ and $\mu(E \triangle \theta E) = 0$ for every $E \in \Sigma$.

(*b*) Every complete decomposable measure space has a multiplicative lifting. [IONESCU TULCEA 69, Ch. IV.]

A6N **Conditional expectation**

(*a*) Let (X, Σ, μ) be a probability space, $T \subseteq \Sigma$ a σ-subalgebra. Then there is a natural embedding of $L^1(X, T, \mu \restriction T)$ into $L^1(X, \Sigma, \mu)$ defined by the inclusion $\mathscr{L}^1(X, T, \mu \restriction T) \subseteq \mathscr{L}^1(X, \Sigma, \mu)$. At the same time there is a map $P: L^1(X, \Sigma, \mu) \to L^1(X, T, \mu \restriction T)$ defined by writing $P(f^\cdot) = g^\cdot$ whenever $f \in \mathscr{L}^1(X, \Sigma, \mu)$ and $g \in \mathscr{L}^1(X, T, \mu \restriction T)$ are such that

$$\int_E g \, d\mu = \int_E f \, d\mu \forall E \in T.$$

P (i) If $f \in \mathscr{L}^1(X, \Sigma, \mu)$, then by the Radon–Nikodým theorem there is a $g \in \mathscr{L}^1(X, T, \mu \restriction T)$ such that $\int_E f = \int_E g$ for every $E \in T$. (ii) For $g_1, g_2 \in \mathscr{L}^1(X, T, \mu \restriction T)$, $g_1^\cdot = g_2^\cdot$ iff $\int_E g_1 = \int_E g_2$ for every $E \in T$. So we have a function from $\mathscr{L}^1(X, \Sigma, \mu)$ to $L^1(X, T, \mu \restriction T)$. (iii) It is now elementary to check that this induces a map $P: L^1 \to L^1$ as required. **Q**

Regarding $L^1(X, T, \mu \restriction T)$ as a subspace of $L^1(X, \Sigma, \mu)$, P becomes a projection, the **conditional expectation** projection. It is easy to check that P is linear, positive, and that

$$\| Pu \|_1 \leq \| u \|_1 \; \forall u \in L^1(X, \Sigma, \mu),$$

$$Pu \in L^\infty(X, T, \mu \restriction T) \text{ and } \| Pu \|_\infty \leq \| u \|_\infty \; \forall u \in L^\infty(X, \Sigma, \mu).$$

(*b*) In the special case $X = \{0, 1\}^I$ with its usual measure [A6K*c*], we see that for any $J \subseteq I$ we have a σ-subalgebra

$$\Sigma_J = \{E : E \in \Sigma, \pi_J^{-1}[\pi_J[E]] = E\}$$

where $\pi_J(x) = x \restriction J \in \{0, 1\}^J$ for $x \in X$. In this case, there is a natural bijection between $\mathscr{L}^1(X, \Sigma_J, \mu \restriction \Sigma_J)$ and $\mathscr{L}^1(\{0, 1\}^J)$ leading to a natural identification of $L^1(X, \Sigma_J, \mu \restriction \Sigma_J)$ with $L^1(\{0, 1\}^J)$. Thus the associated conditional expectation projection P_J can be regarded as a map from $L^1(\{0,1\}^I)$ to $L^1(\{0,1\}^J)$.

A6O κ-additive measures

(*a*) Let (X,Σ,μ) be a measure space. We say that μ is **κ-additive** if whenever $\mathscr{E}\subseteq\Sigma$ is disjoint and $\#(\mathscr{E})<\kappa$, then $\bigcup\mathscr{E}\in\Sigma$ and $\mu(\bigcup\mathscr{E})=\sum_{E\in\mathscr{E}}\mu E$.

(*b*) Let (X,Σ,μ) be a κ-additive measure space. If $\mathscr{E}\subseteq\Sigma$ and $\#(\mathscr{E})<\kappa$ then $\bigcup\mathscr{E}\in\Sigma$ and

$$\mu(\bigcup\mathscr{E})=\sup\{\mu(\bigcup\mathscr{E}_0):\mathscr{E}_0\in[\mathscr{E}]^{<\omega}\}.$$

P Enumerate $\mathscr{E}$ as $\langle E_\xi\rangle_{\xi<\lambda}$ where $\lambda<\kappa$. Set $F_\xi=E_\xi\setminus\bigcup_{\eta<\xi}E_\eta$ for $\xi<\lambda$. By induction on ξ, show that $F_\xi\in\Sigma$ for every $\xi<\lambda$, so that $\bigcup\mathscr{E}=\bigcup_{\xi<\lambda}F_\xi\in\Sigma$. Now

$$\mu(\bigcup\mathscr{E})=\sum_{\xi<\lambda}\mu F_\xi=\sup\{\sum_{\xi\in I}\mu F_\xi:I\in[\lambda]^{<\omega}\}$$
$$\leq\sup\{\mu(\bigcup\mathscr{E}_0):\mathscr{E}_0\in[\mathscr{E}]^{<\omega}\}\leq\mu(\bigcup\mathscr{E}).\quad \mathbf{Q}$$

(*c*) Let (X,Σ,μ) be a complete locally determined measure space. Suppose that the ideal of negligible sets is κ-additive [12A*i*]. Then μ is κ-additive. **P** Let $\mathscr{E}$ be a disjoint subset of Σ of cardinal less than κ. Let F be any set of finite measure. Then $\mathscr{E}(F)=\{E:E\in\mathscr{E},\mu(E\cap F)>0\}$ is countable, and $F'=\bigcup\{F\cap E:E\in\mathscr{E}\setminus\mathscr{E}(F)\}$ is a union of fewer than κ negligible sets, so is negligible. Because (X,Σ,μ) is complete, $F\cap\bigcup\mathscr{E}=(F\cap\bigcup\mathscr{E}(F))\cup F'$ is measurable, and

$$\mu(F\cap\bigcup\mathscr{E})=\sum_{E\in\mathscr{E}(F)}\mu(E\cap F)\leq\sum_{E\in\mathscr{E}}\mu E.$$

As F is arbitrary and (X,Σ,μ) is locally determined, $\bigcup\mathscr{E}$ is measurable and $\mu(\bigcup\mathscr{E})\leq\sum_{E\in\mathscr{E}}\mu E\leq\mu(\bigcup\mathscr{E})$. **Q**

A6P Miscellaneous definitions

(*a*) Let (X,Σ,μ) be a measure space, $\mathscr{A}\subseteq\mathscr{P}X$. Then μ is **inner regular** for $\mathscr{A}$ if

$$\mu E=\sup\{\mu F:F\in\mathscr{A}\cap\Sigma,\ F\subseteq E\}$$

for every $E\in\Sigma$.

(*b*) A measure space (X,Σ,μ) of finite magnitude is **perfect** if whenever $f:X\to\mathbf{R}$ is measurable.

$$\sup\{\mu f^{-1}[K]:K\text{ compact},\ K\subseteq f[X]\}=\mu X.$$

(*c*) If (X,Σ,μ) is a probability space, a family $\langle E_\iota\rangle_{\iota\in I}$ of subsets of

X is **independent** if

$$\mu(\bigcap_{\iota\in J}E_\iota)=\prod_{\iota\in J}\mu E_\iota$$

for every non-empty finite $J\subseteq I$.

(*d*) An **atomlessly measurable cardinal** is an uncountable cardinal κ such that there is an atomless κ-additive probability with domain $\mathscr{P}\kappa$. (In the language of JECH 78, this is a 'real-valued-measurable cardinal $\leq\mathfrak{c}$'.) If there is an atomlessly measurable cardinal, then (i) there is a measure on $[0,1]$, extending Lebesgue measure, and defined on every subset of $[0,1]$ [use A6I*b*] (ii) the continuum hypothesis is wildly false [cf. 35B*a*]. See **B2D** and SOLOVAY 71, DRAKE 74 or JECH 78, §27.

(*e*) If $\mathfrak{A}$ is a Boolean algebra, a functional $\nu:\mathfrak{A}\to\mathbf{R}$ is **additive** ('finitely additive') if $\nu(a\cup b)=\nu a+\nu b$ whenever a, $b\in\mathfrak{A}$ and $a\cap b=0$.

A7 Topological measure spaces

Once again, I start from FREMLIN 74, with references to SCHWARTZ 73.

A7A Definitions

(*a*) **A topological measure space** is a quadruple $(X,\mathfrak{T},\Sigma,\mu)$ where (X,Σ,μ) is a measure space and $\mathfrak{T}$ is a topology on X such that $\mathfrak{T}\subseteq\Sigma$ (i.e. every Borel set is measurable).

(*b*) Let (X,Σ,μ) be a measure space and $\mathfrak{T}$ a topology on X. Then μ is **locally finite** if every point of X has a neighbourhood of finite (outer) measure, and **effectively locally finite** if whenever $\mu E>0$ there is a measurable open set G such that $\mu G<\infty$ and $\mu(G\cap E)>0$. μ is **τ-additive** if $\mu(\bigcup\mathscr{G})=\sup_{G\in\mathscr{G}}\mu G$ whenever $\mathscr{G}$ is an upwards-directed family of measurable open sets and $\bigcup\mathscr{G}$ is measurable. μ is **completion regular** if it is inner regular for the zero sets [A6P*a*].

(*c*) Let X be a topological space. **A Borel measure** on X is a measure with domain precisely the algebra of Borel subsets of X.

(*d*) A **quasi-Radon measure space** is a topological measure space $(X,\mathfrak{T},\Sigma,\mu)$ such that (i) (X,Σ,μ) is complete and locally determined [A6B], (ii) μ is effectively locally finite, τ-additive, and inner regular for the closed sets.

(*e*) A **Radon measure space** is a topological measure space $(X,\mathfrak{T},\Sigma,\mu)$ such that (i) (X,Σ,μ) is complete and locally determined, (ii) $\mathfrak{T}$ is

Hausdorff, (iii) μ is locally finite and inner regular for the compact sets. [The measures of SCHWARTZ 73, p. 13, Definition R_3 are, in my terminology, the locally finite Borel measures inner regular for the compact sets. My Σ corresponds to Schwartz' 'μ-measurable' sets. See A7C*c*.]

A7B **Quasi-Radon measure spaces** [FREMLIN 74, §72]

(*a*) Let X be a topological space, $\mathscr{K}$ a family of closed subsets of X such that $H \cup K = \mathscr{K}$ and $K \cap F \in \mathscr{K}$ whenever $H, K \in \mathscr{K}$ and $F \subseteq X$ is closed. Let $\rho : \mathscr{K} \to [0, \infty[$ be a function such that (i) $\rho(F) = \rho(H) + \rho_*(F \setminus H)$ whenever $F, H \in \mathscr{K}$ and $F \subseteq H$, (ii) whenever $\mathscr{A} \subseteq \mathscr{K}$ is downwards-directed and $\bigcap \mathscr{A} = \varnothing$ then $\inf_{K \in \mathscr{A}} \rho(K) = 0$, (iii) for every $H \in \mathscr{K}$ there is an open $G \supseteq H$ such that $\rho_*(G) < \infty$, where $\rho_*(A) = \sup\{\rho(K) : K \in \mathscr{K}, K \subseteq A\}$. Then there is a quasi-Radon measure on X, extending ρ, which is inner regular for $\mathscr{K}$. [Follow the proof of 72E in FREMLIN 74.]

(*b*) Let X be a topological space, μ a Borel measure on X. Then μ extends to a quasi-Radon measure $\bar{\mu}$ on X iff μ is effectively locally finite and τ-additive, and

$$\mu G = \sup\{\mu F : F \subseteq G,\ F \text{ closed}\}$$

for every open set G of finite measure. [To construct $\bar{\mu}$, use A7B*a* with

$$\mathscr{K} = \{F : F \subseteq X \text{ closed}, \exists \text{open } G \supseteq F,\ \mu G < \infty\}$$

and $\rho = \mu \restriction \mathscr{K}$. To show that $\bar{\mu}$ extends μ, show first that $\mu E = \sup\{\mu(E \cap G) : G \text{ open}, \mu G < \infty\}$ for every Borel set E.]

(*c*) Let $(X, \mathfrak{T}, \Sigma, \mu)$ be a σ-finite quasi-Radon measure space. There is a sequence $\langle G_n \rangle_{n \in \mathbb{N}}$ of open sets of finite measure such that $X \setminus \bigcup_{n \in \mathbb{N}} G_n$ is negligible. **P** Apply A6G*a* to the collection of open sets of finite measure to obtain a sequence $\langle G_n \rangle_{n \in \mathbb{N}}$ of open sets of finite measure such that $\mu(G \setminus \bigcup_{n \in \mathbb{N}} G_n) = 0$ for every open set G of finite measure. Now because μ is effectively locally finite, $\mu(X \setminus \bigcup_{n \in \mathbb{N}} G_n) = 0$. **Q** So X can be covered by a sequence of Borel sets of finite measure.

(*d*) Let $(X, \mathfrak{T}, \Sigma, \mu)$ be a quasi-Radon measure space, and $\langle G_n \rangle_{n \in \mathbb{N}}$ a sequence of open sets of finite measure. If E is a measurable subset of $\bigcup_{n \in \mathbb{N}} G_n$, then $\mu E = \inf\{\mu G : G \supseteq E, G \text{ open}\}$. **P** Given $\varepsilon > 0$, then for each $n \in \mathbb{N}$ there is a closed set $F_n \subseteq G_n \setminus E$ such that $\mu F_n \geq \mu(G_n \setminus E) - 2^{-n}\varepsilon$. Set $G = \bigcup_{n \in \mathbb{N}} (G_n \setminus F_n)$. Then $G \supseteq E$ and $\mu(G \setminus E) \leq 2\varepsilon$. As ε is arbitrary, this proves the result. **Q**

(*e*) (i) Let X be a regular topological space and μ a Borel measure on X. Then μ extends to a quasi-Radon measure on X iff μ is effectively locally finite and τ-additive. [Use (*b*).] (ii) Let X be a topological space and μ a Borel measure on X satisfying the conditions of (*b*) above and also σ-finite. Then the completion $\hat{\mu}$ of μ [A6D] is a quasi-Radon measure. **P** By (*b*) there is a quasi-Radon measure $\bar{\mu}$ extending μ. Use (*c*)–(*d*) to see that $\bar{\mu}^* = \mu^*$ so that $\bar{\mu}$ and μ have the same negligible sets and $\bar{\mu} = \hat{\mu}$. **Q** (iii) Let X be a regular hereditarily Lindelöf space and μ a locally finite Borel measure on X. Then the completion of μ is a quasi-Radon measure. **P** Because X is Lindelöf and μ is locally finite, X can be covered by a sequence of open sets of finite measure; so μ is effectively locally finite and σ-finite. Next, μ is τ-additive because X is hereditarily Lindelöf, and satisfies the last condition of (*b*) because also X is regular. So we can use (ii). **Q**

(*f*) Let (X, Σ, μ) be a σ-finite measure space such that Σ is the σ-subalgebra of $\mathscr{P}X$ generated by a countable set $\mathscr{A} \subseteq \mathscr{P}X$. Then there is a second-countable topology $\mathfrak{T}$ on X such that the completion $\hat{\mu}$ of μ is quasi-Radon with respect to $\mathfrak{T}$. **P** We can suppose that $\mathscr{A}$ includes a sequence of sets of finite measure covering X. Let $\mathfrak{T}$ be the topology generated by $\mathscr{A} \cup \{X \setminus E : E \in \mathscr{A}\}$. Then $\mathfrak{T}$ is second-countable and zero-dimensional, therefore regular, and Σ is precisely the algebra of $\mathfrak{T}$-Borel sets. Because X is covered by members of $\mathfrak{T}$ of finite measure, μ is a locally finite Borel measure with respect to $\mathfrak{T}$, which is hereditarily Lindelöf. So we can apply (*e*)(iii) above. **Q**

(*g*) Let $(X, \mathfrak{T}, \Sigma, \mu)$ be a quasi-Radon measure space. I shall say that a set $E \subseteq X$ is **supporting** if $\mu^*(E \cap G) > 0$ whenever G is an open set meeting E. If $E \in \Sigma$, set

$$F = E \setminus \bigcup \{G : G \in \mathfrak{T},\ \mu(E \cap G) = 0\};$$

then $\mu(E \setminus F) = 0$ [Fremlin 74, 72G*d*], and F is the unique relatively closed subset of E which is supporting and such that $\mu(E \setminus F) = 0$. In particular, any non-negligible measurable set includes a non-negligible supporting measurable set. The **support** of μ is

$$X \setminus \bigcup \{G : G \in \mathfrak{T},\ \mu G = 0\}.$$

(*h*) A quasi-Radon measure space is decomposable. [Fremlin 74, 72B.] Consequently its measure algebra is Dedekind complete.

(*i*) If μ and ν are two quasi-Radon measures on the same topological space $(X, \mathfrak{T})$, and if $\mu \restriction \mathfrak{T} = \nu \restriction \mathfrak{T}$, then $\mu = \nu$ (by which I mean to imply that $\operatorname{dom}(\mu) = \operatorname{dom}(\nu)$). [Fremlin 74, 72F*b*].

(*j*) A hereditarily Lindelöf quasi-Radon measure space is σ-finite (because the measure is effectively locally finite; see (*e*)(iii) above).

(*k*) A quasi-Radon measure space $(X, \mathfrak{T}, \Sigma, \mu)$ is atomless iff all singleton sets in X are negligible. **P** (i) If $\mu^*(\{x\})>0$ then $\inf\{E^{\cdot}:x\in E\in\Sigma\}$ is an atom in the measure algebra $\mathfrak{A}$ of (X, Σ, μ). (ii) If $E^{\cdot}$ is an atom in $\mathfrak{A}$ then there is an open $G\subseteq X$ such that $\mu G<\infty$ and $\mu(G\cap E)>0$; now there is a supporting $F\subseteq E\cap G$ with $\mu F>0$ [(*g*) above]; if $x\in F$ then

$$\mu^*(\{x\}) = \inf\{\mu H: H \text{ open}, x\in H\} = \mu E > 0$$

[(*d*) above]. **Q**

(*l*) If $(X, \mathfrak{T}, \Sigma, \mu)$ is a properly $\mathfrak{c}$-based atomless σ-finite quasi-Radon measure space, it has a $\mathfrak{c}$-Sierpiński set. **P** Set $\mathcal{N} = \{E: E\in\Sigma, \mu E = 0\}$. Let $\mathcal{K}\subseteq\mathcal{P}X$ satisfy A6L*a*(i)–(iii) with $\kappa=\mathfrak{c}$. Set $\mathcal{L} = \{\bar{K}: K\in\mathcal{K}\}$; then $\mathcal{L}$ clearly satisfies A6L*a*(i), and it also satisfies A6L*a*(ii) because μ is inner regular for the closed sets. Let Σ_1 be the σ-subalgebra of Σ generated by $\mathcal{L}$, so that $\#(\Sigma_1)\leq\mathfrak{c}$. Write $\mathcal{N}_1 = \mathcal{N}\cap\Sigma_1$. For $E\in\Sigma$, choose a maximal disjoint family $\mathcal{L}_E$ in $\mathcal{L}\cap\mathcal{P}E$; as $\mu L>0$ for every $L\in\mathcal{L}$, and (X, Σ, μ) is σ-finite, $\mathcal{L}_E$ is countable. Set $H_E = \bigcup\mathcal{L}_E\in\Sigma_1$. Then $E\setminus H_E\in\mathcal{N}$, because every set in $\Sigma\setminus\mathcal{N}$ includes a member of $\mathcal{L}$, and $\mathcal{L}_E$ was maximal. Similarly $(X\setminus E)\setminus H_{X\setminus E}\in\mathcal{N}$, and we have $H_E\subseteq E\subseteq X\setminus H_{X\setminus E}$ while $(X\setminus H_{X\setminus E})\setminus H_E\in\mathcal{N}_1$. This shows that, writing $\mu_1 = \mu\restriction\Sigma_1$, (X, Σ, μ) is the completion of (X, Σ_1, μ_1). In particular, $\bigcup\mathcal{N}_1 = \bigcup\mathcal{N} = X$, by (*k*) above. Now A3F*a* tells us that there is a $(\Sigma_1, \mathcal{N}_1, \mathfrak{c})$-Lusin subset of X which by A3E*b* is a $(\Sigma, \mathcal{N}, \mathfrak{c})$-Lusin set, i.e. a $\mathfrak{c}$-Sierpiński set for μ. **Q** [Compare A3F*b*.]

A7C **Radon measures** [FREMLIN 74, §73]

(*a*) A Radon measure space is quasi-Radon. [FREMLIN 74, 73B.] A quasi-Radon measure space $(X, \mathfrak{T}, \Sigma, \mu)$ is Radon iff $\mathfrak{T}$ is Hausdorff and μ is locally finite and inner regular for the compact sets. A quasi-Radon probability space $(X, \mathfrak{T}, \Sigma, \mu)$ is Radon iff $\mathfrak{T}$ is Hausdorff and $\sup\{\mu K: K\subseteq X, K \text{ compact}\} = 1$.

(*b*) Let X be a Hausdorff space, $\mathcal{K}$ the set of compact subsets of X. Let $\rho:\mathcal{K}\to[0,\infty[$ be such that (i) whenever $F, H\in\mathcal{K}$ and $F\supseteq H$ then $\rho(F) = \rho(H)+\rho_*(F\setminus H)$, (ii) for every $x\in X$ there is a neighbourhood U of x such that $\rho_*(U)<\infty$, where $\rho_*(A) = \sup\{\rho(K): K\in\mathcal{K}, K\subseteq A\}$. Then ρ has a unique extension to a Radon measure on X. **P** Note first that

$$\rho_*(G_0\cup\ldots\cup G_n)\leq\sum_{i\leq n}\rho_*(G_i)$$

for any open sets $G_0,\ldots,G_n$. Hence, condition (iii) of A7B*a* is satisfied and there is a quasi-Radon measure μ extending ρ and inner regular for $\mathcal{K}$.

From (ii) above, μ is locally finite, so is a Radon measure. By A7B*i*, μ is unique. **Q**

(*c*) Let X be a Hausdorff space and μ a Borel measure on X. Then the following are equivalent: (i) μ extends to a Radon measure on X, (ii) μ is locally finite and inner regular for the compact sets, (iii) μ is effectively locally finite and locally finite and $\mu G = \sup\{\mu K : K \text{ compact}, K \subseteq G\}$ for every open set $G \subseteq X$. [For (iii)$\Rightarrow$(i), use A7B*b*.]

(*d*) Let $(X, \mathfrak{T}, \Sigma, \mu)$ be a K-analytic locally finite quasi-Radon measure space. Then it is a Radon measure space. **P** As μ is locally finite and inner regular for the closed sets,

$$\infty > \mu K = \inf\{\mu G : K \subseteq G \in \mathfrak{T}\}$$

for every compact set $K \subseteq X$. Consequently $\mu^* : \mathscr{P}X \to [0, \infty]$ is a 'Choquet capacity' in the sense of DELLACHERIE 80. So $\mu F = \sup\{\mu K : K \text{ compact}, K \subseteq F\}$ for every K-analytic subset F of X [DELLACHERIE 80, I.19 and I.27]; in particular, for every closed $F \subseteq X$ [A5D*b*]. Accordingly, μ is inner regular for the compact sets. But this is all we need. **Q**

(*e*) Let X be an analytic space, μ a locally finite Borel measure on X. Then the completion $\hat{\mu}$ of μ is a Radon measure on X. **P** By SCHWARTZ 73, p. 125, Theorem 11, μ is inner regular for the compact sets. It follows that $\hat{\mu}$ is inner regular for the compact sets. Because X is Lindelöf and μ is locally finite, there is a cover of X by a sequence of (open) sets of finite measure, and $\hat{\mu}$ is σ-finite, therefore locally determined [A6B]. Of course $\hat{\mu}$ is still locally finite so it is a Radon measure. **Q**

(*f*) In particular, Lebesgue measure on $\mathbf{R}^n$ is a Radon measure.

(*g*) Let $(X, \mathfrak{T})$ be an analytic space, $\mathfrak{S}$ a Hausdorff topology on X coarser than $\mathfrak{T}$. Let μ be a totally finite measure on X. Then the following are equivalent: (i) μ is quasi-Radon for $\mathfrak{T}$, (ii) μ is Radon for $\mathfrak{T}$, (iii) μ is quasi-Radon for $\mathfrak{S}$, (iv) μ is Radon for $\mathfrak{S}$. **P** By A5F*e*, $\mathfrak{S}$ and $\mathfrak{T}$ have the same algebra $\mathscr{B}$ of Borel sets. Now the conditions (i)–(iv) all imply that the domain of μ includes $\mathscr{B}$ and that μ is the completion of $\mu \restriction \mathscr{B}$. **Q**

(*h*) If X is a zero-dimensional compact Hausdorff space and $\mathscr{E}$ the algebra of open-and-closed subsets of X, then any non-negative additive functional $\mu : \mathscr{E} \to \mathbf{R}$ has a unique extension to a Radon measure on X. [Apply A7C*b* with $\rho(K) = \inf\{\mu E : K \subseteq E \in \mathscr{E}\}$ for compact $K \subseteq X$.]

A7D **Subspaces**

(*a*) Let $(X, \mathfrak{T}, \Sigma, \mu)$ be a quasi-Radon measure space, $Y \subseteq X$ any set.

Then $(Y, \mathfrak{T}_Y, \Sigma_Y, \mu_Y)$ is a quasi-Radon measure space, where $\mathfrak{T}_Y$ is the induced topology on Y, and Σ_Y and μ_Y are described in A6J. [Use A7B*h* to reduce to the case $\mu X < \infty$.]

(*b*) Let $(X, \mathfrak{T}, \Sigma, \mu)$ be a Radon measure space and $Y \in \Sigma$. Then $(Y, \mathfrak{T}_Y, \Sigma_Y, \mu_Y)$ is a Radon measure space. [Cf. SCHWARTZ 73, p. 20.]

A7E **Products**

(*a*) Let $(X, \mathfrak{T}, \Sigma, \mu)$ and $(Y, \mathfrak{S}, T, \nu)$ be quasi-Radon measure spaces. Then there is a unique quasi-Radon measure $\bar{\theta}$ on $X \times Y$ extending the c.l.d. product measure θ of A6K*a*. [Apply A7B*a* with

$$\rho(K) = \int \nu\{y:(x, y) \in K\}\mu(dx)$$

for closed sets K included in a product of open sets of finite measure.] If μ and ν are both Radon measures, so is $\bar{\theta}$. [Cf. SCHWARTZ 73, p. 63, Theorem 17.]

(*b*) Let $\langle(X_\iota, \mathfrak{T}_\iota, \Sigma_\iota, \mu_\iota)\rangle_{\iota \in I}$ be a family of quasi-Radon probability spaces. Then there is a unique quasi-Radon measure $\bar{\theta}$ on $X = \prod_{\iota \in I} X_\iota$ extending the c.l.d. product measure θ of A6K*b*. [Show first that if $\mathscr{G}$ is a collection of basic open sets in X and $\bigcup \mathscr{G} = X$, then $\sup\{\theta(\bigcup \mathscr{G}_0):\mathscr{G}_0 \in [\mathscr{G}]^{<\omega}\} = 1$. Now use A7B*a* with ρ the restriction of θ^* to the closed subsets of X.] If every μ_ι is a Radon measure, and if the support of μ_ι is compact for all but countably many $\iota \in I$, then $\bar{\theta}$ is a Radon measure.

(*c*) If in (*b*) each X_ι is a compact metric space, and the support of each μ_ι is X_ι itself, then $\theta = \bar{\theta}$ [CHOKSI & FREMLIN 79, Theorem 3]. In particular, there need be no dispute over what the 'usual' measure on $\{0, 1\}^I$ should be, since the c.l.d. product measure of A6K*b*–*c* is already a Radon measure. It is sometimes convenient to transfer this measure to $\mathscr{P}I$; thus the 'usual' measure on $\mathscr{P}I$ (which I shall call **Haar** measure) will be that Radon measure μ such that

$$\mu\{A:J \subseteq A \subseteq I\} = 2^{-\#(J)} \; \forall J \in [I]^{<\omega},$$

$\mathscr{P}I$ being given its 'usual' topology, which is generated by

$$\{\{A:A \cap J = K\}:K \subseteq J \in [I]^{<\omega}\}.$$

A7F **Separable $\mathbf{L}^1$ spaces**

(*a*) Let $(X, \mathfrak{T}, \Sigma, \mu)$ be a second-countable quasi-Radon measure space. Then $L^1(X)$ is separable. **P** (X, Σ, μ) is σ-finite because μ is effectively locally finite and X is hereditarily Lindelöf. So if $\mathscr{B}$ is the algebra of Borel subsets of X, $L^1(X, \mathscr{B}, \mu\restriction\mathscr{B})$ is separable [A6G*c*]. But (X, Σ, μ) is just the completion of $(X, \mathscr{B}, \mu\restriction\mathscr{B})$ so $L^1(X, \Sigma, \mu)$ is separable. **Q**

(*b*) If $(X, \mathfrak{T}, \Sigma, \mu)$ is an analytic Radon measure space [cf. A7C*d–e*], then $L^1(X)$ is separable. **P** Because μ is locally finite and X is Lindelöf, (X, Σ, μ) is σ-finite. Because μ is inner regular for the compact sets, there is a sequence $\langle K_n\rangle_{n\in\mathbf{N}}$ of compact sets such that $\mu(X\setminus\bigcup_{n\in\mathbf{N}}K_n)=0$. Each K_n is metrizable [A5F*d*], therefore second-countable. Let $\mathfrak{S}$ be a second-countable topology on X, coarser than $\mathfrak{T}$, agreeing with $\mathfrak{T}$ on each K_n, containing each $X\setminus K_n$, and including a cover of X by open sets of finite measure. Then $(X, \mathfrak{S}, \Sigma, \mu)$ is a second-countable quasi-Radon measure space so $L^1(X)$ is separable, by (*a*) above. **Q**

(*c*) In particular, Lebesgue measure on $\mathbf{R}^n$ is a quasi-Radon measure with a separable L^1 space.

A7G **Almost continuous functions**

(*a*) Let $(X, \mathfrak{T}, \Sigma, \mu)$ be a topological measure space, Y any topological space. A function $f:X\to Y$ is **almost continuous** if μ is inner regular for $\{E: E\subseteq X, f\restriction E \text{ is continuous}\}$. (Schwartz 73 calls such functions 'Lusin μ-measurable' when X is a Radon measure space.)

(*b*) Let $(X, \mathfrak{T}, \Sigma, \mu)$ be a quasi-Radon measure space, Y a topological space, $f:X\to Y$ an almost continuous function. Then f is measurable. **P** Let $G\subseteq Y$ be open, and E a set of finite measure. For each $n\in\mathbf{N}$, there is a set $F_n\subseteq E$ such that $\mu(E\setminus F_n)\le 2^{-n}$ and $f\restriction F_n$ is continuous. Now $f^{-1}[G]\cap F_n$ is relatively open in F_n, so $f^{-1}[G]\cap F_n\in\Sigma$ for each $n\in\mathbf{N}$. As $E\setminus\bigcup_{n\in\mathbf{N}}F_n$ is negligible, and (X,Σ,μ) is complete, $f^{-1}[G]\cap E$ is measurable. As (X, Σ, μ) is locally determined, $f^{-1}[G]$ is measurable. As G is arbitrary, f is measurable. **Q**

(*c*) Let $(X, \mathfrak{T}, \Sigma, \mu)$ be a Radon measure space, $(Y, \mathfrak{S})$ a Hausdorff space, $f:X\to Y$ an almost continuous function. Set $T=\{F: F\subseteq Y, f^{-1}[F]\in\Sigma\}$, $(\mu f^{-1})(F)=\mu(f^{-1}[F])$ for $F\in T$. Then $(Y, \mathfrak{S}, T, \mu f^{-1})$ is a topological measure space. If μf^{-1} is locally finite, then $(Y, \mathfrak{S}, T, \mu f^{-1})$ is a Radon measure space. [Schwartz 73, pp. 31–2.]

A7H **Hyperstonian spaces**

(*a*) A **hyperstonian** space is an extremally disconnected compact Hausdorff space Z such that for every non-empty open set $G\subseteq Z$ there is a Radon measure ν on Z such that $\nu G>0$ and $\nu F=0$ for every nowhere dense set $F\subseteq Z$.

(*b*) If (X, Σ, μ) is any totally finite measure space, its associated hyperstonian space is the Stone space Z of the measure algebra $\mathfrak{A}$ of (X, Σ, μ). There is a canonical Radon measure ν on Z defined by saying that

$$\nu(\hat{E}^{\cdot}) = \mu E \; \forall E \in \Sigma,$$

where for $E \in \Sigma$ I write $E^{\cdot}$ for its equivalence class in $\mathfrak{A}$ and $\hat{E}^{\cdot}$ for the corresponding open-and-closed set in Z. [Use A7C*h*.] Observe that this measure ν is such that $\nu G > 0$ for every non-empty open $G \subseteq Z$, but $\nu F = 0$ for every nowhere-dense $F \subseteq Z$.

(*c*) If (X, μ) is a compact Radon measure space, with hyperstonian space Z, then there is a canonical continuous function $h: Z \to X$ defined by saying that

$$\{h(z)\} = \bigcap \{F : F \subseteq X \text{ closed}, \, z \in \hat{F}^{\cdot}\} \; \forall z \in Z.$$

The image $h[Z]$ is the support of μ. If Z is given its usual measure, as in (*b*) above, h is inverse-measure-preserving. [Compare 12M*b*.]

A7I **Radon spaces**

(*a*) A Hausdorff space X is **Radon** if every totally finite Borel measure on X is inner regular for the compact sets, i.e. extends to a Radon measure on X. [See SCHWARTZ 73, pp. 117ff.]

(*b*) If X is a compact Hausdorff space and every Radon measure on X is completion regular, then X is a Radon space. **P** Let μ be a totally finite Borel measure on X. By the Riesz representation theorem [FREMLIN 74, 73D], there is a Radon measure ν on X such that $\int f \, d\nu = \int f \, d\mu$ for every $f \in C(X)$. If $H \subseteq X$ is a zero set, then

$$\nu H = \inf \left\{ \int f \, d\nu : f \in C(X), f \geq \chi H \right\} = \mu H.$$

Now if $E \subseteq X$ is Borel,

$$\nu E = \sup \{\nu H : H \subseteq E, \, H \text{ is a zero set}\} \leq \mu E,$$

because ν is completion regular. It follows at once that ν is an extension of μ. As μ is arbitrary, X is Radon. **Q**

A7J **The split interval**

Let $S \subseteq [0, 1]$ be any set. By '[0, 1] split on S' I shall mean the set

$$X = \{s^+ : s \in S\} \cup \{s^- : s \in S\} \cup ([0, 1] \setminus S),$$

located in $\mathbf{R}^2$ by identifying s^+ with $(s, \frac{1}{2}), s^-$ with $(s, -\frac{1}{2})$, and u with $(u, 0)$ for $u \in [0, 1] \backslash S$. Give X the order inherited from the lexicographic total order of $\mathbf{R}^2$, so that, for $s, t \in S$ and $u \in [0, 1] \backslash S$,

$$s^+ \leq t^- \Leftrightarrow s < t, \quad s^+ \leq u \Leftrightarrow s < u, \quad s^- < s^+$$

etc. Then X is Dedekind complete, with greatest and least members. So if X is given its order topology, it is a compact Hausdorff space. [KELLEY 55, Problem 5C.] We have

$$w(X) = \max(\omega, \#(S)),$$

$$\pi(X) = L(X) = hL(X) = c(X) = hc(X) = t(X) = d(X) = hd(X) = \chi(X) = \omega.$$

Let $h: X \to [0, 1]$ be the vertical projection, so that $h(s^+) = h(s^-) = s$ if $s \in S$, $h(u) = u$ if $u \in [0, 1] \backslash S$. Then h is continuous and irreducible. If $G \subseteq X$ is open, then $h^{-1}[h[G]] \backslash G$ is countable; consequently a set $E \subseteq X$ is Borel iff there is a Borel set $F \subseteq [0, 1]$ such that $E \triangle h^{-1}[F]$ is countable.

Let μ_L be Lebesgue measure on $[0, 1]$, and Σ_L its domain. Let Σ_0 be $\{h^{-1}[E]: E \in \Sigma_L\}$ and define μ_0 on Σ_0 by writing $\mu_0 h^{-1}[E] = \mu_L E$ for each $E \in \Sigma_L$; this is well defined because h is surjective, and is clearly a measure. Let (X, Σ, μ) be the completion of (X, Σ_0, μ_0). Then $\mu\{x\} = 0$ for every $x \in X$, so every Borel set of X belongs to Σ. Because h is continuous, X is compact, and μ_L is inner regular for the compact subsets of $[0, 1]$, μ_0 and μ are inner regular for the compact subsets of X; so μ is a Radon measure. h is inverse-measure-preserving for μ_0 and μ_L, therefore also for μ and μ_L. The map $f \mapsto fh: \mathscr{L}^1([0, 1]) \to \mathscr{L}^1(X, \Sigma_0, \mu_0) \subseteq \mathscr{L}^1(X, \Sigma, \mu)$ induces an isomorphism between $L^1([0, 1])$ and $L^1(X)$, so $L^1(X)$ is separable.

A7K Miscellaneous definitions

(*a*) Let X be a Hausdorff space. A set $E \subseteq X$ is **universally measurable** if it is measurable for every Radon measure on X; it is **universally negligible** if it is negligible for every atomless Radon measure on X. If Y is a topological space, a function $f: X \to Y$ is **universally measurable** if it is measurable [A6H*a*] for every Radon measure on X. (Note that this definition can disagree with SCHWARTZ 73 if Y is not metrizable.)

(*b*) Let X be a topological space. Let Σ be the σ-algebra of subsets of X generated by the zero sets. X is **measure-compact** or **almost Lindelöf** if every totally finite measure with domain Σ is τ-additive [A7A*b*]. If X is completely regular, it is measure-compact iff every totally finite measure with domain Σ has an extension to a quasi-Radon measure on X.

A8 Functional analysis

My starting point is KÖTHE 69.

A8A Linear topological spaces

For simplicity, all linear spaces in this book are taken to be over the real field; of course it makes no difference if you prefer to work with **C**.

(*a*) If E and F are linear spaces in duality, I write $\mathfrak{T}_s(E, F)$ for the weak topology on E determined by the action of F [KÖTHE 69, §10.3, uses '$\mathfrak{T}_s(F)$'].

(*b*) If E is a linear topological space, I write E' for the space of continuous real-valued linear functionals on E. A subset of E is **weakly compact** if it is compact for $\mathfrak{T}_s(E, E')$. E is **weakly compactly generated** if there is a weakly compact $K \subseteq E$ such that E is the closed linear subspace generated by K.

(*c*) Let E be a locally convex linear topological space. A set $V \subseteq E$ is **absorbent** if $E = \bigcup_{s>0} sV$. E is **barreled** if every closed convex absorbent set is a neighbourhood of 0.

(*d*) Let E be a Hausdorff linear topological space. A closed linear subspace F of E is **complemented** ['$\mathfrak{T}$-complemented' in KÖTHE 69, §10.7] if there is a continuous linear projection $P: E \to E$ such that $P[E] = F$. If E is a Fréchet space, a closed linear subspace F of E is complemented iff there is a closed linear subspace G of E such that $F \cap G = \{0\}$, $F + G = E$ [KÖTHE 69, 15.12(6)].

(*e*) If E and F are linear topological spaces, an **isomorphic embedding** of E into F is a linear map $T: E \to F$ such that T is a homeomorphism between E and $T[E]$.

(*f*) A **Fréchet** space is a locally convex linear topological space E in which the topology can be defined by a metric under which E is complete.

A8B Function and sequence spaces

(*a*) Let X be any set. I write $\ell^\infty(X)$ for the Banach space of functions $f: X \to \mathbf{R}$ such that $\|f\|_\infty = \sup_{x \in X} |f(x)| < \infty$; $\ell^2(X)$ for the Hilbert space of functions $f: X \to \mathbf{R}$ such that $\|f\|_2 = (\sum_{x \in X} |f(x)|^2)^{1/2} < \infty$; and $\ell^1(X)$ for the Banach space of functions $f: X \to \mathbf{R}$ such that $\|f\|_1 = \sum_{x \in X} |f(x)| < \infty$.

(*b*) If X is a topological space, I write $C(X)$ for the linear space of continuous real-valued functions on X. In this context I write $\mathfrak{T}_p$ for the linear space topology on $C(X)$ induced by that of $\mathbf{R}^X$ [cf. KÖTHE 69, §24.5].

(*c*) If $\langle E_\iota \rangle_{\iota\in I}$ is any family of Banach spaces, their ℓ^1**-sum** $(\bigoplus_{\iota\in I} E_\iota)_1$ is the Banach space of all functions $f \in \prod_{\iota\in I} E_\iota$ such that $\|f\| = \sum_{\iota\in I} \|f(\iota)\| < \infty$.

(*d*) I write $c_0(\mathbf{N})$ for the closed linear subspace of $\ell^\infty(\mathbf{N})$ consisting of sequences converging to 0.
[See also A6C.]

A8C **Vector-valued measurable functions**

(*a*) Let (X, Σ, μ) be a measure space, E a Banach space, and $f: X \to E$ a function. I say that f is **scalarly measurable** if $gf: X \to \mathbf{R}$ is measurable for each $g \in E'$. f is **Pettis integrable**, with **indefinite Pettis integral** $\varphi: \Sigma \to E$, if

$$\int_S gf \, d\mu \text{ exists} = g(\varphi(S))$$

for every $S \in \Sigma$ and $g \in E'$. f is **Bochner integrable** if it is measurable [A6H*a*], there is a separable $F \subseteq E$ such that $\mu f^{-1}[E \setminus F] = 0$, and $\int \|f(x)\| \mu(dx) < \infty$; in this case it is Pettis integrable, and its **Bochner integral** is equal to its Pettis integral over X. [See TALAGRAND *a* and DUNFORD & SCHWARTZ 57, §III.2. The 'μ-integrable' functions of the latter coincide with what I am calling 'Bochner integrable' functions if (X, Σ, μ) is complete and locally determined.]

(*b*) A Banach space E has the **Radon–Nikodým property** if for every probability space (X, Σ, μ) and every $\varphi: \Sigma \to E$ which is countably additive (i.e. $\sum_{n\in\mathbf{N}} \varphi(S_n)$ exists and is equal to $\varphi(\bigcup_{n\in\mathbf{N}} S_n)$ for every disjoint sequence $\langle S_n \rangle_{n\in\mathbf{N}}$ in Σ), there is a Bochner integrable $f: X \to E$ such that φ is the indefinite (Pettis) integral of f.

A8D $\mathbf{L^1}$**-spaces**

Let (X, Σ, μ) be a probability space.

(*a*) The natural duality between $L^1(X)$ and $L^\infty(X)$ given by writing

$$(f^{\cdot} | g^{\cdot}) = \int f \times g \;\; \forall f \in \mathscr{L}^1(X),\, g \in \mathscr{L}^\infty(X)$$

identifies $L^\infty(X)$ with $L^1(X)'$ [FREMLIN 74, 64B and 64F*a*].

(*b*) A linear subspace F of $L^1(X)$ is a **Riesz subspace** if $(|f|)^{\cdot} \in F$ whenever $f \in \mathscr{L}^1(X)$ and $f^{\cdot} \in F$. If F is any closed Riesz subspace of $L^1(X)$, there is a σ-subalgebra Σ' of Σ such that

$$F = \{f^{\cdot} : f \in \mathscr{L}^1(X), f \text{ is } \Sigma'\text{-measurable}\}.$$

Consequently there is a conditional-expectation projection onto F [A6N. See LACEY 74, §17, Theorem 3 and its corollary.]

A8E **Lemma**

Let (X, μ) be a totally finite measure space and $K \subseteq \mathbf{R}^X$ a convex compact set consisting entirely of measurable functions, such that if f, g are distinct members of K then $\mu\{x : f(x) \neq g(x)\} > 0$. Then K is metrizable and there is a countable set $Y \subseteq X$ separating the points of K.

Proof By KHURANA 79, Theorem 1, K is metrizable. Accordingly $C(K)$ is separable under $\| \quad \|_\infty$. For $x \in X$, define $\hat{x} \in C(K)$ by writing $\hat{x}(f) = f(x)$ for each $f \in K$. There is a countable $Y \subseteq X$ such that $\{\hat{y} : y \in Y\}$ is $\| \quad \|_\infty$-dense in $\{\hat{x} : x \in X\}$, and now Y separates the points of K because X does. [See also TALAGRAND *a*.]

A8F **Theorem** [cf. TORTRAT 76, Theorem 1.3]

Let $(E, \mathfrak{T})$ be a Fréchet space with dual E'. Let μ be a totally finite measure on E which is quasi-Radon for $\mathfrak{T}_s(E, E')$. Then μ is Radon for $\mathfrak{T}$.

Proof Let $\langle U_n \rangle_{n \in \mathbf{N}}$ run over a base of neighbourhoods of 0 for $\mathfrak{T}$ and for each $n \in \mathbf{N}$ set

$$U_n^0 = \{f : f \in E', f(x) \leq 1 \; \forall x \in U_n\}.$$

Let X be the support of μ, so that X is $\mathfrak{T}_s(E, E')$-closed in E, and write μ_X for the subspace measure $\mu \restriction \mathscr{P}X$ [A7D]. Then if f and g are distinct members of $C(X)$, $\{x : f(x) \neq g(x)\}$ is a non-empty open set in X, and has positive measure.

For each $n \in \mathbf{N}$, set $K_n = \{f \restriction X : f \in U_n^0\}$. Because U_n^0 is convex and $\mathfrak{T}_s(E', E)$-compact, K_n is convex and compact in $\mathbf{R}^X$. By A8E, there is a countable $Y_n \subseteq X$ separating the points of K_n. Write $Y = \bigcup_{n \in \mathbf{N}} Y_n$. Then we see that if $f, g \in E'$ and $f \restriction Y = g \restriction Y$, $f \restriction X = g \restriction X$ (since there is an $n \in \mathbf{N}$ such that both f and g belong to U_n^0, so that both $f \restriction X$ and $g \restriction X$ belong to K_n). In particular, if $f \in E'$ and $f(y) = 0$ for every $y \in Y$, then $f(x) = 0$ for every $x \in X$. It follows that X is included in the closed linear span of Y [KÖTHE 69, §20.8]. Consequently, as Y is countable, X is separable. But now X is Polish under the topology $\mathfrak{T}_X$ induced by $\mathfrak{T}$. So, applying A7C*g*, we see that μ_X is $\mathfrak{T}_X$-Radon; it follows at once that μ is $\mathfrak{T}$-Radon. [See also TALAGRAND *a*.]

A8G **Banach lattices**

A **Banach lattice** is a Banach space E with a lattice structure such that (i) $x \leq y \Rightarrow x + z \leq y + z$, (ii) $0 \leq x \Rightarrow 0 \leq sx$ for every $s \geq 0$, (iii) $\| x \vee (-x) \| = \| x \|$ for every $x \in E$, (iv) $0 \leq x \leq y \Rightarrow \| x \| \leq \| y \|$. The norm is **order-continuous** if whenever A is a non-empty downwards-directed subset of E and $\inf A = 0$, then $\inf_{x \in A} \| x \| = 0$. [See FREMLIN 74, §24.]

APPENDIX B

Consistency results

So many questions are resolved by Martin's axiom that it is possible to exaggerate its power. It may therefore save some waste of time and energy if I list those results which have been shown to be undecided by MA. I put them in the form 'A is relatively consistent with ZFC $+B$', meaning that if there is a proof in ZFC (Zermelo–Fraenkel set theory including the axiom of choice) that $B \Rightarrow$not-A, then there is a proof in ZFC that $\varnothing = \{\varnothing\}$.

B1 Variants of Martin's axiom

I give precise statements of the principal results showing that the different cardinals of Chapter 1 are indeed different [B1F], with descriptions of four similar cardinals [B1B–E], and some particular cases in which $\mathfrak{p}$ or $\mathfrak{m}_K$ cannot be substituted for $\mathfrak{m}$ [B1G–J].

B1A Classes of partially ordered sets

If Φ is any class of partially ordered sets, we can seek to define $\mathfrak{m}_\Phi$ as the least cardinal such that

> there is a non-empty $P \in \Phi$ and a family $\mathcal{Q}$ of cofinal subsets of P such that $\#(\mathcal{Q}) = \mathfrak{m}_\Phi$ and there is no upwards-centered subset of P meeting every member of $\mathcal{Q}$.

($\mathfrak{m}_\Phi$ is defined iff there is a $P \in \Phi$ such that $\{q : q \geq p\}$ is not upwards-directed for any $p \in P$.) Thus if Φ is the class of all partially ordered sets, $\mathfrak{m}_\Phi = \omega_1$ [11F]; if Φ is the class of upwards-ccc partially ordered sets, $\mathfrak{m}_\Phi = \mathfrak{m}$ [13A]; if Φ is the class of partially ordered sets satisfying Knaster's condition upwards, $\mathfrak{m}_\Phi = \mathfrak{m}_K$; and if Φ is the class of partially ordered sets which are σ-centered upwards, $\mathfrak{m}_\Phi = \mathfrak{p}$ [14C]. Evidently $\mathfrak{m}_\Psi \leq \mathfrak{m}_\Phi$ whenever $\Phi \subseteq \Psi$ and $\mathfrak{m}_\Phi$ is defined.

B1B Countable partially ordered sets

Taking Φ in B1A to be the class of countable partially ordered sets, we obtain a cardinal $\mathfrak{m}_{\text{countable}}$ with $\mathfrak{p} \leq \mathfrak{m}_{\text{countable}}$. Reviewing the arguments of 22D, 22G and 22L we see that (i) the union of fewer than $\mathfrak{m}_{\text{countable}}$ compact Lebesgue negligible sets is Lebesgue negligible (so that $\mathfrak{m}_{\text{countable}} \leq \mathfrak{c}$), (ii) if $\mathfrak{m}_{\text{countable}} = \mathfrak{c}$ then $\mathbf{R}$ has a $\mathfrak{c}$-Lusin subset and there is no atomlessly measurable cardinal. Other known facts are that $\mathfrak{m}_{\text{countable}}$ is the smallest cardinal such that $\mathbf{R}$ is the union of $\mathfrak{m}_{\text{countable}}$ nowhere-dense sets [FREMLIN & SHELAH 79], and that $\text{cf}(\mathfrak{m}_{\text{countable}}) > \omega$ [MILLER 82].

B1C Precaliber ω_1

If P is a partially ordered set, we say that ω_1 is an **up-precaliber** for P if every uncountable $R \subseteq P$ has an uncountable upwards-centered subset. [Compare 43Po(ii).] Taking Φ in B1A to be the class of partially ordered sets for which ω_1 is an up-precaliber, we obtain a cardinal $\mathfrak{m}_{\text{pc}(\omega_1)}$ such that $\mathfrak{m}_K \leq \mathfrak{m}_{\text{pc}(\omega_1)} \leq \mathfrak{p}$. Note that an S-respecting partial order has ω_1 as an up-precaliber iff it satisfies Knaster's condition upwards [31Cb(iv)]; so that we can substitute $\mathfrak{m}_{\text{pc}(\omega_1)}$ for $\mathfrak{m}_K$ in 31G and 35H–J. Making trivial changes in the wording of the proofs we can do the same in 31H–J and 34J–K. Observe also that the principle H of 41L says just that every upwards-ccc partially ordered set has ω_1 as an up-precaliber, in which case $\mathfrak{m}_{\text{pc}(\omega_1)} = \mathfrak{m}$; while if $\mathfrak{m}_K > \omega_1$ then every partially ordered set satisfying Knaster's condition has ω_1 as a precaliber [31B], so that $\mathfrak{m}_{\text{pc}(\omega_1)} = \mathfrak{m}_K$.

B1D σ-linked sets

A partially ordered set P is **σ-linked upwards** if it is expressible as $\bigcup_{n \in \mathbf{N}} P_n$ where each P_n is upwards-linked in P. Let $\mathfrak{m}_{\sigma\text{-linked}}$ be the cardinal associated with the class of σ-linked partially ordered sets; then $\mathfrak{m}_K \leq \mathfrak{m}_{\sigma\text{-linked}} \leq \mathfrak{p}$. We can replace $\mathfrak{m}_K$ by $\mathfrak{m}_{\sigma\text{-linked}}$ in 32E–N and 33B–D.

B1E Stable partially ordered sets

(*a*) A partially ordered set P is **upwards-stable** if for every countable $Q \subseteq P$ there is a countable $R \subseteq P$ such that for every $p \in P$ there are a $p' \geq p$ and an $r \in R$ such that, for every $q \in Q$, p' and q have a common upper bound in P iff r and q have a common upper bound in P. [AVRAHAM & SHELAH 82.]

Let $\mathfrak{m}_{\text{stable}}$ be the cardinal associated with the class of upwards-ccc upwards-stable partially ordered sets.

(*b*) Countable partially ordered sets are stable. So $\mathfrak{m} \leq \mathfrak{m}_{\text{stable}} \leq \mathfrak{m}_{\text{countable}}$. All Aronszajn trees are upwards-stable; so $\mathfrak{m}_{\text{stable}}$ can be substituted for $\mathfrak{p}$ in 25G and for $\mathfrak{m}$ in 41D, 41F–G and 41J, at the cost of rewriting the proofs. If $S \subseteq X \times Y$, the S-respecting partial order on $[X]^{<\omega} \times [Y]^{<\omega}$ is stable if, for every countable $Y_0 \subseteq Y$, $\{S[\{x\}] \cap Y_0 : x \in X\}$ is countable; analysing the proofs of 31G and 35H–J, we find that the latter will be true if $\mathfrak{m}_{\text{stable}} > \omega_1$. The partially ordered set of 34G is upwards-stable, so $\mathfrak{m}_{\text{stable}}$ can be substituted for $\mathfrak{m}_K$ in 34J–K. The partially ordered set of 31H is not itself stable, but a simple modification produces a stable partially ordered set for which the same proof works; so $\mathfrak{m}_K$ can be replaced by $\mathfrak{m}_{\text{stable}}$ in 31H.

(*c*) If $\mathfrak{m}_{\text{stable}} > \omega_1$, then $2^{\omega_1} = \mathfrak{c}$ [AVRAHAM & SHELAH 82].

(*d*) **Problem** If $\kappa < \mathfrak{m}_{\text{stable}}$, does it follow that $2^\kappa \leq \mathfrak{c}$?

B1F **Theorem**

Each of the following is relatively consistent with ZFC:

(*a*) $\mathfrak{m} = \mathfrak{m}_{\text{stable}} < \mathfrak{m}_K$ [B1E, B1I].
(*b*) $\mathfrak{m}_K = \mathfrak{m}_{\sigma\text{-linked}} < \mathfrak{m}_{\mathrm{pc}(\omega_1)} = \mathfrak{p}$ [HERINK 77, 4.3].
(*c*) $\mathfrak{m}_{\mathrm{pc}(\omega_1)} < \mathfrak{m}_{\sigma\text{-linked}}$ [HERINK 77, 4.9].
(*d*) $\omega_1 = \mathfrak{p} < \mathfrak{m}_{\text{stable}} = \mathfrak{m}_{\text{countable}}$ [AVRAHAM & SHELAH 82].

[See also ROITMAN 79, STEPRANS *a*.]

B1G **Proposition**

It is relatively consistent with ZFC $+ \mathfrak{c} = \omega_2 < 2^{\omega_1}$ that the union of ω_1 meagre sets in **R** should always be meagre [MILLER *a*]. Observe that in this case $\mathfrak{m}_{\text{stable}} = \mathfrak{p} = \omega_1$ [21C, B1Ec] and $\mathfrak{m}_{\text{countable}} = \mathfrak{c}$ [B1B].

B1H **Theorem**

It is relatively consistent with ZFC $+ \mathfrak{p} = \mathfrak{c} > \omega_1$ that **R** should be expressible as the union of ω_1 Lebesgue negligible sets [MILLER *a*]. Observe that in this case $\mathfrak{m}_K = \mathfrak{m}_{\sigma\text{-linked}} = \omega_1$ [32F].

B1I **Theorem**

It is relatively consistent with ZFC $+\mathfrak{m}_K = \mathfrak{c} = \omega_2$ that a Souslin tree should exist [KUNEN & TALL 79]. Observe that in this case $\mathfrak{m} = \mathfrak{m}_{\text{stable}} = \omega_1$ [41D].

B1J **Theorem**

It is relatively consistent with ZFC $+\mathfrak{m}_K > \omega_1$ that there should be an $S \subseteq [\omega_1]^2$ such that (i) $[A]^2 \cap S \neq \emptyset$ for every uncountable $A \subseteq \omega_1$ (ii) if $\eta < \xi < \omega_1$ then

$$\{\zeta : \zeta < \eta,\ \{\zeta, \eta\} \in S \text{ and } \{\zeta, \xi\} \in S\}$$

is finite [MILLER *a*]. Observe that in this case $\mathfrak{m} = \omega_1$ [42Lc].

Notes and comments

The recipe of B1A can be used to formulate a very large number of problems. I have concentrated in this book on the three classes of ccc partially ordered sets, partially ordered sets satisfying Knaster's condition, and σ-centered partially ordered sets; but any of the classes of B1B–E could turn out to be as important as Knaster's condition; and there are many other possibilities: e.g. the condition 'σ-f.c.c.' of 32P*a* [see HERINK 77, BAUMGARTNER & TAYLOR 82*b*].

Note that in B1A I use the phrase 'upwards-centered subset of P' rather than the more usual 'upwards-directed'. In the most familiar cases this makes no difference [13A(ii). 14C(iii)], but with $\mathfrak{m}_{\text{stable}}$ there may be a problem, and it is the 'upwards-centered' form that is more important. I think that the only point in this book where directed sets appear essential is Theorem 43E, and there only for a few of its applications; see the notes to §43.

The proofs which have been given of the different parts of B1F are all independent of each other, but all rely on the same idea: start with a model in which there is a partially ordered set P belonging to $\Phi \backslash \Psi$ and a family $\mathcal{Q}$ of cofinal subsets of P witnessing that $\mathfrak{m}_{\{P\}} = \omega_1$; then use iterated forcing to construct a model in which $\mathfrak{m}_{\Psi} = \mathfrak{c} > \omega_1$, but there is still no upwards-centered subset of P meeting each member of $\mathcal{Q}$, so that $\mathfrak{m}_{\Phi} = \omega_1$. It seems that in many, if not all, cases the forcing partial order can be chosen to belong to Ψ. I do not know if there is a useful metatheorem along these lines. The same idea lies behind B1H–J.

From the point of view of the structure of this book, even knowing that $\mathfrak{m} < \mathfrak{m}_K$ and $\mathfrak{m}_K < \mathfrak{p}$ are possible, I still want to be sure that the principal results are in the right chapters. The burden of B1G–J is that some of them certainly are, though a number of questions remain open. A.W. Miller informs me that the results of B1G–H and B1J are not his, and should be regarded as folklore.

I include $\mathfrak{m}_{\text{stable}}$ [B1E] as a way of emphasizing that the pattern dominating this book, $\omega_1 \leq \mathfrak{m} \leq \mathfrak{m}_K \leq \mathfrak{p} \leq \mathfrak{c}$, is by no means the only one present, and that many other arrangements of the material would be equally reasonable. The principles H and K and the cardinal $\mathfrak{l}$ of 41L suggest that the ideas of B1A, although fertile, are not adequate for all the discriminations that we might wish to make; while the most natural divisions of Chapter 2 also seem to be along different lines (see the notes to §24).

B2 Set theory

I give independence results concerning the cofinality of $\mathfrak{m}$ [B2A], stationary sets [B2B–C], ω_1-saturated σ-ideals [B2D], and Aronszajn and Kurepa trees [B2E–G]. B2H describes a variant of $\diamondsuit$. B2I touches on σ-ideals of $\mathscr{P}\omega_1$, and B2J is a strengthening of Szentmiklóssy's Lemma [42B].

B2A Proposition

It is relatively consistent with ZFC that $\mathfrak{m}=\omega_{\omega_1}$ and $\mathfrak{c}=\omega_{\omega_1+1}$. [KUNEN *a*.] Observe that in this case $\mathfrak{m}_K=\mathfrak{m}$ and $\mathfrak{p}=\mathfrak{c}$ [41C*c*, 21K].

B2B Proposition

It is relatively consistent with ZFC $+\mathfrak{m}=\mathfrak{c}\geq\omega_3$ that there is a family $\mathscr{F}$ of closed unbounded sets in ω_1 such that $\#(\mathscr{F})=\omega_2$ and every closed unbounded set in ω_1 includes some member of $\mathscr{F}$. [BAUMGARTNER, HAJNAL & MÁTÉ 75, §4; see KUNEN 80, EX.7.H1.]

B2C Proposition

It is relatively consistent with ZFC $+\mathfrak{m}=\mathfrak{c}>\omega_1$ that for every uncountable regular cardinal κ there should be a set $E\subseteq\kappa$ such that (i) $\mathrm{cf}(\xi)=\omega$ for every $\xi\in E$, (ii) E is stationary in κ (iii) $E\cap\zeta$ is not stationary in ζ for any non-zero limit ordinal $\zeta<\kappa$. [FLEISSNER 78.]

B2D Theorem

If one of the following is relatively consistent with ZFC, so are the others:

(i) there is a two-valued-measurable cardinal;
(ii) there is an atomlessly measurable cardinal;
(iii) there is a proper ω_1-saturated σ-ideal of $\mathscr{P}\mathfrak{c}$ containing singletons;
(iv) MA $+$(iii).

[SOLOVAY 71 and MARTIN & SOLOVAY 70, p. 163; see JECH 78, §34.]

B2E Theorem

The following statement and its negation are both relatively consistent with ZFC $+\mathfrak{m}=\mathfrak{c}>\omega_1$: Let T and T' be Aronszajn trees. Then there is a closed unbounded set $F\subseteq\omega_1$ such that $\{t:t\in T,\ \mathrm{rank}(t)\in F\}$ is isomorphic, as partially ordered set, to $\{t:t\in T',\ \mathrm{rank}(t)\in F\}$. [AVRAHAM & SHELAH *a*, TODORČEVIĆ 83*c*, 5.10.]

B2F Theorem

It is relatively consistent with ZFC $+\mathfrak{m}=\mathfrak{c}>\omega_1$ that there should be a Kurepa tree. [DEVLIN 78.]

B2G The weak Kurepa hypothesis

(*a*) Consider the statement:

wKH: There is a tree of cardinal and height both ω_1 which has more than ω_1 uncountable branches.

(*b*) **Theorem** If it is relatively consistent with ZFC that an inaccessible cardinal should exist, then not-wKH is relatively consistent with ZFC $+\mathfrak{m}=\mathfrak{c}=\omega_2$. [TODORČEVIĆ 81*a*; see BAUMGARTNER *a*, §8, and BAUMGARTNER 83, 7.10.]

(*c*) **Remark** Not-wKH implies that there is no Kurepa tree.

B2H **Variants of $\diamondsuit$**

(*a*) For a cardinal κ and a set $E\subseteq\kappa$ consider the statement

$\diamondsuit_\kappa(E)$: There is a family $\langle S_\xi\rangle_{\xi\in E}$ of sets such that, for any $A\subseteq\kappa$, $\{\xi:\xi\in E,\ A\cap\xi=S_\xi\}$ is stationary in κ.

(*b*) **Proposition** If $E=\{\xi:\xi<\mathfrak{c},\ \mathrm{cf}(\xi)=\omega\}$, then $\diamondsuit_\mathfrak{c}(E)$ is relatively consistent with $\mathfrak{m}=\mathfrak{c}=\omega_2$. [HAJNAL & JUHÁSZ 79.]

(*c*) **Proposition** [$\mathfrak{p}=\mathfrak{c}=\omega_2$ +not-wKH] If E is a stationary subset of ω_2 and $\mathrm{cf}(\xi)=\omega_1$ for every $\xi\in E$, then $\diamondsuit_{\omega_2}(E)$ is true. [BAUMGARTNER 83, Theorem 7.13.]

B2I **Proposition**

(*a*) It is relatively consistent with ZFC $+\mathfrak{m}=\omega_3$ that every uniform σ-ideal of $\mathcal{P}\omega_1$ should be ω_3-saturated. [BAUMGARTNER & TAYLOR 82*a*.]

(*b*) It is relatively consistent with ZFC $+\mathfrak{m}_{\sigma\text{-linked}}>\omega_1$ that there should be a uniform ω_2-saturated σ-ideal of $\mathcal{P}\omega_1$. [BAUMGARTNER & TAYLOR 82*b*.]

B2J **The thinning-out principle**

(*a*) Consider the following statements:

TOP: Let $\langle S_\xi\rangle_{\xi<\omega_1}$ be a family of subsets of ω_1. Suppose that $A\subseteq\omega_1$ is an uncountable set such that for every uncountable $C\subseteq A$ there is an $\alpha<\omega_1$ such that $\{C\cap\alpha\cap S_\xi:\xi\geq\alpha\}$ has the finite intersection property. Then there is an uncountable $C\subseteq A$ such that $C\cap\xi\setminus S_\xi$ is finite for every $\xi<\omega_1$.

TSL: Let X and Y be sets and $S\subseteq X\times Y$ a relation with countable vertical sections. Let P be $[X]^{<\omega}\times[Y]^{<\omega}$ with the S-respecting partial order [31C]. If P is not upwards-ccc there is a family $\langle y_\xi\rangle_{\xi<\omega_1}$ of distinct points in Y and a disjoint family $\langle K_\xi\rangle_{\xi<\omega_1}$ in $[X]^{<\omega}$ such that $y_\eta\in S[K_\xi]$ whenever $\eta<\xi<\omega_1$.

(*b*) **Theorem** TOP is relatively consistent with ZFC $+\mathfrak{m}=\mathfrak{c}=\omega_2$. [BAUMGARTNER 83, 5.1.]

(*c*) **Remark** (i) TOP$\Rightarrow$TSL. [Use 42B(ii).]

(ii) [TSL $+\mathfrak{m} > \omega_1$] $\omega_1 \rightarrow (\omega_1, \alpha)^2$ for every $\alpha < \omega_1$. [TODORČEVIĆ 81*c*, TODORČEVIĆ *d*. Compare 42L*c*.]

(iii) [TSL] Let X be an upwards-ccc partially ordered set of cardinal less than $\mathfrak{m}$. Then *either* there is an uncountable $Y \subseteq X$ such that every countable subset of Y has an upper bound in X *or* X is expressible as $\bigcup_{n\in\mathbf{N}} X_n$ where each X_n is a weak antichain. [See BAUMGARTNER 83, 8.10, or DEVLIN STEPRĀNS & WATSON *a*.]

Notes and comments

B2A marks an interesting difference between $\mathfrak{m}$ and $\mathfrak{p}$ [see 21K]. B2B and B2C were presented by their discoverers as lemmas on the way to other results; I give them here as they may turn out to have further uses. B2D shows that although MA destroys atomlessly measurable cardinals [22L], it need not destroy ω_1-saturated σ-ideals. B2E–G show that MA does not decide fundamental questions about the structure of Aronszajn trees and the existence of Kurepa trees. The model for B2G has further interesting properties, some of which are given in B3I. Note that, in particular, Theorem 41K becomes unsatisfactory under these conditions [B3I*a*]. B2H shows that although $\mathfrak{m} > \omega_1$ is thoroughly incompatible with $\diamondsuit$ [21J], it is not incompatible with some similar principles. B2I is connected with the work of §35; see especially 35N*b* and 35M*c*.

For a short general discussion of techniques for proving that propositions are compatible with $\mathfrak{m} = \mathfrak{c} > \omega_1$, see ABRAHAM & TODORČEVIĆ 83, §3. One of the most important is the proper forcing axiom, described in BAUMGARTNER 83. This unifies a large number of results, notably B2G, B2J*b*, B3A*b*(i) and B3C. The two principles TOP and TSL of B2J are extracted from work of S. Todorčević.

B3 Topology

In this section I give independence results on 'gaps' and subalgebras of $\mathscr{P}\mathbf{N}/[\mathbf{N}]^{<\omega}$ [B3A–B]; the order types of subsets of $\mathbf{R}$ [B3C–D]; hereditarily ccc spaces [B3E–F]; the structure of coanalytic sets [B3G]; Dowker spaces [B3H]; and Baire spaces [B3I*b*, B3J].

B3A Gaps in $\mathscr{P}\mathbf{N}/[\mathbf{N}]^{<\omega}$

(*a*) Consider the statement Gap(κ, λ): There exist indexed families $\langle A_\xi \rangle_{\xi<\kappa}$, $\langle B_\xi \rangle_{\xi<\lambda}$ of subsets of $\mathbf{N}$ such that

for $\xi, \eta < \kappa$, $A_\xi \setminus A_\eta$ is finite iff $\xi \leq \eta$;
for $\xi, \eta < \lambda$, $B_\xi \setminus B_\eta$ is finite iff $\xi \leq \eta$;
$A_\xi \cap B_\eta$ is finite whenever $\xi < \kappa$, $\eta < \lambda$;

there is no $H \subseteq \mathbf{N}$ such that $A_\xi \setminus H$ and $B_\eta \cap H$ are finite whenever $\xi < \kappa$, $\eta < \lambda$.

Thus 21L asserts that Gap(ω_1, ω_1) is true; 21A implies easily that Gap(ω, κ) is false if $\kappa < \mathfrak{p}$; 31J asserts that Gap(κ, λ) is false if $\max(\kappa, \lambda) < \mathfrak{m}_K$ and $\mathrm{cf}(\kappa) \neq \omega_1$. It follows easily that Gap$(\kappa, \mathfrak{c})$ is true if $\omega \leq \kappa < \mathfrak{m}_K = \mathfrak{c}$ and $\mathrm{cf}(\kappa) \neq \omega_1$.

(*b*) **Theorem** (i) It is relatively consistent with ZFC $+\mathfrak{m} = \mathfrak{c} = \omega_2$ that Gap$(\omega_1, \mathfrak{c})$ and Gap$(\mathfrak{c}, \mathfrak{c})$ should both be false. [KUNEN *b*; see BAUMGARTNER 83, 4.3.]

(ii) It is relatively consistent with ZFC $+\mathfrak{m} = \mathfrak{c} > \omega_1$ that Gap$(\kappa, \mathfrak{c})$ should be true for every infinite $\kappa \leq \mathfrak{c}$. [HERINK 77.]

B3B **Remark**

If $\mathfrak{p} = \mathfrak{c}$ and Gap$(\mathfrak{c}, \mathfrak{c})$ is false, then the following hold.

(*a*) There is a Boolean algebra $\mathfrak{A}$, of cardinal $\mathfrak{c}$, which cannot be embedded into $\mathscr{P}\mathbf{N}/[\mathbf{N}]^{<\omega}$. [MILL 83, 2.3.2; BAUMGARTNER 83, 4.6.]

(*b*) If Y is the set of non-$p(\mathfrak{c})$-points in $\beta\mathbf{N} \setminus \mathbf{N}$, then every real-valued continuous function on Y has a continuous extension to $\beta\mathbf{N} \setminus \mathbf{N}$. [DOUWEN *a*.]

B3C (*a*) **Theorem**

It is relatively consistent with ZFC $+\mathfrak{m} = \mathfrak{c} = \omega_2$ that (i) every uncountable set in $\mathscr{P}\mathbf{N}$ should include either an uncountable chain or an uncountable weak antichain (ii) any two subsets of $\mathbf{R}$ each of which meets every non-empty open interval in a set of cardinal ω_1 should be isomorphic.

(*b*) **Remark** In this case every uncountable Boolean algebra has an uncountable weak antichain. [BAUMGARTNER 80; see BAUMGARTNER 83, §6, and 41P*b*.]

B3D **Entangled sets**

(*a*) If $k \geq 1$, say that a set $A \subseteq \mathbf{R}$ is **k-entangled** if whenever $\langle s_\xi^i \rangle_{\xi < \omega_1, i < k}$ is an indexed family of distinct members of A, and $I \subseteq k$, there are distinct $\xi, \eta < \omega_1$ such that $I = \{i : i < k, s_\xi^i \leq s_\eta^i\}$.

(*b*) **Theorem** It is relatively consistent with ZFC $+\mathfrak{m} = \mathfrak{c} = \omega_2$ that there should be for every $k \geq 1$ an uncountable k-entangled set $A \subseteq \mathbf{R}$.

(*c*) **Remarks** (i) If $A \subseteq \mathbf{R}$ is 2-entangled then no two disjoint uncountable subsets of A can be order-isomorphic.

(ii) $[\mathfrak{m} > \omega_1]$ If $A \subseteq \mathbf{R}$ is uncountable, there is some $k \geq 1$ such that A is not k-entangled.
[AVRAHAM & SHELAH 81, AVRAHAM, RUBIN & SHELAH *a*.]

B3E Theorem

Each of the following is relatively consistent with ZFC $+\mathfrak{m} = \mathfrak{c} > \omega_1$: (*a*) there is a hereditarily Lindelöf regular Hausdorff space which is not separable; (*b*) there is a first-countable hereditarily separable regular Hausdorff space which is not Lindelöf [ABRAHAM & TODORČEVIĆ 83; see also JUHÁSZ 80*a*.]

B3F Theorem [TSL $+\mathfrak{m} > \omega_1$] (see B2J) Let X be a hereditarily ccc topological space.

(*a*) If $\mathscr{G}$ is an open cover of X there is a countable $\mathscr{G}_0 \subseteq \mathscr{G}$ such that $X = \bigcup\{\bar{G} : G \in \mathscr{G}_0\}$.

(*b*) If X is regular it is hereditarily Lindelöf.

(*c*) If X is Hausdorff then $\psi(x, X) \leq \omega$ for every $x \in X$ and $\#(X) \leq \mathfrak{c}$. [TODORČEVIĆ 81*b*. TODORČEVIĆ 81*d*, TODORČEVIČ *d*; see ROITMAN 83, 7.2, and JUHÁSZ 83, 2.1.]

B3G Theorem

Let **L** be the statement of 23J. Then if one of the following is relatively consistent with ZFC, so is the other:

(i) there is a weakly compact cardinal;

(ii) $\mathfrak{m} = \mathfrak{c} > \omega_1$ and **L** is false.

[SOLOVAY 70, p. 4; HARRINGTON & SHELAH *a*.]

B3H Proposition $[\mathfrak{p} = \mathfrak{c}]$

Let E be $\{\xi : \xi < \mathfrak{c}, \mathrm{cf}(\xi) = \omega\}$. If $\diamondsuit_{\mathfrak{c}}(E)$ is true [B2H*b*], there is a separable locally compact locally countable Dowker space. [WEISS 81*a*; see RUDIN 83, 3.1(iii). Compare 25N*c*.]

B3I Proposition $[\mathfrak{m} > \omega_1]$

(*a*) Suppose that wKH [B2G] is false. Let X be a compact Hausdorff space satisfying the condition of 41N*d* above, with $c(X) \leq \omega_1$. Then it has the following stronger property: if $\langle \mathscr{G}_\xi \rangle_{\xi < \omega_1}$ is an indexed family of collections of open sets in X such that $\bigcup \mathscr{G}_\xi$ is dense for every $\xi < \omega_1$, then there is a non-empty open $H \subseteq X$ such that for every $\xi < \omega_1$ there is a $G \in \mathscr{G}_\xi$ such that $H \subseteq G$. [TODORČEVIĆ 81*a*; see WEISS 83, 8.7.]

(*b*) Again suppose that wKH is false. If X is a non-empty regular

Baire space with $\#(X) \leq \omega_1$ and $\pi(X) \leq \omega_1$ then X has an isolated point. [DAVIES 79; see WEISS 83, 8.9, and TODORČEVIĆ 81*a*.]

B3J **Proposition**

It is relatively consistent with ZFC $+\mathfrak{m} = \mathfrak{c} > \omega_1$ that there should be a regular Hausdorff Baire space of cardinal and π-weight ω_1 which does not have an isolated point. [WEISS 83, 8.9.]

Notes and comments

B3A sorts out problems left over from 31J, and B3B*a* indicates an obstacle to some natural conjectures arising from 26G–I. Note that these results are significant only if the continuum hypothesis is false, since in B3A it is clear that CH implies both Gap$(\omega, \mathfrak{c})$ and Gap$(\mathfrak{c}, \mathfrak{c})$, while in B3B we have Parovičenko's theorem [see 26 *notes*] to tell us that every Boolean algebra of cardinal ω_1 can be embedded in $\mathscr{P}\mathbf{N}/[\mathbf{N}]^{<\omega}$.

B3C*a*(i) is a strengthening of 42F. B3C*a*(ii) and B3D*c*(i) show that Martin's axiom does not settle whether there can be subsets of **R** of cardinal ω_1 which are non-isomorphic for non-trivial reasons. B3C*b* (which is a consequence of B3C*a*(i)–(ii) only) also looks interesting; but see 41P*b*.

B3E–F show that $\mathfrak{m} > \omega_1$ is not enough to decide whether there is a hereditarily ccc regular Hausdorff space which is not Lindelöf, and is not enough to make all hereditarily ccc regular Hausdorff spaces separable. B3G deals with a question discussed in the notes to §23. B3H–I give applications of principles introduced in B2G–H. B3I*b* and B3J are connected with 23M*d*(iii).

Many further results may be found in AVRAHAM, RUBIN & SHELAH *a*, which reached me too late for proper discussion here.

Bibliography

Each reference is followed by a list of the paragraphs in which it is referred to.

Aarts J.M. & Lutzer D.J. [74] 'Completeness properties designed for recognizing Baire spaces', *Dissertationes Math.* **116** (1974). 43D*c*, 43O*f*.

Abraham U. [= Avraham U.]

Abraham U. & Todorčević S.B. [83] 'Martin's axiom and first-countable *S* and *L* spaces', in KUNEN & VAUGHAN 83 44 *notes*, B2 *notes*, B3E.

Allen K.R. [see Stavrakis N.M.]

Alster K. & Przymusiński T.C. [76] 'Normality and Martin's axiom', *Fund. Math.* **91** (1976) 123–31. 25L, 25M*d–h*, 25N*a*.

Alster K. & Zenor P. [77] 'On the collectionwise normality of generalized manifolds', *Top. Proc.* **1** (1976) 125–7. 44N.

Antonovskii M.Ja. & Chudnovsky D.V. [76] 'Some questions of general topology and Tikhonov semifields II', *Russ. Math. Surv.* **31** (1976) 69–128. 11 *notes*, 24M, 24O*e*.

Argyros S.A. [82] 'On non-separable Banach spaces', *Trans. Am. Math. Soc.* **270** (1982) 193–216. 33G, 33H, 33J*a*.

Argyros S.A. & Zachariades Th. [*a*] 'A result on the isomorphic embeddability of $\ell^1(\Gamma)$', submitted to *Studia Math.* 33H.

Arhangel'skii A.V. [71] 'On bicompacta hereditarily satisfying Suslin's condition. Tightness and free sequences', *Soviet Math. Dokl.* **12** (1971) 1253–7. 43Pd.

[72] 'On cardinal invariants', pp. 37–46 in NOVAK 72. 43N.

[76] 'Martin's axiom and the structure of homogeneous bicompacts of countable tightness', *Soviet Math. Dokl.* **17** (1976) 256–60. 24O*a*, 31L, 43N.

Arias de Reyna J. [80] 'Dense hyperplanes of first category', *Math. Ann.* **249** (1980) 111–14. 32Q*c*.

Avraham U. [Abraham U.] [81] 'Free sets for nowhere dense set mappings', *Israel J. Math.* **39** (1981) 167–76. 42J.

Avraham U., Rubin M. & Shelah S. [*a*] 'On the consistency of some partition theorems for continuous colourings, and the structure of $\aleph_1$-dense real order types', *Ann. Math. Logic.* B3D, B3 *notes*.

Avraham U. & Shelah S. [81] 'Martin's axiom does not imply that every two $\aleph_1$-dense sets of reals are isomorphic', *Israel J. Math.* **38** (1981) 161–76. B3D.

[82] 'Forcing with stable posets', *J. Symbolic Logic* **47** (1982) 37–42. B1E, B1F*d*.

[*a*] 'Isomorphism types of Aronszajn trees', *Israel J. Math.* B2E.

Balcar B., Frankiewicz R. & Mills C.F. [80] 'More on nowhere dense closed *P*-sets', *Bull. Acad. Polon. Sci. (Math.)* **28** (1980) 295–9. 21N*l*, 32N*c*.

Balcar B. & Vojtas P. [77] 'Refining systems on Boolean algebras', pp. 45–58 in LACHLAN, SREBNY & ZARACH 77. 35L*c–d*.

Balogh Z. [*a*] 'Locally nice spaces under Martin's axiom', preprint. 42J, 44N, 44O*c*, 44O*k*, 44P*a*.

Barwise J. [75] *Admissible Sets and Structures*. Springer, 1975. 13D*a*.
[77] (ed.) *Handbook of Mathematical Logic*. North-Holland, 1977.
Baumgartner J.E. [80] 'Chains and antichains in $P(\omega)$', *J. Symbolic Logic* **45** (1980) 85–92. 42J, B3C.
[83] 'Applications of the Proper Forcing Axiom', in KUNEN & VAUGHAN 83. 31L, 31N*a*, 41N*b*, B2G, B2H*c*, B2J, B3A, B3B*a*, B3C.
[*a*] 'Iterated forcing', in MATHIAS *a*. 11E, 11 *notes*, 41N*b*, B2G.
Baumgartner J.E. & Hajnal A. [73] 'A proof (involving Martin's axiom) of a partition relation', *Fund. Math.* **78** (1973) 193–203. 21M, 21O*b*.
Baumgartner J.E., Hajnal A. & Máté A. [75] 'Weak saturation properties of ideals', pp. 137–58 in HAJNAL, RADO & SOS 75. 35L*e*, B2B.
Baumgartner J.E., Malitz J. & Reinhardt W. [70] 'Embedding trees in the rationals', *Proc. Nat. Acad. Sci.* **67** (1970) 1748–53. 41M.
Baumgartner J.E. & Taylor A.D. [78] 'Partition theorems and ultrafilters', *Trans. Am. Math. Soc.* **241** (1978) 283–309. 26L*a*.
[82*a*] 'Saturation properties of ideals in generic extensions I', *Trans. Am. Math. Soc.* **270** (1982) 557–74. 31L, 35M*c*, B2I*a*.
[82*b*] 'Saturation properties of ideals in generic extensions II', *Trans. Am. Math. Soc.* **271** (1982) 587–609. 35N*b*, B1 *notes*, B2I*b*.
Bell M.G. [81*a*] 'Compact ccc non-separable spaces of small weight', *Top. Proc.* **5** (1980) 11–25. 24O*m*.
[81*b*] 'On the combinatorial principal $P(c)$', *Fund. Math.* **114** (1981) 149–57. 14E, 25N*c*.
Bell M.G. & Ginsburg J. [80] 'First countable Lindelöf extensions of uncountable discrete spaces', *Can. Math. Bull.* **23** (1980) 397–9. 24N*k*.
[82] 'Uncountable discrete sets in extensions and metrizability', *Can. Math. Bull.* **25** (1982) 472–7. 44O*d–e*.
Bell M.G., Ginsburg J. & Todorčević S. [82] 'Countable spread of exp Y and lambda Y' *Top. Appls.* **14** (1982) 1–12.
Benda M. & Ketonen J. [74] 'Regularity of ultrafilters', *Israel J. Math.* **17** (1974) 231–40. 35K.
Bennett H.R. & McLaughlin in T.G. [76] *A Selective Survey of Axiom-Sensitive Results in General Topology*. Texas Technical University (Lubbock, Texas), 1976. 13B, 25 *notes*.
Berberian S.K. [62] *Measure and Integration*. Macmillan, 1962. A6K*a*.
Bierstedt K.-D. & Fuchssteiner B. [80] (eds.) *Functional Analysis: Surveys and Recent Results II*. North-Holland, 1980.
Blass A. [72] 'Theories without countable models', *J. Symbolic Logic* **37** (1972) 562–8. 41O*d*.
[73] 'The Rudin-Keisler ordering of *P*-points', *Trans. Am. Math. Soc.* **179** (1973) 145–66. 26J, 26L*b*.
Blaszczyk A. & Szymański A. [80] 'Concerning Parovičenko's theorem', *Bull. Acad. Polon. Sci.* (*Math.*) **28** (1980) 311–14. 26 *notes*.
Boffa M., Dalen D. van & McAloon K. [79] (eds.) *Logic Colloquium* '78. North-Holland, 1979.
Booth D. [70] 'Ultrafilters on a countable set', *Ann. Math. Logic* **2** (1970–1) 1–24. 11H, 21N*a*, 26J, 26L*a*, A3C*b*.
[71] 'Generic covers and dimension', *Duke Math. J.* **38** (1971) 667–70. 25M*a*.
[74] 'A Boolean view of sequential compactness', *Fund. Math.* **85** (1974) 99–102. 22M, 24M.
Bourbaki N. [66] *General Topology*. Hermann, 1966. A4C*a*, A4M*a*, A4S*d*.
Broverman S., Ginsburg J., Kunen K. & Tall F.D. [78] 'Topologies determined by σ-ideals on ω_1', *Can. J. Math.* **30** (1978) 1306–12. 21J, 21M.

Burgess J.P. [77] 'Forcing', pp. 403–52 in BARWISE 77. 11E.
[78] 'On the Hanf number of Souslin logic', *J. Symbolic Logic* **43** (1978) 568–70. 21O*i*.
Burke D.K. & Davis S.W. [81] 'Separability in nearly compact symmetrizable spaces', *Top. Proc.* **5** (1980) 59–69. 24O*c*.
Burke D.K. & Hodel R.E. [76] 'The number of compact subsets of a topological space', *Proc. Am. Math. Soc.* **58** (1976) 363–8. 44P*e*.

Carlson T.J. [80] 'Strongly meagre and strong measure zero sets of reals', *Abstracts Am. Math. Soc.* **1** (1980) 389. 33H.
Čech E. & Pospišil B. [38] 'Sur les espaces compacts', *Publ. Fac. Sci. Univ. Masaryk Brno* **258** (1938) 3–7. A4J.
Chang C.C. & Keisler H.J. [73] *Model Theory*. North-Holland, 1973. 35L*a*, 41O*d*, A3D.
Choksi J. & Fremlin D.H. [79] 'Completion regular measures on product spaces', *Math. Ann.* **241** (1979) 113–28. 32Q*m*, A7E*c*.
Cohen P.E. [78] 'Iterated forcing without Boolean algebras', *Z. Math. Logik Grundlagen Math.* **24** (1978) 323–4. 11E.
Comfort W.W. & Negrepontis S. [74] *The Theory of Ultrafilters*. Springer, 1974. 26A, 26L*a*, A2J, A2N*c*–*d*, A3C*b*.
[82] *Chain Conditions in Topology*. Cambridge University Press, 1982. 12M*a*, 41 *notes*, 43P*i*, A2L.
Csaszar A. [78] *General Topology*. Hilger, 1978. A4S*d*.
[80] (ed.) *Topology*. North-Holland, 1980 (*Colloq. Math. Soc. Janos Bolyai* **23**).

Dalen D. van [see Boffa M.]
Davies P.C. [79] 'Small Baire spaces and σ-dense partial orders', PhD thesis, University of Toronto, 1979. B3I*b*.
Davis S.W. [see Burke D.K.]
Davis S.W., Reed G.M. & Wage M.L. [76] 'Further results on weakly uniform bases', *Houston J. Math.* **2** (1976) 57–63. 25N*a*.
Dellacherie C. [80] 'Un cours sur les ensembles analytiques', pp. 184–316 in ROGERS 80. A5H, A7C*d*.
Devlin K.J. [76] 'An alternative to Martin's axiom', pp. 65–76 in MAREK, SREBNY & ZARACH 76. 41M.
[78] '$\aleph_1$-trees', *Ann. Math. Logic* **13** (1978) 267–330. B2F.
Devlin K.J. & Shelah S. [78] 'A weak version of $\diamondsuit$ which follows from $2^{\aleph_0} > 2^{\aleph_1}$', *Israel J. Math.* **29** (1978) 239–47. 31M*e*.
Devlin K.J., Steprāns J. & Watson S. [*a*] 'The number of directed sets', preprint. B2J*c*.
Douwen E.K. van [77*a*] 'Density of compactifications', pp. 97–110. in REED 77. 14E, 14G.
[77*b*] 'Hausdorff gaps and a nice countably paracompact non-normal space', *Top. Proc.* **1** (1976) 239–42. 25L.
[80] 'The product of two countably compact topological groups', *Trans. Am. Math. Soc.* **262** (1980) 417–27. 21N*i*, 24O*g*.
[83] 'The integers and topology', in KUNEN & VAUGHAN 83. 11G*e*, 14G, 21N*k*, 24N*k*, 24 *notes*, 24O*j*–*k*.
[*a*] 'There can be C^*-embedded dense proper subspaces in $\beta\omega\backslash\omega$', preprint. B3B*b*.
[*b*] 'The product of two normal initially κ-compact spaces', *Trans. Am. Math. Soc.* 24O*f*.
Douwen E.K. van & Kunen K. [82] '*L*-spaces and *S*-spaces in $P(\omega)$', *Top. Appls.* **14** (1982) 143–50. 42J.
Douwen E.K. van, Lutzer D.J., Pelant J. & Reed G.M. [80] 'On unions of metrizable subspaces', *Can. J. Math.* **32** (1980) 76–85. 25N*a*.
Douwen E.K. van & Mill J. van [80] 'Subspaces of basically disconnected spaces or

quotients of countably complete Boolean algebras', *Trans. Am. Math. Soc.* **259** (1980) 121–7. 21*Oe*.
[*a*] 'There can be C^*-embedded dense proper subspaces in $\beta\omega\backslash\omega$; preprint.
Douwen E.K. van & Przymusiński T.C. [80] 'Separable extensions of first countable spaces', *Fund. Math.* **105** (1980) 147–58. 26J, 26K*g*.
Douwen E.K. van & Wage M.L. [79] 'Small subsets of first countable spaces', *Fund. Math.* **103** (1979) 103–10. 24N*k*, 25N*a*.
Dow A. [82] 'Weak *p*-points in compact ccc *F*-spaces', *Trans. Am. Math. Soc.* **269** (1982) 557–66. 24O*n*.
Drake F.R. [74] *Set Theory*. North-Holland, 1974. 23 *notes*, 41 *notes*, A2N*d*, A6P*e*.
Dunford N. & Schwartz J.T. [57] *Linear Operators I*. Wiley, 1957. 25J, A6D, A6K*b*, A8C*a*.

Edgar G.A. [80] 'A long James space', pp. 31–7 in KÖLZOW 80. 22O*e*.
[*a*] Letter of 11 June 81. 32P*m*.
Eidswick J.A. [76] 'The undecidability of a fundamental problem in cluster set theory', *Proc. Am. Math. Soc.* **60** (1976) 116–18. 25N*a*.
Eklof P.C. [76] 'Whitehead's problem is undecidable', *Am. Math. Monthly* **83** (1976) 775–88. 34A.
[77] 'Homological algebra and set theory', *Trans. Am. Math. Soc.* **227** (1977) 207–25. 34M*b*.
[80] *Set-theoretic Methods in Homological Algebra and Abelian Groups*. Montreal University Press, 1980. 34L.
Eklof P.C. & Huber M. [80] 'On the rank of Ext', *Math. Zeitschrift* **174** (1980) 159–85. 34M*a*.
Ellentuck E. [74] 'A new proof that analytic sets are Ramsey', *J. Symbolic Logic* **39** (1974) 163–5. 23N*a*.
Ellentuck E. & Rucker R.V.B. [72] 'Martin's axiom and saturated models', *Proc. Am. Math. Soc.* **34** (1972) 243–9. 21M, 26J.
Engelking R. [77] *General Topology*. Polish Scientific Publishers, 1977 (*Monogr. Mat.* **60**). 26A, §A4.
Erdös P., Galvin F. & Hajnal A. [75] 'On set-systems having large chromatic number and not containing prescribed subsystems', pp. 425–513 in HAJNAL, RADO & SOS 75. 31N*b*, 41O*c*.
Erdös P., Hajnal A. & Máté A. [73] 'Chain conditions on set mappings and free sets', *Acta Scientiarum Mathematicarum* **34** (1973) 69–79. 21N*g*, 42L*a*.
Erdös P., Kunen K. & Mauldin R.D. [81] 'Some additive properties of sets of real numbers', *Fund. Math.* **113** (1981) 187–99. 22N*f*, 32Q*g*.
Erdös P. & Shelah S. [72] 'Separability properties of almost-disjoint families of sets', *Israel J. Math.* **12** (1972) 207–14. 21M.

Fedorčuk V.V. [75] 'A compact Hausdorff space all of whose infinite closed subsets are *n*-dimensional', *Math. USSR Sbornik* **25** (1975) 37–57. 26 *notes*.
Figiel T., Ghoussoub N. & Johnson W.B. [81] 'On the structure of non-weakly-compact operators on Banach lattices', *Math. Ann.* **257** (1981) 317–34. 22O*n*.
Flachsmeyer J., Frolík Z. & Terpe F. [80](eds.) *Proceedings of the Conference on Topology and Measure II* (Rostock-Warnemunde 1977), Ernst-Moritz-Arndt University, Greifswald 1980, part 1.
Fleissner W.G. [75] 'When is Jones' space normal?', *Proc. Am. Math. Soc.* **50** (1975) 375–8. 25L.
[78] 'Separation properties in Moore spaces', *Fund. Math.* **98** (1978) 279–86. 25N*a*, B2C.

[80] 'Martin's axiom implies that de Caux's space is countably metacompact', *Proc. Am. Math. Soc.* **80** (1980) 495–8. 31M*g*.

[82] 'If all normal Moore spaces are metrizable, there is an inner model with a measurable cardinal', *Trans. Am. Math. Soc.* **273** (1982) 365–373. 25 *notes.*

[83] 'The normal Moore space conjecture and large cardinals', in KUNEN & VAUGHAN 83. 25 *notes.*

Fleissner W.G. & Kunen K. [78] 'Barely Baire spaces', *Fund. Math.* **101** (1978) 229–40. 23M*d*.

Fleissner W.G. & Reed G.M. [78] 'Paralindelöf spaces and spaces with a σ-locally countable base', *Top. Proc.* **2** (1977) 89–110. 23N*j*, 31M*d*.

Frankiewicz R. [77*a*] 'On the inhomogeneity of the set of *P*(*m*)-points of ω*', pp. 169–79 in LACHLAN, SREBNY & ZARACH 77. 26L*a*.

[77*b*] 'To distinguish topologically the space m*', *Bull. Acad. Polon. Sci.* (*Math. Astron. Phys.*) **25** (1977) 891–3. 26L*o*.

[78] 'Assertion Q distinguishes topologically ω* and m* when m is regular and m > ω' *Colloq. Math.* **38** (1978) 175–7. 26L*o*.

[81] 'Non-accessible points in extremally disconnected compact spaces I', *Fund. Math.* **111** (1981) 115–23. 26L*d*–*h*, 32P*e*, 32Q*n*.

[*a*] 'Six theorems', *Abstracts Am. Math. Soc.* **1** (1980) 588.

[See also Balcar B.]

Frankiewicz R. & Gutek A. [81] 'Some remarks on embeddings of Boolean algebras and the topological spaces I', *Bull. Acad. Polon. Sci.* (*Math.*) **29** (1981) 471–6. 33I*f*.

Fremlin D.H. [72] 'Tensor products of Archimedean vector lattices', *Am. J. Math.* **94** (1972) 777–98. 41O*b*.

[74] *Topological Riesz Spaces and Measure Theory.* Cambridge University Press, 1974. §A6, §A7, A8D*a*, A8G.

[77*a*] '*K*-analytic sets with metrizable compacta', *Mathematika* **24** (1977) 257–61. 23N*d*.

[77*b*] 'Uncountable powers of **R** can be almost Lindelöf', *Manuscripta Math.* **22** (1977) 77–85. 32Q*l*.

[81] 'Measurable functions and almost continuous functions', *Manuscripta Math.* **33** (1981) 387–405. 32O, 32P*k*, 32R*a*.

[82] 'Measurable selections and measure-additive coverings', pp. 425–8 in KOLZOW & MAHARAM-STONE 82. 32Q*k*, 43Q*b*, A6L*d*.

[*a*] 'On a problem of A. Bellow', Note of 11 May 82, privately circulated. 22O*d*.

[*b*] 'Two new versions of MA', Note of 5 Jan. 84. 13D*b*–*c*, 41L.

[see also Choksi J.]

Fremlin D.H., Hansell R.W. & Junnila H.J.K. [*a*] 'Borel functions of bounded class', *Trans. Am. Math. Soc.* **277** (1983) 835–49. 23N*h*, 23O*a*.

Fremlin D.H. & Shelah S. [79] 'On partitions of the real line', *Israel J. Math.* **32** (1979) 299–304. 23N*e*, 44N, B1B.

Fremlin D.H. & Talagrand M. [79] 'A decomposition theorem for additive set-functions, with applications to Pettis integrals and ergodic means', *Math. Zeitschrift* **168** (1979) 117–42. 22O*a*–*c*, A6L*c*.

Frolik Z. [see Flachsmeyer J.]

Fuchs L. [70] *Infinite Abelian Groups I.* Academic Press, 1970. 34A, 34B, 34M*a*.

[73] *Infinite Abelian Groups II.* Academic Press, 1973. 34A.

Fuchssteiner B. [see Bierstedt K.-D.]

Galvin F. [77] 'On Gruenhage's generalization of first-countable spaces', *Notices Am. Math. Soc.* **24** (1977) A–22. 24O*o*.

[78] 'Indeterminacy of point-open games', *Bull. Acad. Polon. Sci.* (*Math. Astron. Phys.*) **26** (1978) 445–9. 22O*i*.

[80] 'Chain conditions and products', *Fund. Math.* **108** (1980) 33–48. 24O*o*, 41 *notes.*
[see also Erdös P.]

Galvin F. & Miller A.W. [*a*] 'γ-sets and other singular sets of real numbers', *Top. Appls.*

Gardner R.J. & Pfeffer W.F. [80] 'Some undecidability results concerning Radon measures', *Trans. Am. Math. Soc.* **259** (1980) 65–74. 32P*i*.

Ghoussoub N. [see Figiel T.]

Ginsburg J. [see Bell M.G., Broverman S.]

Göbel R. & Wald B. [80] 'Martin's axiom ensures the existence of certain slender groups', *Math. Zeitschrift* **172** (1980) 107–21. 21O*i*.

Gödel K. [40] *The Consistency of the Axiom of Choice and the Generalized Continuum Hypothesis with the Axioms of Set Theory.* Princeton University Press, 1940 (Annals of Mathematical Studies 3). *Intro.*

Gruenhage G. [79] 'A note on the product of Fréchet spaces', *Top. Proc.* **3** (1978) 109–115. 24O*b*.

[80] 'Paracompactness and subparacompactness in perfectly normal locally compact spaces', *Russ. Math. Surv.* **35** (1980) 49–55. 24M, 44N.

Grzegorek E. [81] 'On some results of Darst and Sierpiński concerning universal null and universally negligible sets', *Bull. Acad. Polon. Sci.* (*Math.*) **29** (1981) 1–5. 22 *notes.*

Gutek A. [see Frankiewicz R.]

Hajnal A. [see Baumgartner J.E., Erdös P.]

Hajnal A. & Juhász I. [71] 'A consequence of Martin's axiom', *Indag. Math.* **33** (= *Konink. Nederl. Akad. Wetensch. Proc.* (A) 74) (1971) 457–63. 31 *notes*, 41M, 43N.

[79] 'Weakly separated subspaces and networks', pp. 235–245 in Boffa, Dalen & McAloon 79. B2H*b*.

[82] 'When is a Pixley–Roy hyperspace ccc?', *Top. Appls.* **13** (1982) 33–41. 31M*i–j*, 43 *notes.*

Hajnal A. & Máté A. [75] 'Set mappings, partitions and chromatic numbers', pp. 347–79 in Rose & Shepherdson 75. 31L.

Hajnal A., Rado R. & Sos V.T. [75] (eds.) *Infinite and Finite Sets.* North-Holland, 1975. (Colloq. Math. Soc. Janos Bolyai 10)

Halmos P.R. [50] *Measure Theory.* Van Nostrand, 1950. §A6.

[60] *Naive Set Theory.* Van Nostrand, 1960. A2A.

[63] *Lectures in Boolean Algebras.* Van Nostrand, 1963. 12A, 12D.

Hansell R.W. [see Fremlin D.H.]

Harrington L. & Shelah S. [*a*] 'Equiconsistency results', submitted to *Notre Dame J. Formal Logic.* 21O*h*, B3G.

Hausdorff F. [36] 'Summen von $\aleph_1$ Mengen', *Fund. Math.* **26** (1936) 241–55. 21M, 23 *notes.*

Haydon R.G. [78] 'On dual L^1-spaces and injective bidual Banach spaces', *Israel J. Math.* **31** (1978) 142–52. 33 *notes.*

[80] 'Non-separable Banach spaces', pp. 19–30 in Bierstedt & Fuchssteiner 80. 33J*b*.

Heath R.W. [64] 'Screenability, pointwise paracompactness and metrization of Moore spaces', *Can. J. Math.* **16** (1964) 763–70. 25 *notes.*

Hechler S.H. [71] 'Classifying almost-disjoint families with applications to $\beta\mathbf{N} - \mathbf{N}$', *Israel J. Math.* **10** (1971) 413–32. 21M, 21N*c–d*.

[72*a*] 'Directed graphs over topological spaces: some set theoretical aspects', *Israel J. Math.* **11** (1972) 231–48. 22M.

[72*b*] 'Short complete nested sequences in $\beta\mathbf{N}\backslash\mathbf{N}$ and small maximal almost-disjoint families', *Gen. Top. Appls.* **2** (1972) 139–49. 21O*i*.

[75*a*] 'On some weakly compact spaces and their products', *Gen. Top. Appls.* **5** (1975) 83–93. 24M, 24N*i*–*j*.

[75*b*] 'On a ubiquitous cardinal', *Proc. Am. Math. Soc.* **52** (1975) 348–55. 24N*a*.

[78] 'Generalizations of almost disjointness, *c*-sets, and the Baire number of βN − N', *Gen. Top. Appls.* **8** (1978) 93–110. 26L*k*.

Herink C.D. [77] 'Some applications of iterated forcing', PhD thesis, University of Wisconsin (Madison), 1977. 11 *notes*, 21N*l*, B1F, B3A.

Hodel R.E. [see Burke D.K.]

Hoffman–Jørgensen J. [78] 'How to make a divergent series convergent by Martin's axiom', Aarhus Math. Inst. Preprint Series 21 (1978). 22O*l*, 33I*h*.

Holščevnikova N.N. [81] 'On the sum less than continuum of closed *U*-sets' (Russian), *Vestnik Mosk. Univ.* (*Mat. Mekh. Astr.*) part 1 (1981) 51–5. 22O*h*.

Horn A. & Tarski A. [48] 'Measures in boolean algebras', *Trans. Am. Math. Soc.* **64** (1948) 467–97. 32O.

Huber M. [see Eklof P.C.]

Ionescu Tulcea A. & C. [69] *Topics in the Theory of Lifting*. Springer, 1969. A6B, A6M*b*.

Ismail M. & Nyikos P.J. [80] 'On spaces in which countably compact sets are closed, and hereditary properties', *Top. Appls.* **11** (1980) 281–92. 24M, 24N*b*–*c*.

[82] 'Countable small rank and cardinal invariants II', *Top. Appls.* **14** (1982) 283–304. 44P*e*.

Jayne J.E. & Rogers C.A. [80*a*] 'Fonctions et isomorphismes boreliennes du premier niveau', *C. r. hebd. Séanc. Acad. Sci., Paris* **291** (1980) A351–A354. 23N*k*.

[80*b*] '*K*-analytic sets', pp. 1–183 in Rogers 80. A3G, §A5.

Jech T.J. [71] *Lectures in Set Theory, with Particular Emphasis on the Method of forcing*. Springer, 1971 (Lecture Notes in Mathematics 217). 11E, 11H.

[78] *Set Theory*. Academic, 1978. 11E, 35B, 41M, §A2, A3H*b*, A3J, A3M*a*, A6P*e*, B2D.

Johnson W.B. [see Figiel T.]

Juhász I. [70] 'Martin's axiom solves Ponomarev's problem', *Bull. Acad. Polon. Sci.* (*Math. Astron. Phys.*) **18** (1970) 71–4. 41M, 44N.

[71] *Cardinal Functions in Topology*. North-Holland, 1975 (*Math. Centre Tracts* **34**). 12L, 13B, 43P*l*, §A2.

[77] 'Consistency results in topology', pp. 503–22 in Barwise 77. 43N, 43O*e*, 43R*c*.

[80*a*] 'A survey of *S* and *L* spaces', pp. 675–88 in Csaszar 80. B3E.

[80*b*] *Cardinal Functions in Topology–Ten Years Later*. Mathematisch Centrum, Amsterdam, 1980 (Math. Centre Tracts 123). 22O*j*, §A4.

[83] 'Cardinal functions II', in Kunen & Vaughan 83. B3F*c*.

[see also Hajnal A.]

Juhász I., Kunen K. & Rudin M.E. [76] 'Two more hereditarily separable non-Lindelöf spaces', *Can. J. Math.* **28** (1976) 998–1005. 23 *notes*, 44J*b*.

Juhász I., Nagy Zs. & Weiss W. [79] 'On countably compact, locally countable spaces', *Per. Math. Hung.* **10** (1979) 193–206. 24O*j*–*l*.

Juhász I. & Weiss W. [78] 'Martin's axiom and normality', *Gen. Top. Appls.* **9** (1978) 263–74. 24N*k*, 25L, 25M*e*–*g*.

Junnila H.J.K. [see Fremlin D.H.]

Kanamori A. [76] 'Weakly normal filters and irregular ultrafilters', *Trans. Am. Math. Soc.* **220** (1976) 393–9. 35K.

Kanovei V.G. [79] 'A consequence of the Martin's axiom', *Mathematical Notes* **26** (1979) 548–53. 21O*h*.

Keisler H.J. [see Chang C.C.]

Kelley J.L. [55] *General Topology*. Van Nostrand, 1955. A4M*a*, A7J.

Ketonen J. [see Benda M.]

Khurana S.S. [79] 'Pointwise compactness and measurability', *Pacific J. Math.* **83** (1979) 387–99. A8E.

Knaster B. [45] 'Sur une propriété charactéristique de l'ensemble des nombres réels', *Rec. Math. Moscou* **16/58** (1945) 281–90. 11G*f*, 11H.

Kölzow D. [80] (ed.) *Measure Theory, Oberwolfach* 1979. Springer, 1980 (Lecture Notes in Mathematics 794).

Kölzow D. & Maharam–Stone D. [82] (eds.) *Measure Theory, Obserwolfach* 1981. Springer, 1982 (Lecture Notes in Mathematics 945).

Köthe G. [69] *Topological Vector Spaces I*. Springer, 1969. A8A, A8B*b*, A8F.

Koumoullis G. [83] 'On the almost Lindelöf property in products of separable metric spaces', *Compositio Math.* **48** (1983) 89–100. 32Q*l*.

Krawczyk A. & Pelc A. [80] 'On families of σ-complete ideals', *Fund. Math.* **109** (1980) 155–61. 35M*b*.

Kucia A. & Szymański A. [76] 'Absolute points in $\beta\mathbf{N}\setminus\mathbf{N}$', *Czech. Math. J.* **26** (1976) 381–7. 26J, 26K*e*.

Kunen K. [68] 'Inaccessibility properties of cardinals', PhD thesis, Stanford University, 1968. 14E, 21M, 22N*p*.

[76] 'Some points in $\beta\mathbf{N}$', *Math. Proc. Camb. Phil. Soc.* **80** (1976) 385–98. 26L*a*, 26L*c*, 26L*g*, 32O, 32Q*n*.

[77*a*] 'Luzin spaces', *Top. Proc.* **1** (1976) 191–9. 41O*a*.

[77*b*] 'Strong *S* and *L* spaces under MA', pp. 265–8 in Reed 77. 44P*b*.

[78] 'On paracompactness of box products of compact spaces', *Trans. Am. Math. Soc.* **240** (1978) 307–16. 25N*f*.

[80] *Set Theory*. North-Holland, 1980. 11E, 24 *notes*, 26J, 26K*f*, 32P*h*, 41 *notes*, 44J*a*, §A2, A3H*b*, B2B.

[81] 'A compact *L*-space under CH', *Top. Appls.* **12** (1981) 283–7. 22 *notes*, 32 *notes*, 44J*c*.

[*a*] 'On MA(κ)', handwritten notes (Dec. 1980). B2A.

[*b*] '(κ, λ^*)-gaps under MA', unpublished notes. 31L, B3A.

[see also Broverman S., Douwen E.K. van, Erdös P., Fleissner W.G., Juhász I.]

Kunen K. & Roitman J. [77] 'Attaining the spread at cardinals of cofinality ω', *Pacific J. Math.* **70** (1977) 199–205. 22O*j*.

Kunen K. & Tall F.D. [79] 'Between Martin's axiom and Souslin's hypothesis', *Fund. Math.* **102** (1979) 173–81. 11H, 13C*a*, 41M, 41N*f*, 41P*a*, B1I.

Kunen K. & Vaughan J.E. [83] (eds.) *Handbook of Set Theoretic Topology*. North-Holland, 1984.

Kuratowski K. [66] *Topology I*. Academic Press, 1966. A3G*b*, §A5.

Lacey H.E. [74] *The Isometric Theory of the Classical Banach Spaces*. Springer, 1974. A6E–F, A8D*b*.

Lachlan A., Srebny M. & Zarach A. [77] (eds.) *Set Theory and Hierarchy Theory V*. Springer, 1977 (Lecture Notes in Mathematics 619).

Lane D.J. [80] 'Paracompactness in perfectly normal locally compact locally connected spaces', *Proc. Am. Math. Soc.* **80** (1980) 693–6. 44N.

Larson J.A. [79] 'An independence result for pinning for ordinals', *J. Lond. Math. Soc.* (2) **19** (1979) 1–6. 21O*c*.

Laver R. [75] 'Partition relations for uncountable cardinals $\leq 2^{\aleph_0}$, pp. 1029–42 in Hajnal, Rado & Sos 75. 21N*h*, 21O*a–b*, 41M, 42L*a*, A2K.

[77] 'On the consistency of Borel's conjecture', *Acta Math.* **137** (1976) 151–69. 22N*q–r*.

[82] 'Saturated ideals and non-regular ultrafilters', pp. 297–305 in METAKIDES 82. 35K, 35N*b*.

Levy A. [79] *Basic Set Theory*. Springer, 1979. 31L, 41 *notes*, §A2, A3H*b*.

Loats J.T. [79] 'Hopfian Boolean algebras of power less than or equal to continuum', *Proc. Am. Math. Soc.* **77** (1979) 186–90. 21O*d*.

Loats J.T. & Roitman J. [81] 'Almost rigid Hopfian and dual Hopfian atomic Boolean algebras', *Pacific J. Math.* **97** (1981) 141–50. 21O*d*.

Loomis L.H. [53] *An Introduction to Abstract Harmonic Analysis*. van Nostrand, 1953. 34K.

Louveau A. [72] 'Sur un article de S. Sirota', *Bull. Sci. Math.* (2) **96** (1972) 3–7. 21O*f*.

[74] 'Une démonstration topologique de théorèmes de Silver et Mathias', *Bull. Sci. Math.* (2) **98** (1974) 97–102. 23N*a*.

[80] 'σ-ideaux engendrés par des ensembles fermés et théorèmes d'approximation', *Trans. Am. Math. Soc.* **257** (1980) 153–69. 23N*k*.

Lutzer D.J. [see Aarts J.M., Douwen E.K. van]

Maharam–Stone D. [see Kölzow D.]

Maitra A., Rao B.V. & Rao K.P.S. Bhaskara [79] 'A problem in the extension of measures', *Illinois J. Math.* **23** (1979) 211–16. 32Q*j*.

Malitz J. [see Baumgartner J.E.]

Malyhin V.I. [75a] 'Sequential bicompacta; Cech–Stone extensions and π-points', *Moscow Univ. Math. Bull.* **30** (1975) 18–23. 21O*i*, 26L*p*.

[75*b*] 'Extremally disconnected and similar groups', *Soviet Math. Dokl.* **16** (1975) 21–5. 21O*f*.

[77] 'The equivalence of Martin's axiom and one purely topological statement' [Russian], *Bull. Acad. Polon. Sci.* (*Math. Astron. Phys.*) **25** (1977) 895–900. 13B.

Malyhin V.I. & Sapirovskii B.E. [73] 'Martin's axiom and properties of topological spaces', *Soviet Math. Dokl.* **14** (1973) 1746–51. 24M, 24N*g*, 26L*j*, 31L, 43N, 43P*c–e*, 43P*h*.

Marczewski E. [47] 'Séparabilité et multiplication cartésienne des espaces topologiques', *Fund. Math.* **34** (1947) 127–43. 12L, 12M*a*.

Marek W., Srebny M. & Zarach A. [76] (eds.) *Set Theory and Hierarchy Theory, a Memorial Tribute to Andrzej Mostowski*. Springer, 1976 (Lecture Notes in Mathematics 537).

Martin D.A. [*a*] 'Projective sets and cardinal numbers', unpublished MS. 23 *notes*.

Martin D.A. & Solovay R.M. [70] 'Internal Cohen extensions', *Ann. Math. Logic* **2** (1970) 143–78. 11G*c*, 11H, 12L, 13B, 13 *notes*, 21M, 21O*h*, 22M, 23J, 23L, 23N*b*, 23N*g*, 32O, 32Q*h*, B2D.

Máté A. [see Baumgartner J.E., Hajnal A., Erdös P.]

Mathias A.R.D. [77] 'Happy families', *Ann. Math. Logic* **12** (1977) 59–111. 23N*a*, 26K*b*.

[*a*] (ed.) *Surveys in Set Theory*. Cambridge University Press (LMS Lecture Notes).

Mathias A.R.D., Ostaszewski A.J. & Talagrand M. [78] 'On the existence of an analytic set meeting each compact set in a Borel set', *Math. Proc. Camb. Phil. Soc.* **84** (1978) 5–10. 23L, 23N*c*.

Mauldin R.D. [77] 'On rectangles and countably generated families', *Fund. Math.* **95** (1977) 129–39. 23N*i*.

[78] 'Some effects of set-theoretical assumptions in measure theory', *Adv. Math.* **27** (1978) 45–62. 22N*e*, 23M*e*, 32P*g*, 32R*e*.

[*a*] 'Measurable constructions of preference orders', unpublished MS. 32Q*i*.

[see also Erdös P.]

McAloon K. [see Boffa M.]

McKenzie J.R. & Monk J.D. [75] 'On automorphism groups of Boolean algebras', pp. 951–88 in HAJNAL, RADO & SOS 75. 21O*d*.

McLaughlin T.G. [see Bennett H.R.]
Mekler A.H. [80] 'How to construct almost free groups', *Can. J. Math.* **32** (1980) 1206–28. 34L.
Metakides G. [82] (ed.) *Patras Logic Symposium.* North-Holland, 1982.
Mill J. van [82] 'Weak *p*-points in Cech–Stone compactifications', *Trans. Am. Math. Soc.* **273** (1982) 637–78. 22O*k*.
[83] 'An introduction to $\beta\omega$', in KUNEN & VAUGHAN 83. 26L*n*, 26 *notes*, B3B*a*.
[see also Douwen E.K. van]
Miller A.W. [79] 'On the length of Borel hierarchies', *Ann. Math. Logic* **16** (1979) 233–67. 13C*d*, 23M*a–b*, 23N*f*, 32P*f*.
[82] 'The Baire category theorem and cardinals of countable cofinality', *J. Symbolic Logic* **47** (1982) 275–88. B1B.
[*a*] Letter of 24 Feb. 81. B1G–J.
Mills C.F. [*a*] 'An easier proof of the Shelah *p*-point independence theorem', *Trans. Am. Math. Soc.* 26 *notes.*
[see also Balcar B.]
Monk J.D. [see McKenzie J.R.]
Moschovakis Y.N. [80] *Descriptive Set Theory.* North-Holland, 1980. 23 *notes*, A5H.
Müller G.H. & Scott D.S. [78] (eds.) *Higher Set Theory.* Springer, 1978 (Lecture Notes in Mathematics 669).
Munroe M.E. [53] *Introduction to Measure and Integration.* Addison-Wesley, 1953. 25J, A6K*a*.

Naber G. [77] *Set-Theoretic Topology.* University Microfilms International, 1977.
Nagy Zs. [see Juhász I.]
Navy C. [81] 'Paralindelöfness', PhD thesis, University of Wisconsin (Madison), 1981. 25N*a*.
Negrepontis S. [see Comfort W.W.]
Normann D. [76] 'Martin's axiom and medial functions', *Math. Scand.* **38** (1976) 167–76. 22O*l*.
Novak J. [72] (ed.) *General Topology and its Relations to Modern Analysis and Algebra III*, Academia (Prague), 1972.
Novak J. [83] (ed.) *Proceedings of the Fifth Prague Topological Symposium.* Heldermann, 1983.
Nyikos P.J. [78] 'A compact nonmetrizable space *P* such that P^2 is completely normal', *Top. Proc.* **2** (1977) 359–63. 25N*b*.
[80] 'A provisional solution to the normal Moore space problem', *Proc. Am. Math. Soc.* **78** (1980) 429–35. 25 *notes.*
[81] 'Tunnels, tight gaps and countably compact extensions of N', *Top. Proc.* **5** (1980) 223–9. 24N*n*, 24P*e*, 25N*e*.
[*a*] 'Variations on Martin's axiom', handwritten notes (1981). 41N*f*.
[*b*] 'Notes on hereditarily normal spaces', 1981. 24O*h*.
[see also Ismail M.]

Ostaszewski A.J. [74] 'Martin's axiom and Hausdorff measures', *Proc. Camb. Phil. Soc.* **75** (1974) 193–7. 33I*d*.
[76] 'On countably compact, perfectly normal spaces', *J. Lond. Math. Soc.* (2) **14** (1976) 505–16. 21J, 24 *notes.*
[see also Mathias A.R.D.]
Oxtoby J.C. [61] 'Cartesian products of Baire spaces', *Fund. Math.* **49** (1961) 157–66. 43O*f*.
[71] *Measure and Category.* Springer, 1971 (Graduate Texts in Mathematics 2). 25J, A5B.

Pelant J. [see Douwen E.K. van]
Pelc A. [see Krawczyk A.]
Pfeffer W.F. [see Gardner R.J.]
Pospíšil B. [see Cech E.]
Price R.A. [79] 'CH (and less) solves a problem of Cech', PhD thesis, University of Wisconsin (Madison), 1979. 21O*g*.
Prikry K. [70] 'On a problem of Gillman and Keisler', *Ann. Math. Logic* **2** (1970) 179–87. 35 *notes*.
[76] 'Kurepa's hypothesis and a problem of Ulam on families of measures', *Monatshefte Math.* **81** (1976) 41–57. 35 *notes*.
[*a*] 'A measure extension axiom (MEA)', lecture notes, 1980. 22N*o*.
Pryce J.D. [71] 'A device of R.J. Whitley's applied to pointwise compactness in spaces of continuous functions', *Proc. Lond. Math. Soc.* (3) **23** (1971) 532–46. A4G*b*.
Przymusiński T.C. [73] 'A Lindelöf space X such that X^2 is normal but not paracompact', *Fund. Math.* **78** (1973) 291–6. 25L.
[77] 'Normality and separability of Moore spaces', pp. 325–337 in Reed 77. 25L, 25N*a*, 25N*d*.
[80] 'On the equivalence of certain set-theoretic and topological statements', pp. 999–1003 in Csaszar 80. 25M*k*.
[81] 'A note on Martin's axiom and perfect spaces', *Colloq. Math.* **44** (1981) 209–15. 23L, 23N*j*, 25L, 25N*a*.
[see also Alster K., Douwen E.K. van]
Przymusiński T.C. & Tall F.D. [74] 'The undecidability of the existence of a non-separable normal Moore space satisfying the countable chain condition', *Fund. Math.* **85** (1974) 291–7. 25L.

Rado R. [see Hajnal A.]
Rančin D.V. [77] 'Tightness, sequentiality and closed coverings', *Soviet Math. Dokl.* **18** (1977) 196–200. 24N*d*.
Rao B.V. [see Maitra A.]
Rao K.P.S. Bhaskara [see Maitra A.]
Reed G.M. [75] 'On the productivity of normality in Moore spaces', pp. 479–84 in Stavrakis & Allen 75. 25O*b*.
[76] 'On subspaces of separable first countable T_2-spaces', *Fund. Math.* **91** (1976) 189–202. 25M*k*.
[77] (ed.) *Set-Theoretic Topology*. Academic Press, 1977.
[80] 'On normality and countable paracompactness', *Fund. Math.* **110** (1980) 145–52. 25N*a*.
[see also Davis S.W., Douwen E.K. van, Fleissner W.G.]
Reed G.M. & Zenor P. [76] 'Metrization of Moore spaces and generalized manifolds', *Fund. Math.* **91** (1976) 203–10. 44O*l*.
Reinhardt W. [see Baumgartner J.E.]
Rogers C.A. [80] (ed.) *Analytic Sets*. Academic Press, 1980.
[see also Jayne J.E.]
Roitman J. [79] 'Adding a random or a Cohen real; topological consequences and the effect on Martin's axiom', *Fund. Math.* **103** (1979) 47–60. B1F.
[83] 'Basic *S* and *L*', in Kunen & Vaughan 83. 44Pb, B3F*b*.
[see also Kunen K., Loats J.T.]
Rose H.E. & Shepherdson J.C. [75] (eds.) *Logic Colloquium* '73. North-Holland, 1975.
Rosenthal H.P. [74] 'The heredity problem for weakly compactly generated Banach spaces', *Compositio Math.* **28** (1974) 83–111. 22O*m*.
Rothberger F. [48] 'On some problems of Hausdorff and Sierpiński', *Fund. Math.* **35** (1948) 29–46. 11H, 14E, 21M, 22M, 23L, 24M.

Rubin M. [see Avraham U.]
Rucker R.V.B. [see Ellentuck E.]
Rudin M.E. [75] *Lectures on Set-Theoretic Topology.* American Mathematical Society 1975. 21N*f*, 44O*l*.
[79] 'The undecidability of the existence of a perfectly normal nonmetrizable manifold', *Houston J. Math.* **5** (1979) 249–52. 44N.
[83] 'Dowker spaces', in KUNEN & VAUGHAN 83. B3H.
[see also Juhász I.]
Rudin M.E. & Zenor P. [76] 'A perfectly normal nonmetrizable manifold', *Houston J. Math.* **2** (1976) 129–34. 44 *intro.*
Rudin W. [67] *Fourier Analysis on Groups.* Interscience, 1967. 34K.

Sapirovskii B.E. [72*a*] 'On separability and metrizability of spaces with Souslin's condition', *Soviet Math. Dokl.* **13** (1972) 1633–8. 43N, 43P*l*.
[72*b*] 'On discrete subspaces of topological spaces; weight, tightness and Souslin number', *Soviet Math. Dokl.* **13** (1972) 215–19. A4K.
[80] 'Special types of embeddings in Tychonoff cubes. Subspaces of Σ-products and cardinal invariants', pp. 1055–86 in CSASZAR 80. 43N.
[see also Malyhin V.I.]
Schwartz J.T. [see Dunford N.]
Schwartz L. [73] *Radon Measures on Arbitrary Topological Spaces and Cylindrical Measures.* Oxford University Press, 1973. A4C*a*, A5F, A6B, A6H*a*, §A7.
Scott D.S. [71] (ed.) *Axiomatic Set Theory.* American Mathematical Society, 1971 (Proc. Symp. Pure Mathematics XIII).
[see also Müller G.H.]
Shelah S. [74] 'Infinite abelian groups, Whitehead's problem and some constructions', *Israel J. Math.* **18** (1974) 243–56. 34 *notes.*
[75] 'A compactness theorem for singular cardinals, free algebras, Whitehead problem and transversals', *Israel J. Math.* **21** (1975) 319–49. 34 *notes.*
[78] 'A weak generalization of MA to higher cardinals', *Israel J. Math.* **30** (1978) 297–306. 11 *notes.*
[79] 'On uncountable abelian groups', *Israel J. Math.* **32** (1979) 311–30. 34M*c–d*.
[82] *Proper Forcing.* Springer, 1982 (Lecture Notes in Mathematics 940). 31 *notes.*
[*b*] 'Lifting problem of the measure algebra', *Israel J. Math.* **45** (1983) 90–6. 32 *notes.*
[see also Avraham U., Devlin K.J., Erdös P., Fremlin D.H., Harrington L.]
Shepherdson J.C. [see Rose H.E.]
Shinoda J. [73] 'Some consequences of Martin's axiom and the negation of the continuum hypothesis', *Nagoya Math. J.* **49** (1973) 117–25. 22N*m*, 22N*q*, 23M*e–f*.
Shoenfield J.R. [75] 'Martin's axiom', *Am. Math. Monthly* **82** (1975) 610–17. 22M, 24M.
Sikorski R. [64] *Boolean Algebras.* Springer, 1964. 12A, 12B.
Silver J. [70] 'Every analytic set is Ramsey', *J. Symbolic Logic* **35** (1970) 60–4. 11H, 23N*a*.
Simon P. [78] 'A somewhat surprising subspace of $\beta\mathbf{N}\setminus\mathbf{N}$', *Commentationes Math. Univ. Carol.* **19** (1978) 383–8. 26L*m*.
Solomon R.C. [77] 'Families of sets and functions', *Czech. Math. J.* **27** (1977) 556–9. 26L*a*.
Solovay R.M. [70] 'A model of set theory in which every set of reals is Lebesgue measurable', *Ann. Math.* **92** (1970) 1–56. B3G.
[71] 'Real-valued measurable cardinals', pp. 397–428 in SCOTT 71. 35K, A6P*e*, B2D.
[see also Martin D.A.]
Solovay R.M. & Tennenbaum S. [71] 'Iterated Cohen extensions and Souslin's problem', *Ann. Math.* **94** (1971) 201–45. *Intro.*, 11E, 11H, 41M.
Sos V.T. [see Hajnal A.]

Souslin M. [20] 'Problème 3', *Fund. Math.* **1** (1920) 223. 41 *notes.*

Srebny M. [see Lachlan A., Marek W.]

Stark W.R. [80] 'Martin's axiom in the model theory of LA', *J. Symbolic Logic* **45** (1980) 172–6. 13D*a*.

Stavrakis N.M. & Allen K.R. [75] (eds.) *Studies in Topology.* Academic, 1975.

Steprāns J. [*a*] 'Strong Q-sequences and variations on Martin's axiom'. B1F.
[see also Devlin K.J.]

Stern J. [78] 'Partitions of the real line into $\aleph_1$ closed sets', pp. 455–60 in MÜLLER & SCOTT 78. 23M*g*.

Szentmiklóssy Z. [80] '*S*-spaces and *L*-spaces under Martin's axiom', pp. 1139–45 in CSASZAR 80. 42J, 44N, 44P*c*.

Szymański A. [77] 'The existence of $P(a)$ points of $\mathbf{N}^*$ for $\omega < a < c$', *Colloq. Math.* **37** (1977) 179–84. 26L*a*.
[79] 'Some Baire category theorems for $U(\omega_1)$', *Commentationes Math. Univ. Carol.* **20** (1979) 519–28. 26L*i*.
[80*a*] 'Undecidability of the existence of regular extremally disconnected spaces', *Colloq. Math.* **43** (1980) 61–7. 44P*d*.
[80*b*] 'On *m*-Baire and *m*-Blumberg spaces', pp. 151–61 in FLACHSMEYER, FROLIK & TERPE 80. 26K*d*.
[*a*] Note of October 1981. 21M.
[see also Blaszczyk A., Kucia A.]

Szymański A. & Zhou H.-X. [*a*] 'The behaviour of ω_2^* under some consequences of Martin's axiom', in NOVAK 83.

Talagrand M. [80*a*] 'Compacts de fonctions measurables et filtres non mesurables', *Studia Math.* **67** (1980) 13–43. 22O*d*, 22O*g*, 32O, 32Q*a*, 32Q*e*.
[80*b*] 'Sur les mesures vectorielles definies par une application Pettis-integrable', *Bull. Soc. Math. France* **108** (1980) 475–83. 32Q*f*.
[80*c*] 'Hyperplans universellement mesurables', *C. r. hebd. Seanc. Acad. Sci., Paris* **291** (1980) A501–A502. 22O*f*.
[80*d*] 'Un nouveau $C(K)$ de Grothendieck', *Israel J. Math.* **37** (1980) 181–91. 26 *notes.*
[82] 'Closed convex hull of set of measurable functions, Riemann measurable functions and measurability of translations', *Ann. Inst. Fourier* **32** (1982) 39–69. 32Q*d*, A6L*e*.
[83] 'Filtres: mesurabilité, rapidité, propriété de Baire forte', *Studia Math.* **74** (1982) 283–91. 32Q*b*.
[*a*] 'Measure theory and Pettis integration', in preparation for *Mem. Am. Math. Soc.* 22O*a*–*d*, 32Q*d*–*e*, A6L, A8C, A8E, A8F.
[see also Fremlin D.H., Mathias A.R.D.]

Tall F.D. [74*a*] 'The countable chain condition versus separability – applications of Martin's axiom', *Gen. Top. Appl.* **4** (1974) 315–39. 43N, 43P*d*, 43P*k*, 43P*o*, 43Q*a*, 44N, 44O*a*.
[74*b*] 'On the existence of normal metacompact Moore spaces which are not metrizable', *Can. J. Math.* **26** (1974) 1–6. 24M.
[76] 'The density topology', *Pacific J. Math.* **62** (1976) 275–84.
[77*a*] 'Set-theoretic consistency results and topological theorems concerning the normal Moore space conjecture and related problems', *Dissertationes Math.* **148** (1977). 25L.
[77*b*] 'First countable spaces with caliber $\aleph_1$ may or may not be separable', pp. 353–8 in REED 77. 25L.
[78] 'Normal subspaces of the density topology'. *Pacific J. Math.* **75** (1978) 579–88. 25L, 25M*l*, 32P*i*.
[79] 'Applications of a generalized Martin's axiom', *Comptes Rendus Math. Rep. Acad. Sci. (Canada)* **1** (1979) 103–6. 11 *notes.*

[*a*] 'An alternative to the continuum hypothesis and its uses in general topology', privately circulated. 26K*d*.
[*b*] 'Some applications of a generalized Martin's axiom', *Trans. Am. Math. Soc.* 11 *notes.*
[see also Broverman S., Kunen K., Przymusiński T.C.]
Tarski A. [see Horn A.]
Taylor A.D. [79] 'Regularity properties of ideals and ultrafilters', *Ann. Math. Logic* **16** (1979) 33–55. 35K, 35L*b–c*, 35M*a*.
[80] 'On saturated sets of ideals and Ulam's problem', *Fund. Math.* **109** (1980) 37–53. 35 *notes.*
[see also Baumgartner J.E.]
Tennenbaum S. [see Solovay R.M.]
Terpe F. [see Flachsmeyer J.]
Todorčević S.B. [81*a*] 'Some consequences of MA + ¬wKH', *Top. Appls.* **12** (1981) 187–202. B2G, B3I.
[81*b*] 'On the *S*-space problem', *Abstracts Am. Math. Soc.* **2** (1981) 394. B3F*b*.
[81*c*] '$\omega_1 \rightarrow (\omega_1, \omega + 2)^2$ is consistent', *Abstracts Am. Math. Soc.* **2** (1981) 462. B2J.
[81*d*] 'On the cardinality of Hausdorff spaces', *Abstracts Am. Math. Soc.* **2** (1981) 529. B3F*c*.
Todorčević S.B. [83*a*] 'Real functions on the family of all well-ordered subsets of a partially ordered set', *J. Symbolic Logic* **48** (1983) 91–6.
[83*b*] 'On a conjecture of R. Rado', *J. Lond. Math. Soc.* (2) **27** (1983) 1–8. 41O*f*.
[83*c*] 'Trees and linearly ordered spaces', in KUNEN & VAUGHAN 83. A3J, B2E.
[*a*] Letter of 11 July 82. 25M*m*.
[*b*] 'Martin's axiom and antichains in partially ordered sets', note of February 1981. 41O*e*.
[*c*] '$\mathrm{MA}_{\omega_1} \Rightarrow \omega_1 \rightarrow (\omega_1, \omega + 2)^2$', note of December 1981. 42L*c*.
[*d*] 'Forcing positive partition relations', *Trans. Am. Math. Soc.* B2J, B2 *notes*, B3F.
[see also Abraham U.]
[*e*] 'Some results concerning partially ordered sets', *J. Symbolic Logic.*
Tortrat A. [76] 'τ-regularité des lois, separation au sens de A. Tulcea et propriété de Radon–Nikodým', *Ann. Inst. Henri Poincaré* (B) **12** (1976) 131–50. A8F.

Vaughan J.E. [79] 'Discrete sequences of points', *Top. Proc.* **3** (1978) 237–65. 24N*l*.
[83] 'Countably compact and sequentially compact spaces', in KUNEN & VAUGHAN 83. 24 *notes.*
[*a*] '*R*-spaces', abstract for Birmingham Topology Conference (1980). 25N*e*.
[see also Kunen K.]
Vojtas P. [see Balcar B.]

Wage M.L. [76] 'Countable paracompactness, normality and Moore spaces', *Proc. Am. Math. Soc.* **57** (1976) 183–8. 25N*a*.
[79] 'Almost disjoint sets and Martin's axiom', *J. Symbolic Logic* **44** (1979) 313–8. 41 *notes*, 42J, 42L*b*.
[see also Davis S.W., Douwen E.K. van]
Wald B. [see Göbel R.]
Watson S. [see Devlin K.J.]
Weiss W. [78] 'Countably compact spaces and Martin's axiom', *Can. J. Math.* **30** (1978) 243–9. 24M, 44N.
[81*a*] 'Small Dowker spaces', *Pacific J. Math.* **94** (1981) 485–92. B3H.
[81*b*] 'The equivalence of a generalized Martin's axiom to a combinatorial principle', *J. Symbolic Logic* **46** (1981) 817–21. 11 *notes.*

[83] 'Versions of Martin's axiom', in KUNEN & VAUGHAN 83. 11 *notes*. 21N*j*, 41O*a*, B3I–J.
[see also Juhász I.]
Wheeler R.F. [76] 'The Mackey problem for the compact-open topology', *Trans. Am. Math. Soc.* **222** (1976) 255–65. 24O*i*.
White H.E. Jr. [75] 'An example involving Baire spaces', *Proc. Am. Math. Soc.* **48** (1975) 228–30. 25L.
Williams S.W. [*a*] 'Box products', in KUNEN & VAUGHAN 83. 25N*f*.
Wimmers E. [82] 'The Shelah *p*-point independence theorem', *Israel J. Math.* **43** (1982) 28–48. 26 *notes*.

Zachariades Th. [see Argyros S.A.]
Zarach A. [see Lachlan A., Marek W.]
Zenor P. [see Alster K., Reed G.M., Rudin M.E.]
Zygmund A. [59] *Trigonometric Series*. Cambridge University Press, 1959. 22O*h*.

This bibliography was prepared with the help of the University of Essex Computing Service.

Index to special symbols and abbreviations

References in **bold** type are to definitions; references in *italics* are to passing references. Cross references (*see*, *see also*) are to the general index as well as this index.

General index

References in **bold** type are to definitions; references in *italics* are to passing references.